**About this Book**

The nature of time has haunted man through the ages. Some conception of time has always entered into man's ideas about mortality and immortality, and permanence and change, so that concepts of time are of fundamental importance in the study of religion, philosophy, literature, history, and mythology. How man experiences time physiologically, psychologically, and socially enters into the research of the behavioral sciences, and time as a factor of structure and change is an essential consideration in the biological and physical sciences. In one aspect or another, time study cuts across all disciplines and it is the goal of the International Society for the Study of Time to begin the interdisciplinary comparative study of time. These Proceedings record the substance of an international meeting where a distinguished group of scholars and scientists came together to present the results of research on time in their respective disciplines.

# The Study of Time

*Proceedings of the First Conference of the
International Society for the Study of Time
Oberwolfach (Black Forest) — West Germany*

*Edited by*

J. T. Fraser, F. C. Haber, G. H. Müller

*With 65 Figures*

Springer-Verlag
Berlin Heidelberg New York 1972

Dr. J. T. FRASER, P. O. Box 164, Pleasantville, NY 10570/USA

Dr. F. C. HABER, Professor, Department of History, University of Maryland, College Park, Maryland 20742/USA

Dr. G. MÜLLER, Professor, Mathematisches Institut, Universität Heidelberg, D–6900 Heidelberg, West Germany

ISBN 3-540-05824-9 Springer-Verlag Berlin Heidelberg New York
ISBN 0-387-05824-9 Springer-Verlag New York Heidelberg Berlin

This work is subject to copyright. All rights are reserved, whether the whole or part of the material is concerned, specifically those of translation, reprinting, re-use of illustrations, broadcasting, reproduction by photocopying or similar means, and storage in data banks
Under § 54 of the German Copyright Law where copies are made for other than private use, a fee is payable to the publisher the amount of the fee to be determined by agreement with the publisher
© by Springer-Verlag, Berlin · Heidelberg 1972. Library of Congress Catalog Card Number 72-80472. Printed in Germany. The use of general descriptive names, trade names, trade marks, etc. in this publication, even if the former are not especially identified, is not to be taken as a sign that such names, as understood by the Trade Marks and Merchandise Marks Act, may, accordingly be used freely by anyone.

# Foreword

The First Conference of the International Society for the Study of Time was held at the Mathematisches Forschungsinstitut at Oberwolfach in the Black Forest, Federal Republic of Germany from Sunday, 31 August to Saturday, 6 September, 1969. The origin of this conference and the formation of the Society goes back to a proposal due to J. T. Fraser that was discussed at a conference on "Interdisciplinary Perspectives of Time" held by the New York Academy of Sciences in January, 1966. It was unanimously agreed than that an international society should be formed on an interdisciplinary basis with the object of stimulating interest in all problems concerning time and that this object could best be attained by means of conferences held at regular intervals. J. T. Fraser was elected Secretary, S. Watanabe Treasurer, and I was elected President. It was agreed, at my suggestion, that the organization of the first conference of the newly formed Society be left to a committee of these three officers, on the understanding that they would invite authorities on the role of time in the various special sciences and humanities to form an Advisory Board to assist them.

One of the main difficulties in seeking support for an interdisciplinary conference is that most foundations confine their interest exclusively either to the sciences or to the humanities. On the other hand, although various universities expressed their interest and were willing to provide accommodation for an international interdisciplinary conference on time, they were unable to offer any financial support. It was therefore doubly fortunate that, when Dr. Fraser explored the possibility of holding a conference in the Federal Republic of Germany, he came into contact with Professor Gert Müller of the Mathematical Institute of the University of Heidelberg. Professor Müller not only took up the idea with great enthusiasm, but was able to elicit a favourable response from the authorities in the Federal Republic whom he approached for support. In particular, his personal association both with the journal *Studium Generale,* published by the Springer-Verlag, of which he is the Executive Editor, and with the Mathematisches Forschungsinstitut, Oberwolfach, ensured that our proceedings would have a sympathetic publisher as well as generous financial support. Moreover, it made it possible to hold the conference under ideal conditions.

Thanks to the untiring energy and enthusiasm of Dr. Fraser, on whom the main burden of organizing the conference fell and without whom it would never have taken place, and of Professor Müller, the first conference of our Society was a resounding success and we owe to both of them a great debt of gratitude.

The Conference brought together nearly fifty scholars and scientists, covering a wide range of subjects, all of whom had a professional interest in some aspect or aspects of time. A central feature was a one-day symposium on Dysrhythmia organized and chaired by Dr. Fraser. Generous financial support for this symposium was provided by Pan American Airways corporation. It is unlikely that such a thorough-going and wide-ranging discussion on this topic has taken place previously. Apart from the Presidential Address with which the Conference opened and the Secretarys' General Survey and Report with which it closed, the papers that were read, other than those contributed to the special symposium, were organized by subject.

In the main, the choice of papers was the responsibility of the Chairmen of the respective sessions, although in some cases they were chosen by the President and Secretary. The Programme of papers follows this Foreword.

One of the traditional features of conferences at the Mathematisches Forschungsinstitut is that the seating of participants at the different tables in the dining room is changed from one meal to the next. This arrangement had a particularly favourable effect on our Conference, composed as it was of representatives of a wide range of disciplines, and without doubt contributed to the general success of the Conference. Since the printed page can only convey that part of our proceedings which was committed to paper, the benefits arising from the many informal discussions, some of which went on far into the night, can only be appreciated by those who took part in them.

It is with great sorrow that I have to report that three members of the Conference, who took a prominent part in the proceedings, have since died. ARTHUR PRIOR, who was Chairman of the Logic session and read a paper which stimulated much interest, was a world authority on logic and in recent years had done pioneer work on the logic of time. He was a fellow of the British Academy and of Balliol College, Oxford. He died suddenly in Norway in October 1969, barely a month after our Conference ended. ADRIAN DOBBS, of the Ministry of Defence, London, despite the claims of his official post, had found time over the years to maintain a professional interest in both physics and psychology, and his paper gave rise to considerable discussion. His accidental death in London in April 1970 came as a great shock. Equally unexpected was the sad news of the death in October 1971 of SAMUEL GEORGE FREDERICK BRANDON, who was Chairman of one of the History sessions and read a paper that was one of the most valuable of the whole conference. He had been Professor of Comparative Religion in the University of Manchester since 1951. An international authority on the religions of antiquity, he made important contributions to our subject, notably in his books "Time and Mankind" and "History, Time and Deity". We mourn the untimely passing of these valued colleagues.

Finally on behalf of all who attended the Conference, I should like not only to express our most grateful thanks to Pan American Airways Corporation for the financial support already mentioned, but also to the Bundesministerium für Wissenschaftliche Forschung at Bonn and to the Kultusministerium des Landes Baden-Württemberg at Stuttgart for the generous support that we received for board and accomodation and for part of our travel expenses. We are all grateful to the Mathematisches Forschungsinstitut, Oberwolfach, for acting as our host and to the Springer Verlag, Berlin-Heidelberg-New York, for their generous help in the local organization of the Conference. In particular, I should like to express our thanks to Dr. Petersen of the Bundesministerium, Oberregierungsrat Dieffenbacher of the Kultusministerium and Professor Dr. Barner, Director of the Mathematisches Forschungsinstitut, for their sympathetic co-operation.

London, July 1971

G. J. WHITROW
First President
International Society for the Study of Time

# Contents*

Presidential Address:

Reflections on the History of the Concept of Time. G. J. WHITROW. [Stud. Gen. 23 (1970) 498–508] . . . . . . . . . . . . . . . . 1

## I. Time and the Physical Sciences

The Time Coordinate in Einstein's Restricted Theory of Relativity. J. D. NORTH. [Stud. Gen. 23 (1970) 203–223] . . . . . . . . . 12

General Relativity and Time in the Solar System. G. C. McVITTIE. [Stud. Gen. 23 (1970) 197–202] . . . . . . . . . . . . . 33

Zeit als physikalischer Begriff. F. HUND. [Stud. Gen. 23 (1970) 1088–1101] 39

Time in Particle Physics. J. G. TAYLOR. [Stud. Gen. 23 (1970) 1102–1107] 53

Time in Statistical Physics and Special Relativity. P. T. LANDSBERG. [Stud. Gen. 23 (1970) 1108–1158] . . . . . . . . . . . . . . 59

The Myth of the Passage of Time. D. PARK. [Stud. Gen. 24 (1971) 19–30] . 110

Time Asymmetry, Time Reversal, and Irreversibility. M. BUNGE. [Stud. Gen. 23 (1970) 562–570] . . . . . . . . . . . . . . . . . 122

No Paradox in the Theory of Time Anisotropy. O. COSTA DE BEAUREGARD. [Stud. Gen. 24 (1971) 10–18] . . . . . . . . . . . . . . . . 131

Pierre Curie's Principle of One-Way Process. L. L. WHYTE. [Stud. Gen. 23 (1970) 525–532] . . . . . . . . . . . . . . . . . . . 140

In Defence of *the* Direction of Time. K. G. DENBIGH. [Stud. Gen. 23 (1970) 234–244] . . . . . . . . . . . . . . . . . . 148

Creative Time. S. WATANABE. [Stud. Gen. 23 (1970) 1057–1087] . . . 159

## II. Time in the Life Sciences

Temporal Order as the Origin of Spatial Order in Embryos. B. C. GOODWIN. [Stud. Gen. 23 (1970) 273–282] . . . . . . . . . . . . . . 190

Time in the Evolutionary Process. J. MAYNARD SMITH. [Stud. Gen. 23 (1970) 266–272 . . . . . . . . . . . . . . . . . . . . 200

The Measurement of Perceptual Durations. R. EFRON. [Stud. Gen. 23 (1970) 550–561] . . . . . . . . . . . . . . . . . . . 207

Oscillations as Possible Basis for Time Perception. E. PÖPPEL. [Stud. Gen. 24 (1971) 85–107] . . . . . . . . . . . . . . . . . . . 219

Processing of Temporal Information and the Cognitive Theory of Time Experience. J. A. MICHON. [Stud. Gen. 23 (1970) 249–265] . . . . . 242

---

* All papers previously published in "Studium Generale" as indicated under the titles.

The Psychophysical Structure of Temporal Information. P. MEREDITH. [Stud. Gen. 24 (1971) 70–84] . . . . . . . . . . . . . . . . . 259

The Dimensions of the Sensible Present. H. A. C. DOBBS. [Stud. Gen. 24 (1971) 108–126] . . . . . . . . . . . . . . . . . . . . . 274

Time, Time Stance, and Existence. B. S. AARONSON. [Stud. Gen. 24 (1971) 369–387] . . . . . . . . . . . . . . . . . . . . . . 293

Personality and the Psychology of Time. R. H. KNAPP. [Stud. Gen. 24 (1971) 44–51] . . . . . . . . . . . . . . . . . . . . . . 312

III. Time, Philosophy and the Logic of Time Concept

The Notion of the Present. A. N. PRIOR. [Stud. Gen. 23 (1970) 245–248] . 320

Instants and Intervals. C. L. HAMBLIN. [Stud. Gen. 24 (1971) 127–134] . 324

The Fiction of Instants. M. ČAPEK. [Stud. Gen. 24 (1971) 31–43] . . . . 332

On the Reality of Becoming. E. CASSIRER. [Stud. Gen. 24 (1971) 1–9] . . 345

Whitehead and the Philosophy of Time. W. MAYS. [Stud. Gen. 23 (1970) 509–524] . . . . . . . . . . . . . . . . . . . . . . . . 354

IV. Time and Culture

The Deification of Time. S. G. F. BRANDON. [Stud. Gen. 23 (1970) 485–497] 370

The Darwinian Revolution in the Concept of Time. F. C. HABER. [Stud. Gen. 24 (1971) 289–307] . . . . . . . . . . . . . . . . . . . 383

Temporal Attitudes in Four Negro Subcultures. H. B. GREEN. [Stud. Gen. 23 (1970) 571–586] . . . . . . . . . . . . . . . . . . . . 402

On Hegel – A Study in Sorcery. E. VOEGELIN. [Stud. Gen. 24 (1971) 335–368] . . . . . . . . . . . . . . . . . . . . . . . . 418

Time and the Modern Self: Descartes, Rousseau, Beckett. G. SEBBA. [Stud. Gen. 24 (1971) 308–325] . . . . . . . . . . . . . . . . 452

Time and the Modern Self: A Change in Dramatic Form. T. UNGVÁRI. [Stud. Gen. 24 (1971) 326–334] . . . . . . . . . . . . . . . . 470

The Study of Time. J. T. FRASER . . . . . . . . . . . . . . . . . 479

V. Special Session on Flight Dysrhythmia

Introduction J. T. FRASER . . . . . . . . . . . . . . . . . . 503

The Psychological Effects of Rapid Shifts in Temporal Referents. S. J. BLATT and D. M. QUINLAN. [Stud. Gen. 23 (1970) 533–549] . . . . . . . 506

Evaluation of Circadian Dyschronism during Transmeridian Flights. A. REINBERG. [Stud. Gen. 23 (1970) 1159–1168] . . . . . . . . . 523

Some Factors in the Production of Dysrhythmia and Disorientation Associated with Rapid Latitudinal Transfer. W. GOODDY. [Stud. Gen. 24 (1971) 52–65] 533

Discussion Notes on the Lecture by Dr. Gooddy. G. SCHALTENBRAND. [Stud. Gen. 24 (1971) 66–69] . . . . . . . . . . . . . . . . 547

# Reflections on the History of the Concept of Time

G. J. WHITROW*

*Summary.* The central feature of our current mode of thought about the physical universe and human society is the idea of time as linear advancement without cyclical repetition. We tend to regard this idea as intuitively obvious and a necessity of thought, but a general survey of the historical evidence reveals that this is not the case.

The origins of our concept of time are shrouded in mystery, but from our knowledge of ancient civilizations and also of surviving primitive races it is reasonable to assume that the lives of our remote ancestors were far less consciously dominated by time than are ours. For example, although the children of Australian aborigines are of similar mental capacity to white children, they have great difficulty in telling the time by the clock. They can read off the position of the hands of the clock as a memory exercise but they are quite unable to relate it to the time of day. There is a cultural gap between their conception of time and ours which they find difficult to cross. Nevertheless, all primitive peoples have some idea of time and some method of reckoning, usually based on astronomical observations. The Australian aborigine will fix the time for a proposed action by placing a stone in the fork of a tree, or some such place, so that the Sun will strike it at the agreed time.

Primitive man's sense of rhythm was a vital factor in his intuition of time. Moreover, before he had any explicit idea of time, he seems to have been aware of temporal associations dividing time into intervals like bars in music. The principal transitions in nature were thought to occur suddenly, and similarly man's journey through life was visualized as a sequence of distinct stages – later epitomized in Shakespeare's 'Seven ages of man'. Even in so culturally advanced a civilization as the ancient Chinese different intervals of time were regarded as separate discrete units, so that time was in effect discontinuous. Just as space was decomposed into regions, so time was split up into eras, seasons and epochs. In other words, time was 'boxed'.[1] Even in medieval Europe the development of the mechanical clock did not spring from a desire to register

---

* G. J. Whitrow, M. A., D. Phil., Department of Mathematics in the University of London at the Imperial College of Science and Technology, Exhibition Road, London S.W. 7, Great Britain.

1 Granet, M.: *La Pensée Chinoise*. Paris: 1934, p. 330; but see also, for some modification of this view, Needham, J.: *Time and Eastern Man*. Royal Anthropological Institute Occasional Paper, No. 21, 1965, especially p. 8.

the passage of time but rather from the monastic demand for accurate determination of the hours when the various religious offices and prayers should be said.

It was a long step from the inhomogeneity of magical time as generally imagined in antiquity and the middle ages, with its specific holy days and lucky and unlucky secular days, to the modern scientific conception of homogenous linear time. Indeed, man was aware of different times long before he formulated the idea of time itself. This distinction is particularly well illustrated by the Maya priests of pre-Columban central America, who, of all ancient peoples, were the most obsessed with the idea of time.[2] Whereas in European antiquity the days of the week were regarded as being under the influence of the principal heavenly bodies, e. g. Saturn-day, Sun-day, Moon-day etc., for the Mayas each day was itself divine. Every monument and every altar was erected to mark the passage of time. The Mayas pictured the divisions of time as burdens carried on the backs of a hierarchy of divine bearers who personified the respective numbers by which the different periods – days, months, years, decades, etc. – were distinguished. There were momentary pauses at the end of each prescribed period, for example at the end of a day, when one god with his burden (in this case representing the next day) replaced another god with his. A remarkably precise astronomical calendar was developed embodying correction formulae that were even more accurate than our present leap year *correction* which was introduced about a thousand years later by Pope Gregory XIII in 1582. Our correction is too long by 3 days in ten thousand years, whereas the corresponding Maya correction was 2 days too short in ten thousand years.[3] Despite this astonishing technical achievement, the Mayas never seem to have grasped the idea of time as the journey of one bearer and his load. Instead, each god's burden came to signify the particular omen of the division of time in question – one year the burden might be drought, another a good harvest and so on.

Unlike the Mayas, the ancient Greeks were not obsessed by the temporal aspect of things. At the dawn of Greek literature two contrasting points of view are found in Homer and Hesiod. In the *Iliad* Olympian theology and morality are dominated by spatial concepts, the cardinal sin being *hubris*, that is, going beyond one's assigned province. The whole conception, in the words of Cornford, is "static and geometrical; everything has its limited field with bounds that must not be passed."[4] In Homer's epics time is never the subject of a verb and is therefore not even personified. He was not interested in the origin of things and had no cosmogony. On the other hand, Hesiod, in his *Works and Days* gave an account of the origin of the world, but although his poem can be regarded

---

[2] Thompson, J. Eric S.: *The Rise and Fall of Maya Civilization*. London: 1956, p. 149.
[3] Morley, S. G.: *The Ancient Maya*. 2nd edn., 1947, p. 305.
[4] Cornford, F. M.: *From Religion to Philosophy*. London: 1912, p. 181.

as a moralistic study based implicitly on the time concept, the word 'time' never appears in it.

Two centuries or more later the Ionian pioneers of natural philosophy in the sixth century B. C. visualised the world as a geometrical organism or a space-filling substance with a life cycle. Heraclitus, however, believed the world to be a soul involved in an endless cycle of death and rebirth, the very essence of the universe being transmutation.[5] In place of the space-like concept of *Moira* (necessity or destiny) that we find in Homer, he advocated the time-like *Way of Justice* which admits no barriers between different regions and passes through all phases of the world-cycle. A similar emphasis on time and soul characterized the Orphic religion which appears to have provided the mythical background of Pythagoreanism. According to Plutarch, when asked what Time was, Pythagoras replied that it was the soul, or procreative element, of the universe.[6]

The extent to which Pythagoras and his followers were influenced by Oriental ideas is a matter for speculation, but the Orphic theogonical conception of *Kronos* has common features with the Iranian idea of *Zurvān Akarana* (unending time), and each was in fact depicted as a multi-headed winged serpent. In her Messenger Lectures on *Time in Greek Tragedy* Jacqueline de Romilly has suggested that "tragedy was born precisely when the consciousness of time became ripe and the idea of time important"[7], and that before the fifth century B. C. and the writings of Aeschylus the idea of time was unimportant for the Greeks. This was the century when the Hellenic world was brought into violent contact with the Iranian at Marathon and Salamis. Professor de Romilly stresses that for the Greeks time was generally thought of as a threat, and not as a continued evolution. Unlike writers of our own day, no Greek author would have dreamt of writing a play about time. Although the Greeks are known to have used waterclocks in the law courts in order to limit speeches[8], there is no evidence that the performance of plays was ever restricted in this way (despite a dubious reference in Aristotle's *Poetics*). Nor did they even like to show the action of time on moods and feelings. When Euripides allowed one of his characters, Iphigeneia, to change her decision within a short time, Aristotle was shocked.[9]

---

5 "Homer was wrong in saying: 'Would that strife might perish from among the gods and men', He did not see that he was praying for the destruction of the universe; for, if his prayer were heard, all things would pass away." (Heraclitus, Fragment 43, transl. Burnet, J., *Early Greek Philosophy*, 3rd ed., 1920, p. 136. The reference is to the *Iliad*, Book XVIII, line 107).

6 Plutarch. *Morals*. Rev. and corrected by W. W. Godwin. Boston: 1870. Vol. 5. Platonic Question VIII (transl. R. Brown), p. 440.

7 Romilly, J. de: *Time in Greek Tragedy*. Ithaca, N. Y.: 1968, p. 5.

8 Fyfe, W. H.: *Aristotle's Art of Poetry*. Oxford: 1940, p. 22, footnote 3.

9 Romilly, J. de: op cit., p. 25.

The general outlook of the Greeks in natural philosophy was dominated by the idea of the permanence of the *cosmos*, and in so far as they thought of time in this context they associated it with the regular alternation of things and not with concepts of progress and perpetual transformation. Even their idea of memory differed sharply from ours. They believed that through memory we do not try to grasp the past but rather we seize an eternal and divine truth.[10]

From about the middle of the fifth century B. C. many of the most acute Greek thinkers became sorely perplexed by time and sought to evade the concept as far as possible because it was difficult to reconcile with their severely rational outlook. In particular, Parmenides, the founding father of logical disputation, argued that time cannot pertain to anything that is truly real. The essence of his difficulty was that time and change imply that the same thing can have contradictory properties – it can be both red, say, and not-red depending on the time – and this conflicted with the idea that nothing can possess logically incompatible attributes.

The difficulties involved in producing a logically satisfactory theory of time were emphasized by Parmenides's pupil Zeno of Elea in his famous paradoxes. For, although these paradoxes were primarily concerned with the problem of motion, they raised difficulties both for the idea of time as continuous or infinitely divisible and for the idea of temporal atomicity. Unlike the Pythagoreans, who tended to identify the chronological with the logical, Parmenides and Zeno argued that they are incompatible.

The influence of Parmenides and Zeno on Plato is evident in the different treatment of space and time in Plato's cosmological dialogue the *Timaeus*. Space exists in its own right as a given frame for the visible order of things, whereas time is merely a feature of that order based on an ideal timeless archetype of realm of static geometrical shapes (Eternity) of which it is the 'moving image', being governed by a regular numerical sequence made manifest by the motions of the heavenly bodies. Plato's intimate association of time with the universe led him to regard time as being actually produced by the revolutions of the celestial sphere.

This conclusion was not accepted by Aristotle who rejected the idea that time can be identified with any form of motion. For, he argued, motion can be uniform or non-uniform and these terms are themselves defined by time, whereas time cannot be defined by itself. Nevertheless, although time is not identical with motion, it seemed to him to be dependent on motion. Possibly influenced by the Pythagoreans, he argued that time is a kind of number, being the numerable aspect of motion. Time is therefore a numbering process founded on our perception of 'before' and 'after' in motion: "Time is the number of motion with respect to earlier and later." Aristotle regarded time and motion as

---

10 Vernant, J. P.: *Mythe et Pensée chez les Grecs, études de Psychologie Historique*. Paris: 1965. 51–78.

reciprocal. "Not only do we measure the movement by the time, but also the time by the movement, because they define each other. The time marks the movement, since it is its number; and the movement the time."[11] Aristotle recognized that motion can cease whereas time cannot, but there is one motion that continues unceasingly, namely that of the heavens. Clearly although he did not agree with Plato, he too was profoundly influenced by the cosmological view of time. Moreover, although he began by rejecting any association between time and a particular motion in favour of one between time and motion in general, he came to the conclusion that time is closely associated with the circular motion of the heavens, which he regarded as the perfect example of uniform motion.

Unlike the pre-Parmenideans, Aristotle interpreted the first cause of anything not as its method of origin but as its sustaining principle. Instead of *physis* signifying the whole process of growth, it now came to mean the form of the fully developed thing, "for the process of generation exists for the sake of the complete being, and not this for the sake of the process".[12] Moreover, the order of the universe was declared to be eternal and ungenerated. Consequently, Aristotle objected to the Orphic and similar theogonies because they made the development of the world to a certain extent evolutionary. "The most perfect was not in existence at the beginning, but appeared at a later stage. In his own terms, they made the potential prior in time to the actual, and that for Aristotle was the greatest heresy."[13]

Belief in the cyclic nature of time was widespread in antiquity and found its apotheosis in the concept of the Great Year, which the Greeks may have inherited from the Babylonians. The idea had two distinct interpretations. On the one hand it was simply the period required for the Sun, Moon and planets to attain the same positions in relation to each other as they had at a given time. This appears to be the sense in which Plato used the idea in the *Timaeus*. On the other hand, for Heraclitus it signified the period of the world from its formation to its destruction and rebirth. The two interpretations were combined by the Stoics, who believed that, when the heavenly bodies return at fixed intervals of time to the same relative positions as they had at the beginning of the world, everything would be restored just as it was before and the entire cycle would be renewed in every detail. "Socrates and Plato and each individual man will live again, with the same friends and fellow citizens. They will go through the same experiences and the same activities. Every city and village and field will be restored, just as it was. And this restoration of the universe takes place not once, but over and over again – indeed to all eternity without end. Those of the gods who are not subject to destruction, having observed the

---

11 Aristotle, *Physica* (transl. R. P. Hardie and R. K. Gaye) in *The Works of Aristotle* (ed. W. D. Ross) Vol. II, Oxford: 1930, Book IV, 220 b.
12 Aristotle, *De Partibus Animalium.* 640 a, 17–19.
13 Guthrie, W. K. C.: *Orpheus and Greek Religion.* London: 1935, p. 245.

course of one period, know from this everything which is going to happen in all subsequent periods. For there will never be any new thing other than that which has been before down to the minutest detail."[14]

Essentially the same philosophy of time can be found in such different works as the Book of Ecclesiastes ("Is there anything whereof it may be said, See this is new? It hath been already of old time which was before us.") and Virgil's Fourth Eclogue, in which occurs the famous passage: "A second Tiphys shall then arise and a second Argo to carry chosen heroes; a second warfare too shall there be, and again a great Achilles be sent to Troy."

A cardinal factor that caused time to become a concept of primary importance was the spread of Christianity. Its central doctrine of the Crucifixion was regarded as a *unique* event not subject to repetition and so implied that time must be linear rather than cyclic. Before the rise of Christianity only the Hebrews and Zoroastrian Iranians appear to have developed any teleological conceptions of the universe or had any idea that history is progressive. The historical view of time, with particular emphasis on the non-repeatability of events was, however, the very essence of Christianity. The contrast with the Hebrew view is clearly brought out in the *Epistle to the Hebrews*, chapter 9, verses 25–6: "Nor yet that he should offer himself often, as the high priest entereth into the holy place every year with the blood of others; For then must he often have suffered since the foundation of the world; but now once in the end of the world hath he appeared to put away sin by the sacrifice of himself".

Prominent among the early Church Fathers who, in their struggles against rival doctrines competing for the spiritual conquest of the Greco-Roman world, vigorously disputed the traditional cyclical view of time was St. Augustine. In many passages, both in his *Confessions* and in *The City of God*, we find him passionately concerned with the problem of the nature of time. In the latter work he wrote: "The pagan philosophers have introduced cycles of time in which the same things are in the order of nature constantly being restored and repeated, and have asserted that these whirlings of past and future ages will go on unceasingly ... From this mockery they are unable to set free the immortal soul, even after it has attained wisdom, and believe it to be proceeding unceasingly to false blessedness and returning unceasingly to true misery ... It is only through the sound doctrine of a rectilinear course that we can escape from I know not what false cycle discovered by false and deceitful sages."[15]

Nevertheless, despite this emphasis on linear advance, the idea of denominating the years serially in a single era count, such as the Olympic dating from 776

---

[14] Nemesius, Bishop of Emesa (4th cen.) in J. von Arnim: *Stoicorum Veterum Fragmenta*. II, frag. 625; transl. E. Bevan, *Later Greek Religion*. London: 1927. See also Goldschmidt, V.: *Le Système Stoicien et l'idée de Temps*. Paris: 1953, p. 188, footnote 5.

[15] St. Augustine, *The City of God*, Book XII, Ch. XIII (quoted in translation from Baillie, J.: *The Belief in Progress*. Oxford: 1950, p. 75).

B. C. and the Seleucid from 311 B. C., did not originate in the Christian era until it was introduced by Dionysius Exiguus in 525 A. D., and the B. C. sequence extending backwards from the birth of Christ was only introduced by D. Petavius in 1627.[16] In medieval Europe, just as in medieval China and pre-Columban America, time was not conceived as a continuous mathematical parameter but was split up into separate seasons, divisions of the Zodiac, and so on, each exerting its specific influence. Despite the fact that the Incarnation was a unique event, not subject to repetition, the theory of cycles and of astral influences was accepted by most Christian thinkers down to the seventeenth century. In other words, magical time had not yet been superseded by scientific time. Throughout the whole medieval period, there was a conflict between the cyclic and linear concepts of time. The scientists and scholars, influenced by astronomy and astrology, tended to emphasize the cyclic concept. The linear concept was fostered by the mercantile class and the rise of a money economy. For, as long as power was concentrated in the ownership of land, time was felt to be plentiful and was associated with the unchanging cycle of the soil. With the circulation of money, however, the emphasis was on mobility. In other words, men were beginning to believe that "time is money" and that one must try to use it economically and thus time came to be associated with the idea of linear progress, an idea that was reinforced by religious tendencies in the Reformation.

Although medieval scholars were not concerned with machines, they became more and more interested in clocks, particularly because of their connection with astronomy. For it was generally believed that a correct knowledge of the relative positions of the heavenly bodies was necessary for the success of most earthly activities, and many early clock incorporated elaborate astronomical representations. The most celebrated was the Strasbourg clock set up in 1350, but the most elaborate was that designed by Giovanni de'Dondi of Padua between 1348 and 1364.[17] His famous contemporary Nicole Oresme likened the universe to a vast mechanical clock created and set moving by God so that 'all the wheels move as harmoniously as possible'.[18]

The great leaders of the scientific revolution of the seventeenth century were also much concerned with horological questions and metaphors. Early in the century Kepler specifically rejected the old quasi-animistic magical conception of the universe and asserted that it was similar to a clock, and later the same analogy was drawn by Robert Boyle and others. Thus the invention of the mechanical clock played a central role in the formulation of the mechanistic conception of nature that dominated natural philosophy from Descartes to Kelvin. An even more far-reaching influence has been claimed for the mechanical

---

16 Bickerman, E. J.: *Chronology of the Ancient World*. London: 1968, p. 10.
17 Bedini, S. A., Maddison, F. R.: *Mechanical Universe – the Astrarium of Giovanni de Dondi*. Trans. Am. Phil. Soc. 56, Part 5 (1966) 5–69.
18 White, L. Jr.: *Medieval Technology and Social Change*. Oxford: 1962, p. 125.

clock by Lewis Mumford who has argued that it "dissociated time from human events and helped to create the belief in an independent world of mathematically measurable sequences: the special world of science."[19]

Moreover, the invention of a satisfactory mechanical clock had a tremendous influence on the concept of time itself. The oldest modes of timekeeping were essentially discontinuous. For, instead of depending on a continuous succession of temporal units, they merely involved the repetition of a concrete phenomenon occurring within a unit – as, for example, where Homer in the Twenty-first Book of the *Iliad,* makes one of Priam's sons say to Achilles "This is the twelfth dawn since I saw Ilion." Even the sundials, sandreckoners and waterclocks of antiquity tended to be more or less irregular in their operation. It was not until a successful pendulum clock was invented by the Dutch scientist Christiaan Huygens in the middle of the seventeenth century that man was at last provided with an *accurate* timekeeper that could tick away continually for years on end. This must have greatly influenced belief in the homogeneity and continuity of time.

These characteristics were implicit in the idea of time put forward by Galileo in the dynamical part of his book *Two New Sciences,* published in 1638. Although he was not the first to represent time by a geometrical straight line, he became the most influential pioneer of this idea through his theory of motion. Nevertheless, for the first explicit discussion of the concept of geometrical time it seems that we must go to the *Geometrical Lectures* of Isaac Barrow written about thirty years after the publication of Galileo's book. Barrow, who occupied the chair of Mathematics in Cambridge in which he was succeeded by Newton in 1669, was much impressed by the kinematic method in geometry that had been developed with great effect by Galileo's pupil Torricelli. He realized that to understand this method it was necessary to study time, and he was particularly concerned with the relation of time and motion. "Time does not imply motion, as far as its absolute and intrinsic nature is concerned; not any more than it implies rest; whether things move or are still, whether we sleep or wake, Time pursues the even tenour of its way."[20] However, he argues, it is only by means of motion that time is measurable. "Time may be used as a measure of motion; just as we measure space from some magnitude, and then use this space to estimate other magnitudes commensurable with the first; i.e. we compare motions with one another by the use of time as an intermediary." He regarded time as essentially a mathematical concept which has many analogies with a line "for time has length alone, is similar in all its parts and can be looked upon as constituted from a simple addition of successive instants or as from a continuous flow of one instant; either a straight or a circular line". The reference here to "a circular line" shows that Barrow was not completely

---

19 Mumford, L.: *Technics and Civilization.* London: 1934, p. 15.
20 Barrow, I.: *Lectiones Geometricae* (transl. E. Stone), London: 1735, Lect. 1, p. 35.

emancipated from traditional ideas. Nevertheless, his statement goes further than any of Galileo's, for Galileo only used straight line *segments* to denote particular intervals of time. Barrow was very careful, however, not to push his analogy between time and a line too far. Time, in his view, was "the continuance of anything in its own being."

Barrow's views greatly influenced his illustrious successor, Isaac Newton. In particular, Barrow's idea that, irrespective of whether things move or are still, whether we sleep or wake, "Time pursues the even tenour of its way" is echoed in the famous definition at the beginning of Newton's *Principia:* "Absolute, true and mathematical time, of itself, and from its own nature, flows equably without relation to anything external."

Newton's views, in turn, made a great impression on the philosopher John Locke. In his famous *Essay concerning Human Understanding* (1690), we find the clearest statement of the scientific conception of time that was evolved in the seventeenth century: "duration is but as it were the length of one straight line extended *in infinitum*, not capable of multiplicity, variation or figure, but is one common measure of all existence whatsoever, wherein all things, whilst they exist equally partake. For this present moment is common to all things that are now in being, and equally comprehends that part of their existence as much as if they were all but one single being; and we may truly say, they all exist in the same moment of time."[21]

Thus by the end of the seventeenth century we seem to have arrived at the idea of time that dominated physical science until the advent of Einstein's special theory of relativity – an idea that can be baldly summarized in the symbol $t$ denoting the continuous independent variable of classical dynamics. But the revolution in human thought that this idea implied took effect only gradually and traditional ideas lingered for a long time. For example, even Newton himself adhered to the old cyclical conception of the cosmos based on the conviction that the world was coming to an end. He was convinced that the comet of 1680 had just missed hitting the Earth and in his commentaries on *Revelations* and the *Book of Daniel* he indicated that the end of the world could not be long delayed. Robert Boyle too believed that "the present course of nature shall not last always, but that one day this world, or at least, this vortex of ours, shall either be abolished by annihilation, or, which seems far more probable, be innovated, and, as it were transfigured, and that, by the intervention of that fire, which shall dissolve and destroy the present frame of nature."[22]

The symbol $t$ in Newtonian dynamics carries no arrow, for the equations of motion are unaltered by a reversal in the direction of time. Only with the

---

21 Locke, J.: *Essay concerning Human Understanding*. 1690, Book II, Chapter 15, Paragraph 11.
22 Boyle, R.: *The Excellence of Theology compared with Natural Philosophy*. (1665). London: 1772, p. 11.

rise of thermodynamics in the nineteenth century did the problem of this direction enter physics. Nevertheless, the concept of linear progress in history was the concomitant of the rise of modern science, particularly as prognoticated by Francis Bacon, one of whose works, dating from 1603, bore the significant title *Temporis Partus Masculus (The Masculine Birth of Time)*. In this he wrote: "Science is to be sought from the light of nature, not from the darkness of antiquity. It matters not what has been done; our business is to see what can be done."[23] The scornful rejection by Bacon of the doctrine that the ancients had encompassed all knowledge was echoed by, among others, John Wilkins, who, in 1638, in *The Discovery of a New World* – in which he attempted to show that the Moon is inhabited – wrote: "there are yet many secret Truths which the Ancients have passed over, that are yet left to make some of our Age famous for their Discovery." And, two years later, in his *A Discourse Concerning a New Planet,* in which he advocated the Copernican theory, he wrote, even more significantly for our purpose: "Antiquity does consist in the Old Age of the World, not in the Youth of it. In such learning as may be increased by fresh Experiments and new Discoveries; 'tis we are the Fathers, and of more Authority than former Ages; because we have the advantage of more Time than they had, and Truth (we say) is the Daughter of Time."[24] The spirit of optimism with which, as the seventeenth century progressed, men came to regard the possible achievements of scientists was ecstatically expressed by Henry Power in his *Experimental Philosophy* (1664): "the Solutions of all those former Difficulties are reserved for you (most noble Souls, the true lovers of Free and Experimental Philosophy) to gratifie Posterity withall. You are the enlarged and Elasticall Souls of the world who, removing all former rubbish, and prejudicial resistances, do make way for the Springy Intellect to flye out into its desired Expansion. When I seriously contemplate the freedom of your Spirits, the excellency of your Principles, the vast reach of your Designs, to unriddle all Nature; me-thinks, you have done more than men already, and may be well placed in a rank Specifically different from the rest of groveling Humanity."[25]

The influence of the scientific revolution of the seventeenth century on the change in men's attitude to time, by directing attention away from the past, with its cyclical assumptions, to the future, with its prospect of linear advancement, has not so far been treated with the attention to detail that its importance merits. Although the spirit of philosophical optimism that this change encouraged took a severe knock from the great Lisbon earthquake of 1755, the eighteenth century enlightenment continued to develop on the basis of the new forward-looking view of time. Moreover, in this same century leaders of thought finally abandoned the biblical chronology that excluded the possibility

---

23 Farrington, B.: *The Philosophy of Francis Bacon.* Liverpool: 1964, 69.
24 Wilkins, J.: *Mathematical and Philosophical Works.* 1708, p. 146.
25 Power, H.: *Experimental Philosophy.* 1664, p. 191.

of slow processes of transformation over immense periods of time. In his *Theory of the Earth*, first published in 1785, the geologist James Hutton declared "We find no vestige of a beginning – no prospect of an end."

The linear view of time as continual progression finally prevailed through the influence of the nineteenth century biological evolutionists. Meanwhile, everyday life, at least in the countries influenced by European and American technical developments, tended to become more and more subservient to the tyranny of time. "The clock", as Lewis Mumford rightly declares, "not the steam-engine, is the key-machine of the modern industrial age."[26] The popularization of time-keeping that followed the mass-production of cheap watches in the nineteenth century accentuated the tendency for even the most basic functions of living to be regulated chronometrically: "one ate, not upon feeling hungry, but when prompted by the clock: one slept, not when one was tired, but when the clock sanctioned it."[27] This regulation of our lives by the clock meant that the abstract concept of absolute time as something existing in its own right came to be endowed with a spurious concrete reality. Even the division of the Earth's surface, in 1885, into separate time-zones did little to undermine this widespread belief. When, in 1916, daylight saving (now called British Standard Time) was first introduced in the United Kingdom, the popular novelist Marie Corelli was not the only one who was scandalized by this hubristic interference with what she called 'God's Own Time'! But before we preen ourselves on being above that sort of nonsense, let us not forget that Einstein's penetrating criticism, in 1905, of the classical concept of universal simultaneity that precipitated the final downfall of Newtonian absolute time was at first rejected out of hand by many leading scientists and most philosophers.

Our idea of time as linear advancement without cyclical repetition is the central feature of our current mode of thought about the nature of things. It is a sophisticated concept and, as the historical evidence reveals, it is far from being a necessity of thought. The fact that we have, nevertheless, an ingrained tendency to regard our concept of time as intuitively obvious, and therefore presumably less worthy of detailed consideration than other apparently more exciting and contentious topics, may help to account for the curious fact that this in the first international conference devoted to the study of time in all its aspects.

---

26 Mumford, L., *op. cit.*, p. 14.
27 Mumford, L., *op. cit.*, p. 17.

# The Time Coordinate in Einstein's Restricted Theory of Relativity

J. D. NORTH*

*Summary.* After drawing some preliminary divisions between different conceptions of time, distinguished as "proper time", "Kantian time", "psychological time", "physical time", and "mathematical time", there follows a discussion of time dilatation and the clock paradoxes of Einstein's restricted theory of relativity. Beginning with Lorentz's time transformations, and some remarks on the different attitudes of Lorentz, Poincaré, and Einstein to the transformation equations generally, we turn to the clock paradox, often held to involve Einstein in inconsistency. A new notation is introduced, which not only simplifies the manipulation of the Lorentz transformation equations, but which makes it natural to preserve certain terms which are commonly suppressed unknowingly. The notation makes clear what has long been appreciated by some, but denied by others, namely, that the time dilatation equations, correctly expressed, are consistent under a suitable interpretation (and not merely mathematically consistent). Different attempts to engender contradiction within Einstein's theory are considered, in particular by taking more than two observers, moving uniformly. Necessarily we introduce the notion of an aggregate proper time of two observers, which may not satisfy those looking for a solution to the problem of actual space travellers. There are also difficult problems of specifying what is to count as a journey, under this interpretation. Nevertheless, there is here no sign of the inconsistency in the interpretation of Einstein's theory for which some writers have argued. Where there is an asymmetry in the aggregate times experienced, there is also an asymmetry in the observers.

It was predictable enough that, after the first flush of enthusiastic hero-worship, historians would begin to question Einstein's authorship of the several principles on which the restricted theory of relativity is founded. Given a little more time, and no doubt we shall see their ancestry traced back to William the Conqueror. There are few, even so, who are prepared to deny that Einstein alone derived from his few basic principles, borrowed or not, the theory as it is now generally accepted. One might imagine, therefore, that the interpretation of the theory is no longer a matter for dispute, but this is not so, and even in recent years a good deal of controversy has been concerned with interpretations of Einstein's time coordinate, and of those equations which suggest the idea of "time-dilatation". It is on time-dilatation and the so-called "clock paradox" that I want to dwell. In particular I shall consider versions of the paradox which, by introducing three or more clocks, avoid accelerations. Little of what I have to say is in substance new, and I should make it plain at the outset that I shall not argue either for or against the overall empirical viability of Einstein's

---

* Dr. J. D. North, 28 Chalfont Road, Oxford, Great Britain.

restricted theory under any interpretation. Amongst other things, however, I wish to show that there are certain charges of inconsistency from which the theory may be saved, although not necessarily by following Einstein himself to the letter.

What subsequently became known as the "clock paradox" was given a prominent enough place in Einstein's 1905 paper, and yet at first it provoked less discussion than did Einstein's denial of absolute simultaneity, and his finding that there are events – those outside the observers' light cones, in Minkowski's parlance – whose order in time is different for different observers in relative motion. Once these ideas on simultaneity were widely accepted, it became impossible to assume uncritically an objective and universally acceptable time order. But if the old absolutes had gone, there was a new absolute soon to take their place in the form of an absolute of events. No-one did more to fuse space and time into such an absolute than Minkowski, with the much quoted opening words of his famous paper of 1908, *Raum und Zeit*.[1] Before long, Weyl and Eddington took up the cry with great eloquence, speaking of the human consciousness as something which ranged over a universe pre-existing in space and time, and which singled out its own unique section of experience.[2] The objectivity not so much of time as of time *in isolation* had been called in question. Local, personal or proper time, the time appropriate to any place as taken from a clock at that place, continued to be held meaningful; but all else was inevitably a construction from such local times and from spatial coordinates. When we come to discuss time dilatation and the clock paradox, it will be useful to make a strict division – strict, but founded on an epistemologically unpretentious thesis – between the time of local and distant events. But this is a distinction which cuts across certain others which are more fundamental in the sense that they do not concern a particular scientific theory or group of theories. These more fundamental distinctions I wish to consider first.

---

1 Eleven important papers on relativity, including *Zur Elektrodynamik bewegter Körper* (Einstein, 1905), *Raum und Zeit* (Minkowski, 1908), and two earlier papers by Lorentz (1895 and 1904) are readily available in English translation in "The Principle of Relativity", London, Methuen, 1923, reprinted by Dover subsequently. The translation was made from the German edition "Das Relativitätsprinzip", 4th edition, Teubner, 1922. References below are to the English edition, abbreviated as P. R.

2 Hermann Weyl's *Raum, Zeit, Materie* is available in English translation, unfortunately from the fourth German edition of 1921, rather than the fifth, with the title *Space – Time – Matter*, London, Methuen, 1921, subsequently reprinted by Dover. For A. S. Eddington, see especially *The Mathematical Theory of Relativity*, 2nd edition, Cambridge University Press, 1924, p. 26: "the contraction [of a moving rod] and retardation [of a moving clock] do not imply any absolute change in the rod and clock. The 'configuration of events' constituting the four-dimensional structure which we call a rod is unaltered; all that happens is that the observer's space and time partitions cross it in a different direction".

*Some Preliminary Divisions*

Grandiose metaphysical schemes are a thing of the past, and few would wish it otherwise. One thing to be said in their favour, however, was that their authors were as often as not careful to say precisely what they meant in speaking of the reality (or unreality) of time. Today, "time is real" may mean such disparate things as that "clocks are real", and that "the relative rate of two given clocks is single-valued". We can do worse, under the circumstances, than turn to Kant, who, to say the least, drew his distinctions more finely than we are prone to do.

Time, for Kant, is the form of the intuitions of the self and of our internal state, rather than of outward phenomena.[3] It does not subsist of itself or inhere in things as something objective.[4] It is in fact a presupposition made prior to our appreciation of the coexistence and succession of things in time, and may be thought of independently of phenomena, whereas the converse is not the case.[5] Time was for Kant a necessary condition of all our internal experience, and the reality of the objects of experience was not in doubt. We notice that in all this he confined his attention to time at the observer – proper time. Had there been a generally accepted theory of time relations analogous to the Euclidean theory of space, there is little doubt but that at this point Kant would have argued that time necessarily has a corresponding structure, exactly as he held space to be necessarily Euclidean. He came near to such an argument when he wrote:

> And precisely because the internal intuition presents to us no shape or form, we endeavour to supply this want by analogies, and represent the course of time by a line progressing to infinity, the content of which constitutes a series which is only of one dimension ...[6].

It seems that he has taken time to be a form without structure except insofar as it is pulled into shape by certain apodeictic (clearly demonstrated) principles, or "axioms of time in general", examples of which are, according to him, "Time has only one dimension", and "Different times do not coexist, but are in succession".[7] Such axioms are valid as rules through which experience is possible; and they are not conceived of as coming from experience, for in that case they would give "neither strict universality nor apodeictic certainty". This is a weak point of Kant's argument. As G. J. Whitrow among others has demonstrated[8], there are axioms governing temporal experience of a value which can only be judged by experience, and then only in conjunction with other principles, if

---

3 *Kritik der reinen Vernunft*, 2nd edition, 1787, I. ii. 7 (b).
4 *Ibid.*, I. ii. 7 (a).
5 *Ibid.*, I. ii. 5 (1–2). This did not mean that it had absolute reality in the sense of inhering in things as a condition or property. *Ibid.*, I. ii. 7 (c).
6 *Ibid.*, I. ii. 7 (b).
7 *Ibid.*, I. ii. 5 (3).
8 Whitrow, G. J.: *The Natural Philosophy of Time*. London: Nelson, 1961, chapter 4.

anything of scientific moment is to be derived. The old problem of the viable alternative, so familiar in relation to Euclidean geometry, presents itself again here, to the detriment, as I believe, of Kant's position. On the other hand, we might consider Kant to have made, as it were, a distillation of human experience of space and time, a distillation so pure that *ordinary* experience is always likely to include it as an ingredient.

Kant's philosophy of space and time is often regarded as concealing an essentially psychological thesis on the nature of the human mind. Almost anything one chooses to say about time – indeed about almost anything – may be translated into a statement of psychology, and I have no intention of pursuing the idea here. Time, proper time, treated as an *a priori* necessity, we may call "Kantian time". Time for which axioms provide a structure, and which "instruct us in regard to experience", to use Kant's phrase[9], we might call "physical time". (The two sorts of time might overlap, but need not necessarily do so.) The difference between the modern physicist and Kant is that the physicist tends not to regard his axioms as apodeictic. Einstein's claim that such axioms must involve space as well as time, however, is of little consequence in any reformulation of Kant's philosophical position.

Kant did not consider the possibility that the truth of an interpreted set of axioms may be preserved by varying the interpretation placed on them, but neither did he consider the axioms themselves, in isolation from their interpretation. The uninterpreted axioms embrace what we might call "mathematical time", by analogy with the term "space" as used by the pure mathematician without any thought as to its physical interpretation. (William Rowan Hamilton perceived this, when he argued that as geometry is the mathematics of space, so algebra must be the mathematics of time.)[10] Any particular species of mathematical time may be said to "exist" as long as the axioms describing it are consistent.

Although few mathematicians have arrogated to themselves the word "time" as they have done the word "space", many a physicist defending the truth of Einstein's restricted theory has considered his task done when he has underlined ("proved" would be too strong a word) the consistency of the theory. I think it was Eddington who once said that the Astronomer Royal's time was in the position of a vested interest, and was therefore deemed to require no philosophical support. *Plus ça change, plus c'est la même chose.* Today it is Einstein's time which

---

9 See note 7.
10 *Theory of conjugate functions, or algebraic couples; with a preliminary and elementary essay on algebra as the science of pure time* (read in 1833 and 1835), Transactions of the Royal Irish Academy, Vol. 17 (1837) pp. 293–422. Cf. p. 297: "The notion of intuition of ORDER IN TIME is not less but more deep-seated in the human mind, than the notion or intuition of ORDER IN SPACE; and a mathematical Science may be founded on the former, as pure and demonstrative as the Science founded on the latter". Since Hamilton hoped to derive algebra from a prior intuition of time, this was not strictly mathematical in the sense used here.

is the vested interest. Although, as I have said, I shall myself accept it in order to find some of its implications, I hope to make it clear by adding "mathematical time" to my list that Einsteinian time needs to be much more than a term in a consistent mathematical system.

When a physicist speaks of "time observed" or "time experienced", he is usually referring in the last analysis to clock readings, rather than to that special sort of physical time (in our earlier sense) which we might reasonably call "psychological time". The mind plays tricks with time – which is another way of saying that under a psychological interpretation we cannot expect our physical postulates to prove acceptable. This problem is not without relevance to a strict analysis of the "twins" version of the clock paradox, but there are problems enough when we suppose time to be registered locally by a well-behaved and impersonal clock, without going into this further complication. Initially it is necessary to take this and a good deal more for granted, both of a physical and a philosophical sort. Is it not to beg a great many questions to suppose that mechanical clocks must needs keep the same time under all circumstances as clocks based on electromagnetic principles? And what meaning are we to give to the statement that a clock beats at a constant rate? Although not perhaps without solutions, it is as well to remember that such problems are not trivial. Henceforth, nevertheless, the idea of local or proper time will be accepted as a datum, and it will be assumed that no other immediate interpretation of the time coordinate in terms of experience is possible.

*Ideas of Time Dilatation before Einstein*

Formulae for the transformation of the time coordinate had appeared on several occasions in scientific writings of the decade ending in 1905. From some time before 1895, Lorentz had been investigating transformation equations for electromagnetic systems moving within the aether. He had made use of the idea of a "local time" ($t'$) appropriate to a moving observer, this being defined in terms of the true and absolute time ($t$) by the equation $t = t' + vx'/c^2$. (Here $v$ is the velocity of the moving observer, and $x'$ the spatial coordinate with respect to an observer on the Earth, both being reckoned in the direction of the motion.) At first Lorentz ignored quantities of higher order than $v/c$, and not until 1904 did he find the precise form of the transformation under which the Maxwellian equations of the aether are invariant.[11] Following Poincaré, we call the full set of equations by Lorentz's name. (They are quoted in full subsequently.) In 1900 Joseph Larmor had extended Lorentz's original analysis so as to include quantities of the second order in $v/c$. In the course of doing so he gave extensive consideration to the transformation $t'' = t' - \varepsilon v x'/c^2$, where

---

11 For an ample bibliography of relativity, see the references given in E. T. Whittaker's *History of the Theories of Aether and Electricity,* Vol. 2. London: Nelson 1953, chapter 2. Lorentz's paper of 1904 (not 1903, as given in error by Whittaker) is printed in P. R.

$1/\varepsilon = (1 - v^2/c^2)^{\frac{1}{2}}$, and where the symbols $t''$ and $t'$ are retained in deference to Larmor's notation.[12] On one or two occasions after 1905, Lorentz admitted that he had never considered his "local time" as having anything to do with real and absolute time. He considered his time transformation, he said, "only as a heuristic working hypothesis", a convenient grouping of physically meaningful terms, as we might say.[13] Larmor similarly made nothing of the paradox which his transformation equation might have been thought to entail, although he fully appreciated that "The change of the time variable in the comparison of radiations in the fixed and moving systems, involves the Doppler effect on the wavelength".[14] Time dilatation as an idea associated with important physical effects had to wait for Einstein.

The complete Lorentz equations were derived on the basis of the idea of a fixed aether, an extended form of Maxwell's theory of electromagnetism, and an "electric theory of matter". As is well known, Einstein's basic assumptions were somewhat different. Accepting the principle of relativity itself, that "the same laws of electrodynamics and optics will be valid for all frames of reference for which the equations of mechanics hold good", and the postulate that "light is propagated in empty space with a definite velocity $c$ which is independent of the state of motion of the emitting body", he placed no reliance on Lorentz's electrical theory of matter. He nevertheless accepted Maxwell's theory for stationary bodies, and he made use of the concept of a rigid body. His derivation of the Lorentz equations was inevitably different from Lorentz's, while others subsequently offered derivations different from his, though usually based on comparable principles. By common consent, however, a great conceptual gulf lies between Einstein on the one hand, and Poincaré and Lorentz on the other.

The different approaches of the three men cannot be adequately dealt with in a few sentences. It is sufficient here to point out that Lorentz to the end of his life denied that the Lorentz-Fitzgerald contraction is only an appearance. He is difficult to follow on this point, however, since he was capable of adding these words to his denial: "On the contrary, the contraction could actually be observed by $A$ as well as by $B$, nay, it could be photographed by either observer ...".[15] This seems an odd way of denying that the contraction is only an appearance. Similarly he emphasized that time dilatation is real. When $A$ says that $B$'s clock goes slower, this, he believed, "expresses a real phenomenon". Since $B$ may make a similar statement, is it not perverse of Lorentz to say that the phenomenon is real? He is in fact saved from inconsistency – at least in regard to

---

[12] *Aether and Matter,* Cambridge University Press, 1900, chapter 11, especially pp. 167–77.
[13] See in particular Astrophysical Journal, Vol. 68 (1928), pp. 385–8, and *Lectures on Theoretical Physics*, Vol. 3, London: McMillan 1931, pp. 181 to the end. These lectures on relativity date from 1910–12.
[14] *Op. cit.* (note 11), p. 177.
[15] *Op. cit.* (1931, note 12), p. 303.

time, which is all that concerns us here — by the realization that the statements of $A$ and $B$ "express entirely different things, although both observers use the same words". I shall shortly have more to say of these "different things" and their nature.

As for a more fundamental distinction between the writings of Einstein and the other two, suffice it to say that Lorentz and Poincaré seem to have thought of the impossibility of detecting absolute motion as an empirical matter, whereas for Einstein the principle of relativity was, if not exactly accorded the status of *a priori* certainty, at least not one he would have willingly sacrificed in the case of a clash between the theory as a whole and observation. In his 1905 paper, indeed, Einstein spoke of raising the conjectured principle of relativity to a *postulate*, and in due course it became customary for him and exponents of his theory to say that it was actually *meaningless* to assert that a certain body has a certain absolute motion. However philosophically enlightening this kind of remark may be, it cannot, as some people seem to think, save the restricted theory of relativity as a whole, should sufficiently damning experimental evidence turn up. What it can do, however, is attract the attentions of armchair scientists, and in particular those who would like to prove the theory as a whole incompatible with it.

*Einstein's Clock Paradox*

By far the best known of the inconsistencies — if that is the right word — are those which grew around section 4 of Einstein's 1905 paper.[16] I shall take it that the clock paradox, as it was to be called, or paradox of the twins, following Langevin's anthropomorphic version, is in outline well known, for it has drawn to itself an immense literature. Two synchronized clocks separate and re-unite, and one is held to be retarded in relation to the other, although motion is apparently inherent in neither, but is relative, involving both equally.[17] Here, they argued, is a contradiction, for which "paradox" is nothing more than a polite word.

How did Einstein reconcile himself to the idea in the first place? He phrased the problem rather strangely, subjecting the "moving" clock, not (as is now usual) to a single deceleration between outward and inward journeys undertaken at constant velocity, but to a constant acceleration, the whole journey being in a curved path (covered at constant speed). This device he must later have realized to be illegitimate, for he did not use it again. Later, presenting the paradox in its now usual form, he ascribed the predicted discrepancy between clock readings to a real asymmetry between clocks: a force, he said, was required to move

---

16 P. R., pp. 48–50.
17 That the criticism did not pass unnoticed is suggested by Bishop Barne's *Scientific Theory and Religion*, Cambridge University Press, 1933, p. 114, for example.

one of the clocks, that is, to change its inertial frame.[18] This explanation of the asymmetry has since become fairly standard, although there are several variants of the simple idea. It may be said, for instance, that the moving clock may be identified in absolute terms because it changes its disposition with respect to a very special inertial frame, namely that at rest with regard to the universe as a whole. But such an absolute, it has been held, is not in the spirit of Einstein's restricted theory. Without pursuing the question as to whether in 1905 Einstein showed a hidden preference for one set of space and time measurements rather than another, I will simply point out that in envisaging two clocks, one at the Earth's equator and the other at one of the poles, it was that at the *equator* which he took to go more slowly. The implications are obvious.

The kinematic part of the restricted theory of relativity deals, of course, with a limited class of physical situations, referring them to coordinate frames not in relative acceleration. If it is to be proved capable of leading to inconsistency, then clearly this cannot be by a consideration of problems outside that restricted class. On the other hand, we plainly say nothing for the restricted theory if we choose to solve the problem of the space-travelling twin by recourse to other theories, such as Einstein's general theory of relativity. The paradox, if it is indeed a paradox, is one only with regard to the expectations of the restricted theory. It has been said that if the restricted theory contains a genuine inconsistency, then the very foundations of Einstein's general theory crumble, and hence the fact that the general theory offers a solution involving accelerations is of no consequence. This is not unduly pessimistic. Certainly there is no inconsistency in kicking over the ladder by which one has climbed; but the fact that the general theory reduces to the special in the absence of gravitating matter suggests that inconsistencies of the special theory are inevitable in the general. If they exist, then they are not to be taken lightly.

*The Lorentz Transformation Equations, and Some of Their Consequences*

Although it is impossible to give an entirely verbal analysis of the problem, I shall reduce the mathematics to a minimum, and leave the details to my footnotes. Our first problem is one of notation, which if unchecked may itself easily become a source of error. At the expense of simplicity, I shall represent by the symbol $t_{AB}^{E_B}$ the time assigned by an observer $A$ to an event $E_B$ which occurs at a place $B$. Where there is no cause for ambiguity, I shall denote this more simply by the symbol $t_A^B$. The merit of this notation is that it removes the need for such circumlocutions as are commonly found introducing the formulae of the theory, where for instance we read of events "viewed from the system $S$", "observed from the perspective of an imaginary observer $O$" and so forth. Equally valuable is the fact that the notation allows us to distinguish at a glance

---

[18] Naturwiss., Vol. 6 (1918), p. 697.

between local time, written in the form $t_A^A$, and the time assigned to events at a distance, written in the form $t_A^B$, where $B$ and $A$ are different.

Using this notation the inherent form of the "Lorentz transformation equations" and their inverses is plainly evident, restricting attention to one spatial dimension only, as is not unusual. I shall take the two $x$-axes of coordinates in opposite directions, in order to make the symmetry of the two sets of equations fully apparent:

$$x_B^E = \beta(vt_A^E - x_A^E) \tag{1}$$

$$t_B^E = \beta(t_A^E - \frac{v}{c^2} \cdot x_A^E) \tag{2}$$

$$x_A^E = \beta(vt_B^E - x_B^E) \tag{3}$$

$$t_A^E = \beta(t_B^E - \frac{v}{c^2} \cdot x_B^E). \tag{4}$$

Here $E$ is an arbitrary event, $v$ is the relative velocity of each of the two coordinate frames with respect to the other, $c$ is the velocity of light, and $\beta = (1 - v^2/c^2)^{-\frac{1}{2}}$. The notation for $x$-coordinates follows that for $t$-coordinates, and clearly for any $A$, $x_A^A = 0$, whence follow the Eqs. (1')–(4'):

$$x_B^A = \beta vt_A^A \tag{1'}$$

$$x_A^B = vt_A^B \tag{1''}$$

$$t_B^A = \beta t_A^A \tag{2'}$$

$$t_B^B = \beta(t_A^B - \frac{v}{c^2} \cdot x_A^B) \tag{2''}$$

$$x_A^B = \beta vt_B^B \tag{3'}$$

$$x_B^A = vt_B^A \tag{3''}$$

$$t_A^B = \beta t_B^B \tag{4'}$$

$$t_A^A = \beta(t_B^A - \frac{v}{c^2} \cdot x_B^A). \tag{4''}$$

The Eqs. (3), (4), and their derivatives need not have been given at all, of course, but those unfamiliar with Einstein's theory will perhaps take comfort in the reciprocal nature of the two sets of equations, which in this respect at least conform with the principle of relativity.

In order to illustrate the use of the Lorentz equations with this notation, consider the usual derivation of the so-called "Lorentz-Fitzgerald contraction" of moving rods. In general, if $P$ is an event at one end of a rod, and $Q$ an event at the other end, a double application of Eq. (1) gives, after subtraction,

$$(x_B^P - x_B^Q) = -\beta(x_A^P - x_A^Q) + \beta v(t_A^P - t_A^Q).$$

The last bracket is usually suppressed from the outset, on such grounds as that "we may consider the case where $t=0$". If we do in fact equate the bracket to zero, however, we arrive at an equation which is flatly inconsistent with the similar equation which we have an equal right to derive from Eq. (3): referring to the absolute value of the difference of $x$-coordinates by the symbol $\varDelta$, from (1) we have $\varDelta_B = \beta \varDelta_A$, whilst from (3) we have $\varDelta_A = \beta \varDelta_B (\beta \neq 1)$. The mistake was in taking the events at the two ends of the rod as simultaneous *according to both* A *and* B. There is of course nothing new in the idea that the notion of simultaneity enters into the concept of a rigid rod, and it is a simple matter to show that if events at the two ends of the rod are taken as simultaneous with respect to either $A$ or $B$, but not both, then the two derived equations are no longer inconsistent, but are indeed one and the same equation. All this foreshadows a problem to come, namely that of laying down a definition of a journey in which two observers separate.

Problems of a less tractable sort begin with Eq. (2′) or (4′). Written in the conventional way, $t = \beta t'$, it appears that (2′) makes an absolute distinction between "observers" $A$ and $B$. But "from another point of view", goes the typical argument, "we can write $t' = \beta t$ (i.e. (4′)), and therefore relativity is upheld". A glance at (2′), however, shows that the principle of relativity is never in jeopardy, so long as we are prepared to distinguish between local time ($t_A^A$) and time as a coordinate of distant events ($t_B^A$). It is well known that Einstein introduced such time-at-a-distance into his theory by means of a definition in the first section of his 1905 paper. He imagined a system of clocks made synchronous with any given clock by means of an ideal experiment involving the reflection of light from any event. The time of the event was by definition to be the arithmetic mean of the times of despatch and return of the synchronizing signal. It is for time defined in this way that the symbol $t_B^A$ is here reserved.

At least five years ago H. Dingle introduced a somewhat unusual paradox of his own making with the purpose of proving that although the special theory of relativity is mathematically self-consistent, it "requires clocks to behave in an impossible manner, for one set of synchronized clocks must concomitantly go both faster and slower than another set, and this is impossible".[19] Dingle observed that when Einstein deduced the rate-ratio of two clocks he chose to compare times assigned to a very special pair of events, namely a pair at the origin of one of the two systems. Taking the first event as that of coincidence of $A$ and $B$ (in our previous notation), and the second as any event at $B$, then it seems that the clock at $A$ runs faster than that at $B$. But, as Dingle pointed out, why not take the second event on the other clock? When we do so, we shall decide that the clock at $B$ runs faster than that at $A$. This is what Dingle meant by saying that clocks are required to behave in an impossible manner.

---

19 British Journal for the Philosophy of Science, Vol. 15 (1964), p. 46.

If we denote the interval between the two events by "$dt$", then Dingle's two supposedly conflicting propositions, when expressed in the notation of (2') and (4'), are as follows:

$$\frac{dt_A^B}{dt_B^B} = \beta \qquad \begin{array}{l}\text{– } A \text{ judges the interval to be greater than does } B, \\ \text{and therefore "clock } B \text{ runs faster than clock } A\text{";}\end{array} \qquad \text{(i)}$$

$$\frac{dt_B^A}{dt_A^A} = \beta \qquad \text{– "clock } A \text{ runs faster than clock } B\text{"}. \qquad \text{(ii)}$$

The propositions might well seem flatly contradictory, and yet it is obvious that the verbal forms contain only a part of the meaning of the equations which precede them. I am afraid that I simply cannot see how clocks have been required to behave "in an impossible manner", although Dingle has certainly drawn attention to the dangers of expressing Einstein's theory too casually. Einstein's own interpretation was confused, or how, without further explanation could he have supposed an equatorial clock to go faster than a polar one? And as Dingle remarks, almost all subsequent textbooks have interpreted the rate-ratio in the same way as Einstein, who took the second calibration event to be at the "moving" clock.

The moral of all this is that we should not speak of relative clock rates at all without specifying the events with respect to which they are measured. Einstein was inconsistent to the extent that he spoke as though the ratio were *objective*, but he was under no compulsion to do so. Dingle's purpose was to discredit Einstein's interpretation, and not to indicate ways of saving it. He maintained, however, that "unless [a certain equation leading to (ii) above] is held to rule out [an equation leading to (i)], it may be proved – the proof is withheld for want of space – that the special relativity theory cannot fulfil its purpose of justifying the classical electromagnetic equations".[20] I am unable to envisage this proof, since (i) and (ii) are absolutely identical in form. In considering the kinematic interpretation of relativity, however, one should certainly keep an open mind as to possible inconsistencies in their electromagnetic consequences.

Turning to the clock paradox in its traditional guise, there are certain necessary distinctions which make it desirable that we pay even more attention than before to matters of notation. Suppose that identical clocks $A$ and $B$, synchronized and reading zero at the moment they coincide (the event $E_0$), separate at constant velocity $v$ (in the estimation of both) to some distance. Obviously

---

20 *Loc. cit.* (note 18).

we must distinguish between the event at $B$ ($E_b$, say) when the journey is reckoned by an observer at $B$ to cease, and the event at $A$ ($E_a$) when he reckons the journey to cease. It is a simple matter to show[21] that

$$t_A^{E_b} = \beta t_B^{E_b}$$

and

$$t_B^{E_a} = \beta t_A^{E_a}.$$

If we were to make the mistake of supposing that there was a unique, that is universally valid, moment at which the journey ended, and of identifying the time of either $E_a$ or $E_b$ as estimated by the two observers, then each equation is self-inconsistent, even in isolation, since $\beta \neq 1$ *ex hypothesi*. It goes without saying that to suppose the events simultaneous as judged from a single place (that is, to write $t_A^{E_a} = t_A^{E_b}$ and $t_B^{E_a} = t_B^{E_b}$) would give rise to a pair of inconsistent equations. These difficulties are easily avoided by saying that in each equation the left hand side is conventional or coordinate time, rather than a proper time, and therefore the equations do not compare real with real. To do so, and obtain the clock paradox proper, it is necessary to bring the clocks back into coincidence, and this we cannot do without introducing accelerations into the problem. As H. E. Ives[22] and others have indicated, however, the problem may be re-cast by the introduction of another moving clock. Before adopting this approach it is necessary to look more closely at the meanings which can be assigned to the phrase "time of the journey".

Suppose that $A$ and $B$ are in agreement when at relative rest as to their scales of length and time. The event $E_b$, marking the end of $B$'s journey, may then be regarded as the event when $B$ reaches a point (say the end of a rod) at distance $\Delta$ (measured by $A$) from $A$. Earlier I said that $E_a$ was an event at $A$ when he reckoned this same journey to cease. This might mean at least two different things. Henceforth I reserve $E_a$ to denote the event at $A$ when, in their relative motion, $A$ reaches a distance $\Delta$ (measured by $B$) from $B$. By $E_{a'}$, on the other hand, I denote an event at $A$ which occurs at the instant when $A$ judges $E_b$ to happen at $B$. In short, $t_A^{E_{a'}} = t_A^{E_b}$, by definition. $E_b$ and $E_{b'}$ are defined by analogy. By "time of the (outward)) journey" it seems that we may mean at least four sorts of time in the estimation of each observer, conceptually if not all numerically different. The relations between them are

---

[21] It is a consequence of (4') that

$$\int_{E_0}^{E_b} dt_A^{E_B} = \beta \int_{E_0}^{E_b} dt_B^{E_B},$$

whence (since $t_A^{E_0} = t_B^{E_0} = 0$, by definition) the two equations of the text follow.

[22] *Nature*, Vol. 168 (1951), p. 246.

summarized in the following table, the entries in which are justified more fully elsewhere:[23]

| time | $t^E_A a$ | $t^E_A b'$ | $t^E_A b$ | $t^E_A a'$ | $t^E_B a$ | $t^E_B b'$ | $t^E_B b$ | $t^E_B a'$ |
|---|---|---|---|---|---|---|---|---|
| scale factor | $t/\beta$ | $\beta t$ | $t$ | $t$ | $t$ | $t$ | $t/\beta$ | $\beta t$ |

It is of some interest to consider the status of the different sorts of "journey time" in this table. We may call $t^E_A a$ "real", for it is the local time of some event which could in principle be directly and locally observed – that of passing the end of a rod fixed in relation to $B$, for instance. At first sight, $t^E_A a'$ is just as real, but on closer examination it emerges that the event $E_{a'}$, although local to the clock at $A$, was simply postulated to be such that $t^E_A a' = t^E_A b$. It is no less conventional, therefore, than the quantity to which it is equated, which ultimately takes its meaning from Einstein's definition of the time at distant events. The fourth quantity, $t^E_A b'$, is doubly conventional, for it is the time of a distant event $E_{b'}$, itself postulated in the same way as was $E_{a'}$. The last four sorts of journey time to be listed in the table all have a counterpart in the four already discussed.

*Four Clocks and No Paradox*

Consider now the following symmetrical movement of four clocks. Clocks $A$ and $B$ separate at constant velocity $v$ under the same circumstances as before. A clock $C$ later passes $A$ (the event being $E_a$) and a clock $D$ passes $B$ (the event being $E_b$), the clocks $C$ and $D$ approaching along the same line as $A$ and $B$ formerly separated, and with the same relative velocity. For simplicity let it be supposed that $C$ and $D$ meet when they both read zero, and that as before each measures $x$-distances in the direction of the relative velocity of the other. They will of course assign negative $t$-coordinates to the moments at which they pass $A$ and $B$ respectively. If we allow ourselves to speak of the "joint experience" of $A$ and $C$ (or $B$ and $D$), and suppose it to be given

---

23 As a matter of definition (see the text),
$$t^E_A a' = t^E_A b \quad \text{and} \quad t^E_B b' = t^E_B a .$$
By Eqs. (4′) and (2′), respectively,
$$t^E_A b = \beta t^E_B b \quad \text{and} \quad t^E_B a = \beta t^E_A a .$$
These equations may be repeated with $E_{b'}$ and $E_{a'}$, giving two more relations.
$$t^E_B a = t^E_A b .$$
This follows from the fact that, by hypothesis, scales of length and time have been agreed upon, and each side of the last equation represents the time $B$ and $A$ respectively assigned to a journey of distance $\Delta$ at velocity $v$. (This is merely to affirm Lorentz transformation equations (3′) or (3″) of the text.)

by the total proper time of the round trip, then we may write these totals of proper time as

$$t_A^{E_a} - t_C^{E_a} \quad \text{and} \quad t_B^{E_b} - t_D^{E_b}.$$

Reference to our tables above shows that the positive terms in the two expressions are equal. (Each of $A$ and $B$ covers a distance – the length of the other's rod – which he judges to be equal to the estimate made by the other of his own rod; and the velocities are the same.) But what of the remaining terms, representing times as measured from $C$ and $D$? By what argument could we prove them to be equal or otherwise? We could say that at the event $E_a$ (at $C$) there is associated with $D$ a rod which just reaches from $D$ to $C$, and at the event $E_b$ (at $D$) there is a rod from $C$ to $D$, but we cannot assume without proof that these two rods are equal in length, for we cannot, as in classical kinematics, say that the two events $E_a$ and $E_b$ are simultaneous. It is tempting to produce an argument by "running time backwards", by appealing to symmetry, or by a *reductio ad absurdum*, which amounts to the same thing; but this would be to pre-suppose the very self-consistency of the theory which is in doubt. On the hypothesis that Einstein's restricted theory of relativity is consistent, however, it seems to me that, with the present formulation of the problem, the aggregate proper time predicted for one pair of clocks is equal to that for the other pair. The clock paradox has refused to appear, but up to a point it has been ruled out by *fiat*. We notice, however, that no hypothesis was made as to the relative motion of $C$ and $A$ (or $B$ and $D$), and there is no reason why, for example, $A$ and $C$ should not be taken as relatively stationary, in which case we are left with a three-clock problem. We might equally well stipulate that the return velocity of $C$ with respect to $A$ is equal to that of $D$ with respect to $B$.

The four-clock problem was introduced here partly in an attempt to engender contradiction – and the attempt has, so far, failed – and partly to find a symmetrical analogue to the traditional problem of *two* clocks (or twins), but without introducing accelerations. It falls down as a satisfactory analogue, however, in the absence of absolute simultaneity. With only two clocks, there are no problems of definition in regard to the *limits* of the journey of separation. A way of overcoming this problem with four clocks will be explained shortly.

The version involving three clocks may initially be characterized very briefly. According to $B$, he requires a time of $\Delta/\beta v$ to reach the end of $A$'s rod, while according to $D$, the return journey along $A$'s rod takes the same length of time. The total of proper time kept by $B$ and $D$ together ($2\Delta/\beta v$) is less than the time assigned by $A$ to the round trip.[24] With a little elaboration, the argument

---

24 The total "time assigned" to the journey by $A$ and $C$ is $t_A^{E_b} - t_C^{E_b}$. (It is to be remembered that clocks $C$ and $D$ are zeroed at the end of their journey, making $t_C^{E_b}$ etc. negative.) Applying Eq. (4′) of the text, this is $\beta$ times greater than the time kept locally by $B$ and $D$ jointly.

soon gives us the asymmetrical aging of twins, and so forth, in which for a variety of reasons the great majority today apparently believe. But one way of avoiding this solution is to say that there is an error in equating the "time assigned" by $A$ (and $C$) to the aggregate time kept on the two clocks $A$ and $C$. The time assigned is certainly not the time kept on a clock or clocks, but a coordinate time deriving from Einstein's definition of time at a distance. We are under no obligation, therefore, to accept this "asymmetrical aging" as real without further proof, but such proof is forthcoming, as I shall show.

We return to a fully symmetrical statement of the problem in terms of four clocks. If we stipulate that $A$ and $B$ separate with a relative velocity of $v$, that $C$ and $D$ approach with the same relative velocity, and that each has the same velocity relative to the clock it first passes, then it is easily shown that this last velocity turns out also to be $v$. In other words, $C$ and $B$ are relatively stationary, as are $D$ and $A$. It is this fact which makes for a simple description of the double journey. $A$ holds, as before, that he takes time $\Delta/\beta v$ to reach to a distance which $B$ claims to be $\Delta$. ($B$ maintains that $A$ takes time $\Delta/v$ and $C$, being at rest with respect to $B$, agrees.) In reaching $D$, which we may imagine to be at the end of $A$'s rod, again of length $\Delta$, $C$ reckons time $\Delta/\beta v$.[25] The time jointly experienced during the round trip by $A$ and $C$ will be simply $2\ \Delta/\beta v$, and it is usual to add that $B$ and $D$ will assess that time as $2\ \Delta/v$. But repeating the argument for the other pair, $B$ and $D$ will likewise jointly experience a time $2\ \Delta/\beta v$, which it may be said $A$ and $C$ will assess as $2\ \Delta/v$. Overlooking the assessments of times by distant observers, there is no paradox, for the aggregates of time experienced by the two pairs are the same. It is nevertheless imperative, if Einstein's restricted theory is to be saved from contradiction, that the assessments of times $2\ \Delta/v$ be explained away. There is no longer the obvious asymmetry of the three-clock version for us to fall back on. One alternative would be to examine more critically the nature of the times assessed by distant observers, which as already emphasized are not proper times kept on clocks, but coordinate times stemming from Einstein's definition of time at a distance. In this particular example these "mere coordinate times" are a somewhat incongruous mixture. The first, for instance, is a time at $A$ assessed by $B$ or $C$, added to a time at $C$ assessed by $A$ or $D$. What claims can a time of this sort have to physical significance, even when $C$ is reckoned to be $A$'s *alter ego*, as it were, undergoing the return journey on his behalf? Clearly what is needed is that such times assigned to distant events be in some way correlated with local events, which may in turn be compared with a local clock. Unfortunately, when this is done, the "mere coordinate times" have an uncomfortably real look about them, for they simply turn into proper times. (I considered the (proper) times of receipt of continuous

---

25 The $v$ here is $C$'s velocity with respect to $A$, not with respect to $D$. Given that the latter were different from $v$ so would the former be different. $C$ and $B$ (and $A$ and $D$) would no longer be relatively stationary, and this simple statement of the problem would not be possible.

light signals.) There is, after all, no easy way out of the dilemma for those who would like to make a category distinction between "proper time" and "assigned time", and then say that there can be no conflict between the two expressions ($2\Delta/\beta v$ and $2\Delta/v$) for the time. I shall come back to this point at the end. First, however, the four clock problem will be re-considered in terms which make it unnecessary to stipulate equal velocities of outward and return journeys.

The argument from four clocks in the general form given at the outset had certain weaknesses. Even overlooking $C$ and $D$, in what sense do $A$ and $B$ perform the *same* journey? The assumption that they do was tacitly made when we compared $B$'s estimate (rather than $D$'s or that of some other observer) of $A$'s journey with his own, and so on. The reason was, of course, that the "clock paradox" is generally explained in terms of only two observers, and then only a single division into outward and return journeys is required. It is a mistake to suppose that in the four clock version we must necessarily be describing a journey uniquely when we say, in effect, that $A$ goes to the end of $B$'s rod while $B$ goes to the end of $A$'s. There are two unique events involved, and the Lorentz equations do not themselves contain any means whereby the two may be identified with anything which could be termed the end of the journey of separation of $A$ and $B$. To claim to identify them is to claim to have some absolute standard of simultaneity, and this, we know, Einstein's theory does not have. (The whole problem is reminiscent of that of saying what we mean by "the contraction of a moving rod".) The most we can say is that $A$ and $D$ can agree that the events are simultaneous, as can $B$ and $C$. This does not mean that the theory cannot be re-interpreted in terms of a universal time – and the most natural way of doing so would seem to be to begin from the methods of "kinematic relativity" of the late E. A. Milne and our President Dr. Whitrow. Since, however, the object of the present exercise is to remain more or less within Einstein's interpretation of his theory, the idea of absolute simultaneity must here be rejected. It should, none the less, be possible to introduce a standard of simultaneity achieving much the same purpose, by simply supposing a neutral observer ($O$, say), positioned between $A$ and $B$ in such a way that all three are simultaneously concurrent, and such that $O$ subsequently judges each of the others to have the same x-coordinate and velocity ($w$) with respect to him. Some of the consequences of this supposition will be outlined.

An immediate consequence is that the time-relations of the Lorentz transformation equations are symmetrical about $O$:

$$t_A^O = \gamma t_O^O = t_B^O,$$

$$\gamma t_A^A = t_O^A, \quad t_O^B = \gamma t_B^B,$$

and $$t_B^A = \beta t_A^A, \quad \beta t_B^B = t_A^B,$$

where $$1/\beta = (1-v^2/c^2)^{\frac{1}{2}} \quad \text{and} \quad 1/\gamma = (1-w^2/c^2)^{\frac{1}{2}}.$$

It is a simple matter to show that once $v$ is given, $w$ is uniquely determined, and that $w$ and $\gamma$ are simple functions of $v$ and $\beta$.[26] $O$ is then determined uniquely in velocity and position with respect to both $A$ and $B$ for all times. From Eq. (1''), since by hypothesis $x_O^A = -x_O^B$ (taking $O$'s x-coordinate to be positive in the directions of $A$), $wt_O^A = wt_O^B$. Since $w$ is not zero, $t_O^A = t_O^B$, and therefore from the last group of equations we find also $t_A^A = t_B^B$, and thus $t_B^A = t_A^B$. In other words, there is a way of defining events at $A$ as simultaneous with events at $B$, and this defined simultaneity involves not merely the equality of coordinate times, but also of proper times ($t_A^A$ and $t_B^B$). (The actual existence of the auxiliary "observer" $O$ is, of course, totally irrelevant to the validity of this definition.) To this extent, the definition supplements Einstein's definition of simultaneity at different places. We notice, furthermore, that we may also equate $x_B^A$ and $x_A^B$, and we may therefore take as the two "simultaneous" events the $E_a$ (at $A$) and $E_b$ (at $B$) of our first example, that involving four space-travelling clocks, $A$, $B$, $C$ and $D$. It should by now be clear that in this example, where $C$ and $D$ meet after passing $A$ and $B$ respectively at events $E_a$ and $E_b$, the aggregate proper time recorded by $A$ and $C$ will be equal to that recorded by $B$ and $D$. For this to be true, $C$ and $D$ need not approach with the same velocity as that at which $A$ and $B$ separated, although for $O$ to be the same on both outward and return journeys, $C$ and $D$ must have the same speeds with respect to $A$ and $B$ respectively. There is no obvious sign of any clock paradox here.

No doubt it will be objected that since there is an element of *definition* in all this, nothing conclusive has been proved. This misses the point, however, that the definition laid down is a possible, i.e. consistent, definition of a journey.

*Three Clocks – Asymmetry without Contradiction*

Without changing notation, we consider the problem of three Clocks: $A$ and $B$ separate at constant velocity, and $D$ subsequently passes first $B$ and then $A$ along the same line, also at constant velocity. The time assigned by $A$ to the two parts of the journey will certainly be greater than the sum of those recorded locally by the "travellers", but is the sum of the times assigned by $A$ necessarily equal to the total proper time *recorded* by $A$ between $B$'s departure and $D$'s return? We can easily see how doubt could be cast on their equality. Let us simply

---

26 By Einstein's formula for combining relative velocities, $v = 2w/(1+w^2/c^2)$. Solving this quadratic for $w$ in terms of $v$, we find – after rejecting one root as involving a velocity in excess of the velocity of light –

$$w = \left(\frac{\beta-1}{\beta}\right) \cdot \frac{c^2}{v}, \quad \text{where} \quad \beta = (1+v^2/c^2)^{-\frac{1}{2}}.$$

By definition, $\gamma$ is the same function of $w$ as $\beta$ is of $v$. By substitution of the above value for $w$ in this function, it is found that $\gamma = \sqrt{\left(\frac{\beta+1}{2}\right)}$.

denote by $E_0$, $E_1$ and $E_2$ the events when $B$ leaves $A$, when $B$ meets $D$, and when $D$ meets $A$. When writing down the equation comparing $A$'s assessment of the time of $E_1$ with $B$'s, we are reckoning forward from the event $E_0$, and yet when comparing $A$'s assessment with $D$'s, for the second part of the journey, the time is reckoned backwards from the event $E_2$. The events $E_0$ and $E_2$ are the only events of the journey of $B$ and $D$ of which $A$ can be sure. Considering $A$'s proper time as a line stretching from $E_0$ to $E_2$, how do we know that we are entitled to consider the "time of $E_1$" as represented by a unique point on the line? One way to be sure is to consider $B$ and $D$ as sources of light. If $A$ observes $B$ continually from its departure to $E_1$, he may then immediately observe $D$, also at $E_1$, since according to Einstein the velocity of light is independent of the velocity of its source. $A$ may then observe $D$ until the two meet. It is a simple matter to integrate the time kept locally by $A$ during his observation of $B$ and $D$, and to show that this is precisely the time he assigns to the outward and return journeys of $B$ and $D$, and equal to $\beta$ times the aggregate of their local times during the double journey.[27] (This is so even with a different return velocity.) There can be no doubt but that Einstein's restricted theory of relativity applied to three clocks $A$, $B$, and $D$ as explained, does predict that the proper

---

27 $E$ is the element of emission of photons from a source at $B$, and $R$ is the event of their receipt at $A$. Then

$$t_A^R = t_A^E + \frac{x_A^E}{c},$$

and therefore

$$dt_A^R = dt_A^E \left(1 + \frac{v}{c}\right).$$

(By Eq. (4'), $dt_A^E = \beta dt_B^E$; and equating photons emitted and received, $v_B^B dt_B^E = v_A^B dt_A^R$. Combining, we obtain the Einsteinian Doppler formula for a receding source:

$$v_B^B / v_A^B = \beta\left(1 + \frac{v}{c}\right).)$$

For the return journey of $D$, at velocity $w$, the equation will be

$$dt_A^R = \left(1 - \frac{w}{c}\right) \cdot dt_A^E.$$

Suppose that $B$ goes out to a distance $\Delta$ as measured by $A$, and $D$ returns along the same distance. The total time recorded by $A$ during his observations of $B$ and $D$ will be $\int dt_A^R$,

namely (B) $\quad \left(1 + \dfrac{v}{c}\right)\int dt_A^E = \left(1 + \dfrac{v}{c}\right) \cdot \dfrac{\Delta}{v},$

and (D) $\quad \left(1 - \dfrac{w}{v}\right)\int dt_A^E = \left(1 - \dfrac{w}{c}\right) \dfrac{\Delta}{w}.$

The sum of these quantities is $\Delta\left(\dfrac{1}{v} + \dfrac{1}{w}\right)$, which is the time $A$ would arrive at on the basis of the coordinate time he assigns to the distant event $E_1$.

time recorded by $A$ is (for equal velocities out and back) $\beta$ times the aggregate proper time recorded by $B$ and $D$.[28]

There is an asymmetry here as between recorded local times; but where is the paradox? One might expect that the factor $\beta$ would appear on the other side of the equation if light from $A$ were received first at $B$ and then at $D$. Here certainly, would be a paradox, a true contradiction. But a simple calculation shows that treated in this way the problem yields the same answer as before: the proper time at $A$ is $\beta$ times as great as that at $B$ and $D$. There is no contradiction here, strange as the asymmetry of the result might at one time have appeared to be. It seems at first equally strange that there was no comparable asymmetry in the journey involving four clocks, but the two results are not mutually contradictory. Why need we say more than that the structure of special relativity is very different from that of the corresponding sector of everyday belief?

*On the Cause of the Asymmetry*

The word "perspective" is one of the most useful in the vocabulary of the apologist for Einstein's restricted theory, for we are all of us familiar with the alteration and distortion of spatial experience as a result of changes in our point of view. But what would it mean to a young astronaut to be told that his twin brother's long grey hair and general decripitude were the result of his peculiar time-perspective? Would he be satisfied with the Lorentz equations, or with a Minkowski diagram, and agree not to ask for such an antiquated thing as a causal explanation? One would like to have subjected those brilliant rationalistic twins, Weyl and Eddington, to a real test of this sort. Although it is no argument for the need to provide causal explanations, it is clear enough that most physicists who discuss the paradox of the twins do have a desire to go beyond the restricted theory to a physics where "physical" causes operate. This is as much a pyschological fact as a symptom of concern for a legitimate methodology, and I doubt whether many of those who are prone to distinguish between physical and kinematic explanations would find it easy to justify the distinction. Bishop Barnes, writing in the late 1920s, was untypical only insofar as he was honest enough to say openly that he did not believe that the discrepancy between the time of consciousness of the two men had been satisfactorily explained, although on the previous page he had given the standard Einsteinian explanation.[29] Einstein

---

28 It might be asked why no use is made here of the idea of introducing the quasi-universal time of an auxiliary observer $O$, as in the last section. Would this not lead to a situation in which $A$ and $B$ agreed about the outward journey, and $A$ and $D$ about the return? In fact it would be necessary in this three-clock version to introduce *two* auxiliary observers, $O_1$ and $O_2$, one for the outward and one for the return journey. We should then be committed to combining times recorded by $A$ on two incompatible scales.

29 *Scientific Theory and Religion,* (The Gifford Lectures, 1927–1929), Cambridge University Press, 1933, p. 113.

himself, and most orthodox relativists since, have put the effect down to the acceleration of one and only one observer with respect to an inertial frame. Accelerations of this sort have qualified as causal at least from the time of Newton, a fact which might perhaps have weighed in their favour as explanations of the discrepancies between clocks or twins. But even without accelerations we have seen that similar discrepancies are possible. Are we to ascribe them to the *motion* of the observers $B$ and $D$ in the three-clock example? Clearly not, for that motion is not inherent in them any more than it is inherent in $A$, with respect to whom it is measured. Is *relative* motion the cause? Even allowing the naive desire to find a simple cause, it can hardly be supposed that we have found it in the relative motions alone, for two bodies in relative motion share the motion equally. And was there not relative motion enough in the four-clock examples, where there was no sign of asymmetrical aging? On the other hand, since the relative velocity $v$ is present in the factor $\beta$, the cause-and-effect philosopher might be inclined to speak of a multiple cause of asymmetrical aging involving relative velocity and an asymmetry in the distribution of the moving clocks. (We recall that in the four-clock problem we compared the aggregate time of one pair of relatively moving clocks with that of another pair, whereas in the case of three clocks we compared the aggregate time of such a pair with the time of a *single* clock.)

In passing, we notice that no violence is done to any principle of relativity. In fact it is only when we pass beyond the restricted theory, and treat $B$ and $D$ in that example as one and the same clock, with its motion reversed by deceleration, that even the unrestricted principle of relativity is called in question; for then, the acceleration being relative to $A$, it should therefore be just as capable of making his clock register the lesser time. There is one clear way of avoiding this unacceptable (because contradictory) alternative, and that is to bring back something akin to the much despised aether – under some other name such as "universe" of course. Those who are honest enough to adopt this course openly are not without their causal explanation of the asymmetrical readings of relatively moving clocks. But those who wish to stay within Einstein's restricted theory must be satisfied with the frail causal ingredients mentioned earlier – relative velocity and an asymmetrical situation. Where does the searcher for causes go from there? In all probability he will devise a model displaying all the characteristics of those situations which are deducible from the Lorentz equations. If he is Minkowski's equal, the chances are that his model will differ little from that which Minkowski put forward more than sixty years ago. And in the unlikely event of his feeling that he now has reason for greater confidence in the reality of the situations with which Einstein's theory deals, he is simply deluded.

The syntactic structure of the theory is totally independent of confirmatory experience, and is yet again independent of the rules of correspondence between them. Each of these three things may be modified to a greater or lesser extent.

Since in discussing some of the problems associated with time dilatation I have modified none of them, have kept within the theory and interpretation offered by Einstein, and have ignored all experiment bearing on its plausibility, I realize that I have scarcely done more than scratch the surface of the problem. I hope at least that I have shown special relativity, one of the great corner-stones of twentieth-century physics, to be free from some of the kinematic inconsistencies with which it has been charged. Although I have not found an acceptable answer to the problem of the space twins, that was never my intention. As for the apparently incompatible results obtained with three and four moving clocks, however surprising at first, they seem to me no more difficult to accept than the theorems that two sides of a triangle are together greater than the third, and that two pairs of sides of a rhombus are equal. The unfamiliar is not on that account illogical.

# General Relativity and Time in the Solar System

G. C. McVITTIE*

## 1. Astronomical Times

Astronomers are concerned with the measurement of time rather than with its philosophical definition [5, 9]. They have arrived at several different time-measures which I will briefly describe. A clock is a time-measuring device, and a "natural clock" is provided by a recurrent natural phenomenon such as the periodic passage of the Sun across the observer's meridian. A "man-made clock" is a device constructed by man which produces a repetitive phenomenon. It might be a dial with a rotating pointer that passes over marks made on the rim of the dial. A natural, and a man-made, clock proceed at the same rate if the natural phenomenon recurs at the same reading of the man-made clock.

There are two astronomical times for which the natural clock is the earth rotating about its axis. They are sidereal time and universal time and they proceed at different rates. They need not concern us further here. Of greater interest is ephemeris time [2] in which the natural clock consists of the repetitive motions of all the planets and their satellites in the solar system. But to arrive at a time-measure in this way a theory of these motions is needed, in addition to the observed positions of the bodies of the solar system. The theory[1] employed is said to be "in accordance with the Newtonian law of gravitation, modified by the theory of general relativity" [4]. This theory I shall call "classical celestial mechanics" in order to distinguish it from a theory of solar system motions wholly based on general relativity that may eventually be achieved. Classical celestial mechanics so defined incorporates the following postulates or pre-suppositions: a) space is absolute and Euclidean, so that all distances can be combined by the rules of Euclidean geometry; b) there exists an absolute time, $t_N$, in terms of which all events can be unambiguously dated; c) the equations of motion of a particle state that the rate of change, with respect to absolute time, of the linear momentum of the particle is equal to the force; d) the gravitational force between two particles is given by Newton's inverse square law, in which there occurs the absolute Euclidean distance between the particles at an instant of absolute time; *except that* e) the consequences of the inverse square law are modified by the addition of a small rotation of the line of apsides in a planetary orbit – such as that discovered long ago for the motion of Mercury – which has

---

\* Professor G. C. McVittie, University of Illinois Observatory, Urbana, Illinois, USA.
[1] I am indebted to Dr. G. M. Clemence for his help in drawing up this statement.

the same numerical value as that predicted by Einstein's theory of gravitation. In addition whenever the motion of light has to be taken into account, it is assumed that light moves in vacuo with constant speed along Euclidean straight lines.

The rate at which $t_N$ proceeds is adjusted to produce the best fit with the observations of the constituent bodies of the solar system and it is then called ephemeris time. The rate differs from that of sidereal and of universal time, though connections between the three times can be stated. Classical celestial mechanics also produces the values, at each instant $t_N$, of the Euclidean spherical polar coordinates ($r_N$, $\Theta_N$, $\varphi_N$) of each solar system object. The origin of these space coordinates is at the Sun. It is also worth remarking that this identification of $t_N$, the time-variable in a theory, with the time kept by the solar system natural clock is a matter of definition. Sixty years ago the theoretical time-variable was identified with mean solar time — a variant of universal time — and therefore with the time kept by a different natural clock, namely, the rotating Earth. This led to inconsistencies between the predicted and the observed positions of the Moon. Therefore ephemeris time was introduced and the absolute time inherent in Newton's theory was identified with it. But neither the old nor the new identification proves that either natural clock keeps an "absolute" time, which could thus be called the "real" or "correct" time.

The term "uniform time" occurs frequently in the writings of celestial mechanicians and appears to mean a time which proceeds at a constant and invariable rate. But the term is conventional [10] because there is no way of identifying a natural or a man-made clock that keeps such a time. However it is possible to assert pragmatically that such and such a time is "uniform" with respect to some other. For example, ephemeris time may be asserted to be uniform compared with the time kept by a sundial because the former time gives the most harmonious account of planetary motions, whereas the latter would lead to strange irregularities in their motions, were sundial time be defined as "uniform" time.

## 2. Atomic Time

Ephemeris time is thus based essentially on a theory of gravitation. Another time is derived from atomic theory in which gravitation plays no part but electrical and nuclear forces are dominant. The quantum theory of atomic transitions and the resulting frequencies of the electromagnetic waves, contain implicitly a time-variable and it is this variable which defines atomic time. The variable's somewhat ambiguous character has been emphasized by Zimmerman [12]. There is no a priori reason why atomic and ephemeris time should proceed at the same rate and the question will have to be settled empirically in the course of the next two or three decades. Nor is atomic time "uniform" except as a matter of pragmatic definition.

## 3. General Relativity

In this theory there is no unique time-system as there is in classical mechanics. Nevertheless it is likely that interplanetary radar experiments, which are now technically possible, can be used to elucidate the notion of time in general relativity. The observational result is the time-delay, namely, the lapse of time measured on some terrestrial man-made clock between the emission of the signal from Earth and its return after reflection from the planet. The experimental results can, of course, be interpreted through classical celestial mechanics. But the question can also be asked: Can they be interpreted through general relativity, and what happens if this is done? General relativity is probably a more accurate theory of gravitation than Newton's and the applicable part of general relativity is contained in the Schwarzschild space-time that describes the gravitational field of the Sun. This space-time implies that the geometry of space is non-Euclidean and also that the masses of the planets are infinitesimally small so that their mutual attractions are ignored. It is possible to introduce the analogues of spherical polar coordinates $(r, \Theta, \varphi)$ and a coordinate-time $t$, and then the fundamental formula of the Schwarzschild space-time, namely, its metric is [6]

$$ds^2 = (1 - 2m/r)dt^2 - \{(1 - 2m/r)^{-1} dr^2 + r^2 d\Omega^2\}/c^2,$$
$$d\Omega^2 = d\Theta^2 + \sin^2\Theta \, d\varphi^2. \tag{1}$$

Here the mass of the Sun, $M$, appears as the constant $m = GM/c^2 = 1.5$ km., where $G$ is the universal constant of gravitation and $c$ is the local velocity of light. The coordinate-system $(t, r, \Theta, \varphi)$ is however not unique because coordinate-systems are regarded as methods of labelling events and there can be more than one way of setting up such a labelling system. Nevertheless calculations must be performed in some one coordinate-system which is usually selected by the investigator because he has found it to be the mathematically simplest to use. But with some trouble the computations can be carried out in a number of time and space coordinate systems. I shall then argue that the best system can only be selected *a posteriori* by an appeal to observation.

There is one way of escape from the relativity of time-coordinates. The invariant $s$ can be calculated along the world-line of a moving particle and it is then called the proper-time of the particle. The proper-time is independent of the particular coordinate-system used to calculate it.

## 4. Time-Delay

The motion of the radar signal from Earth to, say, Mercury and back is illustrated in Fig. 1. The outgoing signal travels from the Earth, $E$, at $A$ along $AMC$ to Mercury, $F$, at $C$, is immediately reflected there and returns to Earth at $B$

along $CM'B$. The configuration is assumed to be such that Mercury is seen at some position beyond the Sun, so that general relativity effects are as large as

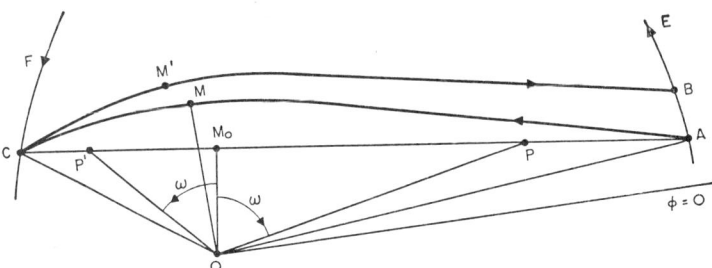

Fig. 1. Diagram of the paths, $AMC$ and $CM'B$, of the radar signals sent between the Earth ($E$) and the planet ($F$). The center of the Sun is at $O$

possible. The paths $AMC$ and $CM'B$ are not Euclidean straight lines but null-geodesics of the Schwarzschild space-time. For technical mathematical reasons I have assumed that the Earth's orbit is "circular" which means that $r=a=$ constant in the equations of the ordinary geodesics that define the orbit. It can then be proved [7] that in terms of terrestrial proper-time the time-delay is

$$s'_e - s_e = 2(\Delta_0/c)(1-\varepsilon_1)(1-\varepsilon_2)(1-\varepsilon_3), \qquad (2)$$

where $\Delta_0$, $\varepsilon_1$, $\varepsilon_2$ and $\varepsilon_3$ are known functions of the $(r, \Theta, \varphi)$ coordinates of $A$ and $C$, respectively, and $\Delta_0$ can be regarded as the "distance" from $A$ to $C$. The three $\varepsilon$'s arise as follows: $\varepsilon_1$ arises from the change of position of the Earth from $A$ to $B$ during the to and fro motion of the signal; $\varepsilon_2$ arises from the null-geodesic theory for the motion of the signals; and $1-\varepsilon_3$ is the conversion factor from $t$-time to terrestrial proper-time. It is thus $\varepsilon_2$ and $\varepsilon_3$ that arise from general relativity.

## 5. Comparison of Theory and Observation

If a formula such as (2) is to be used in the interpretation of measured time-delays the variables it contains must first be identified with measured quantities. This can only be done by definition, as far as I can see. With some confidence, the proper-time of the Earth may be identified with the time measured by terrestrial atomic clocks. The justification lies in Pound and Rebka's [8] demonstration of the change of frequency of radiation as it moves in the Earth's gravitational field. Such a change is interpretable by general relativity if frequencies of atomic vibrations, and those of the resulting electromagnetic waves, are measured by proper-time. Since $c$ is the local velocity of light according to general relativity, it may be identified with that measured in a terrestrial laboratory. But the $(r, \Theta, \varphi)$ coordinates found in the functions that occur in the right-hand side of (2) present a more speculative problem. One can assert that they are identical in value with the $(r_N, \Theta_N, \varphi_N)$ of classical celestial mechanics simply because these are the spherical polar coordinates for which we do have numerical values. For the configuration in which Mercury is at perihelion and the ray $AMC$ grazes the Sun, the $\varepsilon$'s are: –

$$\varepsilon_1 = 4.621 \times 10^{-7}, \quad \varepsilon_2 = -1.503 \times 10^{-7}, \quad \varepsilon_3 = 0.148 \times 10^{-7}, \quad (3)$$

$$2(\Delta_0/c) = (1362.85) \frac{1 \pm 3.3 \times 10^{-6}}{1 \pm 1.4 \times 10^{-6}} \text{ secs.} \quad (4)$$

Thus time-delays measured in atomic time for $(s'_e - s_e)$ would have to be known to an accuracy of one part in $10^8$ in order to check the adequacy of the formula. However the errors shown in (4) present a difficulty: the one in the numerator is that given for the Astronomical Unit [1], the one in the denominator is that assigned to $c$ by Cohen and DuMond [3]. Both errors are larger than the values of the $\varepsilon$'s. However the obstacle might be overcome by regarding $2 \Delta_0/c$ as one of the unknowns to be determined by the time-delay experiments.

Suppose that the time-delay measurements could not in practice be made to fit formula (2). Then recourse could be had to another coordinate system, for example, the one in which a radial coordinate $\bar{r}$ is employed in the Schwarzschild space-time, where

$$r = (1 + m/2\bar{r})^2 \bar{r}. \quad (5)$$

If now $(\bar{r}, \Theta, \varphi)$ are identified with $(r_N, \Theta_N, \varphi_N)$, it can be shown that only $\varepsilon_2$ is affected and that its value is reduced by some 10 per cent as compared with that given in (3). If failure to secure a good fit again occurred, one of the more elaborate coordinate systems that are possible in the Schwarzschild space-time could be employed, and so on until a satisfactory interpretation was achieved. The classical formula for time-delay, namely,

$$t'_N - t_N = 2(\Delta_0/c)(1 - \varepsilon_1), \quad (6)$$

would, of course, be tested also and presumably found wanting. Then it could be concluded: a) that the classical theory for the motion of the signals embodied in (6) was inadequate; b) that general relativity provided a better theory for the motion of the signals; c) that terrestrial proper-time could be identified with atomic time used in the measurements; and d) that the coordinates in the Schwarzschild space-time that most closely agreed with the $(r_N, \Theta_N, \varphi_N)$ coordinates of classical celestial mechanics had been identified.

The pioneer in the theoretical and experimental work on interplanetary radar is I. I. Shapiro [11] who states that the prediction of general relativity has been verified. Unfortunately, it is not clear what theoretical formula he uses nor exactly how he carries out his numerical computations. Neither has he published the time-delays that he has measured. At present (July 1969) we can only await further elucidation by Shapiro of his remarkable investigations.

The work on radar time-delays has been supported by the National Aeronautics and Space Administration, USA, under Grant NGR – 14-005-088.

*References*

1. Allen, C. W.: *Astrophysical Quantities,* 2nd Ed. London: Athlone press, 1962.
2. Clemence, G. M.: *Time Measurement for Scientific Use.* In: The Voices of Time. Ed. J. T. Fraser. 401–414. New York: George Braziller, 1966.
3. Cohen, E. R., DuMond, J. W. M.: *Fundamental Constants in 1965,* Rev. Mod. Phys. *37* (1965) 537–594.
4. Explanatory Supplement to the Astronomical Ephemeris and the American Ephemeris and Nautical Almanac. London: H. M. Stationery Office 1961.
5. Markowitz, Wm.: *Time Measurement.* Encyclopedia Britannica *21* (1968) 1159–1163. Chicago: Encyclopedia Britannica Inc.
6. McVittie, G. C.: *General Relativity and Cosmology* Chp. 5. 2nd Ed. London: Chapman and Hall 1965.
7. — *Interplanetary Radar Time-delays in General Relativity.* Astron. J. (1970) 287–296.
8. Pound, R. V., Rebka, G. A.: *Apparent Weight of Photons.* Phys. Rev. Lett. *4* (1960) 337–341.
9. Sadler, D. H.: *Astronomical Measures of Time.* Q. Jl. R. Astr. Soc. *9* (1968) 281–293.
10. — Private communication (1969).
11. Shapiro, I. I.: *A Fourth Test of General Relativity,* Phys. Rev. Lett. *13* (1964) 789–791.
    — *Testing General Relativity with Radar,* Phys. Rev. *141* (1966) 1219–1222.
    — *Ross-Schiff Analysis of a Proposed Test of General Relativity: A Critique.* Phys. Rev. *145* (1966) 1005–1010.
    — *Fourth Test of General Relativity: Preliminary Results.* Phys. Rev. Lett. *20* (1968) 1265–1269.
12. Zimmerman, E. J.: *Time and the Quantum Theory.* In: The Voices of Time. Ed. J. T. Fraser. 492–499. New York: George Braziller 1966.

# Zeit als physikalischer Begriff

FRIEDRICH HUND*

*Summary.* In daily life "time" has two aspects, and in physics, too. Time as coordinate describes the embedding of events into the universe. Time as consecutive order shows the difference between past and future. This difference is not found in the general laws of physics, but derives from a special fact, a former state of very low entropy; its origin is not conceivable in terms of physics. The second law of thermodynamics, which violates the symmetry of the two directions of time, is a statement about our special real world.

## 1. *Aspekte*

Wir objektivieren unsere Erfahrung als ein in einer eindimensionalen Zeit ablaufendes Geschehen in einem dreidimensionalen Raum. Wir messen nicht nur räumliche, sondern auch zeitliche Abstände. Dabei erscheint uns die Zeit irgendwie unheimlicher als der Raum: der Raum ist da, die Zeit verstreicht und kehrt nicht wieder. Die zeitliche Umkehr des wirklichen Geschehens – wie ein rückwärts laufender kinematographischer Film – gäbe ein unmögliches Geschehen. Diese beiden Aspekte der Zeit, Maß und Dauer einerseits, einsinnige Abfolge andererseits, sind uns allen vertraut. Beispiele sind Sätze wie: Um 8.10 fährt der Zug. Nach einer anstrengenden Wanderung ist man müde. Die beiden Aspekte finden sich auch im physikalischen Zeitbegriff wieder: die *Zeitkoordinate* und die *zeitliche Abfolge*. Der Physiker hat zu untersuchen: welche physikalischen Tatsachen stecken hinter der Koordinate Zeit (Abschnitte 2–4); aus welchen Gründen können Vergangenheit und Zukunft unterschieden werden (Abschnitte 5–8)? Es wird sich zeigen, daß er dabei über den Rahmen der Physik hinausgeführt wird in den Bereich der Kosmologie (die Physik handelt vom Wiederholbaren – im Experiment oder als vielfaches Vorkommen nebeneinander –, die Kosmologie vom einmaligen Universum).

## 2. *Absolute Zeit*

Raum und Zeit sind Grundgegebenheiten der Wirklichkeit. Einfache Vorgänge, die das zeigen, sind Bewegungen von Partikeln, bei denen wir etwa den Ort als Funktion der Zeit angeben, $x(t), y(t), z(t)$, oder Änderungen von Feldgrößen, $\psi(x,y,z,t)$. Eine Welle auf Wasser z. B. läßt beide Auffassungen zu; man kann die Bewegung der Wasserteilchen im Laufe der Zeit verfolgen, man

---

* Professor Dr. Friedrich Hund, D–34 Göttingen, Tuckermannweg 5.

kann aber auch die Höhe der Wasseroberfläche an einer Stelle $x,y$ im Laufe der Zeit betrachten [$h(x,y)$]. Bei beiden Arten von Vorgängen (bzw. beiden Beschreibungsarten) benutzen wir die Zeit als Koordinate.

In großen Abschnitten der naturwissenschatflichen Beschreibung verwenden wir diese Zeitkoordinate im Sinne einer „absoluten Zeit". Mit diesem Begriff, zusammen mit dem des absoluten Raumes, bekommen die Sätze der Mechanik einen Sinn, z. B. der Satz: ein sich selbst überlassener Körper bewegt sich geradlinig und gleichförmig. Von metaphysischen Hintergründen abgelöst, bedeutet der Begriff der absoluten Zeit eine einfache und enge Beziehung zwischen allen Bewegungen: zwischen der Zahl der Schwingungen eines Pendels oder eines elastischen Körpers, der Stellung der Erde bei ihrer Rotation, der Stellung der Planeten, der Menge einer radioaktiv zerfallenden Substanz.

*„Absolute Zeit" ist ein kurzer und zweckmäßiger Ausdruck für einen weltweiten Zusammenhang aller Vorgänge.*

Der Sachverhalt wird deutlicher, weil reichhaltiger, beim Raum. Die einfachen Gesetze der Mechanik gelten nur in nicht rotierenden und nicht beschleunigten Bezugssystemen, den „Inertialsystemen"; diese sind eine unendliche Schar geradlinig und gleichförmig gegeneinander bewegter Systeme. Absolute Rotation ist erkennbar; die Ablenkung der Passatwinde, die Drehung der Schwingungsebene des Foucaultschen Pendels „beweisen" die Rotation der Erde; ein freischwebender im Vergleich zur Kugelform etwas abgeplatteter Flüssigkeitstropfen (etwa der Planet Jupiter) rotiert. An jeder Stelle der Welt existiert ein Etwas, das angibt, in was für Bezugssystemen die einfache Mechanik gilt, in was für Bezugssystemen zusätzlich Trägheitskräfte, z. B. Zentrifugalkräfte, auftreten. Man kann dieses Etwas (eine physikalische Realität) ein Feld nennen, Trägheitsfeld oder Inertialfeld. Man kann in einem Raumstück alle Materie wegdenken, es auch frei von elektromagnetischen Feldern denken; aber das Inertialfeld können wir gedanklich nicht entfernen, wenn wir innerhalb unserer Physik bleiben wollen. Das alte Argument des Aristoteles gegen die Möglichkeit eines Vakuums: in einem solchen wüßte ein Körper nicht, wie er sich bewegen solle, bleibt in etwas verändertem Sinne richtig. Ohne Voraussetzung eines Inertialfeldes haben die Grundgesetze der Mechanik keinen Sinn, „wüßte ein Körper nicht", was geradlinig und gleichförmig ist, wäre für einen freischwebenden Tropfen seine Abplattung nicht bestimmt. Das heißt: ein physikalisches System ist nicht völlig isoliert von der übrigen Welt denkbar. Um physikalische Aussagen über das System machen zu können (etwa aus seinem Zustand auf seine künftige Bewegung zu schließen), muß man angeben, wie es in die übrige Welt eingebettet ist. Abgesehen vom Materie- und Strahlungsaustausch *gibt das Inertialfeld die Art der physikalisch wirksamen Einbettung an.*

Das Inertialfeld ist auf zwei Weisen empirisch feststellbar: einmal durch Kräfte (z. B. Zentrifugalkräfte), also durch den „Trägheitskompaß", zweitens durch den „Sternenkompaß" (Wortprägungen von H. Weyl). In der Praxis des Astronomen wird der Trägheitskompaß bestimmt durch Nachrechnen der Bewegungen aller Planeten und Vergleich mit den beobachteten Planetenörtern. Er definiert die ganze Schar der Inertialsysteme. Die Fixsterne, die wir sehen, die genähert relativ zueinander ruhen, stellen ein Inertialsystem dar, so genau, wie es die oben genannten Planetenbeobachtungen zulassen. Da der Trägheitskompaß die ganze Schar der Inertialsysteme angibt, der Sternenkompaß ein einziges davon auszeichnet, ist die dynamische Einbettung schwächer als der Zustrahlung von den Sternen entspricht; die wirkliche Welt (Anordnung der Sterne) hat weniger Invarianz als die Naturgesetze. Das empirische Faktum, daß der Sternenkompaß mit dem Trägheitskompaß im Einklang ist, legt den Verdacht nahe, daß die Anordnung der Massen in der Welt das Inertialfeld physikalisch bestimme.

Die Fixsterne sind nicht genau gegeneinander ruhend, und im Großen wissen wir von der „Expansion des Universums", der zeitlichen Zunahme der Abstände der Galaxien. Unseren Satz vom Sternenkompaß modifizieren wir: an jeder Stelle der Welt läßt sich ein (lokales) Inertialsystem angeben, in dem die Welt genähert isotrop expandiert.

Die Grundsätze der Einsteinschen Gravitationstheorie formulieren die Verwandtschaft der Trägheitskräfte mit den Gravitationskräften, also des Inertialfeldes mit dem Gravitationsfeld. So kann abgeschätzt werden, wie stark das Inertialfeld an einer Stelle durch umlaufende Massen mitgeführt wird. Man findet: zur völligen Mitführung braucht man ungefähr das ganze Universum. Wir können (vielleicht etwas unvorsichtig) sagen: *Absoluter Raum und absolute Zeit bezeichnen die dynamische Einbettung eines lokalen physikalischen Systems in das Universum.* Raum und Zeit ragen damit in den Bereich der Kosmologie hinein.

## 3. *Zeitmaß und Seinsschichten*

Die Grundgesetze der klassischen Mechanik sind unabhängig von Einheiten der Länge, der Zeit und der Masse. Aus einer möglichen Bewegung wird durch proportionale Änderung aller Längen oder aller Zeiten oder aller Massen wieder eine mögliche. Für unser menschliches Leben sind jedoch „natürliche Einheiten" bestimmend, etwa die Länge unserer Gliedmaßen (ungefähr 1 m), der Pulsschlag oder die Schrittdauer (ungefähr 1 s), das Gewicht eines Armes (einige kp). Die Eigenschaften der Stoffe liegen ebenfalls in einer ganz bestimmten Größenordnung: die Dichten der festen und flüssigen Körper (bei einigen g/cm³), die Schallgeschwindigkeiten (bei einigen km/s), die Energien bei chemischen Umsetzungen, die elektrischen Ladungen bei der Elektrolyse und die Frequenzen der Spektren. Sie sind bestimmt durch drei universelle Naturkonstanten, $\hbar$ (das elementare Wirkungsquantum), $m$ und $e$ (Masse und Ladung des Elektrons). Aus diesen drei Größen lassen sich Einheiten der Masse, der Länge und der Zeit bilden, die man als die für den atomaren Bereich und damit für die Stoffeigenschaften „natürlichen Einheiten" ansehen kann, neben der Masseneinheit $m$ eine Längeneinheit, die die Größenordnung der Atomradien, und eine Zeiteinheit, die die Größenordnung der Umlaufszeiten der Elektronen und damit der Schwingungszeiten der ausgestrahlten Spektrallinien angibt. Diese natürliche Zeiteinheit beträgt einige $10^{-17}$ s.

Die genannten Einheiten sind nicht für die ganze Physik „natürlich". Für die Bausteine der Atome, sowie für Materie unter hohen Drucken und Temperaturen, die nicht mehr aus Atomen bestehen, spielt die elektrische Kraft (die für die „gewöhnliche Materie" sehr wichtig ist) und damit $e$ nur eine geringe Rolle; dafür wird wegen der hohen Geschwindigkeiten die Lichtgeschwindigkeit $c$ wichtig, auch wird die Masse $M$ des Nukleons (Protons und Neutrons, der Bausteine der Atomkerne) wichtiger als die Masse $m$ des Elektrons. Die Größen $\hbar$, $c$, $M$ bestimmen so die „natürlichen Einheiten" der „Hochenergiephysik", der Physik der Elementarteilchen, der Physik der „ungewöhnlichen Materie", wie sie in Sternen vorkommt. Die zugehörige Zeiteinheit beträgt etwa $10^{-24}$ s. Messungen von kürzeren Zeitintervallen als $10^{-24}$ s können wir uns nicht ausdenken.

Daß es für verschiedene Erscheinungsgebiete der Physik verschiedene natürliche Einheiten gibt, deutet darauf hin, daß die physikalische Wirklichkeit verschiedene „Schichten" aufweist. Je nach dem Verhältnis ihrer natürlichen Einheiten zu den menschlichen Größenordnungen sind sie uns in verschiedener Weise zugänglich. Eine Einteilung der Physik nach der Art der Zugänglichkeit entspricht darum auch ungefähr dieser Schichtung: Es gibt die „durchschaubaren" Bewegungen von Körpern, den Gegenstand der klassischen Mechanik; sie hat keine natürlichen Einheiten und umfaßt Bewegungen von Himmelskörpern wie von Staubteilchen. Die elektrischen und magnetischen Erscheinungen stellen einen nicht durchschaubaren Vordergrund von Erfahrungen dar, der durch einen nicht sinnlich wahrnehmbaren, aber makroskopisch in Raum und Zeit, also anschaulich, beschreibbaren Hintergrund, das elektromagnetische Feld, durchschaubar wird. Die darin wesentlich vorkommende Konstante $c$ (die Lichtgeschwindigkeit) setzt Abstände und Zeiten in Beziehung – 300000 km entsprechen 1 s –, legt aber (außer für die Geschwindigkeit) keine natürlichen Einheiten fest. Elektromagnetisches Geschehen gibt es in der weiten Nachbarschaft der Sterne und zwischen kleinen Materieteilchen. Die Wärmeerscheinungen stellen einen nicht durchschaubaren makroskopischen Vordergrund dar, der durch einen mikroskopischen Hintergrund, die kinetische Theorie der Materie und die statistische Thermodynamik durchschaubar wird. Solange diese auf dem Boden der klassischen Mechanik bleibt, legt sie noch keine Atomgrößen fest; die Experimente zeigen aber definierte Größen. Die Chemie und die Stoffeigenschaften, sowie die eben genannten Experimente stellen einen Vordergrund dar, der durch einen nicht mehr anschaulich beschreibbaren Hintergrund, die Quantentheorie, durchschaubar wird; hier gelten die natürlichen Einheiten $\hbar$, $m$, $e$ und die Zeiteinheit von einigen $10^{-17}$s. Das bis heute noch nicht ganz begriffene Gebiet der „schwachen Wechselwirkung" von Elementarteilchen (mit Kräften, die viel schwächer sind als die elektrischen) zeigt charakteristische Zeiten von $10^{-8}$ bis $10^{-10}$s, und für das auch nicht ganz begriffene Gebiet der „starken Kopplungen", die auch für die Physik der Atomkerne wichtig sind, gilt die oben genannte natürliche Zeiteinheit von ungefähr $10^{-24}$s.

Ordnen wir im Rückblick die charakteristischen Zeitdauern der Größe nach, so gibt es eine „*Elementaruhr*" mit der Einheit um $10^{-24}$s, eine „*Atomuhr*" mit der Einheit um $10^{-16}$s, die *Uhr der Lebensdauern* der „nahezu stabilen" Elementarteilchen um $10^{-8}$s, unseren *Pulsschlag*, 1 s; wir können zufügen die „*Planetenuhr*" um $10^8$s (einige Jahre) und die „*paläontologische Uhr*" der Entwicklung des Lebens um $10^{16}$s. Das „*Weltalter*", die Größe $T$ im Gesetz $v \sim r/T$ zwischen Fluchtgeschwindigkeit und Abstand entfernter Galaxien bei der oben genannten Expansion der Welt, liegt zwischen $10^{17}$ und $10^{18}$s. Der Zeitfaktor zwischen den beiden Zweigen der Biologie, der Physiologie und der Paläontologie beträgt $10^{16}$; etwa ebenso groß ist der zwischen den beiden Zweigen der „Hochenergiephysik", dem Gebiet der starken und der schwachen Kopplungen.

## 4. *Die beiden Zeitrichtungen*

Die Zeit ist eindimensional, der Raum ist dreidimensional. Damit ist aber der Unterschied noch nicht ausreichend bezeichnet. Die Grundgesetze der Mechanik lassen sich sinnvoll auf Räume anderer Dimensionszahl übertragen; so kann man das Modell einer Mechanik mit nur einer Raum- und einer Zeitdimension aufstellen. In ihm sind aber Raum und Zeit verschieden. Die Bewegung zweier Partikel unter dem Einfluß einer Abstoßung wird in einem $x,t$-Diagramm etwa durch die „Weltlinie" der Abb. 1 wiedergegeben;

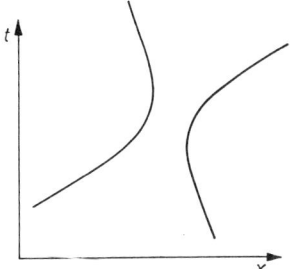
Abb. 1. Möglicher Vorgang

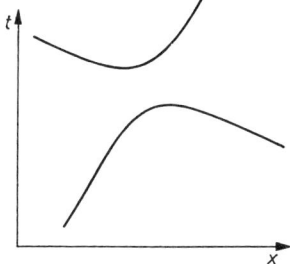
Abb. 2. Unmöglicher Vorgang

Vertauschung von Raum- und Zeitkoordinate ($x$, $t$), Abb. 2, stellt einen unmöglichen Vorgang dar. Es ist der alte Satz: ein Körper kann am gleichen

Ort zu verschiedenen Zeiten sein, aber nicht zu gleicher Zeit an verschiedenen Orten; $dx/dt = 0$ kommt bei Bewegungen vor, niemals aber $dt/dx = 0$. Jeder Punkt einer Weltlinie trennt immer zwei Abschnitte (in Abb. 1 unten und oben); aber er trennt nicht immer zwei Raumabschnitte (in Abb. 1 links und rechts). Bei Bewegungen im dreidimensionalen Raum bleibt das Wesentliche erhalten. *Jeder Punkt der Weltlinie einer Bewegung trennt zwei Zeitabschnitte.* Daß wir sie Vergangenheit und Zukunft nennen, ist hier noch nicht verständlich.

Die (spezielle) Relativitätstheorie verknüpft Raum und Zeit; aber auch sie macht den Unterschied deutlich. Für den vom Bezugssystem unabhängigen raumzeitlichen „Abstand" $ds$ zweier benachbarter Ereignisse, von denen das eine die Wirkung des anderen ist, gilt:

$$ds^2 = c^2 dt^2 - dx^2 - dy^2 - dz^2 \geqq 0; \qquad (1)$$

das Gleichheitszeichen gilt nur, wenn $dx, dy, dz, dt$ zu einer Lichtbewegung gehören; das $>$-Zeichen gilt, wenn sie zu der Bewegung einer Partikel oder einer nicht durch Licht gegebenen Wirkungskette gehören. Der Ungleichung (1) entspricht es, daß das vierdimensionale Gebiet Raumzeit um einen Punkt $P$ in drei getrennte Abschnitte zerfällt (in Abb. 3 ist eine Raumdimension unterdrückt), von denen zwei (das Innere des Kegels in Abb. 3) mit $P$ in Wirkungs-

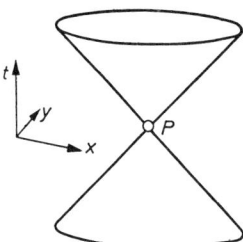

Abb. 3. Zeitrichtungen in der relativistischen Physik

zusammenhang stehen können, der dritte (das Äußere des Doppelkegels in Abb. 3) jedoch nicht. *Jeder Punkt P trennt in Bezug auf Wirkungszusammenhänge zwei Zeitabschnitte.* Wir werden sie später Vergangenheit und Zukunft nennen.

Das Modell einer Physik im zweidimensionalen Raum mit

$$ds^2 = dt^2 - dx_1^2 - dx_2^2 \qquad (2)$$

($c$ haben wir gleich 1 gesetzt), wie sie Abb. 3 entspricht, vergleichen wir mit dem Modell einer „Pseudophysik"

$$ds^2 = dt_1^2 + dt_2^2 - dx^2. \qquad (3)$$

Die in (2) und (3) links von ≧ mit positivem Zeichen auftretenden Koordinaten haben wir $t$, die mit negativem Zeichen auftretenden Koordinaten $x$ genannt. In der Pseudophysik besteht Wirkungszusammenhang zwischen $P$ (Abb. 4) und dem Außengebiet des Doppelkegels. $P$ trennt in Bezug auf Wir-

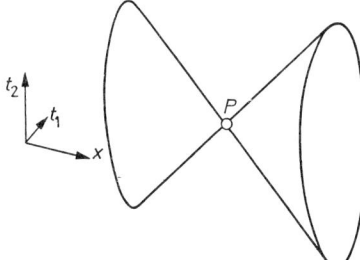

Abb. 4. Pseudophysik

kungszusammenhänge nicht zwei Abschnitte. In dieser Pseudophysik gibt es nichts, das unserer Zeit analog ist. Der Begriff Zeit läßt sich sinnvoll nur in Modellen der Physik mit

$$ds^2 = dt^2 - dx_1^2 - dx_2^2 - \ldots - dx_n^2 \geqq 0 \qquad (4)$$

anwenden, also nur wenn links von ≧ eine einzige Koordinate mit positivem Zeichen vorkommt.

## 5. Vergangenheit und Zukunft

Daß jeder Raumzeitpunkt $P$ in der $3+1$-dimensionalen Raumzeitwelt zwei Zeitabschnitte möglicher physikalischer Verknüpfung (eines Wirkungszusammenhanges) mit diesem Punkt trennt, bedeutet noch nicht, daß die beiden Abschnitte verschieden sind. Zunächst können wir noch willkürlich die eine der beiden Zeitrichtungen positiv nennen, etwa so wählen, daß in Europa die Sonne von Asien her zum Atlantischen Ozean über den Himmel wandert. Die Gleichberechtigung der beiden Zeitrichtungen ist dabei noch nicht verletzt, sowenig wie die Gleichberechtigung der beiden Magnetpole dadurch verletzt wird, daß wir das nach dem Polarstern (α urs. min.) zeigende Ende einer Magnetnadel positiv nennen.

In der wirklichen Welt und in unserem Bewußtsein sind aber die beiden Zeitrichtungen erheblich verschieden. Wir nennen sie Vergangenheit und Zukunft im Verhältnis zu $P$. Die Zeitrichtung, mit der die Sonne von Asien zum Atlantik wandert, ist die gleiche, mit der wir älter werden, mit der unsere

Ideen aufeinanderfolgen, unser Gedächtnisinhalt wächst. Vergangenheit ist für uns das, was unabänderlich feststeht, die Zukunft ist offen. In der Thermodynamik unterscheidet der „zweite Hauptsatz" die beiden Zeitrichtungen. Nach ihm kann die Entropie eines abgeschlossenen Systems mit der Zeit nicht abnehmen, $\Delta S \geq 0$.

Sehen wir uns jedoch die allgemeinen Gesetze der Physik an, so finden wir volle Symmetrie der beiden Zeitabschnitte. Die allgemeinen Sätze der klassischen Mechanik, der Elektrodynamik und der Quantenmechanik ändern sich nicht, wenn man in den Gleichungen $dt$ durch $-dt$ ersetzt. In Bezug auf die allgemeinen Sätze der Physik finden wir keinen Unterschied der beiden Zeitabschnitte. Neben den allgemeinen Sätzen gibt es noch „Materialgleichungen", wie das Ohmsche Gesetz zwischen Stromdichte und elektrischer Feldstärke,

$$i = \sigma E,$$

oder das Gesetz der Wärmeleitung zwischen Dichte des Wärmestroms und Temperaturgefälle,

$$j = -\varkappa \operatorname{grad} T,$$

die bei der Transformation $dt \to -dt$ ($i \to -i$, $j \to -j$) nicht in sich übergehen, also nicht reversible Vorgänge beschreiben. Sie fallen unter den zweiten Hauptsatz. Sie gehören dem an, was oben makroskopischer Vordergrund genannt wurde, und werden im mikroskopischen Hintergrund durch die allgemeinen Sätze, also durch eine Physik reversibler Vorgänge verständlich gemacht. Das geschieht durch statistische Betrachtung von Systemen mit sehr vielen Freiheitsgraden.

*Durch diese statistische Betrachtung des mikroskopischen Hintergrundes der Vorgänge wird der Unterschied von Vergangenheit und Zukunft physikalisch begreifbar.*

Die Zurückführung des zeitlich nicht umkehrbaren makroskopischen Geschehens auf einen mikroskopischen Hintergrund mit Gesetzen, die gegen Zeitumkehr invariant sind, bedarf genauerer Untersuchung. Wir betrachten dazu drei Beispiele. Das erste betrifft die Schwankungen der Teilchenzahl in einem Teilgebiete, das Teilchen mit der Umgebung austauscht (Brownsche Molekularbewegung). Nehmen wir für das ganze Gebiet eine feste Teilchenzahl $N = 10$ an, so werden wir unter symmetrischen Verhältnissen in der einen Hälfte des Gebietes häufig 5, fast so häufig 4 oder 6, weniger häufig 3 oder 7, selten 2 oder 8, noch seltener 1 oder 9, sehr selten 0 oder 10 Teilchen vorfinden. Man kann die Verhältnisse (nach Ehrenfest) simulieren durch „Spiele", in denen bei jedem Spielschritt auf eine Verteilung $(n, 10-n)$ der 10 Elemente auf zwei Kästen eine Verteilung $(n+1, 9-n)$ oder $(n-1, 11-n)$ folgt und durch die Spielregeln dafür gesorgt ist, daß jedes Element mit gleicher Wahrscheinlichkeit in jedem der beiden Kästen ist. Abb. 5 stellt den Verlauf eines solchen

Zeit als physikalischer Begriff 47

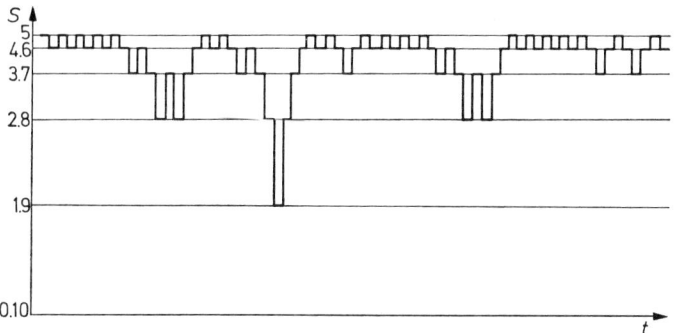

Abb. 5. Entropiespiel

Spieles dar. Als Ordinate ist der Logarithmus der Zahl der Fälle aufgetragen, die zu einer Verteilung $(n, 10-n)$ gehören. Dieser Logarithmus mißt die „Entropie" $S$ der Verteilung. Allgemein wird die Entropie proportional gesetzt dem Logarithmus der Zahl der gleichberechtigten mikroskopischen Fälle, durch die ein makroskopischer Zustand, hier die Verteilung von Elementen auf zwei Kästen, verwirklicht wird. In unserem Beispiel hat die Gleichverteilung (5, 5) die höchste Entropie; die extremen Verteilungen (0, 10) und (10, 0) die niedrigste. Einer Verteilung $n \neq 5$ folgt häufiger ein größerer Wert der Entropie, als ein kleinerer; besonders „wahrscheinlich" ist die Zunahme der Entropie, wenn sie einen kleinen Wert hat. Aber das gilt auch in der anderen Richtung. Häufiger geht ein größerer Entropiewert vorauf; besonders wahrscheinlich ist dies bei niedrigen Entropiewerten. *In Bezug auf die Entropieänderung haben wir immer noch Symmetrie der beiden Zeitrichtungen.*

Im zweiten Beispiel nehmen wir eine große Zahl $N$ an, etwa die große Zahl der Gasmolekeln in einem Gefäß. Die Zahl $n$ der Teilchen in einem Teilgebiet zeigt dann nur relativ ganz kleine Schwankungen um den Wert, der dem Verhältnis der Volumina des Teilgebiets zum Gesamtgebiet entspricht (im Gebiet $B$ der Abb. 6 sind die Schwankungen stark übertrieben). Die Entropie bleibt in der Nähe ihres größtmöglichen Wertes. Wenn wir einen merklich niedrigeren Wert der Entropie vorfinden, so ist es extrem wahrscheinlich, daß die Entropie vorher und nachher höher war. Nun kann man aber durch äußere Eingriffe auch extrem niedrige Entropiewerte herstellen, z.B. das Gefäß leerpumpen und dann von einer Seite her durch eine kleine Öffnung die Gasmolekeln wieder hereinlassen. Das könnte etwa einen Verlauf der Entropie wie bei $A$ (Abb. 6) bewirken. Hier erst finden wir Unsymmetrie der beiden Zeitrichtungen; die Entropie im Gebiet $A$ nimmt (von sehr kleinen Schwankungen abgesehen) zu, wie es der zweite Hauptsatz der Thermodynamik ausspricht. Aber die Unsymmetrie der beiden Zeitrichtungen kommt vom künstlich hergestellten Anfangszustand sehr niedriger Entropie. Von selbst stellt sich solch ein Zustand nicht her; genauer gesagt, die Wahrscheinlichkeit

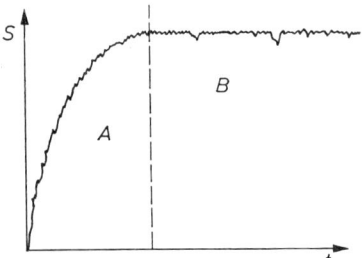

Abb. 6. Entropie bei vielen Teilchen

dafür ist so extrem niedrig, daß man mit der Herstellung nicht zu rechnen braucht (wir rechnen auch nicht mit viel, viel wahrscheinlicheren Vorkommnissen wie Fallen eines Meteorsteins auf unseren Kopf oder kilometerhohen Meereswellen).

So ist die Zunahme der Entropie, der zweite Hauptsatz der Thermodynamik, zu verstehen. *Die Unsymmetrie der beiden Zeitrichtungen kommt von einem außerordentlich unwahrscheinlichen Zustand am Anfang.*

Als drittes Beispiel wählen wir einen großen Ausschnitt aus der wirklichen Welt. Es macht nicht viel Unterschied, ob wir etwa unser Sonnensystem, unsere Galaxie oder die Gesamtheit der beobachtbaren Galaxien wählen. Wir haben in diesem dritten Beispiel ein energetisch nicht abgeschlossenes System. Mit der ausgestrahlten Energie wandert auch Entropie ab. Aber diese Entropieausstrahlung ändert nichts Wesentliches gegenüber dem Abschnitt $A$ des zweiten Beispiels. Der Verlauf der Entropie entspricht dem ausgezogenen Stück der Kurve in Abb. 7. Wir beobachten einen Entropiewert, der viel niedriger

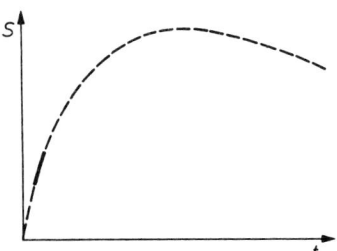

Abb. 7. Entropie eines großen Ausschnittes aus der Welt

ist, als der mögliche Höchstwert. Die Fixsterne haben noch viel Wasserstoff, der in schwerere Elemente umgewandelt werden kann. Wir erwarten, daß die Entropie mit der Zeit zunimmt. Aber wir schließen auch, daß der gegenwärtige Zustand niedriger Entropie aus einem Zustand noch niedrigerer Entropie entstanden ist.

*Der große Ausschnitt aus dem Universum, den wir überblicken, ist in einem Zustand sehr niedriger Entropie – einem „sehr unwahrscheinlichen" Zustand. Er geht in Zustände höherer Entropie, „wahrscheinlichere" Zustände, über. Wir müssen annehmen, daß er aus einem noch unwahrscheinlicheren früheren Zustand hervorgegangen ist. Von den beiden Zeitrichtungen nennen wir die mit den unwahrscheinlicheren Zuständen die Vergangenheit.*

Im ersten der drei Beispiele kommt $\Delta S < 0$, Abnahme der Entropie, vor. Dies im zweiten und dritten erwarten zu wollen, würde der Physik nicht entsprechen. Der zweite Hauptsatz der Thermodynamik ist kein „allgemeines Grundgesetz"; es ist vielmehr eine Aussage, die in unserer besonderen „kosmologischen Situation" niedriger Entropie gilt.

*Eine Unsymmetrie zwischen den beiden Zeitrichtungen ist in den allgemeinen Naturgesetzen nicht zu finden. Aber sie ist ein Faktum der wirklichen Welt; sie ist bedingt durch ein anderes Faktum dieser Welt, nämlich ihre niedrige Entropie. Das erste Entstehen der voraufgegangenen Zustände noch niedrigerer Entropie ist physikalisch nicht zu verstehen.*

Wir können in die Vergangenheit nicht unbegrenzt zurückdenken. Modelle des Weltganzen können frühestens mit der Entropie null beginnen. Die Vergangenheit ist diejenige der beiden Zeitrichtungen, die eine Grenze hat. Das wichtigste Ereignis der Geschichte des Universums oder des großen Ausschnitts aus dem Universum, den wir überblicken, ist ein Ausgangszustand sehr niedriger Entropie. Daß der zweite Hauptsatz der Thermodynamik mit der Unsymmetrie der Zeitrichtungen gilt, liegt daran, daß die Entropie immer noch sehr niedrig ist. Verkürzt können wir sagen: *Vergangenheit und Zukunft sind deshalb so kraß verschieden, weil die Welt noch sehr jung ist.*

## 6. Spezielle Systeme

Es könnte Gebiete des Kosmos geben, in denen die Energiequellen erschöpft sind. Sie sind im Temperaturgleichgewicht mit der Umgebung; die Entropie hat ihren maximalen Wert und behält ihn. Es geschieht nichts. Der zweite Hauptsatz der Wärmelehre ist nicht anwendbar.

Als zweites spezielles System betrachten wir Sonne und Erde. In der Sonne wird Kernenergie in Wärme und Strahlung umgewandelt; dabei wird Entropie erzeugt. Der Erde wird Energie und damit auch Entropie zugestrahlt. Ungefähr ebensoviel Energie, wie sie empfängt, strahlt die Erde auch wieder aus. Die Ausstrahlung von der Erde erfolgt bei tieferen Temperaturen als die Strahlung von der Sonne; darum strahlt die Erde mehr Entropie ab, als sie empfängt. Es ist denkbar und auch ungefähr richtig, daß diese Entropiedifferenz gerade der auf der Erde durch nicht umkehrbare Prozesse entstehenden Entropie entspricht. So ist es möglich, daß nicht nur der Energieinhalt, sondern auch die Entropie der Erde ungefähr gleich bleibt. Es bedeutet also keinen Widerspruch zum zweiten Hauptsatz, wenn die „Ordnung" auf der

Erde, etwa in der Evolution der Lebewesen, zunimmt. Gewiß, Lebewesen sind Gebilde, deren Entropie nur in begrenztem Maße durch Abzählen von „Fällen" definiert werden kann. Aber denken wir daran, daß viele Verrichtungen von Lebewesen auch von Apparaten ausgeführt werden können, so etwa das Trennen roter und weißer Körner aus einem ungeordneten Haufen. Die physikalischen Prozesse in solchen Apparaten sind völlig durchschaubar und genügen natürlich dem zweiten Hauptsatz. Das Herstellen einer Ordnung, eines Zustandes niedriger Entropie, durch ein Lebewesen braucht also keine Verletzung des zweiten Hauptsatzes zu sein.

## 7. Was ist die Zeit?

Die Entropie eines abgeschlossenen Systems nimmt mit der Zeit zu. Aber die Entropie ist doch kein Maß der Zeit. Es gibt Prozesse, bei denen die Entropie langsam wächst (die beinahe reversibel sind), und solche, bei denen sie rasch zunimmt. Entropie und Zeit hängen nicht so fest zusammen, wie Bewegung und Zeit. Also: die Zeit ist nicht die Entropie.

Sie ist aber auch nicht die Expansion der Welt. Wir wissen nicht, ob in der Beziehung zwischen dem Abstand von Galaxien und der Zeit,

$$r = vt,$$

die Geschwindigkeit $v$ konstant bleibt. Modelle des Universums können Verläufe der Abstände zeigen, die der Kurve 1 oder 2 der Abb. 8 entsprechen. Im

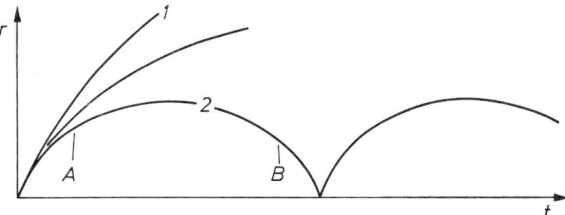

Abb. 8. Weltmodelle

Falle der Kurve 2 verläuft bei $B$ nicht alles umgekehrt wie bei $A$; die Entropie nimmt bei $A$ und bei $B$ zu. Bei $A$ gibt es noch mehr Wasserstoff; der Zustand $B$ ist im Durchschnitt heißer.

*Die Zeit wird nicht durch die Entropie oder die Irreversibilität und nicht durch die Expansion der Welt gemessen. Sie liegt tiefer.*

In der Hochenergiephysik sind kürzlich experimentelle Anzeichen aufgetaucht, daß vielleicht die Invarianz gegen Spiegelung bei gleichzeitiger Vorzeichenumkehr aller ladungsartigen Größen, die PC-Invarianz (parity, charge), verletzt sei. Da aus sehr allgemeinen Voraussetzungen über das Naturgeschehen eine TPC-Invarianz, Invarianz gegen simultane Ausführung von Zeitumkehr, Spiegelung und Ladungsumkehr, folgt, ist die Frage erlaubt, ob die Versuche eine Verletzung der Invarianz allgemeiner Sätze der Physik gegen Zeitumkehr bedeuten. Natürlich denken die Experten daran, daß eine Unsymmetrie zwischen den Zeitrichtungen oder eine Abweichung von der PC-Symmetrie bei den Experimenten von der Einbettung in eine unsymmetrische faktische Umgebung herkommen könne. Die Protonen sind ja positiv geladen, die Welt hat eine begrenzte Vergangenheit. Die Antworten stehen noch aus. Sollte wirklich eine Zeitasymmetrie der allgemeinen Gesetze, nicht nur eine kosmologisch bedingte der Experimente nachgewiesen werden, so wäre eine neue Situation im Denken über die Zeit gegeben.

## 8. *Kausalität*

Wir gebrauchen dieses Wort in zwei Bedeutungen. In der einen drücken wir damit aus, daß der Zustand eines abgeschlossenen physikalischen Systems zu einer Zeit $t=0$, also $\Sigma(0)$, den Zustand zu einer anderen Zeit, $\Sigma(t)$, determiniert; in der klassischen Physik kann $\Sigma(t)$ im Idealfalle genau angegeben werden; in der Quantentheorie werden Wahrscheinlichkeiten für $\Sigma(t)$ ausgesagt. Aber wegen der Symmetrie der physikalischen Sätze im Bezug auf die beiden Zeitrichtungen macht auch die Beziehung zwischen $\Sigma(0)$ und $\Sigma(t)$ keinen Unterschied zwischen $t>0$ und $t<0$. *Kausalität in diesem Sinne zeigt keinen Unterschied von Vergangenheit und Zukunft.*

Diese Betrachtung setzt den Idealfall der vollständigen Kenntnis des Zustandes $\Sigma(0)$ (in der klassischen Physik) oder die „maximale" Kenntnis (in der Quantentheorie) voraus. Da, wo ein makroskopischer Vordergrund durch einen mikroskopischen Hintergrund mit der statistischen Physik erklärt wird, liegen die Dinge anders.

Damit kommen wir zur zweiten Bedeutung der Kausalität: Ein besonderes und auffallendes Ereignis, eine „Ursache", wirkt nur auf die Zukunft. Ein Blitz trifft ein Haus, dieses brennt ab. Die zeitliche Umkehrung kommt nicht vor. Fische sterben in einem Fluß, in den ein Gift geschüttet wurde. Die zeitliche Umkehrung kommt nicht vor. Blitz, Gift sind sehr spezielle Vorkommnisse im Sinne der statistischen Physik, sie kommen aus Gebieten sehr niedriger Entropie. Zerstörte Häuser, tote Fische haben höhere Entropie. Eine „Ursache" hat niedrige Entropie.

Dokumente künden von der Vergangenheit; sie bewahren Zustände geringer Entropie; mit ihrer allmählichen Zerstörung nehmen sie an der Entropievermehrung teil. In den praktischen Anwendungen der Wellentheorie kommen nur die „retandierten Potentiale" vor, die auslaufenden Kugelwellen entsprechen. Der Grund ist, daß Signale aus einem engen Raumgebiet in der Wirklichkeit vorkommen, einlaufende Kugelwellen nur bei künstlichen Vorrichtungen. Beide sind statistisch unwahrscheinliche Vorkommnisse; aber Signale gibt es (ins Wasser fallende Steine, explodierende Sterne).

*Kausalität in diesem zweiten Sinne unterscheidet Vergangenheit und Zukunft, weil sie von speziellen Anfangszuständen handelt, aus dem gleichen Grunde wie der zweite Hauptsatz der Wärmelehre.*

## 9. Zusammenfassung

Wie das tägliche Leben, zeigt auch die Physik zwei Aspekte der Zeit. Zeit als Koordinate beschreibt die Einbettung in das Universum. Zeit als Folgeordnung zeigt den Unterschied von Vergangenheit und Zukunft. Der Unterschied steht nicht in den allgemeinen Gesetzen der Physik, sondern kommt von einem speziellen Faktum, einem früheren Zustand sehr niedriger Entropie, dessen Entstehung physikalisch nicht begreifbar ist. Der zweite Hauptsatz der Wärmelehre, der die Symmetrie der beiden Zeitrichtungen verletzt, ist eine Aussage über unsere besondere Welt.

*Literatur*

Čapek, M.: *The philosophical impact of contemporary physics*, bes. Kap. III, VIII, XI, XVII. Princeton, N. J. 1961.
Costa de Beauregard, O.: *Irreversibility problems*, Proc. Intern. Congr. Logic, Methology, Philos. Sci. 1964.
Dako, M.: *The direction of time*, Stud. Gen. *22* (1969) 965–984.
Ehrenfest, P. u. T.: *Über zwei bekannte Einwände gegen das Boltzmannsche H-Theorem*. Phys. Z. *8*, (1907) 311.
Feynman, R.: *The character of physical law*, bes. Abschn. 5. London 1965.
Frazer, J.T. (ed.): *The voices of time*. Ann. N.Y. Acad. Sci. *138* (1967) 367-915.
Hund, F.: *Die Zeit in der Begriffswelt des Physikers*. Stud. Gen. *8* (1955) 469–476; *Denkschemata und Modelle in der Physik*, Stud. Gen. *18* (1965) 174–183; *Grundbegriffe der Physik*. Mannheim 1969.
Margenau, H.: *The nature of physical reality*, bes. Abschn. 7. New York 1950.
v. Weizsäcker, C. F.: *Der zweite Hauptsatz und der Unterschied von Vergangenheit und Zukunft*. Ann. Physik *36* (1939) 275.
Weyl, M.: *Raum, Zeit, Materie*, bes. § 36 und 39. 5. Aufl. Berlin 1923; 7. Aufl. Darmstadt 1961.
Whitrow, G. J.: *The natural philosophy of time*. London 1961.

# Time in Particle Physics

J. G. TAYLOR[*]

*Summary.* Time and space are shown to have very different properties, both from each other and from that of the macroscopic world, when they are investigated at distances of the size of elementary particles; even the definition of time is in some doubt in such a case. Space reflection is violated maximally, but time reversal is not. Indeed it may not even be violated at all, but if it is not then causality is violated over such short distances. In any case the basic laws of physics will have to be changed; when the details are finally worked out the concept of time will be altered ineradicably from the common sense one.

## 1. *Introduction*

Time as a basic concept has always played a very important role in physics, and in modern physics since Newton's time has been on the same level of importance as spatial extension or distance. Indeed Einstein's theory of special relativity would seem to put these physical concepts together and make of them one concept on which the space-time continuum is based. However recent developments in the physics of elementary particles are beginning to show that time and space have to be given very different properties when events occurring over distances of the order of $10^{-14}$ cms are considered. Such distances are the order of the sizes of the elementary particles; it is processes involving these particles that give us this new picture. I want to try and explain what this picture is without too many technical details in this paper. At the same time I want to describe some conjectures being put forward by various particle physicists which have very important implications for the notion of causality, and so of time. These conjectures will show some of the turmoil that is going through particle physics at present; this turmoil reflects our uncertainty as to some of the basic laws which govern particle physics.

In the remainder of this introductory section let me remark briefly on some general properties of these basic laws. These properties are usually associated with conserved quantities. The most important conserved quantities are the total energy, linear momentum and angular momentum for any isolated system. These conserved quantities do not depend on the time at which they are measured. The description of the motion in terms of them thus allows time to be dispensed with. Indeed one recent approach in particle physics has attempted to remove the notion of time altogether. It does so in terms of the "strict observability" approach of Heisenberg. The directly observable quantities of

---

[*] Professor J. G. Taylor, Department of Physics, University of Southampton, Southampton, England.

particle physics are the energies and momenta of scattered particles. The scattering process is to be thought of as arising from the interaction between a set of particles of known energies and momenta which are termed "incoming" particles. These incoming particles scatter due to their mutual interactions, and after the scattering processes produce "outgoing" particles. The energy and momenta of these outgoing particles are also measured. This scattering process is described, as is usual in quantum mechanics, by a scattering amplitude. The square of this amplitude is equal to the probability that the particular scattering process occurs. Thus the scattering amplitude is directly measurable, up to a constant phase. We see that time doesn't appear explicitly in this description, except in the notion of "incoming" and "outgoing". It enters here in a very weak way: we need to be able to distinguish between the infinite (or near infinite) past and the infinite future. In other words an over-all time separation is required to be known. Once this is given then the scattering amplitude, or S-matrix element as it is usually called, describes a scattering process in a directly observable fashion. This framework for discussing elementary particle processes is called, rather naturally, the S-matrix approach.

In the S-matrix approach it is possible to consider spatial extension in detail, It is not possible to consider time development in detail; the S-matrix approach is only an asymptotic theory in time since it describes the relation between particles either in the infinite past or the infinite future. Intermediate finite times are not accessible to experimental analysis in the general S-matrix framework. It is possible to consider the operation of time reversal T in this framework. This operation is that which changes the sign of time in theories which have intermediate finite times; in the S-matrix framework time reversal is that operation which interchanges the initial states (of the incoming particles of various given energies and momenta) with the final states (of the outgoing particles). Thus the operation of time reversal relates a process, say of the scattering of two incoming particles $A$ and $B$ to produce outgoing particles $C$ and $D$ to the process in which $C$ and $D$ are incoming and $A$ and $B$ outgoing. Thus one reaction can be represented by the equation

$$A + B \longrightarrow C + D \tag{1}$$

whilst the time reversed process is

$$C + D \longrightarrow A + B \tag{2}$$

We can ask if these two processes are related; the theory is said to possess time reversal invariance if the S-matrix elements for the two processes are identical. This corresponds to the condition that it is not possible to distinguish between incoming or outgoing states.

In a similar fashion we can introduce the parity operation $P$, which reflects all processes in the origin of space co-ordinates, so changing all particle momenta to their negatives. The time reversal operation is similar because it changes all the particles' time co-ordinates to their negatives. However the parity operation $P$ doesn't relate two different processes to each other, but invariance under $P$ only imposes conditions on a given process. It was shown in 1957 that the parity operation is violated in particle processes; one can distinguish the mirror world from the actual world without having to be told that we are actually looking in a mirror. In order to understand the details of this a little further we must realise that there are four different sorts of particle interactions or forces. These are: strong or nuclear forces, holding the subnuclear neutrons and protons together in the nucleus; the electromagnetic forces, which are the forces between electrically charged particles; the weak forces, mediating radioactive decays of nuclei; and finally the gravitational force between massive particles. The strength of these forces vary enormously, and are approximately in the ratio $10:1/137:10^{-14}:10^{-40}$. As is to be expected the gravitational forces appear to have little effect in elementary particle processes.

It turns out that both the electromagnetic and strong forces are invariant under parity reflection, but the weak forces are not. Indeed these latter forces appear to violate parity as much as they can. It is still possible to introduce an operation under which the weak forces could be invariant. This is achieved by introducing the operation $C$ of changing the sign of the electric charges of all particles. C is called the charge conjugation operator. Then the results of parity violation discovered in 1957 are still consistent with the invariance of weak forces under the operation $CP$. Thus if you look in a mirror and at the same time change the sign of the charges of all charged particles you see the same world.

There is a famous theorem known as the *CPT* theorem, which may be used to relate these three operations $C, P$ and $T$ together. This theorem states that if each of them is applied then the result is the same world, so that the operation *CPT* is an invariance operation of the world of elementary particles. The *CPT* theorem is evidently important since if we can separately test for the invariance under $C$, $P$ and $T$ we can test whether or not the conditions under which the theorem is valid are themselves true. What are these conditions? The most important one is that of causality. This can be expressed as: no information can travel faster than the speed of light. The other conditions are that the principle of special relativity is valid and that there are only positive energies, both very well tested and accepted. The causality condition is the weakest of the conditions; whilst it appears to be valid in the macroscopic world its validity in the microscopic world is not necessarily valid. It has been proved to be valid for electromagnetic and strong interactions between particles down to about $10^{-15}$ cms. However its validity at shorter distances, and for weak interactions, is not certain. The importance of the *CPT* theorem now becomes clear:

it relates C, P and T and microscopic causality. If the simultaneous operation CPT is violated then causality must be violated; if causality is regarded as sacrosanct then the violation of CP means that T must be violated in such an amount as to make CPT invariant. We will see the importance of this in the next section.

Before we turn to that a final remark on the further introduction of time in particle physics is apposite. In order to follow the most important recent developments in particle physics it is necessary to re-introduce a finite time. This is done by means of quantum field theory. There seem to be many possible quantum field theories. That one which will agree with experiment will tell us how the finite time developments of a system occurs in detail. This finite time development is that chosen to fit all the asymptotic experimental results obtained from scattering experiments (which are the only experiments which can be made). Thus we seem to be back to the position before the S-matrix approach was applied, though we don't know, as yet, what this detailed finite time development is.

## 2. Time Reversal Violation?

The shock of parity violation in 1957 was somewhat mollified when it was suggested, and then shown, that CP is invariant. However the shock was not escaped in this way for long, since in 1964 it was shown that CP itself is not invariant. This violation of CP invariance was found in the decay of the long-lived neutral K-meson. The neutral K-meson has two components, a short lived one $K_S$ and a long lived one $K_L$. It is in the decay of $K_L$ to two pi-mesons that CP is violated, since it may be seen that the CP value of $K_L$ is minus one, that of the two pi-mesons is plus one. The amount of this CP violation is small; only one part in a thousand of the weak interaction violates CP, so there is no "maximal" violation of CP as there was of parity P.[1]

How can this violation of CP be explained? There have been numerous suggestions. One of these is that this violation of CP actually occurs in the strong or electromagnetic interactions, which then contribute internally to the decay of $K_L$ to two pi-mesons. The suggestion that electromagnetism violates CP has caused a great deal of excitement; if true it means that we haven't properly understood the nature of the Coulomb force between charged particles. This suggestion implies that various other evidences of CP violation will show up. And if the CPT theorem is true there should also be violations of time reversal invariance T in electromagnetic processes. These have recently been searched for, and there is some (not conclusive) evidence for the existence of this violation of T. In particular the relation between processes (1) and (2) has

---

[1] For a more detailed discussion of this situation and that of violation of parity and time reversal, see, for example, Taylor, J. G.: *PC and T Violation*, Lectures in Theoretical High Energy Physics (ed. H. H. Aly). John Wiley and Sons, 1968.

been investigated when $A$ and $B$ are photons and deuterons, $C$ and $D$ are protons and neutrons. A recent experiment[2] at Berkeley has shown that the rates of these two processes are different; the amount of this difference seems in conflict with an experiment performed at Harwell[3] by an Oxford group, looking at the case when $A$ was photons, $B$ oxygen, $C$ α particles, $D$ carbon. However the situation is not completely clear since the various analyses of these experiments are model dependent.

If it is true that $T$ is violated, and the Berkeley results are correct, then it is possible that $CPT$ is valid and causality is also satisfied. If $T$ is not violated (and the Oxford experiments are valid) then $CPT$ is violated and causality must be destroyed. In either case another of the 'well-understood' and sacrosanct principles of physics has to go. That is quite certain. Just which one is not so certain. It may also be, of course, that the foundations on which the $CPT$ theorem are based are themselves wrong, in which case we are really at sea in understanding the elementary particles.

There have been other suggestions for violating $CP$ and $T$[1]; one of these[1] is through the existence of Dirac's magnetic charges (the magnetic equivalent of the electron). These magnetic charges, or monopoles as they are usually called, have been searched for but not found. Their existence would not only violate $CP$ and $T$ but also explain why electric charge is quantised; as such it is hoped that the experimental search for them will be ultimately successful.[4]

## 3. Causality Violation?

A model which gives causality violation but which may conserve time reversal has recently been considered.[5] This involves particles which travel faster than light; these particles are called tachyons from the Greek 'tachys' for swift. As is to be expected these particles violate causality, as may be seen from the old limerick:

> There was a young lady called Bright
> Who could travel much faster than light
> She went out one day
> The Einsteinian way
> And returned the previous night.

This means she could have slipped on the soap and sprained her ankle after she had returned the previous night, so preventing her from going out the

---

2 See Casella, R. C.: *Time Reversal and the $K°$ Meson Decays II*. Phys. Rev. Lett., 22 (1969) 554 and references quoted there.
3 Allardyce, B. W., et al.: *A Failure of the Time Reversal Invariance Principle*, contributed talk at the Conference on Nuclear and Elementary Particle Physics, 24–26 September 1969, University of Sussex, England.
4 Kohn, H. H : *Magnetic Monopoles*, Science Journal, September 1968.
5 See Taylor, J. G.: *Particles Faster than Light*, Science Journal, September 1969.

following morning in that Einsteinian way. This violation of causality is not achieved without some cost; the existence of tachyons would imply initially that particles can appear to travel backwards in time. These particles would at the same time have negative energy, and so are to be avoided. It is possible to reinterpret the emission of negative energy particles as the absorption of positive energy ones; this reinterpretation at the same time causes tachyons to only go forward in time. Thus the difficulties presented by tachyons are reduced to that of violating causality; there is now no good reason why tachyons don't exist. They have been looked for[5] but have proved as elusive as the magnetic monopole and the quark. However they may still exist, and as such give rise to acausal effects. It has been recently[6] suggested that unobservable tachyons may exist which also[7] give rise to a violation of causality, so of the *TCP* theorem, but conserve time reversal invariance. This possible model is presently being explored.

## 4. Conclusions

It has been shown that time and space have very different properties in particle physics; space reflection is violated maximally by the weak interactions whilst it is not certain that time reversal is violated, but it is certainly not violated maximally. It is also certain that either causality or time reversal (or both) are violated; the microscopic notion of time is very different from the macroscopic. The possibility of obtaining macroscopic violations of causality or time reversal are very interesting; there may be regions of the universe where such possibilities occur. It will then be easier to perform experiments to determine the true nature of time and space. Till we find these we have to dig deeper and deeper, to smaller and smaller distances, till we understand the correct nature of the laws governing physical processes. In any case these are very different from what we would expect from our macroscopic experience; in the next years the details of this difference will become much clearer. By the time of the next Conference it is to be hoped that the situation is better understood, so that we know what disturbing aspects of time at the microscopic level that we have to live with.

---

6 Lee, T. D., Wick, G. C.: *Nuclear Physics*, B 9, (1969) 209.
7 Taylor, J. G.: *PC and PCT Violation with Indefinite Metric*, Queen Mary College preprint (unpublished) (1969).

# Time in Statistical Physics and Special Relativity

P. T. LANDSBERG*

CONTENTS LIST

1. Introduction ............................................................. 59
2. Time as an Abstraction to Give a Unique Status to Events ................. 62
3. Complications Arising for Large Systems ................................ 62
4. A Random Walk ........................................................ 63
5. The Irrelevance of Time in Some Statistical Problems .................... 65
6. A Reinterpretation of the Random Walk ................................. 67
7. Statistical Entropy: A Time Symmetrical Law for Its Probable Increase .... 67
8. Thermodynamic Entropy: Its Certain Increase for Infinite Systems ........ 69
9. Life as Violating Weak T-Invariance ..................................... 73
10. Life, Entropy and Disorder ............................................. 74
11. The Direction of Life; the Status of Boundary Conditions ................ 75
12. The Cosmological and the Statistical Arrow ............................. 76
13. Two Problems of Relativistic Time ..................................... 79
14. Tachyons ............................................................. 81
15. Concluding Remarks ................................................... 83

*Appendices*

A. Quotations on Irreversibility and Entropy ............................. 84
B. Concepts for Irreversibility Discussions ............................... 87
C. Properties of Entropy and of "Disorder" ............................... 89
D. A Random Walk ....................................................... 92
E. A Brief account of the Thermal History of the Universe ................ 93
F. Conditional probability, Bayes' Rule and Entropy ...................... 95
G. Determinism from Relativity? ......................................... 97
H. Some Properties of Tachyons ......................................... 98
I. Conditions for Signalling to one's Past ............................... 102
J. The Threshold of Classical Cosmology ................................. 105

## 1. Introduction

The effort to meet the Society's request for a talk on this topic has been a humbling experience. The literature is vast; it contains the names of the ancient philosophers, of Newton, Leibniz, Bergson, Planck, Einstein, and of many contemporary men of learning. Yet some deep doubts and differences remain which, in the case of statistical physics, may be at least hinted at by quotations which are at variance with each other (Appendices A, B). A survey of this

---

* Professor P. T. Landsberg, Department of Applied Mathematics and Mathematical Physics, University College, Cardiff, Great Britain.

field is therefore out of the question (but see [5]). Here it will be easier for all concerned if, instead, I present a point of view.

It has been emphasized (Whitrow, 1961) that without a rudimentary idea of time everything could happen at once; I accept this and take as primitive the notion of time as a co-ordinate. It is the direction of time that will occupy us here.

I want to study this question with the aid of a dematerialised intelligence $I$. Its contemplations will become increasingly more complicated. It will proceed in turn to do the following:

1. Study small systems on a planet devoid of life (Poincaré recurrence, $T$-invariance);
2. Study larger systems on a planet devoid of life (Qualitative similarity with 1.)
3. Communicate with humans about large systems (macrostates; time's arrow fixed arbitrarily in the direction of greater probabilities; second law);
4. Study the implications of life processes.

As regards statistical physics, the broad conclusion is that an objective arrow of time can be defined by a statistical tendency towards more probable states. The law of entropy increase is then a deduction from this definition.

The idea of keeping life as we know it out of the analysis as long as possible, is to see precisely how important boundary conditions are in this problem. It will be found that boundary conditions are time-symmetrical so long as we argue from them in both directions of time. Furthermore, that they are not involved explicitly if our intelligence $I$ surveys astronomical objects which are devoid of life. They come in with the experimenter. They are thus what Bridgman would have called paper-and-pencil operations which we insert into nature, but they are not essential to her workings when men are not present. That the past is known and leaves records, that the future is unknown and under investigation are obvious points, relevant to the psychological arrows. But in our treatment the consideration of these matters comes *after* the problems of this lecture have been sorted out. When we say that a boundary condition is an *initial* condition it is this psychological (or at least biological) arrow which imposes a direction of time. Such points are taken to be understood and not in need of elaboration.

The procedure will be to note time-reversal invariance and the recurrence again and again of the same states, if a system is isolated (§ 2). Large systems display the same phenomena, but the recurrence of states is less obvious. The new effect here is the need to introduce statistics (§ 3). That this does not by itself introduce a preferred *direction* of time is illustrated by an example (§ 4). Large systems also lead to the need to consider macrostates, each macrostate consisting of a large number of microstates. This number is the weight of the macrostate. For an isolated equilibrium system, entropy goes up with this weight, and so does the probability of finding the system in a macrostate. The latter condition depends on the assumption of equal à priori probabilities for distinct microstates. Hence, upon withdrawal of a partition, or upon manipulating the system in some other way, the new equilibrium state is one of greater entropy (§ 5). But statistical entropy also increases with separation in time from a known macrostate A in *both* directions of time, if only A and the structure of the system is known (§ 7). One then considers very large systems (strictly: infinite systems) to remove the troubles arising from Poincaré recurrence, and the direction of entropy increase is taken as the positive direction of time for a given system. A further assumption is needed to ensure that the time arrows of various macro-systems are in this limit parallel and not antiparallel. It is suggested, as a primitive notion, that there is a kind of continuity in macroscopic objects in the following sense: if entropy tends to increase with time for one isolated system, then, in a slightly different system, it will not have the opposite tendency (§ 8). In this way a consistent direction of time emerges as a *macroscopic* concept.

These statistical ideas generate the thought that the direction of time is *only* a macroscopic property and that at a microscopic level, when the second law of thermodynamics is irrelevant, then causality and a unique time-direction may be suspended, and the direction of time may possibly fluctuate, in a sense already suggested by certain experiments on elementary particles.

Other characteristic features of the present approach are the insistence on the coarseness of human perception, and the resulting opposition between macrostates and micro-states; the dependence of entropy on measuring apparatus or information available (§ 5)[1]. In the course of the discussion we come to a definition of life (§ 9) and of disorder as distinct from entropy (§ 10), and to important cautions with regard to the use of the entropy concept for biological systems (§ 10) and for the universe (§ 12). A principle of impotence (that a contracting universe cannot be seen), is proposed for discussion, though the writer has an open mind on its validity (§ 12).

In a slightly lighter vein special relativity is considered and in what sense it allows preferred frames of reference and therefore preferred time scales. Two recent and startling proposals are discussed, one that the time transformation implies determinism (§ 13), the other that faster-than-light particles (tachyons) are possible (§§ 14, 15).

The Appendices contain mathematical support for statements in the text and Appendices C, E, G, H, I, J all contain some original work not available elsewhere. The Appendices involve only elementary mathematics in accordance with the interdisciplinary nature of this meeting. Some very interesting mathematical developments in the theory of irreversibility, notably by Zwan-

---

[1] This agrees with a solution of the so-called Gibbs paradox ([51], p. 324). This reference is the first book to give an information theoretical approach to statistical mechanics. See also [45, 84].

zig, and the schools of Prigogine, Truesdell, Meixner and others could therefore not be covered.

## 2. *Time as an Abstraction to Give a Unique Status to Events*

Our intelligence starts with systems of only a few atoms or molecules and is possessed by knowledge concerning their interactions fully up to, but not too far beyond, contemporary human knowledge. It would be interesting to consider how it will introduce the notion of time. But are space and time so different? If it has a notion of space must it not also have the notion of time? For example: $I$ has measuring rods placed end to end; it counts them. Is this not a process which requires knowledge of time? We short-circuit these problems here by postulating that the events noted by $I$ are embedded in a spacetime manifold which assigns a unique quadruplet of numbers to each event in the observable universe. By thus endowing $I$ with a very human intelligence, it may be of course that we are already engaged on what will, in the future, turn out to be an incorrect analysis. It may be essential to incorporate a basic statistical uncertainty or fluctuation in the numbers which make up the quadruplet [2, 20, 41, 60, 61, 76]. It may be that some form of quantisation of space and time is needed which introduces a shortest distance and a shortest interval [12, 22, 36, 77, 78].

Neglecting these possibilities, however, $I$ is possessed of a knowledge of collision processes and of classical and quantum mechanics. Hence the Poincaré recurrence of states (that an equilibrium system with a finite number of states available to it will pass through anyone state again and again) and $T$-invariance of elementary processes (see Appendix B) will be known to it. This maximum available knowledge about the systems which it is studying may be described by saying that $I$ knows all about the *microstates* of these systems. Of course, it must not be too far ahead of humans, though it can be ahead a small amount since $I$ has this benefit: $I$ is not me.

Strictly speaking it discovers $T$-invariance in *many* cases or even in *all* cases. Whether it is "many" or "all" depends on a matter not fully understood by us, namely the universal validity, or otherwise, of the $PCT$ theorem. This theorem states that a certain operation, denoted by $PCT$, produces from known laws for particle interactions the same known laws again, i.e. the law are $PCT$-invariant. The operation is parity reflection $P$, which interchanges left and right, together with charge conjugation $C$, which converts particles into antiparticles (which have opposite charge), together with time reversal $T$ (see Appendix B). If this $PCT$ theorem is universally valid then the violation of the $PC$ symmetry which was discovered in 1964 for the so-called weak interactions implies a violation of $T$-invariance in these cases. If this $PCT$ theorem is not universally valid then the $T$-invariance may always be valid.

## 3. *Complications Arising for Large Systems*

Let us enable our intelligence $I$ to develop. It can now handle, again with maximum knowledge, systems consisting of a large number of particles. This is quantitative change only, and we would not expect anything new. It is true that our intelligence now becomes a Laplacian calculator, a super-human intelligence. However, if it confines its attention to the same detailed description which it gave of the smaller systems earlier, and if it is willing to wait for extremely long recurrence times, there is nothing new. In particular, it has no need of thermodynamics or statistical mechanics. These subjects, if they were explained to it, might be regarded by it as a form of occult sun-worship, an illustration of lack of computer power or of a human need for auxiliary concepts, and in any case as second best; subjects one resorts to when one does not know enough. The notion of a wall is unknown to it, but the positions and vibrations of the atoms which make up the wall are known. The idea of pressure is unknown to it, but the changes in momentum suffered by all the particles colliding with the atoms of a wall are known. The transfer of kinetic energy to molecules by collision with each other or with a wall are known to it, but it does not need the notion of heat. The intelligence $I$ would have to learn that in communication with humans the very fact that its information is so complete is a handicap. It must learn to approximate, to average out, to discard knowledge. Although it knows probability theory (needed in quantum mechanics), it must learn about *statistics*.

We now consider the implications of statistical distributions on the problem of time. We will try to avoid the introduction of a specific law postulating time asymmetry for such distributions [66], interesting though this idea is. To look into this question we take leave from our intelligence $I$ for three sections.

## 4. *A Random Walk*

Consider an infinitely long row of squares and label them by the integers from $-\infty$ to $+\infty$. A token is placed on the square marked zero. Now a coin is tossed and if heads show the token is placed one square to the right, if tails show it is placed one square to the left. The coin is tossed again and the token is placed on a new square. One stops after a definite number $N$ of tosses. The system has a number of states $s_i$ which are labelled by the integers of the square on which the token is placed. Thus the $i^{\text{th}}$ move yields the alternatives for $s_i$ indicated on the line marked $s_i$:

$$
\begin{array}{lcccc}
s_0 & & 0 & & \\
s_1 & & -1 & 1 & \\
s_2 & & -2 & 0 & 2 \\
s_3 & -3 & -1 & 1 & 3 \\
\end{array}
$$

The procedure is in effect a way of generating a sequence of integers

$$\mathbf{S}: (0, -1, 0\ 1, \ldots\ldots\ s_N),$$

or
$$S': (0, -1, -2, -1, \ldots\ldots s_N) \text{ etc., etc.}$$

If one prefers this way of talking, one can say that $S$ and $S'$ are two systems specified by an ordered set of $N$ integers.

One will agree that time does not enter into this problem in any essential way. The probability of finding the token at $s_0 = 0$ is unity for $N = 0$, and decreases as $N$ increases. The greater $N$ the further the probability distribution spreads: the peak at $s_N = 0$ becomes lower and less significant as $N$ increases. If many people play this game so that there are many tokens on the square $s_0 = 0$ at the beginning, then we have what a statistical mechanics student may call a *token-fluid*; it spreads as $N$ increases. The relatively ordered states of small spread around $s_0 = 0$ give way to disordered token arrangements. For large $N$ some tokens begin to occupy squares as far as the eye can see.

The next step is to introduce the statistical entropy, following Gibbs. As information theory has shown, there are merits in associating a statistical entropy function (see Appendix C, and note $\ln p \equiv \log_e p$)

$$S_{\text{stat}} = -k \int [p(x) \ln p(x)] \, dx \quad \text{or} \quad -k \sum p_i \ln p_i$$

with a probability distribution $p(x)$ or discrete probabilities $p_i$. It is not to be assumed that it is exactly the same as the thermodynamic entropy

$$S_{\text{therm}} = \int dQ/T$$

Thus statistical entropy increases with $N$. The reason is easy to see. With every move our ignorance concerning the state of the system increases due to the probability law governing each move. Thus, while all is known and all is certainty for the state $s_0 = 0$, the further we depart from this state the greater is the entropy of the distribution.

One can, of course, imagine a *particle* as generating the states. It moves with infinite speed and presents us with all the states of a particular system $S$ simultaneously. Time can be said to be present *qualitatively* already since the moves are distinguished and counted. Now let us introduce time *quantitatively* by requiring that the particle has a constant incubation period at each site, followed by an instantaneous jump. Then the particle is performing a random walk and the number of moves $N$ is proportional to the time for which we have the patience to follow the game. "Very sensible", a thermodynamics student was heard to mumble, "I see entropy goes up with $N$ and so it increases with time." But one day his friend came to see him: "My token has reached the position $s_0 = 0$ again. Can you tell me about the position $N$ moves prior to this situation?" "The answer is given in terms of a probability distribution

$P(s_N = m)$; and the distribution is, of course, the same whether we calculate backwards in time or forward in time" said the student. "But then the entropy of the distribution goes up as time goes backwards, or *decreases* with forward time" protested his friend Fig. 1, see p. 1117. The tale finishes here and little more was heard of our thermodynamics student.

The notion of entropy as used here is always the Gibbs entropy, i.e., it is based on the probability distribution for the whole system. It is not the Boltzmann entropy which is based on the single particle distribution function.

## 5. *The Irrelevance of Time in Some Statistical Problems*

The story enables me to stress two points already made fully elsewhere ([51], pp. vii, 237, 324). Firstly, the entropy of a system depends on what we know about it and what measurements we can make. It is not characteristic only of the system. One may object that the model does not describe an isolated system. What about the coin, and what about the tosser? I reply: Exactly, let us include these items. Let us suppose we know all about them, so that the tosser becomes a coin-tossing machine and the outcome of each toss is known or calculable. The moves become all known and only *one* set of integers **S** is now appropriate and possible as viewed by a more penetrating intelligence: the set which actually occurs. The moves are certain, the entropy of our game remains zero. What has changed? The game is the same: it is as uninteresting now as it was before. *But somebody more knowledgable is making the calculation.*

The second point is that time is strangely irrelevant to thermodynamics. Nothing ever happens in thermodynamics. I now want to illustrate this second point. Reverting to the less knowledgable description of our game (with probabilities $\frac{1}{2}$), the probability $P(s_N = m)$ of finding the token $s$ on square $m$ after $N$ moves is a normal distribution (for $m \ll N$) and satisfies an interesting differential equation:

$$\frac{\partial P}{\partial N} = \tfrac{1}{2} \frac{\partial^2 P}{\partial m^2} \ (m<<N)$$

No-one is going to worry about the fact that the equation changes when $N$ changes sign – after all we have not defined negative $N$. But suppose that, with the idea of an incubation period, $N$ is put proportional to time $t$ and $m$ to the distance $x$ travelled by the particle in this time? Philosophical handwringing breaks loose as we extract the diffusion equation (which lacks $T$-invariance) from the model (see Appendix D).

$$\frac{\partial P}{\partial t} = D \frac{\partial^2 P}{\partial x^2} \ (D = \text{diffusion constant}, N \propto Dt)$$

The model does not really contain time, and so is certainly $T$-invariant. Why then does the diffusion equation rear its unwanted head? The answer is as follows: The ordered set of integers **S** which specifies our system is presented to us without reference to time. One can, therefore, speak of the distribution $N$ steps *away* from $s_0 = 0$ (not $N$ steps *before* $s_0$ or $N$ steps *after* $s_0$). This is just one case. The difficulty occurs when it is split up into two cases (a) $s_0 = 0$ is the *initial* condition, (b) $s_0 = 0$ is the *final* condition. Mathematically the trouble is seen in the relation $N \propto Dt$ which introduces the time $t$ which is then reversed, thus giving $N$ a negative sign. This is inadmissible in terms of the original problem.

Note that the model allows for the recurrence of the initial state $s_0 = 0$ after $N$ moves. It occurs with probability

$$P(s_N = 0) = \sqrt{(2/\pi N)} \propto (Dt)^{-\frac{1}{2}} \, (N \gg 1)$$

This is an example of Poincaré recurrence (Appendix B). This phenomenon is also expected with a fully deterministic system.

To broaden our discussion, let the notion of a group of macroscopically indistinguishable microstates be introduced. Such a group, appropriately defined, constitutes a *macrostate*. Thermodynamics and much of statistical mechanics deal with such macrostates. Microstates, in contrast, are those giving the maximum possible information about a system. It is clear, therefore, that a number of microstates will be compatible with a given macrostate. This number is the number of distinct ways of realising a macrostate, and can for brevity be called the (statistical) *weight* of the macrostate. The transition to macrostates is called *coarse-graining*.

It should be made clear that in talking about macrostates of systems in equilibrium we are concerned with time averages. They are the real observational material. We shall put them equal to averages over microstates. One is here assuming some form of *ergodic hypothesis*. For the present purpose it is convenient to do so, leaving it to discussions at other places to consider if such a procedure is justifiable.

Next consider two macrostates of a gas in equilibrium in a box of volume $V$:

(A) The gas is in the left-hand half of the box.
(B) The gas fills the whole box.

A usual statement relating these states is "State A goes most probably over into state B irreversibly and statistical entropy increases then with time". This is satisfactory, except when one is speaking philosophically about time. Can we reword the statement to eliminate the disruptive effect of biological time? A proposal is the following principle which does not involve time:

(**P**) *When a macrostate corresponding to equilibrium* (or other given constraints) *has an overwhelmingly larger weight than any other macrostate* (compatible with the given constraints), *the system is found in this state with a corresponding overwhelming* *probability.* Its statistical entropy is an increasing function of the weight of this state.

It follows at once that if the constraint which confines the gas to the left half of the box is removed, the whole box will be filled by the gas with increase of entropy. This is the usual transition: State A goes over into state B. However, unidirectional time does not enter the argument. After all, one might start with state B, and insert a shutter when state A is known to occur. This is not an unlikely situation with 3 or 4 particles. With many particles it is unlikely, but it is not impossible.

We shall return to the important principle $P$ in § 8 after discussing some incidental issues which arise.

## 6. *A Reinterpretation of the Random Walk*[2]

A simple theory of the box of gas fits also into a game. To generate a typical state of the box with the whole volume $V$ available to the $N$ particles proceed as follows. Toss a coin; "heads" means the first particle goes into the left-hand side, "tails" means it goes into the right-hand side. Repeat, and after $N$ tosses the $N$ particles of the gas have been accommodated by the theory. Each particle has the probability $\frac{1}{2}$ of being in one half of the box. Let $N$ be even, and let the number of particles in the left-hand side of the box be $N/2 + m$. The number in the right-hand half is then $N/2 - m$. This macrostate can be specified as $(V, N, m)$. The probability for this state is calculated as the weight of the state, divided by the total number of microstates ($2^N$). The particles may be distinguishable, and the number of microstates can also be worked out if they are indistinguishable. The problem turns out to be mathematically just the random walk problem all over again.

With our new notation we have:

Macrostate $A \equiv (V, N, N/2)$
Macrostate $B \equiv (V, N, O)$

The weights are calculable, and so the entropies of states $A$ and $B$ can be worked out. They can therefore be compared. But time is again irrelevant to these considerations.

## 7. *Statistical Entropy: A Time Symmetrical Law for Its Probable Increase*

Returning now to the intelligence $I$, we explain to it that statistical mechanics and thermodynamics are needed for macroscopic systems owing to the limited information normally at human disposal. In particular, we define for it statistical entropy. The intelligence $I$ responds that with all the information *it* has, the entropy *it* calculates remains constant in time whether the system is in equilibrium or not (see Appendix C). So we explain that statistical mechanics deals with coarse experiments and microscopically incompletely defined situations, and that coarse-graining is the corresponding theoretical tool. Statistical entropy, we explain, is coarse-grained unless a remark is made to the contrary. It slowly transpires that a macroscopic measurement corresponds to rough

---

[2] This section may be omitted on first reading.

information about a system, in what we might again call a macrostate $A$. Thus as one leaves this state $A$ the information about it must become more diffused. A few strong filaments in phase space may, for example, be drawn out into many fine filaments which may be beyond the experimental threshold should another macroscopic measurement be made. The macroscopic nature of the system ensures large Poincaré recurrence times, and for times small compared with these one can say that the information concerning state $A$ becomes, very probably, less and less relevant to experiments which can be performed at increasing *separation in time* from state $A$. This yields highly probably entropy increase. Now what does *separation in time* mean here? In human forward time state $A$ is an *initial* boundary condition and we thus obtain *the law of highly probable statistical entropy increase* in our time. It holds if no quantitative details are known *after* the time of state $A$ (except that the qualitative structure of the system must of course be known).

What does *separation in time mean if time is reversed*? The reversed form of the preceding statements appear to remain valid: In human *reversed* time state $A$ is a *final* boundary condition and we thus obtain *the law of highly probable statistical entropy increase in our reversed time*. It holds if no quantitative details are known *before* the time of state $A$.

An example will illustrate this state of affairs. An isolated equilibrium gas fills in state $A$ half its available box. Predicting into forward time, we naturally say there will be entropy increase (Fig. 2). Retrodicting into the human past we say: State $A$ is by the law of entropy increase a fluctuation from a more normal equilibrium state. This remark implies an increase in entropy in reversed time and illustrates the time symmetry of the law of statistical entropy increase. Note that both predictions are probabilities not certainties. They could be

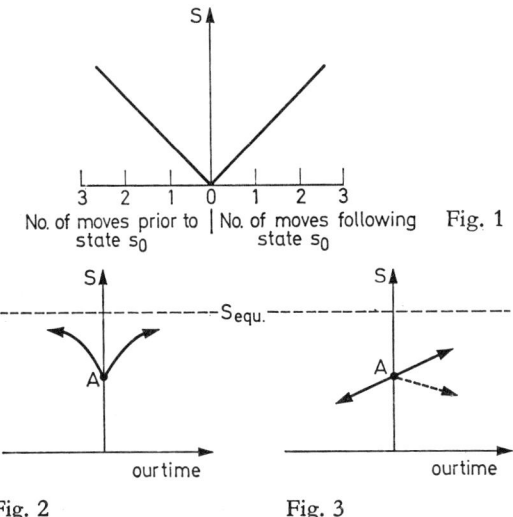

Fig. 1

Fig. 2    Fig. 3

falsified by highly unlikely occurrences such as membrane formation, which would stop expansion of the gas in our forward time. Or, the entropy prior to state $A$ could have been a constant if in state $A$ a membrane had just been destroyed without a trace. This would falsify the retrodiction. We see that in the absence of other information the macrostate $A$ observed corresponds to a temporal entropy minimum.

The retrodiction from state $A$ would be altered were we informed that the system is *approaching equilibrium*. The retrodiction would assume that the strong filaments in phase space would bunch together even more and entropy would decrease into the past (Fig. 3). This case differs from that of Fig. 2 in that information about the states prior to state $A$ had to be used.

Suppose we know not only that the system is approaching equilibrium, but also that we have all required microscopic information. In that case the Gibbs entropy will remain constant, so that the approach to equilibrium must be measured by another quantity, say the Boltzmann entropy. One would expect to find again Fig. 3. Now suppose one achieves time reversal for the gas relative to our time, by reversal at state $A$ of all molecular velocities. Then we would see the system return in our forward time into a low entropy state ([10] dotted in Fig. 3). These cases have been investigated in the spin echo experiments pioneered by E. L. Hahn and in computer experiments [65] on model gases, and the ordered low entropy states have been synthesized in this way.

What is new here is what one might call the deliberate mismatch between microscopic boundary conditions and the *macroscopic* description of the ensuing development of the system. The experiment proceeds simultaneously at the two levels of description. Microscopic boundary conditions are not available in *pure* thermodynamics and that is why Boltzmann was safe when in discussions about velocity reversals in a gas he is reputed to have said: "Go ahead, reverse them". Within the context of the present exposition one may regard these experiments as confirmation of $T$-invariance of microscopic descriptions, and no surprise need be experienced at the results obtained.

## 8. *Thermodynamic Entropy: Its Certain Increase for Infinite Systems*

There are two essential steps to be taken now in order to arrive at the second law of thermodynamics. The first is a technical step. We restrict attention to the idealised case of infinite systems. This lengthens the Poincaré recurrence times to infinity and the highly probable increase of statistical entropy becomes a certain increase (for forward or reverse time). No system on which we experiment is infinite, but this limit is useful in that it facilitates verbal statements, and often it also simplifies quantitative expressions for thermodynamic quantities.

(A "limit $L$" was introduced in 1954 [49] and discussed further in [51]. Asymptotic properties were recently proved for this limit, now called "thermodynamic limit", and for a remark on the name of this limit see Uhlenbeck, 1968).

The statistical entropy goes over into the thermodynamic entropy in the thermodynamic limit; and this entropy increases with certainty in an adiabatically isolated system. The technical proof that

$$\lim_{N, V \to \infty} \{S_{\text{stat}}\} = S_{\text{therm}}$$

is a problem which need not concern us here. Various plausibility arguments designed to establish approximate equality $S_{\text{stat}} \sim S_{\text{therm}}$ for large systems exist in the literature (for example van der Linden and P. Mayer, 1967). We proceed next to the second step needed to establish the law of entropy increase.

It is now important to return to the principle *P* of § 5. In the thermodynamic limit, or at least for very large systems, the ratios of the probabilities of various macrostates of a system (compatible with given constraints) will be calculable and sometimes simple, e.g., a ratio may be very small for certain states. Our intelligence *I* is still contemplating a planet devoid of life and remarks to us: "As you know I perceive a four-dimensional network of world lines of particles. As requested, I have averaged, and I have discarded information in the process. In this way I have been led to perceive ill-defined structures like caves, hills stone walls, which seem to form ingredients of your language. I also now have knowledge of quantities like pressure, heat and entropy, rough measures though they are. In the local universe parts of systems are approximately isolated for intervals of time, and their statistical coarse-grained Gibbs entropies then change. I do not know in what direction you wish to say that time increases; this is obviously pretty arbitrary; or it may depend on your psychology and physiology about which I know very little so far. But your principle *P* implies a choice of available sign already, for you say that more probable states shall have greater entropy. [You might by omission of the negative sign in $S = -k \Sigma p_i \ln p_i$ have achieved the reverse correlation of entropy and probability]. Thus I could order states of systems by their probabilities and give smaller $t$ - values to more probable states." We object immediately that this would be contrary to human conventions. "All right, it makes no difference, except note that you have two procedures available in any case:

*Either* You take the time direction as given and formulate the entropy law as a physical law.

*Or* You define time as increasing with the probabilities whence the entropy law is a deduction from principle *P*".

Suppose we have obtained a time direction for a system in a certain state by looking for the direction of time in which its entropy increases. Will this sense be the same for all initial states of this system? Will it be the same for the vast majority of systems? Nature answers "Yes" and this calls for an inclusion of this answer in any list of natural laws. This particular aspect of entropy increase is, however, often rather neglected, even in books on thermodynamics ([51], p. 81 contains a full discussion). Reichenbach elevates this property to a principle (of the parallelism of entropy increase), and he also refers in this connection to the statistical isotropy of the universe.

Its observational content is clear if we consider the sun as shining on an area of a planet, but interupted for a period by a shadow. What would our intelligence $I$ see? There are four main possibilities (see Fig. 4). The second occurs if $I$ uses reversed human time. But the last two cases are unlikely to be seen. The third case inplies that the system follows reversed human time first and it then follows human time when the sun appears again. The fourth case is equally improbable.

How can this parallelism be understood intuitively? Since we are at the foundation of physical theory, we can at best hope to reduce it to some philosophically appealing notion. One may be reluctant to import the rest of the universe in order to understand this coherence in nature, since one is often dealing effectively with independent systems. But it will be granted as plausible that one usually finds only slightly different macroscopic observations if the macroscopic system is changed only slightly. (However, one would have to be very cautious in the neighbourhood of phase changes.) With this in mind we could imagine any two situations linked by a large number of intermediate situations. In this way the same arrow of time can be expected for all the observational material displayed. Hence one can gain an intuitive appreciation of the parallelism of the various entropy increases. [We cannot expect more in the present context, for we have laid down no axioms, so that we can give no rigorous proofs.]

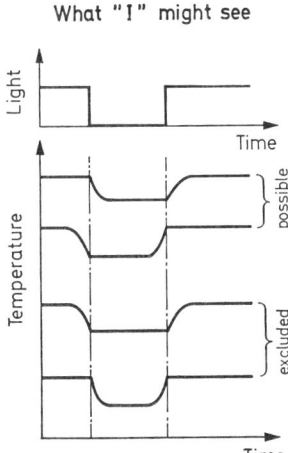

Fig. 4

These views are in contradiction with the opinion that the direction of time in statistical physics derives entirely from human thought habits which appears to have been at least hinted at in [58] and in [88]. At this point it would be tempting to approve selected writings and commentaries, to criticise others, etc. We apologise that this cannot be done here. We would lose ourselves in the vast literature and break our promise that we will present a point of view.

We accordingly agree with the intelligence $I$ that the direction of time – impersonal time ordering with the resulting notion of earlier and later – emerges as a statistical concept resulting from the overlapping entropy increases of almost isolated large systems, as broadly conceived by Boltzmann and by Reichenbach. Human conventions *are* involved:

(C1) Entropy *in*creases with the probabilities of states.
(C2) Time *in*creases with the probabilities of states,
(C3) Statistical physics deals with microscopically ill-defined macrostates.

But the direction of life (the "biological arrow") does not seem to be explicitly needed in this discussion. In equilibrium systems the statistical arrow has died out and an external interference is needed to reactivate it.

The direction of time in statistical physics is thus not related to boundary conditions imposed by human beings in their laboratories; nor is it related to the biological fact that the past is known to us and cannot be altered while the future is unknown, as suggested by some writers. Rather, the direction of time in statistical physics is impersonal: The moving shadow of a crater on the surface of a planet will initiate heat conduction in the normal sense, whether or not it is observed by an intelligence. It is due to the statistical weight of macrostates of systems determining the evolution of these systems. The biological, cosmological and electromagnetic arrows may be related to the statistical one, but it is convenient that, for the moment in any case, and on the basis of the present analysis, these arrows can be regarded as uncoupled from the statistical arrow.

Our intelligence will remark next: "I have knowledge of bundles of world lines which constitute brains. I suppose that to a person's thoughts there corresponds a section at constant proper time of such bundles – a 'brain state', if you like. This picture is that of a block universe [90] and a man's consciousness hops along this bundles from perception to perception. Thus he adds to the impersonal *earlier – later* relationship the (speaker-reflexive) notions of *past*, *present* and *future*. How fast does he hop along?[3] This is given by the proper time intervals between slices which represent perceptions. The notion *becoming* about which your philosophers have written so widely [5, 32] is perhaps more complicated, but in scientific terms there is no more to it than what I have said."

---

3 This is a question raised by Professor David Park at the Conference.

## 9. *Life as Violating Weak T-Invariance*

One day life may be clearly exhibited as resulting from the processes of physics and chemistry. For the moment this is not so and we must try to characterise this (possibly temporary) feature. For this purpose, introduce the following rough definition:

> A complex process is weakly $T$-invariant if its time inverse, though perhaps improbable, does not violate the laws of the most elementary processes in terms of which it is fully understood.

The laws of mechanics and electromagnetism are not only weakly $T$-invariant, they are even $T$-invariant, for they are invariant under time reversal: The earth could travel its orbit in the opposite sense; electromagnetic waves could meaningfully converge to heat the filament of a lamp. The laws of heat conduction or diffusion are not $T$-invariant: the diffusion equation changes under time reversal. However, the most elementary processes of particle collisions in terms of which diffusion and conduction are understood are not violated under time reversal. Anti-diffusion is improbable but possible. A thermostat which is designed to keep a temperature fairly constant changes under time reversal: it takes on the appearance of a device designed to induce temperature fluctuations. However, the elementary processes in terms of which it is fully understood are not violated by time reversal.

Such claims cannot so far be made for all examples of living processes: for example, the response to stimuli and the nature of this response in living things are not yet fully understood in terms of physics and chemistry, i.e in terms of time-reversible elementary processes. Definitions of life are not very profound, and the definition to be given now achieves rather little. Its merit is that it may be new, and that it is a useful focal point in our discussion. Its demerit is that it merely codifies our ignorance. It is this:

**L**: Life is a macroscopic process which violates weak $T$-invariance.

Its usefulness as polarising our ideas is illustrated by the following deductions.

(1) For **L** to be a useful definition, all macroscopic processes in inanimate matter must be weakly $T$-invariant. That this is so has been discussed.

(2) From a definition which makes a rather sharp distinction between inanimate and living matter one can infer the existence of a strong doubt whether the laws of physics and chemistry as they stand now are adequate for the elucidation of living processes. It appears that we have not yet succeeded in breaking down into elements which can be treated quantitatively the obvious macroscopic notions of purpose and consciousness which play so important a part in dividing the biological from the physical sciences. This present lack of understanding must, however, not be taken to imply that physical principles

such as quantum mechanics are in contradiction with life. This would be going too far [1, 53, 92, 93]. New *physical* principles *may* be needed to describe living processes, but this has not been established yet.

(3) The definition focusses attention on the *reduction problem of biology*, viz. the question if living processes can be described fully in terms of physics and chemistry[4].

## 10. *Life, Entropy and Disorder*

It may be said in response to these remarks that one can, and does in fact, apply physical ideas to specific biological notions, e.g. the notion of biological disorder is related to entropy. Complications which are too often overlooked are, however, involved here. First to assign a *thermodynamic* entropy to an object "demands some reversible method of getting to the object from a standard starting point, and this, for an organism, is close to the problem of creating life" [15]. Secondly, the *statistical* entropy concept, "can be applied only to a system whose distinguishable states can be clearly specified and enumerated, and this can be an exceedingly difficult matter in complicated systems" [50]. Thirdly, heeding these cautions, one is still left with an effect which is normally ignored, namely that the number of states $N(t)$ available at time $t$ to an organism with non-zero probability can rise with time, thus introducing an important modification into the usual formulae used in physics, when $N$ is often a given constant. If $N$ increases sufficiently rapidly one can have a situation in which (a) entropy increases, but (b) disorder decreases (see Appendix C). This effect can, therefore, decouple entropy and disorder. We propose (for a related suggestion see [27]) to decouple entropy and disorder by an apparently trivial device. For "disorder" change the base of the logarithm which occurs in entropy and information formulae from e or 10 to $N(t)$ itself. This has the startling result that the new quantity is no longer extensive (like volume or energy, say) but becomes intensive (like temperature or density). Fourthly, it must be remembered that information theory does not take into account the *meaning* of the sentences it handles. In spite of the general value of information theory this can become a serious handicap in biological applications. For example, of four possibilities treated on an equal footing, one may imply the death of the organism; or the end of a message may be "the preceding statements are wrong". Such situations are not yet fully taken into account in the theory.

The neglect of these points casts doubt on existing discussions on the entropy of living things, and explains to some extent the wide variety of views

---

[4] See also *Towards a Theoretical Biology* (Ed. C. H. Waddington) Vol. 2 Sketches, Edinburgh University Press 1969; *Theoretical Physics and Biology* (Ed. M. Marois), North Holland 1969.

encountered (see the correspondence in *Nature* in the years 1965 to 1968 [16, 18, 69, 70, 71, 81, 94, 96, 97]).

The mixing up of genes in propagation of species could at first sight be taken as illustrating the working of the second law, but it has also been noted that in evolution (as well as in breeding) improbable structures are developed, in apparent violation of the second law[5]. With these cautions in mind, neither view can be accepted until the states of the systems contemplated are properly enumerated. It is clear from what has been said that different descriptions can lead to different entropies and to different changes in entropy. The suggested violation of the second law is certainly not correct since evolving systems are open systems. In such systems entropy can decrease if compensations occur elsewhere; photo-chemical reactions in which incident radiation is the low entropy component provide examples.

One should consider, however, if evolution does not provide an example of lack of weak $T$-invariance in biological systems. Another example could be processes satisfying Dollo's law (that identical physical structures are not again acquired by animals which revert after a long gap to their original environment). The discussion of this suggestion requires more elucidation of those biological laws which are irreducible to physics and chemistry. Unfortunately we have no space for this here (see H. Kalmus in [28]).

## 11. *The Direction of Life; the Status of Boundary Conditions*

Our intelligence $I$ has now been acquainted with the human mental equipment, and with some cautions concerning the use of the entropy and information concept in biological contexts. We are, therefore, ready to consider from its point of view the broad coincidence of two arrows:

(1) The arrow in the direction of more probable states of systems, and
(2) The arrow from birth to death.

One can perhaps regard death as a continuing process throughout life. It ends while the last signs of life disappear and the body comes into equilibrium with its surroundings. It can then be argued that the arrows coincide because living things are also physical systems so that the arrow (1) imposes its direction on arrow (2). The argument would represent a violation of our caution concerning the use of entropy in complicated systems, if we were trying to say anything very quantitative. But we are not. We are merely implying broad analogies, for example, between the rusting of iron in the atmosphere and the calcification of the arteries of living things: two systems which approach their more probable states.

---

5 See articles in *Theoretical Physics and Biology* (Ed. M. Marois), North Holland, 1969.

It has been seen, in § 7 for example, that a boundary condition is still time-symmetrical since one can start from it into an unknown past or an unknown future. It is because of the direction of life that some boundary conditions are labelled by humans as initial and others as final. In our development initial and final boundary conditions arise therefore *after* the statistical arrow has been identified and labelled, and they can therefore not be made responsible for the direction of that arrow.

It is sometimes said that "irreversibility in macroscopic systems has its origin in the asymmetry of the initial or boundary conditions that are normally imposed on them" ([30], p. 112. See also pp. 126–128, 233; a more cautious statement occurs in Bondi's Halley Lecture, 1962). The analysis presented here does not support this view. There is also the obvious point that if one is ready to discuss the significance of initial conditions, one must already have an understanding of the direction of life which serves to distinguish initial from final conditions.

## 12. *The Cosmological and the Statistical Arrow*

The notion of the entropy of the universe is subject to the cautions indicated in § 10 in connection with biological systems. The states of the universe cannot be enumerated and it is not clear that it can be treated as either a finite or as a closed system. But the situation is not hopeless. One can make model universes, and if one is in doubt about entropy in general relativity one notes that Newtonian cosmology often yields broadly equivalent results (see Appendix E), and entropy can certainly be used in this context. All other talk about the entropy of the universe is, strictly speaking, meaningless.

As regards the model universes, it is known that in both the expanding and the contracting phases of oscillating models the entropy may remain constant or increase according to Tolman's studies [81, 98]. One must therefore think of the contracting phase as initiated by gravitational effects, but in this phase stars may still shine normally and life's arrow may remain unchanged. Let the "cosmological arrow" be introduced as in one direction for expansion and in the other direction for contraction. Then Tolman's studies suggest that the cosmological and the statistical arrow are decoupled.

Recently it has been suggested (Treder, 1965; Pompe and Schöpf, 1968) that advanced fields are eliminated by the expansion of the universe; and that as a consequence it is impossible for an observer in his subjective time ever to see a contracting universe. The cosmological and the statistical arrow are supposed to swing round together. I should, therefore, like to propose the following *principle of impotence* (Whittaker, 1949) for discussion:

*A contracting universe cannot be seen.*

Any good principle of impotence must be falsifiable. In the present case this falsification would be brought about for example, if the stellar red shifts were to give way to blue shifts in a future epoch. Thus one should show why the arrows of cosmology and of entropy (and therefore of life) are parallel. We cannot here scrutinise this matter, since the relation between cosmology and the retardation in electrodynamics (Wheeler and Feynman; Hoyle and Narliker) is beyond our scope. [It is worth noting the searching investigation by Reichenbach (1946, § 16) who considers the low entropy state of the present universe explicitly as a possible assumption. The notion of "the entropy of the universe" is, however, used by him without regard to the cautions described in § 10 above.]

Two further arguments which fall in this area of the Boltzmann-Reichenbach work should be noted. A possible link between the statistical arrow and the cosmological arrow has been discussed as follows [31], [11]. One regenerates the statistical arrow in an isolated equilibrium system by opening a window to the rest of the universe for a short interval. The temperature unbalance restores an arrow which had disappeared after equilibrium had been reached. Here it is the large scale expansion of the universe, the resulting blackness and coldness (which incidentally saves stellar surfaces from the heat suggested by Olbers' paradox), that is responsible for this regeneration of time's arrow in the system. This argument suggests that the two arrows are parallel, and implies what I shall call a *theory of the dominance of the cosmological arrow*. The trouble with this argument is that it goes through also for a cold system which *gains* energy from the cosmic radiation. It therefore applies also for a hot background as one might have in a contracting universe, and therefore it does not show convincingly if and why the cosmological and the statistical arrow are coupled to be parallel.

Another attempt at effecting this coupling is to suppose a low entropy state at the cosmological origin of the universe. This leads to an arrow of time which points in the direction of an increasing lapse of time from the origin. It leads naturally to the emission of radiation by hot stars, and hence to entropy increase on the stellar and galactic scale. Life, in order to survive has to absorb the radiation from the sun. If the arrow for living things were counter-directed, the sun would become hotter by the absorption of radiation according to human reckoning, and solar radiation could *perhaps* not be utilised by living things. One can therefore see how a cosmological directedness of time could impose its arrow on entropy changes and on life. This is another version of the theory of the dominance of the cosmological arrow. As emphasized by the italicised word (perhaps), it is not convincing just because the status of the above principle of impotence needs further study.

Some writers support a theory of the dominance of the cosmological arrow in a stronger form, namely by doubting the statistical arrow:

"The 'arrow' of time.... does not seem to be of a stochastic character" [69].
"It is somewhat offensive to our thought to suggest that if we know a system in detail then we cannot tell which way time is going, but if we take a blurred view, a statistical view of it, that is to say throw away some information, then we can". [11]
"Surely the fact that we had to deal with the problem in statistical terms rather than compute in detail the behaviour of all constituent parts of our system, that constituted

merely a lack of precision; surely it is not by rejecting information about our system that we can make it reveal to us the sense of time which it would not otherwise show". [31]

I think this view leaves out of account the essentially coarse nature of all human perceptions, which has been emphasized here. One could, in fact, remark similarly: Surely the coarse heat conduction initiated by the moving shadows on a planet's surface cannot be dependent for its one-sidedness on the movement of distant matter. (This view has some support, e.g. Morrison in Ref. 30, p. 143.)

The theory of the dominance of the cosmological arrow draws its greatest strength from the simple fact that any theory of the developing cosmos (for example a general relativistic model universe) must imply an approximate theory of the time development of the material contained within the cosmos. To be satisfactory, such a theory must lead to the observed coincidences of the various arrows of time. If and when available, this theory will then incorporate boundary conditions of some kind. In this sense the cosmological boundary conditions may be expected to impose their arrow of time on the other available time arrows, swinging them around (if necessary) as part of a normal process of deductions from assumptions. A theory of this kind will thus yield coincident arrows. This way of looking at the problem poses an important question: What are the cosmological theories and their boundary conditions which imply the coincidence of the arrows of time? And furthermore is it agreed that, other things being equal, such theories are preferred to those which do not lend themselves to deductions showing the coincidence of the arrows?

The perfect symmetry of the space-time framework can be restored by supposing that a hidden world exists in which entropy decreases in our timereckoning, and in which living things have also a time arrow opposite to our own, so that we would regard them as growing younger. For these beings the law of entropy increase would however still be valid, and they would regard themselves as growing older. Many writers have speculated on this topic and the difficulties of communicating with this world have been emphasized (Wiener). Phenomena in reversed time have been called anti-kinetic (Orban and Bellemans), anti-thermodynamic (Prigogine, Pittsburgh meeting), and matter with a completely reversed time has been given the name faustian (Stannard). These speculations, in spite of their interest, will not be pursued here. They were initiated by Boltzmann on his presumption, now no longer granted, that the universe is broadly in equilibrium and static.

Lastly, we wish to point out that cosmology is a good example of retrodiction as discussed in § 7: We reconstruct the past of the universe from its present macrostate, subject to the additional information that certain physical laws must hold.

A very different analysis must be behind the following remark, which at first sight seems to be in contradiction with our view: "According to the present theory, a complete mathematical description of the universe must unfold from a description of the initial state. It is not possible to reconstruct the past history of the universe by working backward from a complete macroscopic description of the present state" ([57], p. 260).

In the course of his discussion Layzer [57] considers an interesting problem: The early states of the universe are believed to have been much simpler than the present state, whereas normal evolution of isolated systems is from complex initial to simple (equilibrium) final states. Is this consistent? We should like to explain this effect by noting two ways of increasing the entropy of a system: one by spreading the probability distribution more uniformly over the given macro-states. This is the normal evolution of systems. The second is to keep the system under the given constraints, but to bring more accurate instruments to bear on it so that one can distinguish more macro-states. This is what happens as the universe expands: it is as if we were increasing the magnifying power of the microscope we use to look at a picture. As the universe expands the magnifying power of the microscope increases. More and more states are distinguishable [$\dot{N}(t) > 0$] and the entropy goes up. One can see this simply from the formula for the *maximum* entropy (Appendix C):

$$S_{max} = k \ln N(t), \quad \dot{S}_{max} = k \dot{N}/N$$

This effect may also be seen more generally from inequalities for entropies given in Appendix F.

This Appendix also refers to Bayes' rule and the probability of causes or hypotheses. This is seen in Eq. (F. 2) to be an elementary result which is not essential to the understanding of time. The emphasis given to it by some authors (for instance Watanabe [28] can therefore be due only to their desire to *illustrate* rather than to *establish* a point of principle. In the present exposition there is no space for purely philosophical questions. It must be assumed that, with a normal use of language, causes precede effects. Hence questions of causality arise logically *after* the notion of human time has been introduced and fully explored (for a recent reference see [29]).

## 13. *Two Problems of Relativistic Time*

The preceding discussion of those aspects of cosmology which are of interest here has implied a universal time in which the universe is expanding. Such a preferred time scale is derivable by looking around the universe and determining the inertial frame in which the red shifts, number counts of galaxies, etc., have maximum isotropy defined in some convenient way. This inertial frame is preferred since any other will show less isotropy. An observer's proper time in this frame is the preferred (or "cosmic") time which one often uses in cosmology.

The existence of such preferred inertial frames has been illustrated by the discovery in 1965 of what is widely thought to be a relic of the original (hot) big bang. It is believed to exist now as a general background black body radiation at the low temperature of about 3° K, which is so low because of the expansion of the universe which has taken place in the meantime. It is very isotropic, suggesting that the solar system is nearly at rest in it. (See [21, 54] and references quoted there.) Are these preferred frames in contradiction with the special theory of relativity which is sometimes stated to require the *laws of physics* to be the same for all inertial frames? This raises at once the further question whether it *is* a law of physics when we say "red shifts are broadly isotropic from here". Indeed it is not very informative to ask for the laws of physics to be the same for all inertial frames because this leaves in doubt what one means by a law of physics and how one wants it to be formulated. This doubt is well substantiated by the subsequent developments in general relativity, when one wanted laws to remain valid in a uniformly rotating frame of reference and then in a frame in arbitrary motion. This gave rise to the general principle of covariance which required the laws, when properly formulated, to be valid in all co-ordinate systems. A well-known discussion between Einstein and Kretschmann led already in 1918 to the realisation that such a principle had no physical content (see also [24, 46, 26]). Thus, returning to special relativity, it is clearly better to restrict oneself to practically equivalent, but less controversial statements. An example is: Uniform rectilinear motion of a material system as a whole does not affect the processes occurring within it. This leaves intact the possibility of both special relativistic time (mainly for local use) and of cosmic time (for cosmological use).

I shall assume that, for this audience, the clock paradox has long been resolved, and I shall merely refer to two brief notes [52, 56].

The properties of special relativistic time have recently led to the view "that there is determinism". This remarkable result has not so far been discussed, refuted or withdrawn [75], even though its claim to deduce determinism from relativity seems of immense philosophical importance. The argument is, in fact, in error, but I shall expound it here as an illustration of the conceptual difficulties inherent in relativistic time.

Two observers $A$ and $A'$ at a distance $d$ apart synchronise their clocks to $t = t' = 0$ at what I shall call the critical time. They are in uniform relative motion, and one can show from the Lorentz transformation (quite correctly) the following result. Take any event $E$ in $A$'s future $t > 0$. Then, by chosing the velocity of $A'$ appropriately, this event can be arranged to have a negative time co-ordinate for $A'$. Thus at the critical time $t = t' = 0$ the event seems to have already been determined in the sense of *being there* as far as $A'$ is concerned. Nothing $A$ can do, can influence the existence of this future event. It is (pre)-determined. We can find an observer $A'$ for *each* future event $E$, at *all* times on $A$'s world line, and for *all* observers $A$. Thus all events are determined and there is no free will. Now one feels certain that such a philosophical deduction goes far beyond what has been put into special relativity, so that, in a sense, this is a paradox: the *determinism paradox of special relativity*. Its resolution depends on a distinction which is not at once available in this ancient English language of ours, which was designed before relativity. It is that a negative time co-ordinate at $t=0$ means "the past" relative to $t=0$ only in the Newtonian sense. In relativity it may mean the absolute past, or it may mean an event outside the light cone which is not past, present or future. In the case in point one can show that $E$ does not lie for $A'$ in the absolute past at $t'=0$. It therefore lies in the indeterminate region and so the paradox is resolved (see Appendix G).

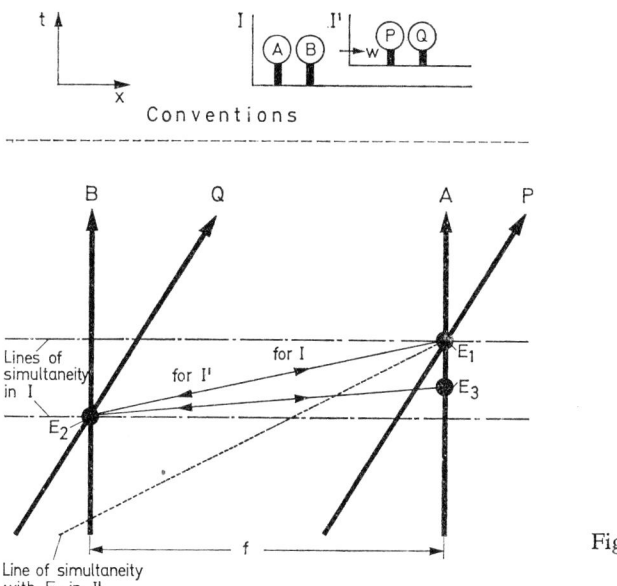

Fig. 5

## 14. *Tachyons*

I now come to a recently discussed case of time inversion, which is illustrated in Fig. 5. Observers $A$ and $B$ are at rest in an inertial frame $I$, while observers $P$ and $Q$ rest in frame $I'$, which has velocity $v$ as shown. We focus attention on three events: $E_1$ occurs when $A$ and $P$ coincide and the event is actually the loss of £ 1 000 by $A$ at a betting table. $P$ at once flashes a message to his friend $Q$ at rest in $P$'s own frame. The motion is so arranged that when $Q$ receives the message about $A$'s misfortune, $Q$ happens to be just in coincidence with $A$'s friend $B$. There is practically instantaneous exchange of information between $Q$ and $B$ since they are at the same spot. This is event $E_2$. At this event $B$ at once sends a signal back to $A$, and it reaches $A$ at event $E_3$. It is not essential that these messages be sent at once: one could tolerate a time delay here, but this would only complicate the situation. Note that the first signal $E_1$ to $E_2$ slopes downwards in Fig. 5, but it is not going backward in time. For it is sent in $I'$ and the line of simultaneity in $I'$ also slopes downwards. The signal takes time in $I'$ so our main requirement is that $E_2$ shall lie above the line of simultaneity. Turning to the second signal, this is sent in $I$ *and our diagram is drawn for frame I.* The lines of simultaneity for $I$ are therefore horizontal, and the second signal slopes upwards because it takes time in $I$. We have shown the arrival $E_3$ of the signal at $A$'s world line *before* $E_1$. This is to make the message really useful. For it is then a warning for $A$ not to place his bet. But if he does not place it, the signal should not have been sent. So we are involved in a causal anomaly. $A$ has signalled to his own past. One feels it ought not to be allowed!

It may be shown that the condition for this sort of signalling is that both signals travel faster than light (Appendix I). Particles with this property have been called tachyons. They have many peculiar properties which can be investigated by simply using standard formulae. For instance, there exists no rest frame for a tachyon, so there is no harm in letting its proper mass be imaginary (for it cannot be observed), while its relativistic mass is real. Like a photon, a tachyon is not localisable (see Appendix H).

The key point involved here is that the signalling to one's past is possible only if the relative velocity between the frames is large enough. This requirement forces the notion of "earlier" and "later" with respect to the events $E_1$, $E_2$, $E_3$ to reverse as one changes frames from $I$ to $I'$ or from $I'$ to $I$. This has a peculiar result: absorbers and emitters interchange roles. The procedure was conceived to follow the sequence $E_1 \rightarrow E_2 \rightarrow E_3$ but it involved a change of frame from $I'$ (for $E_1 \rightarrow E_2$) to $I$ (for $E_2 \rightarrow E_3$). If one were to describe the whole process in $I'$, then both tachyons would *converge* to event $E_2$; if the whole process were to be described in $I$, both tachyons would appear to be emitted at $E_2$. It is with the aid of this observation that the tachyon advocates seek to avoid the causal anomaly which I have indicated [7, 8, 25]. They argue that any one observer will not interpret the closed cycle which has been described as signalling; he will instead entertain certain doubts. When $B$ at $E_2$ seeks to emit a tachyon to signal to $A$, what will encourage him to do so? An incoming tachyon from $A$, of course. But $B$ will not see such a tachyon, he will merely see another tachyon emission and he will say: Dear me, my absorber-emitter has emitted a tachyon, how do I know whether this is a spontaneous emission or a tachyon coming in from my friend $A$? It is these doubts, it is argued, that will undermine the signalling process and resolve the causal anomaly.

I cannot be quite satisfied with this reply (which uses only two observers). The exposition which I have given here uses 4 observers, and this seems to me to make the causal anomaly rather clearer. For it is now not $B$ that is interested in the incoming tachyon. It is $Q$ who watches out for it, and he then tells nearby $B$ what he has seen. The problem of tachyons is an open one; just as neutrinos were first stipulated and later discovered, it could be that a search for tachyons will be successful.

It is fun to take one's own stand in such questions, and so I shall do so. I believe that tachyons cannot be anything like the particles already discovered, for I believe that the above argument shows conclusively that they cannot trigger off macroscopic events so as to contribute to the passage of information. If they exist at all, I can therefore only conceive then as part of a still undiscovered larger domain of sub-quantum phenomena.

## 15. *Concluding Remarks*

Surprisingly enough, it is possible to combine the ideas of statistical physics and of relativity as presented here. Granted time as a primitive notion, the *direction* of time emerged through statistics and entropy. One can infer from this, if our considerations were sound, that the direction of time is a macroscopic concept. This suggests that at an appropriate level of microscopic phenomena time may no longer be unidirectional. The close link between time direction and causality suggests that causality can then also suffer some violations at this level. In this domain, where a multi-directional time is physically important, where the second law of thermodynamics does not apply, and where even causality appears in a new light – in this domain (if there is one) the existence of tachyons seems to me just possible.

In this domain too, one could speculate that a real understanding of the uncertainty relations in quantum mechanics will be found [48], just as Zeno's paradoxes find some clarification in current quantum mechanics [32, 48].

One may combine our ideas also by combining the appropriate constants of nature

$\hbar$ Planck's constant divided by $2\pi$
$G$ the constant of Gravitation
$c$ the velocity of light.

Already Planck inferred from these a natural unit of time [67]

$$\tau \equiv \sqrt{(\hbar G/c^3)} \sim 10^{-44} \text{sec}$$

(except that he used $h$ instead of $\hbar$). This has recently been given a more precise physical meaning [35]. The continuous fluid models of the universe can be valid only for times considerably in excess of $\tau$. This number gives therefore the threshold time for *classical* cosmology (an exposition is given in Appendix J).

Discussions with my colleagues D. A. Evans and J. Hindmarsh which helped to clarify ideas are gratefully acknowledged. I am also grateful to E. R. Harrison for discussions during a pleasant stay at the University of Massachusetts at Amherst in Spring 1969, during which Appendix J was written, and to my colleague K. A. Johns for discussions concerning Appendix H.

## APPENDIX A

### QUOTATIONS ON IRREVERSIBILITY AND ENTROPY

(The selection is rather arbitrary and possibly not typical of the cited author's present views)

#### 1. *Irreversibility Not Yet Understood*

a It is not very difficult to show that the combination of the reversible laws of mechanics with Gibbsian statistics does not lead to irreversibility, but that the notion of irreversibility must be added as an extra ingredient... the explanation of irreversibility in nature is to my mind still open.

P. G. Bergmann, 1967 [6], p. 11

b "...causality plus statistics means irreversibility. I think that is nonsense."

P. G. Bergmann, 1967 [30], p. 190

#### 2. *Entropy Increase due to non-isolation of Systems*

The failure of $S$ to increase with time is due to the fact that we have over-idealized an "isolated" system... The momentum and energy transferred between outside molecules and the system proper then acts as a source of true randomness influencing the dynamical behaviour of the system inside the walls. We maintain that this is the origin of randomness and increasing entropy in statistical mechanics.

J. M. Blatt, 1959 [9], p. 751

#### 3. *Time Direction Due to Measurement*

a In any observation process there must be a signal coming from the observed system to the recording apparatus, and since the propagation of any signal requires a finite time interval, this gives the possibility of defining the arrival of the signal to be 'later' than the time of emission. This specification of the sense of time is perfectly general.

L. Rosenfeld, 1967 [30], p. 193. See also Rosenfeld 1961 [17], p. 3

b ... nur durch die Messung oder einen anderen Prozeß ihrer Wechselwirkung mit dem Makrokosmos tritt die Irreversibilität auf ... Die hier gegebene Analyse, die zu der Einsicht führt, daß nur der Einbau der Objekte in den Kosmos als ganzen zu einem Verständnis der „Auszeichnung" der Zeitrichtung führen kann...

G. Ludwig, 1966 [83], pp. 286, 296

## 4. *Irreversibility Due to Large Systems*

<sup>a</sup> ...irreversible evolution towards equilibrium is an asymptotic property of large systems, for long times, derivable from mechanics alone.

R. Balescu, 1967 [3], p. 434

<sup>b</sup> ...a necessary condition for a rigorous transition from statistical mechanics to thermodynamics consists in taking the so-called thermodynamic limit $N \to \infty$, $V \to \infty$, $N/V$ finite, where $N$ represents the number of particles and $V$ the volume of the system.  E. J. Verboven, 1967 [86], p. 49

## 5. *The Need for Coarse-Graining and Macro-Observables*

<sup>a</sup> Thus we have arrived at the crucial question of how to choose the set of macroscopic variables $A^\nu$. This seems to me the main problem in statistical mechanics of irreversible processes.   N. G. van Kampen, 1961 [44], p. 183

<sup>b</sup> Any really satisfactory demonstration of the second law must therefore be based on a different approach than coarse graining.

E. T. Jaynes, 1965 [42], p. 392

<sup>c</sup> Also, if the root of thermodynamic irreversibility would reside in equality (III.4), it would depend essentially on the precision of observations (as introduced by the distinction between $\varrho$ and $P$).

($\varrho$ and $P$ are fine-grained and coarse-grained probabilities. The inequality relates coarse-grained Gibbs entropies at different times.)

I. Prigogine, 1970 [73], p. 7

## 6. *The Importance of Measurement and Knowledge*

<sup>a</sup> The increase of entropy comes where a *known* distribution goes over into an *unknown* distribution.   G. N. Lewis, 1930 [58], p. 573

<sup>b</sup> This illustrates clearly how the entropy of a system or text depends not only on the system or text, *but also on our knowledge of it, and on the questions we ask about it*.   P. T. Landsberg, 1961 [51], p. 237

<sup>c</sup> For it (entropy) is a property, not of it physical system, but of the particular experiments you or I chose to perform on it.

E. T. Jaynes, 1965 [42], p. 398

ᵈ ...the irreversibility exhibited by this system consists in the information becoming less relevant to the experiments which can be performed on the system.      A. Hobson, 1966 [37], p. 411

## 7. Irreversibility Due to Causality Conditions

... one may say that *irreversibility appears as a special aspect of the physical causality requirement*, which states that the distribution function at a given point is influenced only by the distribution function at points which correspond to earlier times on the trajectory.      I. Prigogine, 1962 [72], p. 296

## 8. Irreversibility Due to Ignorance Concerning Initial Conditions

ᵃ The phenomenon of irreversibility in isolated physical systems has its origin in the absence of microscopic information about initial states. The assumption that *initial* states have this property singles out a direction of time.
      D. Layzer, 1967, [57], p. 258

ᵇ ... I presume that most of us would agree... that the initial conditions generate thermodynamics... The striking asymmetry of the dynamics originates from this asymmetry in the boundary conditions.
      J. A. Wheeler, 1967 [30], p. 233–34

## 9. Irreversibility Due to Smoothing

ᵃ Die Irreversibilität ist eine Folge der Reduktion der exakten mechanischen Gleichung (3) durch Mittelung auf die statistische Gleichung (8)... Diese Mittelung... stellt eine absichtliche „Fälschung" der Mechanik dar, und angesichts dieses Umstandes ist es klar, daß kein Widerspruch zwischen Mechanik und Thermodynamik besteht: sie beruhen auf verschiedenen Grundannahmen.      M. Born, 1948 [13], p. 109

ᵇ The total probability density function $W$, even for a thermodynamically isolated system, does not obey the Liouville equation, $\partial W/\partial t = LW$, since small fluctuations due to its contact with the rest of the universe necessarily "smoothes" $W$, by smoothing the direct many-body correlations in its logarithm. This smoothing is the cause of the entropy increase...
      J. E. Mayer, 1961 [62], p. 1207

## APPENDIX B

CONCEPTS FOR IRREVERSIBILITY DISCUSSIONS

*Quasiperiodicity*

Consider a finite system in classical mechanics in which the forces are not velocity-dependent. The state of a system is then represented by a point in (p,q)-phase space. Choose any state $s(t)$ at a time $t$ and an uncertainty $\varepsilon$. Then there exists a finite time $\tau_\varepsilon$ such that

$$|s(t+\tau_\varepsilon) - s(t)| < \varepsilon$$

This theorem of Poincaré has been widely discussed (Kurth, 1960 Kac, 1959). An analogous theorem holds in quantum mechanics (Ono, 1949; for a recent discussion and additional references [38]). The simplest case arises if $s(t)$ is a specific point in $(p,q)$-space (classical case) or a specific vector in Hilbert space (pure state in quantum mechanics).

Consider next the situation where $s(t)$ is specified precisely but the system develops according to probability laws. This case can arise in quantum mechanics, and the random walk considered in the text falls in this category. Recurrence is possible in a finite time (see Appendix D).

The notion of recurrence must be generalised if the condition of the system at time $t$ is itself specified only statistically, let us say by a set $\{p(t)\}$ of non-zero normalised probabilities. In that case the recurrence of the *distribution* (within a stated approximation) constitutes recurrence. Quasiperiodicity occurs again if the number of non-zero probabilities is finite or enumerably infinite. It does not if the number of non-zero probabilities is non-enumerably infinite. A simple example is a one-dimensional one-particle classical gas whose initial state is not specified precisely [14].

*T-invariance*

The invariance of certain equations under time reversal is known as *T*-invariance. Given a classical one-particle time-independent Hamiltonian $H(r, p)$, write $\dot{r}_1 \equiv dr_1/dt \equiv dx/dt$. The classical time reversal operator $T_c$ and the quantum mechanical analogue $T$ may then be defined by the intuitive properties

$$r_{\text{rev }i}(t) \equiv T_c r_i(t) \equiv r_i(-t) \qquad TrT^{-1} \equiv r$$
$$\dot{r}_{\text{rev }i}(t) \equiv T_c \dot{r}_i(t) \equiv -\dot{r}_i(-t)$$

Hence also

$$p_{\text{rev }i}(t) \equiv T_c p_i(t) = -p_i(-t) \qquad TpT^{-1} = -p$$

Hence also

$$T(\mathbf{r} \times \mathbf{p}) T^{-1} = -\mathbf{r} \times \mathbf{p}$$
$$T\sigma T^{-1} = -\sigma$$

Here $\sigma$ is the Pauli spin vector operator interpreted as similar to an angular momentum $\mathbf{r} \times \mathbf{p}$. It follows that

$$H_{\text{rev}}(\mathbf{r}, \mathbf{p}) = T_c H(\mathbf{r}, \mathbf{p}) = H(\mathbf{r}, -\mathbf{p}) \quad | \quad T H(\mathbf{r}, \mathbf{p}) T^{-1} = H(\mathbf{r}, -\mathbf{p})$$

Here the quantum mechanical Hamiltonian has been assumed to be a real function of its parameters independent of $\sigma$, as the detailed analysis shows [55].

The Hamiltonian equations of motion in classical mechanics

$$\frac{dr_i(t)}{dt} = \frac{\partial H(\mathbf{r},\mathbf{p})}{\partial p_i}, \quad \frac{\partial p_i(t)}{\partial t} = -\frac{\partial H(\mathbf{r},\mathbf{p})}{\partial r_i}$$

become under time reversal

$$\frac{dr_i(-t)}{d(-t)} = \frac{\partial H(\mathbf{r},-\mathbf{p})}{\partial(-p_i)}, \quad -\frac{\partial p_i(-t)}{d(-t)} = -\frac{\partial H(\mathbf{r},-\mathbf{p})}{\partial r_i}$$

This is

$$\frac{dr_{\text{rev}\,i}(t)}{dt_{\text{rev}}} = \frac{\partial H_{\text{rev}}(\mathbf{r},\mathbf{p})}{\partial p_{\text{rev}\,i}}, \quad \frac{\partial p_{\text{rev}\,i}(t)}{\partial t_{\text{rev}}} = -\frac{\partial H_{\text{rev}}(\mathbf{r},\mathbf{p})}{\partial r_{\text{rev}\,i}}$$

They determine the same motion if $H_{\text{rev}}(\mathbf{r}, \mathbf{p}) = H(\mathbf{r}, \mathbf{p})$. This means

$$H(\mathbf{r}, \mathbf{p}) = H(\mathbf{r}, -\mathbf{p})$$

and implies for example, that $H$ must not involve odd powers of $\mathbf{p}$ (velocity independent forces) if there is to be $T$-invariance.

The statement is sometimes made that "irreversibility implies the arrow of time". In the semi-philosophical literature, but also in the physics journals, the kind of irreversibility considered is not always explained [23], [4], [39]. The system under consideration must also be specified for such a statement (see phrases S and B below). The Table contains 48 possible interpretations of our statement. These do not exhaust the possibilities for one might sub-divide B2 into two further cases depending on whether or not the classical Liouville operator has a continuous spectrum. When it has, a necessary condition for a form of irreversibility is satisfied [72]. Or one might divide S2 according to whether the distribution involved is defined at an enumerable number of points, or whether it is continuous. The continuous case yields (I 1)-irreversibility [38]. The (I 3)-irreversibility might also be sub-divided, depending on whether the

Boltzmann or the Gibbs entropy is used (see, for example, [42]). (I 4)-irreversibility means that an initial distribution $\varrho(q, p, 0)$ for the system evolves in time through a function $\varrho(q, p, t)$ into a time independent function $\varrho_{eq}(q, p)$ so that the averages of any smooth function (e.g. $\sum f(q, p) \varrho(q, p, t) dq dp$) in phase span becomes time-independent.

*Irreversibility and the Arrow of Time*

Table. *Possible Interpretations of "Irreversibility implies the arrow of time"* [a]

| | | | |
|---|---|---|---|
| I 1. Absence of Quasiperiodicity | S. 1 in a determinate system free of all statistical uncertainties | B 1. and in a system which is finite | 1. is necessary |
| I 2. Absence of T-invariance | S 2. in a system containing statistical elements in its equations of motion | B 2. and in a system which is infinite | 2. is required with overwhelming probability |
| I 3. Entropy increase | S 3. in a system containing statistical elements in the boundary conditions | | |
| I 4. Approach to equilibrium | | | |

[a] The Table contains 48 possible statements, e.g. I4 followed by S3 followed by B2 followed by 1.

## APPENDIX C

### PROPERTIES OF ENTROPY AND OF "DISORDER"

*(a) Entropy in General*

An isolated thermodynamic system is at a certain time $t$ represented by an ensemble of identical systems. A number of these are in some state 1, others in state 2, etc. From this one can, by using the frequency definition of probabilities, find probabilities $p_i(t)$ of finding the system in state $i$ at time $t$. From these probabilities one can derive the entropy at time $t$:

$$S(N, t) = -k \sum_{i=1}^{N(t)} p_i(t) \ln p_i(t) \tag{C. 1}$$

Here $N(t)$ is the number of distinguishable states for which the probabilities $p_i(t) > 0$. The maximum entropy for given $N(t)$ at time $t$ occurs if

$$p_i(t) = 1/N(t), \text{ i.e. } S_{\max}(N, t) = k \ln N(t). \tag{C. 2}$$

Also
$$\dot{S}_{\max}(N, t) = k \dot{N}(t)/N(t) \tag{C.3}$$

*(b) "Disorder" in General*

Entropy in these formulae satisfies

$$0 \leq S(n, t) \leq k \ln N(t) \tag{C.4}$$

The dimensionless entropy with the basis $N$ (instead of $e$) is a closely related quantity, which we shall call the "disorder" and it satisfies

$$0 \leq D(N, t) \equiv -\sum_{i=1}^{N(t)} p_i(t) \log_{N(t)} p_i(t) \leq 1 \tag{C.5}$$

Thus

$$D(N, t) = S(N, t)/k \log_e N(t) \tag{C.6}$$

and

$$\dot{D}(N, t) = \left\{ \frac{\dot{S}(N, t)}{S(N, t)} - \frac{\dot{N}(t)}{N(t) \log_e N(t)} \right\} D(N, t) \tag{C.7}$$

It follows that even though entropy can increase with time, the disorder will decrease provided

$$\frac{\dot{S}(N, t)}{S(N, t)} < \frac{\dot{N}(t)}{N(t) \log_e N(t)} = \frac{\dot{S}_{\max}(N, t)}{S_{\max}(N, t)} \tag{C.8}$$

*(c) Time-Independence of the Fine-Grained Entropy*

Let $\varrho$ be the density operator of a system in quantum statistics. Its time development is given by the Liouville equation

$$\frac{\partial \varrho}{\partial t} = [H, \varrho], \quad [H, \varrho] \equiv \frac{1}{i\hbar}(H\varrho - \varrho H) \tag{C.9}$$

where $H$ is the Hamiltonian of the system. It follows that

$$\varrho(t) = U(t)^{-1} \varrho(0) U(t), \quad U(t) \equiv \exp(iHt/\hbar) \tag{C.10}$$

Here $\hbar$ is Planck's constant divided by $2\pi$, $U$ is a unitary operator and the function exp(iA) is defined by the exponential series in the usual way. Since the sum of the diagonal elements (or trace) of a product satisfies

$$Tr(AB) = Tr(BA),$$

it follows that for all integral $j$

$$Tr\{[\varrho(t)]^j\} = Tr\{[\varrho(0)]^j\}$$

The trace of a power series in $\varrho$ is therefore also independent of time. It follows that the particular expression

$$Tr[\varrho(t) \ln \varrho(t)] = \text{independent of time} \qquad (C.\ 11)$$

This is the quantum statistical expression for the microscopic (or "fine-grained") entropy which is seen to be independent of time.

*(d) Disorder as Intensive Variable*

Two systems are joined together without any interaction. For example, they may be regarded only conceptually as a single system. Then the number of available states, which may be $N_1$, $N_2$ for the systems separately, becomes $N = N_1 N_2$ for the joint system. The entropy is additive $S = S_1 + S_2$. How do the disorders combine? We have from (C. 6).

$$S = k \ln N^D = S_1 + S_2$$

$$= k \ln (N_1^{D_1} N_2^{D_2}).$$

Hence for the combined system the disorder is with $q_i \equiv \ln N_i / \ln N_1 N_2$,

$$D = q_1 D_1 + q_2 D_2, \quad q_1 + q_2 = 1 \qquad (C.\ 12)$$

In particular, if two identical systems are combined, $D$ behaves like an intensive variable: if $D_1 = D_2$ then the resulting system has the same disorder as the two component systems.

If the systems are not identical,

$$(k \ln N_1 N_2) D = (k \ln N_1) D_1 + (k \ln N_2) D_2.$$

This is of the mathematical form

$$(C_1 + C_2) T = C_1 T_1 + C_2 T_2,$$

where the $C_i$ can be interpreted as constant heat capacities and the $T_i$ as temperatures. Thus the $D$'s are intensive variables.

## APPENDIX D

## A RANDOM WALK

A one-dimensional random walk of $N$ steps, as described in § 4, is a system $S_N$ consisting of a states

$$S_N: (s_0 = 0, s_1, s_2, \ldots s_N), \quad s_{t\pm 1} = s_t \pm 1$$

where the $s_t$ are integers. The probability $P(s_N = m)$ of finding a given state $m$ is known [19] and simplifies for $m \ll N$ to

$$P(s_N = m) = (2/\pi N)^{\frac{1}{2}} \exp(-m^2/2N) \quad (m \ll N) \tag{D. 1}$$

This distribution has the following properties:

(i) If we take $N$ to be a measure of the time $t$ considered, then $m$ may be thought of as the distance $x$ covered by the particle from the position $s_0 = 0$ to $s_N = m$ in this period. Accordingly put

$$m = 2x/a, \quad N = 8 Dt/a^2 \tag{D. 2}$$

where $a$ is a unit of length and $D$ is a constant of dimension $L^2 T^{-1}$ (diffusion constant). Then the probability (D. 1) becomes

$$a (1/4 \pi Dt)^{\frac{1}{2}} \exp(-x^2/4 Dt) \tag{D. 3}$$

Thus the probability per unit distance of finding the particle near $x$ at time $t$ is given by a normal distribution of zero mean:

$$P(x, t) = (1/2 \pi \sigma^2)^{\frac{1}{2}} \exp(-x^2/2 \sigma^2), \quad (x^2 \ll 2 \sigma^2), \tag{D. 4}$$

where the standard deviation is

$$\sigma = (2 Dt)^{\frac{1}{2}} \tag{D. 5}$$

(ii) The distribution (D. 4) satisfies the partial differential equation (diffusion equation)

$$\left[\frac{\partial P}{\partial t}\right]_x = D \left[\frac{\partial^2 P}{\partial x^2}\right]_t \tag{D. 7}$$

(iii) The dimensionless entropy of a normal distribution is ([51], p. 249)

$$S/k = -\int P \ln P \, dx = \tfrac{1}{2} \ln (2 \pi e \sigma^2) = \tfrac{1}{2} \ln (4 \pi e Dt) \tag{D. 8}$$

Here $k$ is Boltzmann's constant.

## APPENDIX E

### A BRIEF ACCOUNT OF THE THERMAL HISTORY OF THE UNIVERSE

The classical equation that mass times acceleration is equal to the force acting on the mass is valid in relativity. The acceleration can, for a conservative force, be derived from a potential $\phi$ (per unit mass):

$$\ddot{r} = \nabla \phi \quad \text{or} \quad \ddot{r} = -d\phi/dr \tag{E. 1}$$

for spherical symmetry with respect to the origin from which $r$ is measured.

Let $\mu$ and $\lambda$ be disposable constants which have to be put equal to zero if one wants strict Newtonian mechanics. One can then write Poisson's equation for a uniform fluid without angular rotation

$$\nabla^2 \phi = 4\pi G \left[\varrho + \mu \frac{3p}{c^2}\right] - \lambda c^2 \tag{E. 2}$$

where $G$ is Newton's gravitational constant, $\varrho$ is the volume density of matter, and $p$ is the hydrostatic pressure. For spherical symmetry the left-hand side is

$$\frac{1}{r^2} \frac{d}{dr}\left(r^2 \frac{d\phi}{dr}\right). \tag{E. 3}$$

If $p$ and $\varrho$ are constant in space at any one time (but not necessarily constant in time), one integration yields

$$\frac{d\phi}{dr} = \frac{4\pi}{3} G \left(\varrho + \mu \frac{3p}{c^2}\right) r - \frac{\lambda c^2}{3} r \tag{E. 4}$$

Putting (E.4) into (E.1)

$$\ddot{r} = -\frac{4\pi}{3} G \left(\varrho + \mu \frac{3p}{c^2}\right) r + \frac{\lambda c^2}{3} r \tag{E. 5}$$

To integrate this equation we need to know the dependence of $\varrho$ on $r$. The time used is cosmic time (see § 13).

Note that with $\lambda = 1$ a particle at $r$ is not only gravitationally attracted to the particle at the origin, but there is also superposed a cosmic repulsion which is proportional to $r$. The $\lambda$-term is therefore the original cosmological term as introduced ad hoc by Einstein into the general theory of relativity. It is here introduced similarly into Newtonian mechanics. The above relations with $\mu = 1$ yield the correct equations for general relativity theory, which is therefore seen to differ from Newtonian mechanics only if the cosmological model has non-zero pressure. The $\mu$-term may be justified in a broadly Newtonian theory by a variety of devices all of which utilise the equivalence $E=mc^2$ which implies that $p/c^2$ is the volume density of mass [63], [34].

It will be assumed that the cosmological fluid is an ideal quantum gas in the technical sense ([51], p. 201)

$$pV = gU, \text{ i.e. } p/c^2 = gU/c^2V = g\varrho \tag{E. 6}$$

where $g$ is a constant and $U$ is the internal energy of a volume $V$ of the gas.

Let $\varepsilon^0$ and $p^0$ be the energy density and the pressure of the cosmological fluid in the local rest frame. Then in that frame the gain of energy of a suitably chosen volume is equal to the compression work done on this volume,

$$d(\varrho R^3) = c^{-2} d(\varepsilon^0 R^3) = -(p^0/c^2) d(R^3) \tag{E. 7}$$

This is a general energy equation for homogeneous cosmological models in Newtonian and general relativistic mechanics. In the latter it can be deduced from the Einstein field equations. $R$ may be interpreted as a dimensionless scale factor which depends only on time and governs the time dependence of the distance between any two particles. Hence any volume $V$ defined by a set of labelled particles is proportional to $R^3$. Using this fact with (E. 6) and (E. 7)

$$(1+g)\varrho d(R^3) + R^3 d\varrho = 0. \tag{E. 8}$$

It follows that the following quantities are independent of time

$$\varrho R^{3(1+g)}, \quad UV^g, \quad p^0 V^g \tag{E. 9}$$

Using (E. 7) and (E. 9) in (E. 5)

$$2\dot{R}\ddot{R} = -\frac{8\pi}{3} G(1+3\mu g) \frac{\varrho R^{3+3g}}{R^{2+3g}} \dot{R} + \frac{2}{3} \lambda c^2 R \dot{R}$$

If $-Kc^2$ be the constant of integration,

$$\dot{R}^2 = \frac{8\pi}{3} \frac{1+3\mu g}{1+3g} G \frac{\varrho R^{3+3g}}{R^{1+3g}} + \frac{1}{3} \lambda c^2 R^2 - Kc^2 \tag{E. 10}$$

Thus the general relativistic result ($\mu = 1$) can generate the Newtonian result ($\mu = 0$) simply if $G$ is replaced by $G/(1+3g)$.

In the early universe $R$ is relatively small and the first term in (E. 10) dominates. Also this includes the era in which radiation dominates, so that $g = \frac{1}{3}$.

Time in Statistical Physics and Special Relativity 95

This means that using (E. 9) and the properties of black body radiation, that the expansion can be treated as a quasistatic adiabatic process with

$$\varepsilon° R^4 = A, \quad TR = B \qquad (T = \text{absolute temperature})$$

where $A$ and $B$ are independent of time. Hence (E. 10) yields for $\mu = 1$

$$R \, dR = \left(\frac{8\pi G}{3c^2} A\right)^{\frac{1}{2}} dt.$$

It follows that

$$t = \left(\frac{3c^2}{8\pi GA}\right)^{\frac{1}{2}} \tfrac{1}{2} R^2 = \left(\frac{3c^2}{8\pi GA}\right)^{\frac{1}{2}} \tfrac{1}{2} \left(\frac{B}{T}\right)^2 \qquad (E.\ 11)$$

The temperature in the early universe is accordingly

$$T = \left(\frac{3c^2}{32\pi GD}\right)^{\frac{1}{4}} t^{-\frac{1}{2}} \qquad (E.\ 12)$$

where

$$D \equiv \frac{A}{B^4} = \frac{\varepsilon° R^4}{T^4 R^4} = \frac{\varepsilon°}{T^4} = \frac{U}{VT^4} = \frac{4}{c}\sigma = \frac{8\pi^5 k^4}{15 c^3 h^3}$$

Here $\sigma$ is Stefan's constant ([51] p. 257) and hence

$$D = 7{,}59 \times 10^{-15} \text{ erg cm}^{-3} \text{ deg}^{-4}.$$

For Newtonian cosmology we replace $G$ by $G/2$, so that if $T$ is in °K and $t$ is in seconds,

$$T \sim \frac{10^{10}}{\sqrt{t}} \quad (\textit{relativistic or Newtonian}) \qquad (E.\ 13)$$

Placing the beginning of the radiation era at $t = 1$ sec and its end at $10^6$ years [33], we see that the temperature must have fallen from $10^{10}$ °K to 2370 °K in this period; but the law (E. 13) may be approximately valid for a greater period.

APPENDIX F

CONDITIONAL PROBABILITY, BAYES' RULE AND ENTROPY

Let $\{A_i\}$ be a set of mutually exclusive and exhaustive events so that we consider a situation in which any event is a combination of the $A_i$'s. Let $B$ be an arbitrary event in this situation, then $P(A_i|B)$ is the (conditional) probability that $A_i$ occurs given that $B$ occurs. The probability that $A_i$ and $B$ both occur is

$$P(A_i \cap B) = P(A_i) P(B|A_i) = P(B) P(A_i|B) \tag{F. 1}$$

Hence Bayes' rule

$$P(A_i|B) = \frac{P(A_i \cap B)}{P(B)} = \frac{P(A_i) P(B|A_i)}{\sum_k P(A_k) P(B|A_k)} \tag{F. 2}$$

It is a symmetrical law, for we obtain correct results by interchanging $A_i$ and $B$.

If the $A_i$ are a set of "causes" or hypotheses of known à priori probabilities we may know the probability $P(B|A_i)$ that $A_i$ will lead to $B$. The formula can then be used to determine the inverse probabilities: given that $B$ has happened, one can estimate the probability of $A_i$, as on the left-hand side of (F. 2).

Replace $B$ by $B_j$, a member of another set of mutually exclusive and exhaustive events, then take logarithms in (F. 2), multiply by $-k P(A_i \cap B_j)$ noting (F. 1), and sum over $i$ and $j$. Then

$$-k \sum_{ij} P(B_j) P(A_i|B_j) \ln P(A_i|B_j) =$$
$$-k \sum_i P(A_i) \ln P(A_i) + k \sum_j P(B_j) \ln P(B_j)$$
$$-k \sum_{ij} P(A_i) P(B_j|A_i) \ln P(B_j|A_i) \tag{F. 3}$$

The left-hand side and the last term on the right-hand side are respectively conditional entropies of $A$ given $B$, and of $B$ given $A$, according to the relations

$$S_B(A) \equiv \sum_j P(B_j) S_{B_j}(A) \equiv -k \sum_{i,j} P(B_j) P(A_i|B_j) \ln P(A_i|B_j)$$

The first two terms on the right hand side of (F. 3) are just the entropies so that (F. 3) is

$$S_B(A) = S(A) - S(B) + S_A(B). \tag{F. 4}$$

The entropy for all possible events $A_i \cap B_j$ can be defined by

$$S(A \cap B) \equiv -k \sum_{i,j} P(A_i \cap B_j) \ln P(A_i \cap B_j)$$

$$= -k \sum_{i,j} \{P(A_i \cap B_j) [\ln P(A_i) + \ln P(B_j|A_i)]\}$$

$$= S(A) - k \sum_{i,j} P(A_i) P(B_j|A_i) \ln P(B_j|A_i)$$

$$= S(A) + S_A(B). \tag{F.5}$$

From (F. 4,5) one finds

$$S(A \cap B) = S(A) + S_A(B) = S(B) + S_B(A). \tag{F.6}$$

If one distinguishes the maximum number of states one has a greater entropy than one has on averaging out some of the states, say those due to the $B_i$. This is expressed by

$$S(A \cap B) > S(A), \tag{F.7}$$

which follows from (F. 6).

One may interpret this inequality as follows. One is reducing the number of states which are deemed distinguishable in passing from $A \cap B$ to $A$. In the limit when one can recognise only one state the entropy is zero. [This is quite different from throwing away information but keeping the number of states constant.] $A \cap B$ can refer to a microscopic description, and $A$ to a macroscopic description of the same system. In this sense:

The micro-entropy exceeds the macro-entropy. (F. 8)


APPENDIX G

DETERMINISM FROM RELATIVITY?

Observers $A$ and $A'$ are at the origins of their reference frames $F$ and $F'$, the latter moving with velocity $v$ in the positive $x$-direction of $F$. The $x'$-axis coincides with the $x$-axis. The clocks of $A$ and $A'$ are adjusted so that when $A'$ is at $x = -d < 0$ we have $t = t' = 0$. We shall talk $y = y' = z = z' = 0$ for all events considered.

We make use of the notion of the pre-relativistic future of $A$ at $t=0$ by defining

$A$'s "future" at $t=0$ means $t>0$

Similarly

$A'$'s "past" at $t=0$ means $t' < 0$.

It follows that an event $E(x, t)$, $x > 0$, in $A$'s "future" has for $A'$ the co-ordinates

$$x' = \gamma [x+d-vt] \tag{G. 1}$$

$$t' = \gamma [t-v(x+d)/c^2] \tag{G. 2}$$

It therefore lies in the "past" of $A'$ if

$$x+d > c^2 t/v \tag{G. 3}$$

It lies in the normal relativistic *absolute past* of $A'$ if it lies in the light cone, i.e.

$$|x'| < |ct'| = -ct' \tag{G. 4}$$

By (G. 1) to (G. 3)

$$\frac{x'}{-ct'} = \frac{c}{v} \frac{x+d-vt}{x+d-c^2t/v} > \frac{c}{v} \frac{x+d-vt}{x+d-vt} > 1 \tag{G. 5}$$

Thus an event in the "future" of $A$ and in the "past" of $A'$ cannot by (G. 4,5) lie in the absolute past of $A'$. In relativistic terms the event $E$ is therefore not pre-determined for $A$ by being already in another observer's "past", for it can lie only in a relativistically indeterminate part of $A$'s history.

APPENDIX H

SOME PROPERTIES OF TACHYONS

*Generalities*

Let $A_\mu$, $A'_\mu$ be a four-vector in inertial frames $I$, $I'$. Then the Lorentz transformation for relative velocity $v$ ($|v|<c$) between the frames (see Fig. 5) and with $\underset{v}{\gamma} = (1-v^2/c^2)^{-\frac{1}{2}}$ yields

$$\frac{A_1'}{A_1} = \underset{v}{\gamma}\left(1 - \frac{v}{c}\frac{A_4}{A_1}\right), \quad \frac{A_4'}{A_4} = \underset{v}{\gamma}\left(1 - \frac{v}{c}\frac{A_1}{A_4}\right), \quad \frac{A_2'}{A_2} = \frac{A_3'}{A_3} = 1$$

Here relative motion parallel to the common $x$-axis has been assumed for simplicity.

The space-time coordinate of an event

$$A_\mu = (\mathbf{r}, ct), \quad A_1 \equiv x, \quad x/ct \equiv u/c$$

and the energy-momentum vector of a system (or particle)

$$A_\mu = P_\mu = (c\mathbf{P}, E), \quad cP_1/E \equiv w/c$$

Time in Statistical Physics and Special Relativity 99

are notable examples. If $(x, t)$ are running coordinates of an object, this object has velocity $u$ in $I$. If $(\mathbf{P}, E)$ refer to a system, $w$ is its centre-of-mass velocity in $I$. We have

$$E^2 - c^2 P_1^2 = (1 - w^2/c^2) E^2,$$

$$c^2 t^2 - x^2 = (1 - u^2/c^2) c^2 t^2.$$

Thus there are two kinds of vectors $P_\mu$

(i) time-like if $(w/c)^2 < 1$. The sign of $E$ is,
the sign of $P_1$ is not, Lorentz-invariant. (H. 1)

(ii) space-like if $(w/c)^2 > 1$. The sign of $P_1$ is,
the sign of $E$ is not, Lorentz-invariant. (H. 2)

Case (i) refers to usual systems, case (ii) to possible super-luminary particles, called tachyons. In that case $P_1$ is the single-particle momentum.

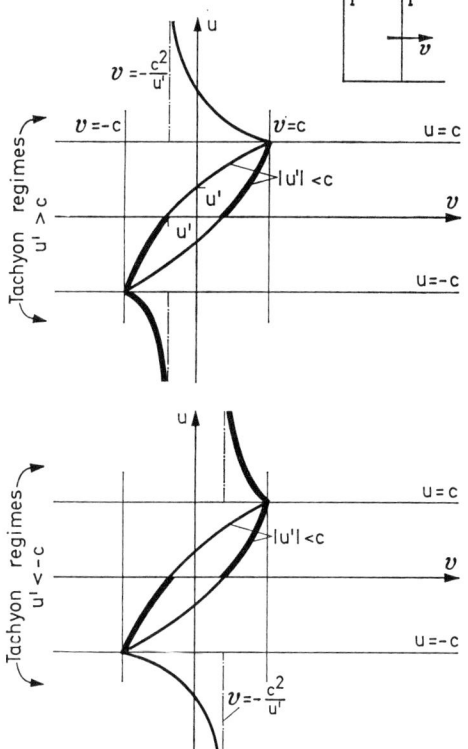

Fig. 6

The simplest frame for usual systems is that for which $\mathbf{P}=0$. This is the rest frame; call it $I_0$ and give quantities measured in $I_0$ a suffix 0. Then in a general frame $I$ (see Fig. 6) in which the system has velocity $w$,

$$E = \underset{w}{\gamma} E_0, \quad \mathbf{P} = (w/c^2) \underset{w}{\gamma} E_0 \qquad (H.3)$$

The simplest frame for superluminary systems is a frame, also to be called $I_0$, in which $E=0$. Then

$$\mathbf{P} = \underset{w}{\gamma} \mathbf{P}_0. \quad E = \underset{w}{\gamma} \mathbf{w} \cdot \mathbf{P}_0 \qquad (H.4)$$

Clearly tachyons have no rest-frame

Relativistic velocity addition for two collinear velocities is shown in Fig. 6. The central strips correspond to normal systems, the top and bottom strips correspond to tachyons. The formula

$$u = \frac{u' + v}{1 + u'v/c^2}$$

shows that velocity reversal is achieved if the velocity $v$ of $I'$ in frame $I$ satisfies

$$\left. \begin{array}{l} -c < v < -u' \text{ if } c > u' > 0 \\ -u' < v < c \text{ if } 0 > u' > -c \end{array} \right\} \text{ normal} \qquad (H.6)$$

$$\left. \begin{array}{l} -c < v < -c^2/u' \text{ if } u' > c \\ -c^2/u' < v < c \text{ if } -c > u' \end{array} \right\} \text{ tachyons} \qquad (H.7)$$

Thick lines in Fig. 6 correspond to cases where $u'$ an $u$ are in opposite directions. The figure shows also that the sub-division of systems into normal systems and tachyons is Lorentz invariant.

For tachyons of velocity $w$ in a frame $I$, the Lorentz factor

$$\underset{w}{\gamma} \equiv (1 - w^2/c^2)^{-\frac{1}{2}}$$

is imaginary and by (H. 5) its rest mass may be assumed imaginary as it cannot be observed:

$$m_0 = i\mu_0 \qquad (H.8)$$

Here $\mu_0$ is real and positive or negative. This makes the relativistic mass,

$$m = \underset{w}{\gamma} m_0, \qquad (H.9)$$

real, as it should be since it could possibly be observed. It follows that

$$m = \frac{i\mu_0}{\sqrt{1-w^2/c^2}} = \pm \frac{\mu_0}{\sqrt{w^2/c^2-1}} \tag{H. 10}$$

The freedom introduced by the observation that multiplication by $i$ will change a square root so as to make it real, is expressed in (H. 10). The sign may be fixed by noting that the $x$-momentum of the tachyon is

$$P_1 = m w_1. \tag{H. 11}$$

By (H. 2) and (H. 7) $m$ must have the sign of $w_1$ so that we may put

$$m = \frac{w_1}{|w_1|} \frac{\mu_0}{\sqrt{w^2/c-1}}. \tag{H. 12}$$

This procedure avoids the amendment of standard Lorentz transformations as suggested by Terletskii (1967, p. 92).

If an observer moves with a speed $v$ such that the tachyon speed becomes infinite, he will find that its mass $m$ tends to zero by (H. 10), and hence it energy $mc^2$ will also tend to zero.

To obtain more insight into the behaviour of a tachyon, consider its motion under a constant force $f$ (in a frame $I$) in the $x$-direction. If $t$ is the time in $I$, the equation of motion is by (H. 11) and (H. 12)

$$\frac{d}{dt} \frac{\mu_0 |w|}{\sqrt{(w^2/c^2-1)}} = f \quad (w_1 \equiv w) \tag{H. 13}$$

Integrating over a period of time during which the direction of $w$ is constant, and introducing a new symbol $d$ in order to denote the constant of integration by $f\theta/c\mu_0$ one finds two possible relations, each of which can be put into two distinct forms

$$\frac{w/c}{\sqrt{(w^2/c^2-1)}} = \frac{f(t+\theta_1)}{c\mu_0} \equiv \tau_1 \begin{pmatrix} w>0 \\ \tau_1>0 \end{pmatrix} \text{ or } -\frac{f(t+\theta_3)}{c\mu_0} \equiv -\tau_3 \begin{pmatrix} w>0 \\ \tau_3<0 \end{pmatrix}$$

$$-\frac{w/c}{\sqrt{(w^2/c^2-1)}} = \frac{f(t+\theta_2)}{c\mu_0} \equiv \tau_2 \begin{pmatrix} w<0 \\ \tau_2>0 \end{pmatrix} \text{ or } -\frac{f(t+\theta_4)}{c\mu_0} \equiv -\tau_4 \begin{pmatrix} w<0 \\ \tau_4<0 \end{pmatrix}$$

The stated inequalities can be tightened to

$$c < w, \quad 1 < \tau_1; \quad c < w, \quad \tau_3 < -1$$
$$w < -c, \quad 1 < \tau_2; \quad w < -c, \quad \tau_4 < -1$$

The equations allow for four different types of trajectories in $(w, t)$ space depending on the signs of $\tau$ and $w$. The four cases are illustrated in Fig. 7.

Treating all four cases together, and using $\tau$ as a measure of time in $I$,

$$|w|/c = |\tau|/\sqrt{(\tau^2-1)}$$

If $x=a$, when $|\tau|=1$ one has

$$|x-a| = (c^2\mu_0/f)\sqrt{(\tau^2-1)}$$

This suggests that at $|\tau| > 1$ four kinds of particles can exist, there being present at each time either no particle or at most two particles subject to equation (H. 13). If two particles coexist their masses $m$ and energies are numerically equal but of opposite signs, while their momenta are equal. The pair of early tachyons are accelerated from $|w|=c$ at $t=-\infty$ to $|w|=\infty$ when they disappear and their velocities become imaginary. The late tachyons are slowed down from their appearance at infinite velocities to $|w|=c$ at $t=\infty$.

## APPENDIX I

### CONDITIONS FOR SIGNALLING TO ONE'S PAST

The procedure to be adopted is illustrated in Fig. 5 drawn for an inertial frame $I$. The four bold arrowed lines are the world lines of observers: $A$ and $B$ at rest at distance $f$ apart; $P$ and $Q$ at rest in another frame $I'$ distance $d'$ apart. Signals and relative velocities are arranged to lie parallel to the common $x$-axis of $I$ and $I'$.

Introduce event $E_1$ as defined by coincidence of $A$ and $P$; at $E_1$ a signal with velocity $c_1 < 0$ is sent from $P$ to $Q$ (in $I'$). The arrival of the signal must

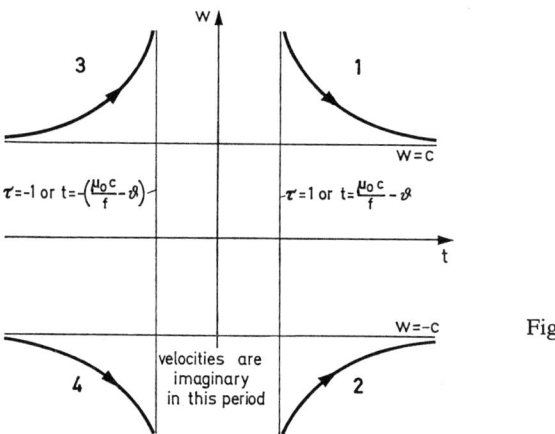

Fig. 7

be shown on our figure, taking the point of view of $I$. In order to do so, let us find the line of events $(x, t)$ in $I$ which correspond to events simultaneous with $E_1$ in $I'$. If $E_1 \equiv (a_1, t_1)$ then in $I'$ the event is

$$E_1' = [a_1', t_1'] = [\gamma(a_1 - vt_1), \gamma(t_1 - \tfrac{v}{c^2} a_1)]$$

Hence in general

$$t_1' = \gamma (t - \tfrac{v}{c^2} x) \text{ i.e } x = \tfrac{c^2}{v} (t - \tfrac{t_1'}{\gamma}).$$

The required line passes therefore through $E_1$ and has slope $v/c^2$. It is drawn dashed in the figure. The event of receipt of the signal must on our diagram lie above this line.

It is arranged that the signal reaches $Q$ at the $(B, Q)$ coincidence. This is event $E_2$ and is marked by a signal sent by $B$ to $A$ (in $I$) with velocity $c_2 > 0$. The signals are shown, and event $E_3$ denotes the arrival of the signal at $A$. $A$ will effectively have signalled to his own past if $t_3 < t_1$ (as drawn). However, we require $t_3 > t_2$ (as shown).

We note the coordinates of the events in an obvious notation with $d' < 0$ and $f > 0$

$$E_2': \left[\gamma (a_1 - vt_1) + d';\ \gamma (t_1 - \tfrac{v}{c^2} a_1) + \tfrac{d'}{c_1}\right]$$

$$E_2: \left[a_1 + (1 + \tfrac{v}{c_1}) \gamma d';\ t_1 + (\tfrac{1}{c_1} + \tfrac{v}{c^2}) \gamma d'\right]$$

$$E_3: \left[a_1 + f + (1 + \tfrac{v}{c_1}) \gamma d';\ t_1 + (\tfrac{1}{c_1} + \tfrac{v}{c_2}) \gamma d' + \tfrac{f}{c_2}\right]$$

We must impose the condition that $E_3$ occurs at $A$, i.e. $E_3(a_3, t_3) = E_3(a_1, t_3)$. It follows that

$$f = -\left(1 + \tfrac{v}{c_1}\right) \gamma d'. \tag{I. 1}$$

Using (I. 1) and defining

$$x \equiv \tfrac{c_1}{c} < 0,\ y \equiv \tfrac{c_2}{c} > 0,\ 1 > z \equiv \tfrac{v}{c} > 0, \tag{I. 2}$$

one finds that

$$\tfrac{c}{\gamma d'} (t_3 - t_1) = \tfrac{1}{xy} (y - x - z + xyz),$$

i.e. $-\tfrac{c(t_3 - t_1)}{\gamma |d'|} = \tfrac{1}{|x| y} (z + x yz - x - y) \tag{I. 3}$

From (I. 1) and (I. 2) we also have

$$f = \left(1 - \frac{v}{|c_1|}\right) \gamma |d'| = \left(1 - \frac{z}{|x|}\right) \gamma |d'| > 0. \tag{I. 4}$$

The condition for signalling to one's past is that $t_3 - t_1 < 0$, so that the quantity (I. 3) must be positive. By (I. 3) and (I. 4)

$$|x| > z > \frac{|x| + y}{1 + |x| y}$$

whence (I. 5)

$$x^2 > 1.$$

Secondly (I. 3) also implies

$$y > \frac{|x| - z}{|x| z - 1} \equiv g(x, z).$$

Note that

$$\left(\frac{\partial g}{\partial z}\right)_x = \frac{1 - x^2}{(|x| z - 1)^2} < 0 \tag{I. 6}$$

by (I. 5). But since $z < 1$

$$g(x, z) > g(x, 1) = 1. \tag{I. 7}$$

Now (I. 6) and (I. 7) show that

$$y > g(x, z) > 1.$$

It has thus been established that $t_3 - t_1 < 0$ implies

$$|c_1|, c_2 > c. \tag{I. 8}$$

Thus two *tachyons* have to be employed for signalling.

For the procedure to work it is clearly necessary that at the time $t_2$ of $E_2$ satisfies $t_2 < t_1$. This implies that

$$v > -\frac{c^2}{c_1} = \frac{c^2}{|c_1|} \quad (c_1 < 0) \tag{I. 9}$$

Inspection of Fig. 5 shows that a negative velocity tachyon which travels from $E_1$ to $E_2$ in $I'$, becomes a positive velocity tachyon which travels from $E_2$ to $E_1$ in $I$ if (I. 9) is satisfied. Hence it is seen that two tachyons appear to be *emitted* in $I$ at the event $E_2$ and absorbed in $I'$ at $E_2$. Similarly as far as as $A$ is concerned a tachyon is *absorbed* at each $E_3$ and $E_1$.

Time in Statistical Physics and Special Relativity 105

APPENDIX J

THE THRESHOLD OF CLASSICAL COSMOLOGY

A homogeneous isotropic expanding non-rotating model of the universe is considered. Then at an early age one can show that

$$6\pi G \rho t^2 = 1 \tag{J.1}$$

where $\rho$ is the volume density of matter at time $t$. Let $d$ be the mean interparticle distance and $\theta d^3$ be the volume per particle, where $\theta$ is a constant of order unity. Suppose the cosmic fluid consists of particles of relativistic mass

$$m = \gamma m_0, \; \gamma \equiv 1/\sqrt{(1-v^2/c^2)}, \; \rho = m/\theta d^3 \tag{J.2}$$

where $v$ is an average velocity in the comoving coordinates. Then

$$\left(\frac{1}{ct}\right)^2 = \frac{6\pi G \rho}{c^2} = \frac{6\pi G m}{\theta c^2 d^3} \tag{J.3}$$

Suppose the fluid consists of *elementary* particles. Hence the continuous fluid model requires $d < ct$. If this condition is violated, and the particles cannot be subdivided then the only remaining possibility is that the particles are distributed into cells of space whose size exceeds on average the particle horizon; this is drastically different from a fluid. Suppose that at $t = t_1$ we have $d = d_1$, $\rho = \rho_1$ and $d_1 = ct_1$. Then by (J.3)

$$ct_1 = d_1 = \frac{6\pi G m}{\theta c^2} \tag{J.4}$$

As one goes back in time the relativistic mass $m$ of a particle increases and the de Broglie wavelength

$$\lambda = \hbar/mv \tag{J.5}$$

decreases to the inter-particle distance, which may be reached when $v = v_2$, $m = \gamma_2 m_0$, and $\lambda = \lambda_2$ (say). This sets another limit to $d$ in the sense that we require

$$d > d_1 \text{ and } d > \lambda_2. \tag{J.6}$$

Our hypothesis is that the two limits $d_1$, $\lambda_2$ are the same. Then Eqs. (J.4) to (J.6) yield

$$\frac{6\pi \gamma G m_0}{\theta c^2} = \frac{\hbar}{\gamma m_0 v} \quad \text{or} \quad \frac{c}{\gamma^2 v} = b^2 \left(\equiv \frac{6\pi G m_0^2}{\theta \hbar c}\right) \tag{J.7}$$

The maximum average velocity $v^*$ determined by this hypothesis is a solution of

$$\left(\frac{v^*}{c}\right)^2 + b^2 \frac{v^*}{c} - 1 = 0.$$

This determines all limiting quantities in terms of $m_0$.

Let $m_0$ be a typical nucleon mass $m_{no}$. Then

$$b = \left(\frac{6\pi}{\theta}\right)^{\frac{1}{2}} \frac{m_{no}}{(\hbar c/G)^{\frac{1}{2}}} \sim 10^{-20} \tag{J. 8}$$

$$\gamma = \gamma^* = 1/b.$$

$$m_n^* = \gamma^* m_{no} = b^{-1} m_{no} = \left(\frac{\hbar c \theta}{6\pi G}\right)^{\frac{1}{2}}$$

Other critical quantities follow. For example by (J. 3)

$$d^* = \frac{6\pi G}{\theta c^2} \left(\frac{\hbar c \theta}{6\pi G}\right)^{\frac{1}{2}} = \left(\frac{6\pi}{\theta} \frac{\hbar G}{c^3}\right)^{\frac{1}{2}} \tag{J. 9}$$

$$= \left(\frac{6\pi}{\theta} \frac{m_n^* G}{c^2} \frac{\hbar}{m_n^* c}\right)^{\frac{1}{2}} = \left(\frac{6\pi}{\theta} \frac{m_n^* G}{c^2}\right)^{\frac{1}{2}} \lambda_n^* \sim 10^{-20} \lambda_n^*$$

Here $\lambda_n^*$ is the de Broglie wavelength for a nucleon travelling very nearly with the speed of light. Also by (J. 2), (J. 8) and (J. 9)

$$\varrho^* = \frac{1}{\theta} \left[\frac{\hbar c}{6\pi G} \frac{\theta^3 c^9}{6^3 \pi^3 \hbar^3 G^3}\right]^{\frac{1}{2}} \tag{J. 10}$$

$$= \frac{\theta c^5}{36 \pi^2 \hbar G^2} \sim 10^{95} \; gm/cm^3$$

The age of the universe below which we can not probe by the fluid-type cosmological models is therefore

$$t^* = \frac{d^*}{c} = \left(\frac{6\pi \hbar G}{\theta c^5}\right)^{\frac{1}{2}} \sim 10^{-44} \; sec. \tag{J. 11}$$

Note that $t^*$ and $m_n^* c^2$ satisfy the uncertainty relation

$$t^* m_n^* c^2 = \hbar. \tag{J. 12}$$

Note Added in Proof: Since this lecture was given and the manuscript was handed in (September 1969), the literature on tachyons has grown further. Some of the references are given below: [99] to [103. Of these references, [102] uses considerations based on a figure close to our Fig. 5, but some of the arguments given here in Appendices H and I are still not available elsewhere. Ref. [104, 105] are relevant to the remarks made at the end of Sect. 12.

## References

1. Ageno, M.: *Does quantum mechanics exclude life.* Nature 205 (1965) 1306–1307.
2. Aron, J. C.: *Quantum laws in connection with Stochastic Processes.* Prog. Theoretical Physics 33 (1965) 726–754.
3. Balescu, R.: *Velocity inversion in Statistical mechanics.* Physica 36 (1967) 433–456.
4. — *Some comments on irreversibility.* Phys. Letters 27A (1968), 249.
5. Beauregard, O. Costa de: *Irreversibility problems. In: Logic. Methodology and Philosophy of Science.* Ed. Y. Bar-Hillel, p. 313–342. Amsterdam: North Holland 1965.
6. Bergmann, P. G.: *Foundations Research in Physics.* In: *Delaware Seminar in the Foundations of Physics* (Ed. M. Bunge) p. 1–14. Berlin: Springer 1967.
7. Bilaniuk, O. M., Sudarshan, E. C. G.: *Particles beyond the light barrier.* Physics Today 22 (May, 1969) 43–51.
8. —, Deshpande, V. K., Sudarshan, E. C. G.: „Meta "relativity.* Am. J. Phys. 30 (1962) 718–723.
9. Blatt, J. M.: *An alternative approach to the ergodic problem.* Prog. Theoret. Phys. 22 (1959) 745–756.
10. Boltzmann, L.: *Vorlesungen über Gastheorie* (Leipzig: Barth 1896). English translation by S. G. Brush, Univ. of California Press 1964.
11. Bondi, H.: *Physics and Cosmology.* Observatory, 82 (1962) 133–143 (Halley Lecture).
12. Bopp, F.: *Der Wechselwirkungsoperator im Gitterraum.* Z. Phys. 200 (1967) 142–157.
13. Born, M.: *Die Quantenmechanik und der zweite Hauptsatz der Thermodynamik.* Ann. Phys. (Lpz) 3 (1948) 107–114.
14. — *Continuity, Determinism, and Reality.* K. Danske Vidensk. Selsk. mat.-fys. Medd. 30 (1955) No. 2.
15. Bridgman, P. W.: *The Nature of Thermodynamics.* Chapter III (Harper Torchbook, p. 213). Harvard 1941.
16. Büchel, W.: *Entropy and information in the universe,* Nature 213 (1967) 319–320.
17. Caldirola, P. (Ed): *Ergodic Theories.* New York: Academic Press 1961.
18. Campbell, B.: *Biological Entropy.* Nature 215 (1967) 1308.
19. Chandrasekhar, S.: *Stochastic Problems in Physics and Astronomy.* Rev. Mod. Phys. 15 (1943) 1–89.
20. Cohn, J.: *Special Relativity in the Quantum Domain.* Am. J. Phys. 36 (1968) 749–751.
21. Conklin, E. K.: *Velocity of the earth with respect to the cosmic background radiation.* Nature 222 (1969) 971–972.
22. Darling, B. T.: *The irreducible volume character of events. I A Theory of the elementary particles and of fundamental length.* Phys. Rev. 80 (1950) 460–466.
23. Earman, J.: *Irreversibility and temporal asymmetry.* Journal of Philosophy 64 (1967) 543–549.
24. Einstein, A.: *Prinzipielles zur allgemeinen Relativitätstheorie.* Ann. Phys. 55 (1918) 241–244.
25. Feinberg, G.: *Possibility of faster-than-light particles.* Phys. Rev. 159 (1967) 1089–1105.
26. Fock, V.: *The theory of Space. Time and Gravitation.* p. 4. 2nd revised ed. Pergamon 1964.
27. Foerster, H. von: *Memory without record.* In: *The Anatomy of Memory.* Ed. D. P. Kimble, Palo Alto 1965.
28. Fraser, J. T. (Ed.): *The Voices of Time.* New York: Braziller 1966.
29. Gale, R. M.: *Why a cause cannot be later than its effect.* Rev. of Metaphysics 19 (1965) 209–234.
30. Gold, T. (Ed.): *The Nature of Time.* Ithaca: Cornell Univ. Press 1967.
31. — *The arrow of time.* 11th International Physics Congress, Solvay 1958. Am. J. Phys. 30 (1962) 403–410 (Richtmyer Lecture).

32. Grünbaum, A.: *Modern Science and Zeno's Paradoxes.* London: Allen and Unwin 1968.
33. Harrison, E. R.: *The early universe.* Physics Today 21 June, (1968) 31—39.
34. — *Cosmology without general relativity.* Annals of Physics 35 (1965) 437—446.
35. — *Quantum cosmology.* Nature 215 (1967) 151—152.
36. Hellund, E. J., Tanaka, K.: *Quantized Space-Time.* Phys. Rev. 94 (1954) 192—195.
37. Hobson, A.: *Irreversibility in Simple Systems.* Am. J. Phys. 34 (1966) 411—416.
38. — *Irreversibility in Finite Quantum-Statistical-Mechanical Systems.* J. Chem. Phys. 46 (1967) 1365—1372.
39. — *Further comments on irreversibility.* Physics Lett. 28A (1968) 183—184.
40. Hoyle, F., Narlikar, J. V.: *Electrodynamics of direct interparticle action I The quantum mechanical response of the universe.* Annals of Physics. 54 (1969) 207—239.
41. Ingraham, R. L.: *Stochastic space-time.* Nuovo Cimento 34 (1964) 182—197.
42. Jaynes, E. T.: *Gibbs vs Boltzmann Entropies.* Am. J. Phys. 33 (1965) 391—398.
43. Kac, M.: *Probability and Related Topics in Physical Sciences.* New York: Interscience 1959.
44. Kampen, N. G. van.: *Fundamental problems in the statistical mechanics of irreversible processes.* In: *Fundamental Problems in Statistical Mechanics I*, p. 173—202. (Ed. E. G. D. Cohen) Amsterdam: North Holland 1962.
45. Katz, A.: *Principles of Statistical Mechanics.* San Francisco: Freeman 1967.
46. Kretschmann, E.: *Physical Significance of the relativity postulate. Einstein's new and original theories of relativity.* Ann. Phys. 53 (1918) 575—614.
47. Kurth, R.: *Axiomatics of Classical Statistical Mechanics.* New York: Pergamon 1960.
48. Landsberg, P. T.: *The uncertainty principle as a problem in philosophy.* Mind 56 (1947) 250—256.
49. — *On Bose-Einstein condensation.* Proc. Camb. Phil. Soc. 50 (1954) 65-76.
50. — *Entropy and the Unity of Knowledge.* Univ. of Wales Press, 1961 p. 18. Inaugural Lecture.
51. — *Thermodynamics with Quantum Statistical Illustrations.* New York: Wiley 1961.
52. — *Two relativistic paradoxes.* Mathematical Gazette 47 (1964) 197—202.
53. — *Does quantum mechanics exclude life?* Nature 203 (1964) 928—930.
54. — Johns, K. A.: *Energy flux from the $3°K$ radiation.* Nature 220 (1968) 1120.
55. — *Time reversibility.* In: *Solid State Theory: Methods and Applications.* (Ed. P. T. Landsberg) New York: Wiley, 1969.
56. — *Special Theory of Relativity.* Nature 220 (1968) 1182—1183.
57. Layzer, D.: *A unified approach to cosmology.* In: Lectures in Applied Mathematics. Vol. 8, p. 237 American Mathematical Society 1967; Ed. J. Ehlers.
58. Lewis, G. N.: *The Symmetry of time in physics.* Science 71 (1930) 569—577.
59. Linden, J. van der, Mazur, P.: *The asymptotic problem of statistical thermodynamics for a real system.* Physica 36 (1967) 491—508.
60. Marshall, T. W.: *Random electrodynamics.* Proc. Roy. Soc. A. 276 (1963) 475—491.
61. — *Statistical electrodynamics.* Proc. Camb. Phil. Soc. 61 (1965) 537—546.
62. Mayer, J. E.: *Approach to thermodynamic equilibrium.* J. Chem. Phys. 34 (1961) 1207—1223.
63. McCrea, W. H.: *Relativity theory and the creation of matter.* Proc. Roy. Soc. A 206 (1951) 562—575.
64. Ono, S.: *The long time behaviour of quantum mechanical systems.* Mem. Fac. Eng. Kyushu Univ. 11 (1949) 125—140.
65. Orban, J., Bellemans, A.: *Velocity-Inversion and irreversibility in a dilute gas of hard discs.* Physics. Lett. 24A (1967) 620—621.
66. Penrose, O., Percival, I. C.: *The direction of time.* Proc. Phys. Soc. 79 (1962) 605—616.
67. Planck, M.: *The Theory of Heat Radiation.* Dover, 2nd Ed. of 1912, published in 1959, 175.
68. Pompe, W., Schöpf, H.-G.: *Kosmologische und statistische Überlegungen zur Irreversibilität.* Annalen d. Phys. 21 (1968) 26—30.
69. Popper, K.: *Time's arrow and entropy.* Nature 207 (1965) 233—234.
70. — *Time's arrow and feeding on negentropy.* Nature 213 (1967) 320.
71. — *Structural information and the arrow of time.* Nature 214 (1967) 322.
72. Prigogine, I.: *Non-equilibrium Statistical Mechanics.* New York: Interscience 1962.
73. — *Dynamic Foundations of Thermodynamics and Statistical Mechanics.* In: A Critical Review of Thermodynamics Ed. E. B. Stuart, B. Gal-Or, A. J. Brainard). Baltimore: Mono Book Corp. 1970

74. Reichenbach, H.: *The Direction of Time*. Berkely: Univ. of California Press 1956.
75. Rietdijk, C. W.: *A rigorous proof of determinism derived from the special theory of relativity*. Phil. of Science 33 (1966) 341—344.
76. Schmidt, H.: *Eine makroskopisch reale Quantentheorie*. Zeit. für Natur. f. 18a (1963) 265—275.
77. Snyder, H. S.: *Quantised space-time*. Phys. Rev. 71 (1947) 38—41.
78. — *The electromagnetic field in quantised space-time*. Phys. Rev. 72 (1947) 68—71.
79. Stannard, F. R.: *Symmetry of the time axis*. Nature 211 (1966) 693—694.
80. Terletskii, Y. P.: *Paradoxes in the theory of relativity*. New York: Plenum 1968.
81. Tolman, R. C.: *Relativity, Thermodynamics and Cosmology*. Oxford 1934.
82. Treder, H. J.: *Kosmische und subjektive Zeit*. Forschungen und Fortschritte. 39 (1965) 325—327.
83. — (Ed.): *Einstein Symposium*. Berlin: Akademie Verlag 1966.
84. Tribus, M.: *Thermostatics and Thermodynamics*. Princeton: van Nostrand 1961.
85. Uhlenbeck, G. E.: *An outline of statistical mechanics*. In: Fundamental Problems in Statistical Mechanics II p. 27. (Ed. E. G. D. Cohen). Amsterdam: North Holland 1968.
86. Verboven, E. J.: *Quantum Thermodynamics of an infinite system of harmonic Oscillators*. In: Statistical Mechanics (Ed. T. A. Bak), p. 49—54. New York: Benjamin 1967.
87. Viswanadham, C. R.: *Entropy, evolution and living systems*. Nature 219 (1968) 653.
88. Weizsäcker, C. F. von: *Der zweite Hauptsatz und der Unterschied von Vergangenheit und Zukunft*. Annalen d. Phys. 36 (1939) 275—283.
89. Weyer, E. M. (Ed.): *Interdisciplinary Perspectives of Time*. Annals of the New York Academy of Sciences 138 (1967) 367—915.
90. Whitrow, G. J.: *The Natural Philosophy of Time*. London: Nelson 1961.
91. Whittaker, E. T.: *From Euclid to Eddington*. Cambridge: Univ. Press 1949.
92. Wigner, E. P.: *The probability of the existence of a self-reproducing unit*. In: The Logic of Personal Knowledge. London: Rutledge and Kegan Paul 1961.
93. — Landsberg, P. T.: *Reply to Ageno*. Nature 205 (1965) 1307.
94. Wilson, J. A.: *Increasing entropy of biological systems*. Nature 219 (1968) 534—535.
95. Wiener, N.: Cybernetics. New York: Wiley 1948.
96. Woolhouse, H. W.: *Negentropy, information and the feeding of organisms*. Nature 213 (1967) 952.
97. — *Entropy and Evolution*. Nature 216 (1967) 200.
98. Zanstra, H.: *On the pulsating or expanding universe and its thermodynamical aspect*. Kon. Nederl. Akad. van Wetenschappen 60 (1957) 285—307.
99. Bilaniuk, O. M., and others: *More about tachyons*. Physics Today (December 1969) 47—52.
100. —,*Causality and space-like signals*. Nature 223 (1969) 386—387.
101. Thouless, D. J.: *Causality and tachyons*. Nature 224 (1969) 506.
102. Root, R. G., and Trefil, J. S.: *An amusing paradox involving tachyons*. Lett. Nuovo Cimento 3 (1970) 412—415.
103. Baltay, C., Feinberg G., Yeh, N., and Linsker, R.: *Search for uncharged faster-than-light particles*. Phys. Rev. D 1 (1970) 759—770.
104. Watanabe, S.: *Conditional Probability in Physics*. Progress of Theoretical Physics Suppl. Extra Number (1965) 135—160.
105. Cocke, W. J.: *Statistical Time Symmetry and two-time boundary conditions in Physics and Cosmology*. Physical Review 160 (1967), 1165—1170.

# The Myth of the Passage of Time

DAVID PARK*

> Le temps s'en va, le temps s'en va, madame.
> Las: le temps, non, mais nous nous en allons.
> Ronsard

> Absolute, true, and mathematical time, of itself and from its own nature, flows equably without relation to anything external.
> Newton

> The objective world simply is, it does not happen. Only to the gaze of my consciousness, crawling upward along the life-line of my body, does a section of the world come to life as a fleeting image in space which continually changes in time.
> Weyl

*Summary.* The main purpose of this paper is to point out that the technique, familiar to physicists, of adopting various representations of physical phenomena for various purposes is convenient (and, I believe, necessary) in order to clarify some of the aspects of time that have been regarded as mysterious. Two representations are distinguished: the one that physics always uses and I have called atemporal, and the temporal one implicit in ordinary language and metaphor and most of the philosophical speculation with which I am familiar. The myth mentioned in the title is the idea that this representation somehow *is* time itself. The representation are explained by the use of pictures, which are to be distinguished from the representations they illustrate, and an attempt is made to clarify some old questions by their use.

In physics, and in other disciplines as well, we start with our sensations. We observe that there are phenomena, and that there are other people who report to us their observations of what seem to be the same phenomena. Further, the phenomena exhibit certain consistencies which enable us to plan our lives and even to make successful physical theories. This encourages us to use words like self and other, imagining a world other than ourselves that we can observe and in which we can participate. Ideas of space and time are a part of this imagined world, and we use them to relate ourselves to it. The resulting picture is very consistent internally. Whether we ought to call it necessary, or one-to-one, or true, or false, or a metaphor, or whether there is finally anything rational that can be said about it, are questions which I believe are of the highest importance and, since I have nothing to say to them, I wish at all costs to avoid them.

Space and time are part of the physicist's theoretical picture of the world, and also of the language in which we all describe our thoughts and sensations. Quite apart from the deep philosophical questions I have alluded to, there are questions bordering on physics and psychology as to the relation between physicists' space and time and those of ordinary experience which I think can

---

* David Park, Professor of Physics at Williams College, Thompson Physical Laboratory, Williamstown, Mass., U.S.A.

be discussed by themselves. This lecture is intended not to answer, but to show what is involved in answering one of these questions: Why is it that though we can voyage pretty freely in space and return to our starting point, we seem to be much less free to explore time? Something about time makes us use such phrases as "not yet" and "too late" which have no counterparts in the description of our experience of space. Time passes, but space does not. (The first of my quotations shows that even this common experience can be expressed in more than one way.) It is curious that although the brain is a physical organism, physics does not contain this special property of time. There are then two possibilities: 1) Physics is wrong as applied to mental processes. This is quite possible, since quite apart from the rudimentary state of our knowledge of the physical mechanism underlying our response to perceptual stimuli, there is the far more disturbing question to which I have alluded above, whether there is any hope at all of explaining our conscious experience in terms of models involving physical structures isolated, or partly isolated, from an outside world. 2) The passage of time does not need to be explained by physics.

I maintain that the latter is the case. I have called the passage of time a myth rather than an illusion, say, because it involves no deception of the senses – we shall see that one cannot perform any experiment to tell unambiguously whether time passes or not. There are no sensory data in the ordinary meaning of the term. I use the word myth because this picture of a time that passes us, or through which we pass, is the dominating source of the private as well as the literary language through which we seek to express our experience of time. In fact, the main difficulties of this paper, for the reader as well as the writer, are to find a way of thinking about phenomena related to time which does not assume that time passes, and to try to understand, from this point of view, why the myth has so strong a hold on our imaginations.

It is usual to think of the present as a special moment, the one during which we are in contact with our surroundings; further, we speak of this moment as one that we all share. In examining anything that purports to be a statement of fact it is useful to consider the alternatives. As an alternative to our being limited to the experience of the present instant, let us consider time travel in the manner of H. G. Wells. Suppose that I were really to travel in time back to my fifth birthday. Here are some children sitting around a table. I am five years old and know nothing of the time machine in my future. If I really go back, then all traces of the intervening years, inside and outside me, are gone. There is nothing remarkable about the birthday party. It is indistinguishable from the original one; in fact it *is* the original one. There are no consequences to time travel. A statement that time travel can, or cannot, or does, or does not take place is unverifiable and therefore, in my logic as a physicist; meaningless. What is usually called time travel should be called lack of time travel; Wells's picture is that I take my present mind back to past events. This I take to be fiction.

As to the assertion that we all share the same present, I must confess that I don't know what it means. If I try to imagine an alternative, I expect it to conflict with experience but instead I find it conflicting with language and logic. I therefore suspect that this assertion is part of language and logic as we use them, but that it does not say anything about experience. It is because these two aspects of time seem rooted in our language and imagination rather then in our experience that I say they are parts of a myth.

I have said earlier that physics does not contain the idea of a present instant or of a time that passes. I must explain this statement, since if it is true, then physics will provide a way of describing our sensations, and to some extent even our consciousness, without the use of these ideas, and such a description may enable us to understand the myth. There is, in fact, no quantitative measure of the rate of the passage of time that can be used scientifically, and I cannot find any statement one can make about it that is capable of experimental proof. How quickly does time pass? At a rate of one second per second? That will get us nowhere. There are assumptions impossible to verify in Newton's definition that I have quoted above; he overcomes the difficulty by never again referring to the concept he has so carefully explained, but we shall not get off so lightly.

The second hint I gained from physics is that physics does not need, and indeed can scarcely express, the idea of "now". A typical formula in which time occurs is the relation between distance $x$ and time $t$ at uniform speed $v$: $x = vt$. For a given value of $v$ there are many pairs $(x, t)$ which satisfy this relation, and ordering them by increasing values of $t$ is something that we may do, though it is in no way required by the equation *or by its interpretation*. And so it is in all of physics. There is no theory, even the theory of the cosmos as a whole, in which "now" is any more than a word that might be used to denote any value of $t$, selected at random. No formula of mathematical physics involving time implies that time passes, and indeed I have not been able to think of a way in which the idea could be expressed, even if one wanted to, without introducing hypotheses incapable of experimental verification.

Most of the general and useful terms of the philosophy of physics, such as space, time, cause, effect, and even self, have counterparts in contemporary physical theory in such a way that, even though all may not be clear, it is reasonably evident to what element of the theory they correspond. I therefore think that we should take warning about the present and the now and the absolute time that flows equably, and should try to understand exactly why they do not occur in physics. It is the more urgent because one encounters works of considerable intellectual sophistication which use the word "now" as if there were no difficulty about giving it an unambiguous definition, and I am uneasy about discussions as to whether "now" is like a point or has width, since the question seems wholly devoid of meaning within my own discipline. I think that the so-called passage of time, however one chooses to define it, has such a questionable status in experience, in logic, and in epistemology that its occur-

# The Myth of the Passage of Time

rence in any philosophical discussion that does not carefully label it as a subjective impression is a mistake. Parenthetically, I am just as skeptical about the word "now" and the word "becoming"; this will become clear even as I stick to my subject.

I have several tasks before me: I must show the superfluity of the idea of the advancing present in those situations in which it is often considered necessary, and express in other language the essential physical content of what Newton said in his famous Scholium on Absolute Time. Finally, since the idea of the passage of time, if not the phenomenon, has an air of reality about it, I must try to see whether the idea refers in any way to an analyzable physical process, in the brain or elsewhere, for this ought to show us how to explain its origin.

I am going to make use of two different representations of ordinary temporal events: an atemporal one and another, involving time, that corresponds more closely to our intuitive temporal way of understanding. These representations are somewhat abstract and must be held separate in the mind with great care. Because of the conceptual difficulties to come I am going to present these representations by the use of diagrams, but the diagrams are not the representations; I could use mathematical or even verbal formulations.

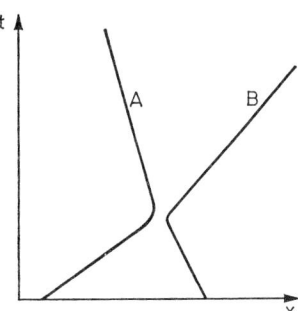

Fig. 1
Two objects move close to each other, interact, and finally move apart. The straight line segments represent motion at constant velocity

The diagram underlying both representations is a graph of events in which time is represented on the vertical axis and a spatial dimension on the horizontal one. It is called a Minkowski diagram.[1] Fig. 1 shows an example, the representation of two objects or people that move towards each other, interact, and then separate. Note that before the interaction A moves faster than B, whereas after it the reverse is true. On this diagram, before and after mean simply below and above with respect to the time axis, in accordance with the usual convention in drawing graphs. A and B on the diagram might be two people encountering each other in the street and then separating again; the lines could then be labelled to show their states of consciousness, the instants at which

---

[1] Minkowski used diagrams of this kind to represent the four-dimensional structure of space-time in his formulation of special relativity. Our considerations, however, have nothing to do with the theory of relativity.

they speak to each other, the passage of the sound waves between them, and so on. The diagram enables one to answer the question "What is the state of affairs at time $t$?" In contains no element corresponding to any particular "now" or to the passage of time. The diagram, as a diagram, is static; the representation which it is designed to help one to visualize is atemporal in that it, too, contains no special present instant and cannot express our intuition that this present instant somehow advances. I distinguish between the *static* diagram and the *atemporal* representation because to call the representation static would be to impose on it, superfluously, the subjective sense of time in the mind of the person considering the representation. The representation has, on the contrary, nothing at all to do with anybody's sense of time.

The temporal word present does not refer to any aspect of the atemporal representation. The Minkowski diagram is simply a display, where we have omitted or coalesced two of the spatial dimensions in order to draw it, of a certain set of events.[2]

Now, let us show this diagram, this visible, tangible, temporary drawing, to some friend and ask him to read from it the course of the events it describes, starting at a particular time. He will start reading upwards on the diagram at an even rate: "Now A is here, now it has gone here," and so on. This is the situation considered by Weyl, in which the speaker's consciousness is described as crawling along the world-line representing his body (as a spark burns along a fuse), although when read carefully, Weyl's often-quoted description seems to fall between my two representations. I call this kind of picture an *animated Minkowski diagram,* and the representation underlying this kind of diagram corresponds to the intuitive idea of advancing time as it is usually imagined and expressed. I will call it the temporal representation.

If we represent our own experience on an animated diagram, the part of it above the level referring to the present instant must be left blank, to be filled in at a uniform rate as our experience continues. If the diagram represents someone else's earlier experience it need not be continuously filled in, since we know what was going to happen to him.

These are my two representations of time. A reading of the famous passage in the *Timaeus*[3] that analyzes the nature of time convinces me that Plato understood and distinguished the two representations. When he says that time is the moving image of eternity, I think his argument makes it quite clear he means it just as we might say that the animated Minkowski diagram is the moving image of the atemporal representation.

---

2 Parenthetically, I would like to deal with a difficulty that is sometimes raised: how does one draw the diagram if one insists on representing the entire 3-dimensional configuration of things as a function of time? If one insists on doing this one must use projective techniques, as when making a drawing of a scene that has depth. It is more convenient to stop drawing graphs and express the content of the atemporal representation mathematically.

3 *Timaeus* 37–38.

I now want to make the vital point that the animated diagram may be more intuitive, or more picturesque, or make better cinema than the atemporal one, but that it contains no more specific, verifiable information. All of the science of dynamics, that is, all we know about how complex systems (including ourselves) behave and interact, is already represented on the atemporal Minkowski diagram. And what can one say about the present instant or the passage of time that is specific or scientifically verifiable?

If one wishes to define now as the moment when we are in contact with our surroundings, there is no experimental way of proving that there is such a moment or any way to put it onto the atemporal diagram. We must use the animated one, and there, if we introduce a second consciousness in communication with the first, we have to explain how they happen to have reached the same level at the same instant so that their now's correspond. In the atemporal representation the question does not occur.

If we animate the Minkowski diagram we introduce, quite properly, a second time scale, for in order to avoid confusion we must realize that the time scale of the animated diagram is purely arbitrary and has no necessary connection with the temporal behavior the diagram represents. In terms of this second scale we can now tell how quickly time advances on the animated diagram: at a rate of so many seconds per second. It has been claimed that this "second time" is of a different kind from the $t$ on the Minkowski diagram; further, the attempt to systematize the situation by introducing a second diagram that shows someone observing the first one, and so on, leads finally to an infinite regress of further times. I think this all rests on a confusion between the animated diagram and the system it represents: the various times, if one wishes to use them, are of exactly the same kind and no new conceptual apparatus has been introduced.

Now let us use the atemporal diagram for a purpose which I think can ultimately be very fruitful, the representation of processes of perception and thought.

I do not know the best way to describe processes in the human brain, but one way, in principle, is to diagram the history of its interacting particles. Even though we may not know how to read the diagram, all the physical aspect of experience is represented there, somehow, and such a diagram, *a complete description* of the perceptive and cerebral processes that underly our conscious life, is atemporal and contains no elements of description corresponding to the ideas of now or the passage of time. Consciousness remains a mystery, but clearly the passage of time and other ideas associated with it are of such uncertain logical status that should at least try to conduct careful discussions of perception and intellectual experience without their use, that is, in the atemporal representation. Mathematics contains such a representation and so does logic: logical notation, as I understand it, is essentially atemporal (though of course it can, like mathematical notation, be used to analyze temporal events) but no expression of temporal ideas such as past, present, future, or becoming in logical terms should be taken as self-explanatory; it takes an open-eyed analysis to bridge the gap from the atemporal to the atemporal. Anyone interested in doing this should start with the *Timaeus,* for Plato sees exactly what the problem is and as good as warns me that if on an atemporal diagram I label the vertical scale $t$ and call it time, I am likely to get us all into trouble. I can give some justification for not having followed his advice, but he may be right even so.

A example of the errors that one can make if one is not careful is a common objection to the representation of events on a static Minkowski diagram: that it assumes a complete determinism, since the world lines as drawn now do not end but extend into the future where, even though we cannot see them, they predict what will happen. I have tried to explain myself earlier in such a way that this objection sounds like nonsense, the result of trying to give a temporal interpretation to a representation that exists out of time. The mistake is to confuse the static diagram, which of course exists in time like any other man-made object, with the atemporal representation it is intended to illustrate. I do not think that this particular objection can be rephrased in a way which makes it any the less nonsense.

I am shortly going to use diagrams to make assertions, or at least formulate questions, on some of the problems I have so far avoided. A diagram is a form of written language; it conveys meaning in something the same way as primitive Chinese calligraphy, part representation and part metaphor. I do not know how to state the difference between representation and metaphor, though it seems to me that some statements about he world are more metaphorical than others. I do not know the connection between symbols and experience, or even how to formulate the relevant questions, since questions involve written or spoken words, diagrams, equations, or gestures, all of which, as they refer to experience, are symbolic. I consider my diagram no more or less symbolic than any other form of description. It is simply a way of making statements. I say this because the atemporal Minkowski diagram represents time by making a picture containing a pictorial element labelled $t$. It spatializes time. As a matter of fact the theory of relativity shows that, within certain limits, no absolute distinction can be made between spatial and temporal components of an interval, on this diagram or in experience. The spatialization is convenient for all of physics and for certain parts of the present discussion, but though I have shown that it is possible to represent some of the experiences of human consciousness in this way, I do not know how the mind works. Thus I cannot exhibit our conscious experience of time, what Bergson calls true duration, on this diagram, but must use the animated form, with its now and its coming to be. That is, I cannot explain how true duration can be spatialized.

Let me stop for a minute to discuss Newton's scholium on absolute time, for it says something about the passage of time that clearly reflects our temporal experience. I think that for Newton the important evidence was the observation of periodic phenomena such as the swing of pendulums and the revolution of the planets. These seem to our senses to proceed at an even rate. Further, the mathematical theory shows that also with respect to the variable $t$ occurring in the equations they proceed at an even rate. Therefore, in some sense, the variable $t$ proceeds with respect to our senses at an even rate. This can be rephrased simply and exactly: the mental and sensory processes by which we estimate intervals of time are ultimately mechanical in nature and are governed by exactly the same dynamical equations as are the motions of planets and pendulums. Since the equations are the same, the variable $t$ occurs in every context with the same fundamental meaning, and it is hardly surprising that we tend, unless disturbed, to keep time with the planets.

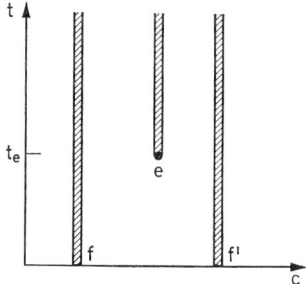

Fig. 2
Schematic diagram of the function of memory. The event $e$, occurring at time $t_e$, leaves a trace in the memory subsequent to its occurrence. The traces $f$ and $f'$ represent traces already present in the memory. (The diagram should not be taken to imply that these traces are localized somewhere in the brain)

I come now to the problem of explaining our experience of time, and since what I have to say is very speculative it can also be brief. I have characterized the atemporal Minkowski picture as the fundamental one because I believe it completely describes the physical processes underlying our experiences. Let us make a static diagram of a brain receiving sensations, Fig. 2. Here $t$ is the time axis, $c$ is some coordinate, not spatial, that distinguishes one mental impression from another. An event $e$ leaves a memory trace extending towards the future; there are of course other memory traces, $f$, $f'$, etc., corresponding to facts learned in the past. Prior to its occurrence, the event $e$ may be quite unanticipated. (Prior means values of $t$ less than $t_e$, "be" is a tenseless present.) Why do the consequences of $e$ stream upwards on the diagram? That at least can be answered: the establishment of a memory trace is an entropy-increasing biochemical process and one can prove from statistical mechanics the second law of thermodynamics which requires that the memory trace, whatever it may actually consist of, extend upwards. In fact, a map of events recorded in our brains is more an entropy map than a time map. That is, there is no record of time as such but only of entropy-increasing processes. Our mental processes are connected, not very remotely, to the evolution of the cosmos, and the memory trace takes its direction from that in a way we understand by following the path of energy from the sun to the earth to plants and animals and coal and human food and warmth. Thus memory, thus the freshness of unexpected events.

Now let us look for the present on a static Minkowski diagram. To define the present as the instant at which we are in contact with our surroundings assumes that there is an instant that is distinguished from other instants by this fact. But my earlier argument on time travel shows that in the world of physics there is no such special instant, and one can see on the Minkowski diagram that we are in contact with our surroundings at all instants while we are alive. Thus the definition is without content because it does not exclude any instant of our lives. The explanation of the passage of time is, I think, that at every moment we are in contact with our surroundings. Our conscious history can be represented on a suitable static diagram, extended alongside other elements of the diagram that represent the various physical events, internal and external to ourselves, by which we measure time. On the afternoon I write these words, certain events are occurring outside my window and inside my body which are peculiar to this afternoon. These events can be taken as marking the time of day and year. I cannot hope to look out and see events identifiable with last year or next year because the consciousness of which I speak is that represented on this part of the diagram and not on some other. Light waves travel very quickly to me from the things I see; their lines on the diagram are almost horizontal and establish a necessary near-simultaneity between my sense impressions and the events causing them. That is the way in which I am bound to this particular instant of time and am not allowed with my present mind to experience any others. I can remember what has already happened but I cannot foresee what will happen next. At every point along my world line on a static diagram the past and the future are similarly unremembered and unforeseen. The atemporal representation is the one in which these relations are evident, but they are not easily explained in words, since the language of ordinary explanation embodies in its structure the temporal assumptions I wish to do without. The language of mathematics, on the other hand, is entirely atemporal. Thus I shall follow the example of Pythagoras, who drew a diagram of a triangle and three squares and wrote below it "Behold!" He left the rest to his readers, as I must do.

I do not know the best way to define the present; it depends on how the definition is to be used, but *in the atemporal representation,* the definition might begin "The present is an arbitrary instant around which we organize our conception of time". Perception is determined by the physical laws governing the brain; it is therefore temporal. These laws are governed by thermodynamics; it is therefore progressive in time. These are the two main qualitative features of our intuitive picture of our own perception of time, and I do not see how more can be said until we have an analyzable physical model of mental processes. This is clearly one of the main concerns of physics in the near future. An obstacle to be overcome is the use of loose and intuitive definitions; part of the confusion that clouds this area would be removed by the adoption of a definition of the word consciousness that takes account of the distinctions to be made with regard to time. Obviously, consciousness must be defined differently in the two representations I have described. I would tend to define it so as to make Weyl's description automatically true – that is, to include its exploratory nature. To transform such a definition into the atemporal representation so that it could then be rigorously used in physics would not be difficult but would take a little thought.

It is clear from what I have said that for some purposes I prefer the atemporal representation to the temporal one. Perhaps I can explain my preference by referring to another case in which there are two ways of picturing the same phenomena, the heliocentric and the geocentric pictures of the solar system. These pictures are of course fully equivalent kinematically, since everybody from Copernicus onward has recognized that our senses enable us to establish the relative motions of sun and planets but provide no absolute standards of rest or motion for either. Still, the heliocentric system is preferred, and the reason for this preference lies in physics: if one adopts the heliocentric picture, the motion of an object with no forces acting on it is (nearly) in a straight line at constant speed, whereas in the geocentric system it is a great looped spiral with a double periodicity of one terrestrial year and one terrestrial day. Other laws of nature are similarly changed and thus the presumably accidental parameters of the Earth's motion appear in the explanation of phenomena that have nothing to do with the Earth. The heliocentric picture is chosen for the simplicity of the dynamical laws by which it is governed, that is, the simplicity with which we can perform in it the calculations of dynamics. The further implications of this choice with respect to the philosophy of man are so wide and deep that for most nonscientific people the matter of dynamical equations would seem quite unimportant by comparison, but the distinction between the two pictures is based there.

My preference for the atemporal representation rests first on its logical economy and second on an exactly analogous situation in physics. There is a class of phenomena which receives an explanation in the atemporal representation that is simpler than and to me preferable to that in the temporal one. I refer to Feynman's discovery[4] in 1949 that a positron (or any antiparticle) can be represented in the atemporal representation as an electron (or other particle) that has a world line whose direction (as determined by the line's continuity with other lies with which it joins) runs downwards on the diagram, from future towards past (Fig. 3). Feynman showed that not only is this picture in accord with the experimental facts, but that also the mathematical theory derivable from it gives the same results as one obtains from the theory based on particles and antiparticles in the temporal representation, and in a far shorter and simpler way.

---

4 Feynman, R. P.: Phys. Rev. 76 (1949) 749. The introductory discussion is accessible to the nonspecialist. The same idea had been suggested but not exploited earlier by E. G. C. Stueckelberg.

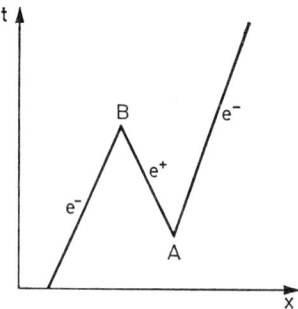

Fig. 3
Feynman diagram of a process in which an electron is present initially, an electron-positron pair is produced somehow in an event A, the positron encounters the original electron and annihilates it in event B, and the newly produced electron goes on alone. If the world line segment BA, directed backwards in time, is interpreted to represent a positron, the entire process can be viewed as the path of a single electron that suffers two interactions

Most phenomena of daily life are explained just the same, for practical purposes, in terms of the geocentric picture as in terms of the heliocentric one. The crucial distinction is made in those phenomena which are not. That the Feynman argument implies a similarly radical distinction between two representations of time was recognized by a few sensitive physicists, among them Y. Nambu,[5] who wrote:

His ingenious method is indeed attractive, not only because of its intuitive procedure which enables one to picture to oneself the complicated interactions of elementary particles, its ease and the relativistic correctness with which one can calculate the necessary matrix elements or transition probabilities, but also because of its way of thinking which seems somewhat strange at first look and resists our minds that are accustomed to causal laws. According to the new standpoint, one looks upon the world in its four-dimensional entirety. A phenomenon that will come into play in this theatre is now laid out beforehand in full detail from immemorial past to ultimate future and one investigates the whole of it at a glance. The time itself loses sense as the indicator of the development of phenomena; there are particles which flow down as well as up the stream of time; the eventual creation and annihilation of pairs that may occur now and then is no creation or annihilation, but only a change of direction of moving particles, from past to future, or from future to past...

It seems to me far more important that the two ways of representing events in time differ, if only at one or two points, than that they are the same at all other points, for this is a clue (unless further research shows it is a false clue) that there exists a way of organizing our experience of time which is different from that we have received from our ancesters and may lead us to insights that were beyond their reach.

I am aware that when people refer to the mysteries of time they may be speaking of any of a great variety of problems. For myself, the mysteries began

---

5 Nambu, Y.: Progr. Theor. Phys. (Kyoto) 5, (1950) 82.

in the private and unverifiable nature of some of our most definite intuitions of time. When I began to understand how these intuitions are psychological by-products, as one might say, of a situation that can actually be explained very simply, they became less mysterious. For example, I found that Aristotle's problem of the sea-battle and what I understand of the logical difficulties surrounding the idea of becoming are problems that belong to the temporal representation. In the atemporal one they are easily settled by appropriate definitions. Probably, for some people, the arguments of this paper will resolve none of the difficulties that confuse them, but for me, now that the work is done, time is exactly as mysterious as space, no more and no less.

# Time Asymmetry, Time Reversal, and Irreversibility

M. BUNGE*

Of all the important ideas those of being and of time seem to have been muddled throughout their history. And of the two, the second seems always to have won the confusion derby. Much of the time muddles come from conceiving of time as something that flows, i.e. from identifying flux and time rather than construing time as the step of becoming. That reification of time, so typical of ordinary knowledge, comes down from archaic thought – and this not only in the Orient and in the Mediterranean. Thus the Mayas, who created a dynamical world view, believed that the gods carried both the sun and time on their backs, and they designated time, the sun and the day, with the same word, namely *kinh* (León-Portilla [8]). The conceptions of time as thing and as process are of course found in a number of locutions in the Indo-European languages, suggesting that time flows, flees and even fleeces us; that it can be lost and found, stolen and gained; that it is ever in a hurry to pass from past to future; that it can cause birth and death. And, at least in Spanish, time can also be killed and must even be given time (*Hay que dar tiempo al tiempo*).

However, confusions and absurdities concerning time are not exclusive of archaic thinking and its fossil record – ordinary language. Time pathologists and teratologists should have a good time examining certain confusions conspicuous in contemporary scientific literature. One such confusion, perhaps the most harmful of all, is the conflation of three quite distinct ideas huddled under the umbrella of the so-called "arrow of time": time asymmetry, non-invariance under time reversal, and irreversibility. Let us try to clear up this confusion even at the risk of error.

## 1. Time Anisotropy

Everyone seems to agree that time is asymmetrical or anisotropic but it is by no means clear what is meant by this. In order to clarify this idea we need a definite theory of time. We shall avail ourselves of a theory of time proposed elsewhere (Bunge [3], [4]) and which formalizes the ancient insight that time is but the pace of becoming. (For forerunners see Whitehead [13], Russell [11] and Noll [9].)

This theory analyzes the concept of local time function $T$ in terms of three undefined concepts that may be elucidated in other contexts, namely those of

---

* Mario Bunge, Professor of Philosophy, McGill University, Montreal, Canada.

event ($E$), reference frame ($K$) and chronometric scale ($S$). The postulates of this theory are:

A 1. *There are events* [changes of state of systems], *physical* [not just geometrical] *reference frames, and chronometric scales* [ways of mapping durations onto numbers].

A 2. *$T$ is a real valued function on the set of all ordered quadruples $(e, e', k, s)$, where $e$ and $e'$ are in $E$, $k$ is in $K$, and $s$ is in $S$.*

A 3. *The set of events is compact* [in the sense that, for any $e$ in $E$, every $k$ in $K$, each $s$ in $S$, and any given $t$ in the real line $R$, there exists a second event $e'$ such that $T(e, e', k, s) = t$].

A 4. Contiguous durations are additive [i.e. for any $e$, $e'$ and $e''$ in $E$, relative to a fixed frame $k$ and on a given scale $s$: $T(e, e', k, s) + T(e', e'', k, s) = T(e, e'', k, s)$].

These axioms determine the concept of local time as anchored in real change and as relative to some frame. They entail, among others, the following result pertinent to the question of time anisotropy:

*Theorem on time direction.* Duration is an oriented interval. [That is, for any two events $e$, $e'$ in a given frame $k$ and on a given scale $s$: $T(e, e', k, s) = -T(e', e, k, s)$]

*Proof:* By Axiom 4. [Set $e'' = e$ and take into account that, by the very same postulate, $T(e, e, k, s) = 0$.]

In short, the local time function is odd in the events: this is all there is to the asymmetry of anisotropy of time. In particular, the above is not the statement that the future is unlike the past because there are irreversible and unique events and processes. Indeed, the local time function $T$ is the same whether the set $E$ of events is composed of reversible or of irreversible events, of repeatable events or unique ones. In particular, the function $T$ is the same for particle mechanics and thermostatics (both of which deal with reversible processes), for continuum mechanics and irreversible thermodynamics. Which is just as well for, if every theory had its own peculiar time concept, inter theory comparison would be hardly possible.

The preceding theorem says that the time lapse between two non simultaneous events $e$ and $e'$ is oriented in the direction opposite to the time interval between $e'$ and $e$. But this does not tell us which of the two events comes first. We can surely add a convention relying the concept of order (of events) to that of directionality (of duration). But any such convention will take it for granted that we have independent criteria for ascertaining which of two non simultaneous events comes first. The standard convention is, of course, the following:

*Definition.* $e$ is *earlier* than $e'$, relative to $k$ and $s$, if and only if $T(e, e', k, s) > 0$.

This is not a law of nature: nothing but tradition prevents us from inverting the inequality. In other words, the asymmetry of time, as expressed by the above theorem, is a fact, but the decision to count time forwards, i.e. in the direction of coming events, is arbitrary. Put in metaphorical terms: nature tells us that time "flows", but not whither. Better: time has no arrow built into it. Arrows must be sought in whole processes not in one of the features of processes.

## 2. Time Reversal

Time reversal consists in the inversion of the sign of the time variable or coordinate. This is a mathematical operation. In order to find out its physical meaning, if any, it may be helpful to clarify the notion of physical coordinate as distinct from the one of mathematical coordinate.

While in elementary physics one introduces coordinates right at the beginning, in the foundations of physics coordinates should appear, if at all, at a late stage, once the nature of the underlying manifolds, and possibly also of the basic laws, have been characterized in a coordinate free way. This method staves off unnecessarily restrictive geometrical assumptions and it de-emphasizes the importance of the viewpoint and of the problem solving techniques while bringing out what belongs to nature rather than to our representations of it. Moreover, if one believes that space and time are neither things nor a priori intuitions but rather certain relations among things and their changes (i.e. events), he will start with reality itself – i.e. the collection of things, or the collection of events – as the object a physical geometry should map.

One may think of reality, or the physical space, as the set $E$ of events or as the collection of elementary physical objects such as point particles and points in an electromagnetic wave front. A way of conceptualizing reality is to map $E$ on a geometrical space $G$. That is, we assume that there is a function $\gamma$ on $E$ to $G$ enabling us to handle the conceptual image $g = \gamma(e)$ of every element $e$ of $E$. (The map $\gamma$ is an injection but not a surjection: different physical points must be assigned different geometrical points but there may be geometrical points without a physical correlate. Hence $\gamma$ has a left inverse. The inverse we are interested in is the representation function $\triangleq : G \to E$ discussed in Bunge [5].)

The geometrization of the world must be made more precise than this: we have to represent the "distance" (spatial, temporal, spatiotemporal, or other) between any two physical points $e$ and $e'$. This can be done by introducing a suitable function $\delta$ assigning every ordered pair $(e, e')$ the difference $g-g'$ of two vectors in $G$, which difference is a third point in $G$. Briefly, $\delta: E \times E \to G$ with $(e, e') \mapsto g-g' \in G$. A few other axioms (e.g. "If $e = e'$ then $g - g' = 0$") will determine this physical separation function $\delta$ uniquely. The geometrization of $E$, sketched so far, can be summed up in the following commutative diagram, where $p$ stands for the projection of $E \times E = E^2$ into any of its factors. Since $\delta = p\gamma$, the axioms for $\delta$ determine the geometrical image $G$ of $E$.

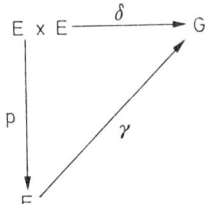

Fig. 1. Mapping pairs of events on geometrical points

The next step consists in fully arithmetizing $E$, in the sense of mapping its geometrical image $G$ on some number field, usually the real line $R$. (More precisely, we introduce a function $\alpha$ on $G$ to $R^4$.) Arithmetization may involve coordinatization. Since there are two spaces involved, $E$ and $G$, there will be two sets of coordinates: physical and geometrical. The $i$th geometrical coordinate is a function $f_i$ that assigns every point of a vector space (like $G$) an element of a field (e.g. $R$) in such a way that $f_i(b_j) = \delta_{ij}$, where $b_j$ is the jth base vector and $\delta_{ij}$ the Kronecker delta. Assuming $E$, hence $G$, to be four dimensional, we have the

*geometrical coordinates* $x_i: G \rightarrow R$ with $i = 1, 2, 3, 4$,

which fix the position of any point in the geometric space $G$. But we have in addition four functions that may be called the

*physical coordinates* $X_i: E \rightarrow R$ with $i = 1, 2, 3, 4$,

which identify any point in the physical space $E$. (Unless $E$ can be assumed to be a vector space, these will not be coordinates in the mathematical sense. Nor are the particle coordinates employed in mechanics: these are real valued functions on $\Sigma \times K \times T$, where $\Sigma$ is the set of particles, $K$ the set of inertial frames, and $T$ the range of the time function.) The $x_i$ and $X_i$ are different functions even though their values coincide at the points of $G$ that happen to have an image in $E$. In factual science, the $x_i$ help to describe what goes on *at* a geometrical point $g \in G$, while the $X_i$ help describing what happens *to* a physical element $e \in E$. (These two modes of description, though different, are equivalent: see Truesdell [10].) The two sets of coordinates are related by the composition $X_i = x_i \gamma$.

The arithmetization can be carried further, by metricizing both $E$ and $G$. There will be two metric forms, one for each space: $d_E$ and $d_G$. The line element $d_E s$ will contain the physical coordinates $X_i$, while the geometrical line element $d_G s$ will contain the geometrical coordinates $x_i$. For example, in special relativity the Lorentz metric concerns the physical space while the Minkowski metric concerns the geometrical space. Mathematicians and physicists often talk through each other about these matters because they talk about different kinds of coordinates, hence different metric forms.

The following diagram summarizes the formalization process sketched heretofore.

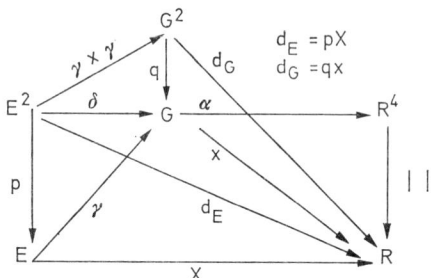

Fig. 2. Coordinatizations: geometrical and physical

The geometrical coordinates $x_i$ have no foot on the space $E$ of events, hence they are physically meaningless. Only the physical coordinates $X_i = x_i \gamma$ are factually meaningful, as they have one leg on $E$, i.e. $X_i \in E \times R$. This holds, in particular, for $x_4$ and $X_4$: only the latter is a time coordinate. For example, in special relativity the fourth coordinate that occurs in the Lorentz metric is a time coordinate; also the proper time is a time coordinate. On the other hand $x_4$ may be regarded as that function whose square, added to the sum of the squares of the spatial coordinates, yields a rotation invariant.

Assuming then that $X_4$ is a time coordinate, what does the inversion of its sign mean? The answer depends on whether we look at the substratum $E$ of $X_4$ or at the role $X_4$ plays in law statements (equations of motion, field equations or constitutive equations). Let us answer each question at a time.

Consider $X_4$ as a mapping on the physical space $E$ to the real line. $E$ may be regarded as a "pointed" set, i.e. one with a distinguished element $e_0$ interpretable as the initial event of a series, and thus as the physical correlate of the zero of the time coordinate. This enables us to identify $X_4$ with the restriction of the local time function $T$ (Section 1) to the set $E \times \{e_0\} \times \{k\} \times \{s\}$, where $k$ and $s$ are the same frame and scale involved in the definition of $X_4$. In simpler terms, $X_4(e) = T(e, e_0, k, s) = t$. Now, by the theorem on time anisotropy, recalled in Section 1, $T(e, e_0, k, s) = -T(e_0, e, k, s)$. Hence the inversion of the sign of $t$ corresponds to an inversion of the order of appearance of the underlying pair events: if formerely $e_0$ preceded $e$, now it is the other way around. In short, *time reversal corresponds to process reversal*.

What about the inversion of $t$ in a law statement $L(t)$ such as an equation of motion or a field equation? There are two possibilities: either $L(-t) = L(t)$ or the equality does not hold. If it does, one says that $L(t)$ is invariant under time reversal, or briefly *T-invariant*. All basic (high level law) microphysical statements known to date are T-invariant, but they may have logical consequences (low level laws) lacking that property. Just as before, a time reversal – with or without $T$-invariance of the law statement concerned – is no indication of the backward "flow" of time. The time inverse of a process described by a $T$-invariant law statement is just the reverse of the original process – e.g. a motion with inverted velocities and spins. Such a process goes "forward in time" as much as the original process does: only, one chooses to describe it as the time inverse of $L(t)$ provided this formula is identical with $L(-t)$.

If a law statement fails to be $T$-invariant then it concerns irreversible processes. But if it is $T$-invariant then it may or may not refer to reversible processes, depending on the accompanying formulas. Thus Maxwell's equations are $T$-invariant but, if conjoined with the condition of outward radiation (no incoming waves), they describe the irreversible propagation of an outgoing (retarded) wave. In short, $T$-invariance has nothing to do with a return to the past and it is not an unambiguous indicator of reversibility. But reversibility belongs in the next Section.

## *3. Irreversibility*

Reversibility is a property of certain processes, most microphysical. A reversible process is, strictly speaking, one in which both the system concerned and its surrounding can be restored to their original condition. Now, every law statement concerns some system $S_1$ of a certain kind under the action of some environment $S_2$ (eventually the null environment or free space). This environment, whether constant or changing, is not supposed to change appreciably under the reaction of the system $S_1$: the law statement focuses on the latter and neglects the feed-back. When the reactions are taken into account, the law statements concern a third system composed of $S_1$ and $S_2$, which will be immersed in a further system $S_3$ that will in turn be regarded as large enough so as to disregard any changes produced by the combined system. In this way a Chinese pill-box is built. The outer box is the universe, but for practical purposes it may be taken to be the immediate environment of the system we happen to be interested in.

This being so it is no wonder that irreversibility may arise even if the system's laws are $T$-invariant. Although the emergence of irreversibility out of smaller scale reversible process has been called a paradox, there is nothing paradoxical about it if one recalls that laws determine processes only partially. Indeed, a process or history is determined jointly by a set of laws and a set of constraints, initial conditions, boundary conditions and other subsidiary hypotheses representing particular circumstances of both the system and its environment. In particular, boundary conditions, so important in continuum physics, field physics and quantum mechanics, constitute really a schematic or black-box representation of the state of the environment. (In the case of materials with memory one may have to add a good stretch of the previous history of the system. Because the stress tensor of any material system is determined by both the present and past deformations of the body, continuum mechanics, electrodynamics and thermodynamics cannot be $T$-invariant in general.) This explains why $T$-invariant theories can account for irreversible processes like radioactivity and wave propagation. With physical systems, just as with persons, the actual career is determined by a bunch of laws and by the *concours de circonstances*.

For this reason the belief that reversibility is more fundamental than irreversibility is erroneous. (On the other hand it does seem that $T$-invariance is more basic than its opposite. At least, there is hope of explaining $T$ noninvariant laws, such as Fourier's equation of heat transfer, in terms of $T$-invariant laws and certain subsidiary conditions.) A particularly mistaken if appealing way of trying to save reversibility and disparage irreversibility is to assume that, in the long run, all processes are reversible: i.e. that irreversibility is a delusion characteristic of short-lived beings. It is fancied that, to a being whose life span were several Poincaré cycles, every process would appear as reversible – the Stoic dream of eternal recurrence popularized by Nietzsche. To begin with, the clause "over sufficiently long time intervals" (or "if one could wait long enough") renders the hypothesis practically irrefutable, for every instance of irreversibility could be dismissed as an evidence that one has not waited long enough. Secondly, the hypothesis (like most of the discussions of the reversibility-irreversibility conundrum) presupposes that everything consists of a bunch of structureless Newtonian particles, so that quantum-mechanical potential barriers, boundary conditions, and constraints such as the one of outward radiation play no role.

This latter assumption is to be blamed for the usual focusing on the $T$-invariance of the basic equations with neglect of the subsidiary assumptions usually responsible for irreversibility. No such mistake might have been made if classical mechanics had been identified with continuum mechanics – where boundary conditions, constraints and often even past histories play a paramount role – or if the existence of irreversible microphysical processes, such as radioactivity, had been recalled.

Another sobering thought is that in most cases a randomly acting environment will produce irreversible processes quite irrespective of the $T$-invariance of the basic laws concerned. (See Bergmann and Leibowitz [1] and Blatt [2].) Since every real system short of the whole universe is under the action of random perturbations, reversibility stands no chance in the long run. Therefore the assumption that if every system were left to itself it would evolve in an entirely reversible manner (Gold [6]) is an idle contrary to fact statement.

In conclusion $T$-invariance and reversibility, though related, are distinct: whereas the former concerns laws (or rather law statements) reversibility can be predicated only of processes. Further, the two are inequivalent: the $T$-invariance of laws is necessary but insufficient for the reversibility of processes. That is, if a process is reversible then it satisfies $T$-invariant laws but not conversely – notwithstanding the authority of Prigogine [9].

We seem to know a few conditions that are severally sufficient for irreversibility: $T$-non-invariant laws, semipermeable boundaries (such as nuclear potential barriers), or the coupling of the system with a randomly acting environment. But we do not know yet of any necessary and sufficient condition for irreversibility, nor do we know whether there is a universal condition of that kind, i.e. one holding for systems of all kinds. And we are not likely to find any necessary, sufficient and universal condition so long as we center our research on the very special and artificial case of Newtonian particles – if only because, for these, the reversibility-irreversibility distinction is a matter of degree (Grad [7]). Nor are we likely to succeed in that search if we stick to entropy increase as the sole irreversibility criterion: first, because there are a number of entropy functions (both in thermodynamics and in statistical mechanics), not all of which are steadily increasing; second, because there are microphysical (hence neither entropic nor nonentropic) processes that are irreversible – e.g. the decay of short-lived elementary particles.

*4. Conclusions*

Here goes the upshot of our discussion:

1(a) The asymmetry or anisotropy of time consists in that durations are oriented intervals – or, equivalently, that the local time function is odd with respect to events.

1(b). The anisotropy of time is independent of causality, $T$-non-invariance, and irreversibility. Both $T$-invariant laws and reversible processes are described with the help of an asymmetric time function.

2(a). $T$-invariance (permanence under time reversal) is a property of certain statements, in particular law statements, containing the time variable.

2(b). $T$-invariance entails the possibility of process reversal (e.g. motion inversion), not the reversal of the direction of time.

3(a). Irreversibility concerns certain processes, not their laws, let alone time.

3(b). If a process is reversible then its laws are $T$-invariant but not conversely: $T$-invariance is only necessary for reversibility. Processes being determined jointly by laws and circumstances, $T$-invariance is compatible with irreversibility.

If 1(a), 2(a) and 3(a) are true, it is misleading to speak of the "arrow of time", for this picturesque expression covers quite distinct ideas. And if 1(b), 2(b) and 3(b) are true, then although time is but an aspect of process – to the point that an uneventful universe would be timeless – the description of a process presupposes some idea of time – the clearer and less metaphorical the better.

*Acknowledgement:* This paper has been supported by the Canada Council Research Grant 69-0300.

*References*

1. Bergmann, P., Leibowitz, J. L.: *New Approach to Nonequilibrium Processes.* Phys. Rev. *99* (1955) 578–587.
2. Blatt, J. M.: *An Alternative Approach to the Ergodic Problem.* Progr. Theor. Phys. *22* (1959) 745–756.
3. Bunge, M.: *Physical time: the objective and relational theory.* Phil. Sci. *35* (1968) 355–388.
4. — *Physique et métaphysique du temps.* Proc. XIVth Intern. Congress of Philosophy, I. Wien: Herder 1968.
5. — Foundations of Physics. Berlin–Heidelberg–New York: Springer 1967.
6. Gold, T.: *The arrow of time.* Am. J. Phys. *30* (1962) 403–410.
7. Grad, H.: *Levels of description in statistical mechanics and thermodynamics.* In: M. Bunge (Ed.), Delaware Seminar in the Foundations of Physics. Berlin–Heidelberg–New York: Springer 1967.
8. León-Portilla, M.: *Tiempo y realidad en el pensamiento maya.* México: Universidad Nacional Autónoma de México 1968.
9. Noll, W.: *Space-time structures in classical mechanics.* In: M. Bunge (Ed.), Delaware Seminar in the Foundations of Physics. Berlin–Heidelberg–New York: Springer 1967.
10. Prigogine, I.: *Introduction to Thermodynamics of Irreversible Processes,* p. 14. New York: Interscience 1961.
11. Russell, B.: *Our Knowledge of the External World.* London: Allen & Unwin 1952.
12. Truesdell, C.: *The Elements of Continuum Mechanics.* Berlin–Heidelberg–New York: Springer 1967.
13. Whitehead, A. N.: *The Principles of Natural Knowledge.* Cambridge: Cambridge University Press 1919.

# No Paradox in the Theory of Time Anisotropy*

O. COSTA DE BEAUREGARD**

*Summary.* Factlike rather than lawlike character of the physical irreversibility principle expressed by means of Bayes' principle, and considered as a boundary condition for integrating the macroscopic evolution equations. The Ritz-Einstein controversy reexamined and J. von Neumann's irreversibility of quantum measurements reworded: one to one connection between the principles of statistical irreversibility and of wave retardation in the theory of quantized waves. Factlike time asymmetry and lawlike time symmetry in the cybernetic context.

Before we formulate (in accord with quite a few recent authors) the answer we feel appropriate to the question in the title, it is useful to understand why it has been felt that the theory of the time anisotropy contains a paradox.

## 1. The Paradox behind the Loschmidt and Zermelo Paradoxes: Time Anisotropy in the "Principle of Probability of Causes"

The well known Loschmidt and Zermelo "paradoxes" in statistical mechanics have merely uncovered the existence of a much older "paradox" inherent in the probability theory itself since the early days of Pascal, Fermat and Bayes, where it came to be named, very significantly, the "principle of probability of *causes*". That is to say, it was obscurely felt that the time anisotropy inherent in the anthropomorphic notion of a "cause" developing aftereffects rather than before effects is somehow connected with the empirical fact that, *even if the transition probabilities between two possible states of a system are symmetric* (as in such classical examples as card shuffling) more probable macroscopic complexions follow in time less probable ones – not the other way.

As Watanabe puts it, the *empirical fact* is that *blind statistical prediction is physical while blind statistical retrodiction is not* – a situation with which probability theory copes by using in retrodictive problems Bayes' formula for conditional probability. But, as the Bayes coefficients are by definition independent of the internal dynamics of the system under study, this amounts to

---

\* This paper is a slightly improved version of the one I read at the Pittsburgh International Symposium on Relativistic and Classical Thermodynamics, April 7–8, 1969, the Proceedings of which are now published under the title *A Critical Review of Thermodynamics*, Baltimore: Mono Book Corp. 1970.
\*\* Professor Olivier Costa de Beauregard, Laboratoire de Physique Théorique Associé au C.N.R.S., Institut Henri Poincaré, 11, rue Pierre Curie, F–75 Paris, France.

saying that the theoretical description of time anisotropy in probability problems is of an extrinsic rather than intrinsic nature. It is, very exactly, a *boundary condition* imposed upon the macroscopic evolution equations. This boundary condition reads "blind retrodiction forbidden", very much like the boundary condition in macroscopic wave theories reads "advanced waves forbidden". The point is that in both cases the boundary condition is an initial, not a final condition. To this we will come back later.

Now we stress the connection between the temporal application of Bayes' principle and the causality concept. To say that the evaluation of Baye's coefficients is extrinsic to the dynamics of the system is to say that they are used to describe an interaction between the system and its surroundings. And to say that the Bayes coefficients must in fact be used in retrodictive but not in predictive problems amounts to saying that the effects of the interaction upon the system are felt after it has ceased, and not before it has begun. But *this* is the very definition of causality, that is, "retarded actions", as opposed to finality or "advanced actions".

A typical example of this general physical law is the ink drop that dissolves in a glassful of water after it has been deposited in it by a pipette; the reversed procedure (including the pipette and the hand holding it) can only be seen by running backwards a movie film. The same can be said of card shuffling[1] or radioactive disintegration.

We thus come to the conclusion that *the time dissymmetry inherent in causality as opposed to finality is of an essentially macroscopic nature,* and that its mathematical formulation consists in the temporal application of Bayes principle expressing that blind statistical retrodiction is forbidden *in physics*. To our cognizance Van der Waals was the first, in 1911, to state that the statistical derivation of Carnot's law is merely a temporal application of Bayes' principle – an idea which is also implicit in an often quoted sentence of Willard Gibbs (1914)[2]. That the mathematical expression of statistical irreversibility is, as Mehlberg (1961) puts it, of a factlike rather than lawlike character has also been expressed in recent years by Watanabe, Reichenbach, E. N. Adams, J. A. Mc Lennan, Wu – Rivier, Grünbaum, C. F. von Weizsäcker, G. Ludwig, O. Costa de Beauregard. The identification of the causality concept with the physical law of increasing probabilities has been especially strongly stressed by Reichenbach, Grünbaum, Terletsky, Costa de Beauregard.

---

1 Of course a deck of cards is said to be "in order" or "in disorder" according to the fact that the sequence of the cards belongs or not to some selected small sub-ensemble of the ensemble of possible permutations of the cards.

2 "It should not be forgotten, when our ensembles are chosen to illustrate the probabilities of events in the real world, that while the probabilities of subsequent events may often be determined from those of prior events, it is rarely the case that probabilities of prior events can be determined from those of subsequent events, for we are rarely justified in excluding the consideration of the anteceding probability of the prior events".

## 2. The Einstein – Ritz Controversy: Retarded Waves and Probability Increase

The existence of a close connection between the two principles of wave retardation and probability increase is strongly suggested by many physical examples such as, for instance, the slowing down of a meteorite in the earth's atmosphere. In these recent years it has been more or less explicitly stated, in various contexts, by quite a few authors among those listed at the end of this paper (Mc Lennan, Penrose – Percival, Costa de Beauregard and others). To our cognizance the discussion started in the late 1900's with the celebrated Einstein-Ritz controversy, in which Ritz insisted for deducing the law of entropy increase from the principle of wave retardation while Einstein maintained that the law of wave retardation should follow from the principle of probability increase.

That Einstein's and Ritz' statements are *reciprocal* should be obvious now that the formulation of the principle of statistical irreversibility has been recognized to be of the nature of a boundary condition.

The aim of this Section is to show that, in a restricted but precise context, wave retardation and probability increase are indeed two names for one and the same principle. At this end we will work with a theory implying *essentially* the two concepts of waves and probability, namely, quantum mechanics.[3]

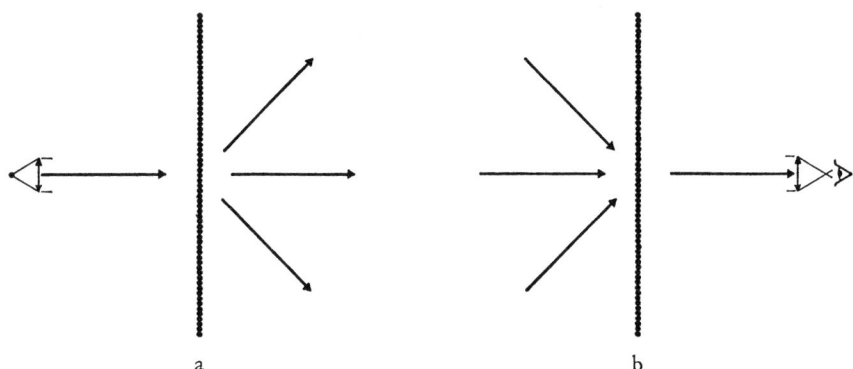

Fig. 1a and b. Physical Irreversibility as a Boundary Condition: the quantized wave and grating thought experiment. (a) Retarded waves and blind statistical prediction. (b) Advanced waves and blind statistical retrodiction

---

3 The connection between the two principles of probability increase and wave retardation is implied in Planck's definition of the entropy of a light beam, where the constant $h$, that is, the photon concept, is essential. It follows from Planck's formula that the entropy of a light beam is increased by scattering or "dif-fusion", a time asymmetric process following from the principle of retarded waves. If advanced waves were macroscopically existent, then phase coherent "in-fusion" would decrease the entropy of the light beam. Of course, in the days of the Einstein-Ritz controversy, the photon concept had already "come to light", but de

Let us first illustrate our statement by using an example. A plane monochromatic wave falling upon a linear grating (the wave planes being parallel to the lines on the grating, which we assume infinite in number for simplicity) generates a finite number $g$ of outgoing plane monochromatic waves; this follows from the necessity of phase coherence and the principle of retarded waves. Now, if one of these outgoing plane waves is received in a collimator and the observer knows nothing else than the presence of the grating (and, of course, the space frequencies of both the wave and the grating), the only thing he can say is that an incoming plane wave falls on the grating and that it is one among a well defined class of $g$ waves (comprising the one considered first). He does *not* conclude that the wave he receives is built up by phase coherence of the $g$ possible incident waves, because this would amount to accept the macroscopic existence of advanced waves.

Now we must remember that our waves are assumed to be quantized, so that we can transpose a discourse on intensities into a discourse on probabilities. If, in a predictive problem, $n$ corpuscles with the same sharply defined momentum fall per time unit upon the grating, then, to use Watanabe's excellent terminology, a "blind statistical prediction" yields $n!/\pi\, (n_i!)$ as the probability $p\,(n_i)$ that $n_i$ corpuscles per time unit come out on each of the admissible outgoing waves; for simplicity we have assumed that the transition probabilities between the $g$ initial and the $g$ final mutually exclusive states are equal; also, that $n$ is low enough for each wave train to carry not more than one corpuscle, so that we can neglect quantum statistical interactions. The above probability $p\,(n_i)$ is maximized when all the $n_i$'s are equal; this is the statistical transposition of the classical computation of intensities using the principle of retarded waves. The point is that, in the "retrodiction problem" where n corpuscles per time unit are received on one outgoing plane wave, a "blind statistical retrodiction" would entail the paradoxical conclusion that the incoming particles were equally distributed among the $g$ admissible incoming waves. But, according to the now accepted view, the principle of statistical irreversibility in physics amounts to the boundary condition that *blind statistical retrodiction is forbidden*.

We thus come to the conclusion that, in the theory of quantized waves, the principle that blind statistical retrodiction is forbidden is just an other wording for the principle that advanced waves are *macroscopically* non existent.

The consideration of this (or any similar) example is a good preparation for understanding the abstract proof of the equivalence in quantum mechanics of the two irreversibility principles that will now be presented. This proof merely consists in a rewording of J. von Neumann's celebrated theorem on statistical irreversibility in the measuring process.

---

Broglie's symmetrical association of waves with matter particles was still lying in the future. Had Einstein and Ritz known that particle and wave scattering go hand in hand, then both of them would have recognized that they were merely issuing *reciprocal* statements.

In a simplified form von Neumann's irreversibility statement boils down to this. Denoting $p_{ij}$ the (intrinsic) transition probabilities between some initial $|i\rangle$ and some final $|j\rangle$ orthonormalized set, $p_i$ and $p_j$ the (extrinsic) statistical weights of the $|i\rangle$'s and the $|j\rangle$'s, the formula

$$p_j = \sum_i p_i \, p_{ij}$$

holds in a predictive calculation. Then, denoting $p$ the largest $p_i$ and using the normalization condition

$$\sum_i p_{ij} = 1$$

the result

$$p_j \leq p \text{ for all } j\text{'s}$$

follows; this is a typical instance of the levelling of statistical frequencies in blind prediction, that is, physical prediction.

*The point is that the principle of retarded waves has been implicitly used* when stating that, *macroscopically speaking in the sense of J. von Neumann's ensembles*, $p_j$ is the predictive and *not the retrodictive* probability in the transition $|i\rangle \rightleftharpoons |j\rangle$.

In other words, as retarded and advanced waves are respectively used, in quantum theory, for statistical prediction and retrodiction, to say that advanced waves are *macroscopically* non existent or that blind statistical retrodiction is forbidden are two different wordings for one and the same statement. Both ways, macroscopic irreversibility is extracted from microscopic time symmetry *via* a boundary condition – very much like one way driving is secured by the appropriate sign post.

Finally we must emphasize that the clarity of the connection we have established between wave retardation and probability increase is such by virtue of a precise, but narrow context. J. von Neumann's micro-entropy (which, for brevity, has not been introduced explicitly) is a much simpler concept than the macro-entropy of thermodynamics. Nevertheless, as we have said, there are so numerous examples of an observable connection between wave retardation and entropy increase that I am confident that this kind of argumentation can be very largely extended. For instance, if between time instants $t_1$ and $t_2$ a physicist moves a piston in the wall of a vessel containing a gas in equilibrium, the fact is that Maxwell's velocity distribution law is altered after time $t_2$, not before time $t_1$. But the fact is also that the perturbation is emitted (not absorbed) by the moving piston as a retarded (not advanced) wave.

## 3. The Factlike Character of Physical Irreversibility and the Information Concept

It is certainly very striking that cybernetics has rediscovered, without having searched for it, the twofold aspect of the old Aristotelian "information" concept, namely (i) gain in knowledge and (ii) organizing power. That (except in a few philosophical circles interested in finality) the second aspect of Aristotle's "information" happened to be almost completely forgotten can be understood as a consequence of physical irreversibility, as will now be explained.

That information is a two faced concept in cybernetics is quite obvious in the characteristic chain.

*information*$_1$ → *negentropy* → *information*$_2$

of communication systems or computers; also, in the characteristic chain

*negentropy*$_1$ → *information* → *negentropy*$_2$

of physical measurements and the classifications they allow such as, for example, in the Maxwell demon problem analyzed by Smoluchowsky, Szilard, Demers, Brillouin, or the von Neumann measuring process in quantum mechanics. The *learning transition*

(i)  *negentropy* → *information*

appears there as symmetrical to the *acting transition*

(ii)  *information* → *negentropy*.

In (i) *observational awareness* follows in time the physical situation which, in accordance with Reichenbach's analyses, it *registers*. In (ii) *willing awareness* precedes in time the physical situation which it contributes to *produce*.

Physical irreversibility consists in the *fact* that the above arrows all point towards lower information or negentropy values. Thus, for instance, the learning transition (i) appears as a generalization of the passive Carnot degradation of negentropy in closed systems. But, as Mehlberg and others have so strongly stressed, physical irreversibility is of a factlike rather than lawlike character. Thus cybernetics implies an invitation to inquire about the lawlike rather than factlike status of our problem – very much like the internal symmetries of the Dirac electron theory have been justified by the discovery of Anderson's positron which, though *de facto* much rarer than the electron, is *de jure* its twin brother.

The irreversibility principle, as stated in the two preceding Sections, amounts to saying that, physically speaking, a low probability complexion can *in fact* be taken as the starting point of a regressing fluctuation rather than the end point of a progressing fluctuation. So the factlike (not lawlike!) irreversibility principle of cybernetics turns out to be that the learning transitions (i) are more frequent than the acting transitions (ii). In terms of awareness, observation is easier, or less tiring, than action.

Now, it should be quite clear that the very value of universal constants in terms of "practical" or anthropomorphic physical units reflects an existential situation. For instance, to say that the value of the light velocity $c$ is "very large" is to say that the ratio of associated length and time units we find convenient is much smaller; this fortuitous circumstance (which, in our opinion, may well stem from the very value of our nervous influx velocity: some small multiple of 1 m/sec) is, as is now well known, at the origin of everybody's (wrong) feeling that "there is an absolute time". Quite similarly, the "smallness" of Boltzmann's constant $k$ and of the conversion coefficient $k$ Ln 2 between an entropy expressed in "practical" thermodynamic units and an information expressed in binary units, according to the equivalence formula

$$negentropy = k \text{ Ln } 2 \times information,$$

may well be taken as a direct expression of the *fact* that observation is for us much easier than action: the conversion rate is such that gaining knowledge is very cheap in negentropy units while producing negentropy costs a lot in *bits*. Going to the limit $k \to 0$ would imply that observation is completely costless and action impossible. This crude approximation to cybernetics has been known as the theory of "epiphenomenal consciousness".

Our final remark will be almost philosophic. Since long ago it has been recognized that progressing fluctuations and advanced waves can be taken as the objective aspect of finality, just as regressing fluctuations and retarded waves are now understood to be *the* physical expression of causality. There should then be no wonder that the finality concept is so elusive in terms of cognitive awareness, for it follows from above that the learning transition being "causal" and the acting transition being "final" (in the above sense), the evidence of causality belongs to cognitive awareness just as the evidence of finality belongs to willing awareness. This, of course, is well known to philosophers, but cybernetics helps understanding why things are so.

*Conclusion*

We have resumed in modern terms the old "paradoxical" problem of deducing physical irreversibility from elementary laws assumed to be time symmetric. We have found, with quite a few recent writers, that strictly speaking there is no paradox at all, but merely a *factlike* state of affairs which is mathematically expressible as an appropriate boundary condition.

As for the lawlike status existing beyond the factlike situation we feel that cybernetics has something to say. Very strikingly, a quantum measurement essentially implies a perturbation on the measured system so that, in this sense, cognizance and action are inseparable. More thinking and more knowledge are needed here, and we feel that quantum biology might well take part in the discussion.

Finally there is of course the possibility that the recent PC violations in elementary particle physics imply T violations that should be superposed as a slight perturbation upon the preceding scheme. But this is part of to-morrows problems.

*References*

Adams, E. N.: *Irreversible processes in isolated systems.* Phys. Rev. 120 (1960) 675.
Aharonov, Y., Bergmann, P. G., Lebowitz, J. L.: *Time symmetry in the quantum process of measurement.* Phys. Rev. 134 (1964) B 1410.
Buchel, W.: *Das H-Theorem und seine Umkehrung. Philosophische Probleme der Physik*, S. 79. Freiburg: Herder 1965.
Costa de Beauregard, O.: *L'irréversibilité quantique, phénomène macroscopique. Louis de Broglie physicien et penseur,* p. 401. A. George (éd.). Paris: Albin Michel 1952.
— *Equivalence entre les deux principes des actions retardées et de l'entropie croissante.* Cah. de Phys. 12 (1958) 317.
— *Equivalence entre le principe de Bayes, le principe de l'entropie croissante et le principe des ondes quantifiées retardées.* Comptes Rendus 251 (1960) 2484.
— *Irreversibility Problems.* – Proceedings of the International Congress for Logic, Methodology and the Philosophy of Science. Y. Bar-Hillel (ed.), p. 313. Amsterdam: North Holland 1964.
Gibbs, W.: *Elementary Principles in Statistical Mechanics,* p. 150. New Haven, Conn.: Yale Univ. Press 1914.
Gold, T.: *The Arrow of Time.* Amer. Journ. Phys. 30 (1962) 403.
Grünbaum, A.: *Temporally asymmetric principles, parity between explanation and prediction and mechanism versus teleology.* Philos. of Science 29 (1962) 146.
— *The anisotropy of time. Philosophical problems of space and time,* p. 209. New York: A. A. Knopf 1963.
Lewis, G. N.: *The Symmetry of Time in Physics.* Science 71 (1930) 569.
Ludwig, G.: *Problematik des Zeitbegriffs in der Physik. Grundlagen der Quantenmechanik,* p. 178. Berlin: Springer 1954.
McLennan, J. A.: *Statistical mechanics of transport in fluids.* 3 (1960) 493.
Mehlberg, H.: *Physical Laws and Time Arrow. Current issues in the Philosophy of Science,* p. 105. H. Feigl & G. Maxwell (eds.). New York: Holt, Rinehart, Winston 1961.
Penrose, O., I. C. Percival.: *The Direction of Time.* Proc. Phys. Soc. 79 (1962) 605.
Planck, M.: *The Theory of Heat.* Parts III & IV. London: MacMillan 1932.
Popper, K.: *The Arrow of Time.* Nature 177 (1956) 538.
Ritz, W., A. Einstein: *Zum gegenwärtigen Stand des Strahlungsproblems.* Phys. Zeits. 10 (1909) 323.
Reichenbach, H.: *The Direction of Time.* Los Angeles: Berkeley. Univ. of California Press 1956.
Schrödinger, E.: *Irreversibility.* Proc. Roy. Irish Acad. 53 (1950) A 189.
Terletsky, J. P.: *Le principe de causalité et le second principe de la thermodynamique.* Journ. de Physique 21 (1960) 680.
Waals, J. D., van der: *Über die Erklärung der Naturgesetze auf statistisch-mechanischer Grundlage.* Phys. Zeits. 12 (1911) 547.
Watanabe, S.: *Reversibility of quantum electrodynamics.* Phys. Rev. 84 (1951) 1008.
— *Réversibilité contre irréversibilité en physique quantique. Louis de Broglie physicien et penseur,* p. 385. A. George (éd.). Paris: Albin Michel 1952.
— *Symmetry of Physical laws.* Phys. Rev. 27 (1955) 26, 40, 179.

— *Le concept de temps dans le principe d'Onsager. Transport processes in statistical mechanics*, p. 285. I. Prigogine (ed.). New York: Interscience 1958.

Weizsäcker, C. F. von: *Der zweite Hauptsatz und der Unterschied von Vergangenheit und Zukunft*. Ann. der Physik. 36 (1939) 275.

Wu, T. Y., D. Rivier: *On the Time Arrow and the Theory of Irreversible Processes*. Helv. Phys. Acta 34 (1961) 661.

Yanase, M. M.: *Reversibilität und Irreversibilität in der Physik*. Annals of the Japan Association for Philosophy of Science 1 (1957) 131.

# Pierre Curie's Principle of One-Way Process

LANCELOT LAW WHYTE*

*Summary.* Curie formulated the principle that an asymmetry in causes can vanish in symmetrical effects. This has been discussed by Enriques, Sellerio, Renaud, and the author, and is important for the role of time in physics. Standard methods in physical theory use symmetrical relations; an alternative would use asymmetrical relations and Curie's principle applied to relaxing structures. This is illustrated by the properties of a skew (deformed) tetrahedron relaxing towards the regular form. Physical theory would benefit from the identification of a 3D geometrical model of electromagnetism using Curie's principle.

In a paper "On the Symmetry of Physical Phenomena" Curie (Curie, 1894) applied to physical processes the considerations on symmetry developed for crystals, and showed that for a given physical effect, such as electric or magnetic fields, a particular type of asymmetry[1] must be present. "C'est la dissymétrie qui crée le phénomène."

He also formulated three rules on cause/effect relations: 1. When certain causes produce certain effects, the elements of symmetry present in the causes must be preserved in the effects. 2. When certain effects show a particular asymmetry, this asymmetry must be present in the causes that produced them. But the converse does not hold, since: 3. An asymmetry in the causes can decrease and vanish, without external causes having to be inferred. I call processes illustrating 3 "one-way"[2] processes, and rule 3 "Curie's principle of one-way process" as he was the first to express the idea with the precision made possible by the study of symmetry in crystals.

Though Curie was writing when spatial and temporal relations were regarded as distinct and his principle gives priority to spatial relations, it repays examination. For it correlates asymmetrical* spatial and temporal relations in the simplest and most general manner: *a spatial asymmetry decreases*[3] (in the internal relations of a class of causally isolable processes).

---

\* Lancelot Law Whyte, 93 Redington Road, London N.W. 3, Great Britain.

1 "Asymmetrical" is here used to mean the absence of any spatial symmetry element, unless otherwise indicated. "Asymmetrical*" and "symmetrical*" are used in the more general sense of the theory of relations.

2 This term was introduced in Whyte (1949), and Whyte (1955). It refers to directly observed 3D spatial relations and their changes in course of time, and avoids the ambiguity of "irreversible".

3 I have called signless decreasing quantities "diminants". Whyte (1955). In limit cases they become invariants. "Causal" is here used to mean providing a complete representation, not in the sense of covariant field theory.

In doing so the principle renders explicit the fact, important for the rôle of temporal relations in physics, that the conception of physical causation expressed in its most general form contains a temporal asymmetry, entropy processes being a subclass of a wider class of one-way processes which includes structural relaxations. If in limit cases there is no initial asymmetry, the principle reduces to the simplest and most general correlation of symmetrical* spatial and temporal relations: *symmetry elements are conserved* (in isolated systems). (Rule 1). Curie's principle is only univocal when the physical system can be represented by a *3D geometrical* model so that symmetry and asymmetry can be given the spatial meanings used by Curie.

The precise scope of Curie's principle is not known, but the following one-way processes illustrate it:

*All processes moving towards a terminal state marked by higher spatial symmetry than earlier states,* whether the system is closed or loses energy.[4] This covers (i) heat (entropy) processes from an initial asymmetrical temperature distribution towards maximal entropy; (ii) structural relaxation and similar processes, surplus energy being converted into radiation or heat (e.g. neutralization of ions, and the vanishing of an electron/positron pair); (iii) the formation of stable structures, such as molecules and crystals, in cases where the system passes over a less symmetrical threshold before relaxing to equilibrium; (iv) frictional or damping processes, terrestrial or astronomical, which reduce asymmetries or irregularities; (v) – a hypothetical case of (ii) to which I shall return – processes in which a skew[5] deformed tetrahedron relaxes toward the regular form. These examples show that entropy processes are not the only ones which show directed tendencies, if open systems losing energy are included.[6]

The asymmetrical* relations of physics have sometimes been neglected by physicists whose attention was directed to symmetrical* relations. One example must suffice. Einstein, in his 1905 paper "On the Electrodynamics of Moving Bodies", stated that "*all* our judgments in which time plays a part are *always*

---

4 I have given the name 'morphic' to processes leading towards greater 3D order or symmetry. Whyte (Forthcoming).

5 There are two main classes of ordered chiral forms, which I call 'skew' (about a centre) and 'screw' (about a line).

6 In addition to the above examples of one-way processes (defined as leading towards higher 3D symmetry), there are also in current physics two processes of which the reverse process cannot occur (which do *not* lead towards higher 3D symmetry): *isotropic expansion from one centre*, of the universe of galaxies and of a divergent spherical EM wave-front. But owing to the heterogenity of the universe, these processes and their reverse processes only occur in a zero-th or first order approximation. Moreover they are both idealisations which neglect electromagnetic (photon/electron/proton) interactions and are non-relativistic phenomena giving a unique status to spatial relations. Thus these two cases of *prima facie* "irreversibility" must be regarded as open to re-interpretation in a unified or more fundamental theory, though certain recent theories, e. g. Hoyle, of cosmic expansion, will *sub-judice*, attempt to correlate the expansion of the universe with fundamental electron and photon effects on a relativistic basis including interactions.

judgments of simultaneity" (my emphasis). Homer nods; Einstein slips. For we have seen that an extensive class of physical processes display a one-way tendency the identification of which involves the *temporal judgment of succession,* or some equivalent temporal relation, not merely the judgment of simultaneity.

Students of Curie's principle have found it difficult to apply, for it presents a radical contrast to the principles of mechanics. The principle uses asymmetrical* internal spatial and temporal relations of systems, and certain open as well as closed systems, while mechanical theories use symmetrical relations of closed systems expressed in external coordinates. Moreover the principle uses the global properties of symmetry and asymmetry, whereas mechanical theories use local or atomic properties. But quantum mechanics uses also spatial and other symmetries, and is in this respect a half-way house to Curie's principle.

However the infertility of Curie's principle to date is due also to a defect, if that is the term for a restriction which it shares with quantum mechanics: it possesses no observational content unless supplemented by rules indicating what model, or energy function, is required for some class of processes. But while quantum mechanics benefits by a long tradition of successful mechanical models, Curie's principle requires geometrical models providing a representation of relaxing deformations of 3 D structures, and this is a relatively new realm. I call such a geometrical model of one-way process a *Curie model.* It provides a representation of a decreasing structural asymmetry which is not statistical, but bears some similarity to the local increase of structural neg-entropy.

The ground has been well prepared for the study of Curie's principle. I select two predecessors and four interpreters of the principle: –

G. T. Fechner (1873). "It may be believed ... that the disposition of any isolated natural system to assume a regular arrangement of its parts and a regular external form is related to the principle of the tendency toward equilibrium".

E. Mallard (1880) "This tendency towards symmetry is one of the major laws of inorganic nature. It is, in fact, merely a manifestation of the more general tendency toward stability".

F. Enriques (1909) regarded Curie's rules as the use of Leibniz's principle of sufficient reason to restrict the types of abstraction and of models permissible in physical theory.

A. Sellerio (1929, '35, '35) discussed the principle and drew a parallel between the statistical entropy tendency and the structural tendency towards increased spatial symmetry. In both the number of operations with respect to which a system is invariant increases. But in the first the operations are in phase space, in the second in ordinary 3-space.

P. Renaud (1935, '37, '39, '51) made an extensive examination of Curie's principle, which he generalized thus: "If a system evolves in such a manner that the causes of its evolution are contained in it, the number of transformations (not only spatial) with respect to which it is invariant can only increase", i.e. can remain constant or increase, and he calls this number the "symmorphy" of the system. Renaud considers that this generalized principle has wide applications.

L. L. Whyte (1931, '44, '49, '49, '55). I shall not refer to my own publications. In contrast to my earlier work, this paper is directed towards quantitative formulations.

I now give my view of the main significance of the principle in the form given to it by Curie.

I consider that Curie's principle has two major consequences: —

*First:* It shows that *the class of processes which can be isolated for causal representation, not requiring the inference of external causes, is wider than the class of energetically closed systems.* One-way processes in which the system loses energy can be *isolable*, in the sense that they can be given a complete representation without taking their environment into account.[7]

*Second:* It suggests the possibility of a *geometrical physics* treating 3 D spatial relations, i.e. angles or lengths, as primary.[8] Just as statistical mechanics, the theory of crystal symmetry, and Group Theory in quantum mechanics, are useful without assumptions about forces, so Curie's principle, with an appropriate model, can determine the path of a one-way process without such assumptions, as we shall see.

In treating complex systems two methods are in principle open to physical theory:

*A. The Standard Method Based on Symmetrical\* Relations.* Here the aim is, by the spatial analysis of systems into smaller parts to reduce complex phenomena to effects which can *to some order* be represented in external coordinates by using symmetrical\* relations. Residual interactions, arbitrary phases, or small terms in a power series, are treated as of no theoretical importance.

The success of this method up to the P. C. T. Theorem of 1955 does not "establish the reversible character of fundamental processes". For these are not yet known, the equations treated as 'fundamental' being, one must assume, limit cases, valid only to some order, of an unknown unifying law. What the success, until recently, of method *A* does prove is that a wide class processes can be so analysed to a finite accuracy, i.e. reduced by spatial analysis to effects covered by symmetrical\* relations. But here caution is necessary, as we shall see.

*B. An Alternative Method Based on Asymmetrical\* Relations,* which can in principle cover the results of method *A* as limit cases. This is a hypothetical method applicable to all levels of structure the consideration of which enables us the better to understand the assumptions underlying method *A*. Here a Curie

---

[7] This formulation is more closely related to the procedures used in measurement than an analysis in terms of hypothetical retrodiction, advanced potentials, etc.

[8] E. Mach appears to have considered this possibility, Mach (1919), see also Hertz (1894).

model appropriate to the particular level is used making a geometrical theory possible, internal relations replace external coordinates, *spatial asymmetries* are treated as primary, and spatial symmetries and temporal invariants as limit cases, valid to some order, in which any asymmetries and one-way changes are, to that order, negligible. A further experimental failure of invariants, e.g. in "elementary" particles, would suggest that method B should be tried. In the meantime the value of Curie's principle is heuristic; its empirical scope has to be determined.

It will be useful to summarize the relation of Curie's principle to some physical concepts.

*Potential Energy.* The rule that potential energy tends spontaneously to decrease unless prevented by constraints, is equivalent to the Curie principle in those cases where a potential energy minimum marks a state of higher symmetry.

*Conservation.* As suggested, temporal invariants can be treated as limit cases of diminant asymmetries.

*Covariance.* This is irrelevant to representations which use internal relations in place of external coordinates.

*Dimensional Analysis.* Most physical theories are *mechanical*, using L, T and M as primary quantities, though some recent theories are *kinematic*, using only L and T. If a Curie model can be found for any class of processes, a *geometrical* theory becomes possible, using only angles or lengths as primary and deriving ratios of times and of masses from changing spatial relations.

*P. C. T. Theorem.* This Theorem may prove of fundamental importance, for the closely linked set of concepts based on symmetrical* relations (symmetry, group theory, conservation, Lorentz-invariance, local field theory, reflection symmetry, reversibility, and the assumption that particles and antiparticles differ only in the sign of their charge[9]) defines the scope of symmetrical* relations as now conceived. To advance further it seems that a theory of the spatial and temporal asymmetries of electromagnetism will be necessary.

How do the two methods, *A* and *B*, approach the task of unifying theory?

The unification of theory under *A* has to be achieved by identifying a *unified field equation,* invariant under T reversal, treating the various local particle-fields as limit cases connected by derivable coupling constants. Under *B* the task is to find *a Curie model of one-way process with similar unifying power.* The failure to construct a unified field theory suggests that it is time to look for Curie models, and first for a model of electromagnetism, the realm best tested and most used in observations.

---

9 This assumption is only valid in a restricted sense excluding the population statistics, or occupation numbers, of particular fields. Electrons are abundant, while positrons are very rare. It is not an adequate interpretation of this and similar facts to postulate the existence of a hypothetical anti-universe somewhere else. The range of this bias over the various levels of structure is not yet known.

Between 1860 and 1900 many physicists searched for a mechanical model underlying the Maxwell's equations. Maxwell himself believed that the universe consists of *matter in motion,* and at one time regarded the electromagnetic quantities as alien intruders which should be shown to be the expression of an underlying material structure serving as a dynamical or kinematic model. He was disappointed in this, no such model gained general support, and by 1900 his equations were widely accepted as fundamental in their own right.

We should now return to the search, but for a geometrical model. *Can a changing 3D geometrical structure be discovered which has the spatial and temporal properties of positive and negative electric charges, with their spins and statistics, or more generally, of electromagnetic interactions?* The history of physics encourages one to rely on geometry, and Minkowski-Einstein space-time is only necessary if external coordinates are used.

I shall now describe a simple model which does not represent known processes, but may indicate the direction in which one should look for a model of electromagnetism.

We assume that *3D systems of permanent particles* (perhaps some class of baryons) are necessary, and represent these by arrangements of discrete points in Euclidean space. Such 3D arrangements are reducible to non-overlapping *tetrahedra,* so we take the tetrahedron as an *indivisible unit of 3D space,* more precisely as a *unit cell of 3D angular relations between discrete points.* To represent the chiral aspects of electromagnetism, we replace the skew tensors and Dirac spinors of field theory by a monoparametric isosceles *skew* tetrahedron, the ortho-rhombic di-sphenoid (Class 222) of crystallography. This can be generated, in two (left and right) forms, by the *deformation of a regular tetrahedron,* the congruent equilateral faces being changed by a continuous deformation into congruent scalene faces. This skew tetrahedron is a member of a *continuous family* of varying skew angle which can serve as a geometrical model of a *temporal process* i.e. of a spontaneous *one-way relaxation* under a decreasing skew asymetry. Under Curie's rule 3 this model can relax towards the regular form spontaneously, while the regular form cannot be deformed without some external interaction. Thus the relaxing model represents an isolable one-way process which is not invariant under T reversal.

This model provides a simple illustration of properties which one may except to find also in more complex systems.

First, the model determines the path of a one-way process without assumptions about forces. It also shows that a single parameter can generate a multiplicity of variables associated with particular geometrical elements, e.g. with pairs of sides or medians. For it demonstrates that the decrease of a single skew deformation parameter can generate contrasted variables representing particular components of the global process, of which the most interesting are: (i) the (scalar) changing ratios of medians and the (polar vector) associated linear displacements, outward and inward, of opposite edges; and (ii) the (axial vector) relative twists or rotations of opposite edges. Moreover the model unifies these scalar and vector quantities by geometrical identities holding between them. More one could not ask from so simple a model.

But for me it also throws light on what has been called the "problem of reversibility and irreversibility", treacherous terms which suggest a precision they lack.

For many decades physicists have been concerned with the contrast between the temporal asymmetry, or directed tendencies, shown by most processes, and the invariance under T reversal of the standard differential equations of classical, relativistic, and quantum mechanics, for example of the equations representing most "elementary" particles to some finite accuracy. One view is that all such directed tendencies of complex systems are statistical in origin and express an entropic effect. But our model shows that statistical-entropic factors are not the only possible source of this contrast, which can also be due to steric structural factors.

I shall first express this as a general principle and then demonstrate it in the model.

The relaxation process of a deformed structure, the representation of which is not invariant under T reversal, does not correspond to a combination of the various differential equations representing its parts taken separately, which are invariant under T reversal, but to *coupled combinations of particular solutions* of these equations, and the coupling term is not in general invariant under T reversal. What we observe is the effect of coupled particular solutions, and the differential equations *representing the parts taken separately* do not cover this coupling. Thus in this case the contrast can be explained as due to a structural factor.

Let us see how the model illustrates this. In the relaxing model the decrease of a particular, say *right*, skew deformation of its internal relations corresponds to combinations of relative rotations and displacements of opposite edges expressed in external coordinates. But these combinations are not arbitrary; they are coupled spatially and temporally and are skew combinations of a particular sense, the relative phasing of the rotations and displacements being such that their joint effect is to *reduce* the *right*-handed skew deformation. To provide an algebraic representation of this geometrical property a coupling or interaction term must be included in the potential energy term which causes the decrease of the particular skew deformation characterizing the model, this coupling term not being invariant under T reversal.

This model proves nothing about any particular particle interactions. But it shows that there is no necessary conflict between (a) the one-way relaxation of a structural deformation, which is not invariant under T reversal, and (b) the reversible differential equations of separate parts, since the observations are of uniquely coupled particular solutions or possibly of coupled boundary values. To prove that skew coupling terms actually play this role in "elementary" particle interactions would require a general quantum mechanical theory of the coupling terms of interacting quantised fields, which has not yet been adequately developed.

Thus the model suggests that an examination of the properties under T reversal of complex systems should demonstrate whether or not non-statistical steric coupling terms, or their equivalent, may not be responsible for such directed tendencies as the systems possess. The beauty of the geometrical model is that the equivalent of a skew coupling term representing a one-way process is an integral part of it; there is no need for the introduction of an *ad hoc* coupling term.

Philosophers of physics normally study past theories. But it is sometimes useful to investigate the properties of a conceivable future theory, as I have done by examining some necessary properties of a geometrical model of electromagnetism. I have paid Curie the compliment of outlining a task for theoretical research in the 1970s based on a paragraph of his paper of 1894, and my remarks may assist our understanding of the role of time in physics.

*References*

Curie, P.: *Les Symmetries en Physique* Œuvres. Paris: Gauthiers-Villars 1908. pp. 118/141. See p. 127. 1894.
Enriques, F.: *Le Principe de Raison Suffisante dans la Construction Scientifique*. Scientia V. Anno III. (1909) No. IX.
Fechner, G. T.: *Einige Ideen zur Schöpfungs- und Entwicklungsgeschichte*, p. 32. Leipzig: Breitkopf & Härtel 1873.
Gray, W. (Ed.): *Organic Structural Hierarchies*. Essay in: Unity and Diversity in Systems. Bertalanffy Festschrift. New York: Braziller. In Press.
Hertz, H.: 1894 *The Principles of Mechanics* (transl.) pp. 15/35. New York – Dover: 1956.
Mach, E.: *Science of Mechanics*, p. 255. Chicago: Open Court 1919.
Mallard, E.: *Sur les proprietés optiques*. Bull. Soc. Mineral. de France. 3 (1880) 16.
Renaud, P.: *Sur une Generalisation du Principe de Symmetrie de Curie*. Comptes Rendus 200 (1935) 521.
— *Analogies entre les Principes de Carnot, Mayer, et Curie*. Exposes 516. Paris: Hermann 1937.
— *Expression analytique du Principe de Curie généralisée*. Rev. Gen. des Sciences (December) (1939).
— *Mise en Usage d'un Principe*. No. 1153. Paris: Hermann 1951.
Sellerio, A.: *Entropia, Probabilita, Symmetria*. Nuovo Cimento VI (1929) No. 5 (May).
— *Les Symmetries en Physique*. Scientia (August) (1935) (i) p. 33.
Whyte, L. L.: *Critique of Physics*, (1931). p. 140 London: Kegan Paul 1931.
— *Next Development in Man*. London: Cresset. Ch. II and Glossary 1944.
— *The Unitary Principle in Physics and Biology*. Historical survey pp. 27/32. (1949) London: Cresset. passim.
— *Tendency towards Symmetry in Fundamental Physical Structures*. Nature 163 (1949) 762.
— *Note on the Structural Philosophy of Organism*. B. J. Phil. Sci. 6 (1955) 107.

# In Defence of *the* Direction of Time

K. G. DENBIGH*

*Summary.* To the question, Does time have an arrow?, it seems that physics alone can give no unequivocal answer at the present. In this situation, and especially if we believe that time should not be reified, it is useful to consider what bearing the facts of consciousness may have on the issue. Perception and cognition, it is pointed out, are irreversible in an absolute sense. Also the criterion of *before* and *after* which is offered by consciousness has a logical primacy over any criterion arising from science. These points give support to the view that time should be regarded as anisotropic. They indicate further that it is reasonable to speak of *the* direction of time. These considerations, it is suggested, should not lightly be set aside for the sake of symmetry in physics. To attribute any degree of reality to a reverse time direction which would make us "unobservers", rather than "observers", would create endless paradox and absurdity.

## *Reversibility or Irreversibility?*

The three main processes, or groups of processes, which have been discussed as possibly offering a physical basis for time's arrow are: 1) those which are the province of thermodynamics and statistical mechanics; 2) the 'expansion'[1] of wave fronts; 3) the supposed 'expansion' of the universe as a whole.

Whether or not these processes actually demonstrate time's arrow is, as we all know, a matter of considerable dispute. On the basis of almost equally convincing arguments, entirely opposing conclusions have been reached. Consider for example the views of Mehlberg and Grünbaum. Mehlberg [6] points to the insensitivity of all fundamental physical theories to a replacement of the time variable $t$ by $-t$ and he argues that no consideration of boundary conditions, no merely local or regional anisotropy of time (such as is suggested by thermodynamics and by geology and evolutionary biology) can have any significant bearing on time's pervasive isotropy, as indicated by these basic theories. Therefore he believes "--- on presently available scientific evidence time should be considered as having no arrow or unique direction, and as involving no intrinsic (observer-independent) distinction between past and future. --- the only plausible way of accounting for the fact that so many well-established and comprehensive laws of nature somehow conceal time's arrow from us is simply to admit that there is nothing to conceal. Time has no arrow."

---

* Dr. K. G. Denbigh, F. R. S., Queen Elizabeth College, Campden Hill Road, London, W 8, Great Britain.

1 Where a conventional usage of words already presupposes a unique temporal direction this will be denoted, where the point might otherwise be overlooked, by single inverted commas.

Against precisely the same background of physical theory Grünbaum [5] argued quite differently. It is, of course, now widely accepted that there is a profound difference between arguing from time to entropy on the one hand, and from entropy to time on the other, and also that the force of reversibility objections to the latter type of argument can be countered only by the invocation of de facto boundary conditions. This indeed is the line of argument used by Grünbaum (as by Reichenbach before him) and he asks by what right does Mehlberg assume that the time-reversible laws, as they are known to us in our limited sample of the universe, have a general cosmic relevance; our warrant, he suggests, for a cosmic extrapolation of the time-symmetric laws is certainly no greater than for a corresponding extrapolation of the factlike conditions making for observed irreversibility. "- - - what is decisive for the anisotropy of time is not whether the non-existence of the temporal inverses of certain processes is factlike or lawlike; instead, what is relevant for temporal anisotropy is whether the required inverses do actually ever occur or not, whatever the reason." And he goes on to argue that the non-occurrence of certain inverse processes, such as were previously discussed by Reichenbach and Popper amongst others, does indeed provide a reliable indication of time's anisotropy.[2]

It seems therefore that the present scientific status of time's arrow turns on the relative amount of significance which is to be attached to the time-reversible laws on the one hand, or to the de facto boundary conditions on the other. For it is readily conceivable, as Grünbaum argues, that the time-reversible laws, as we know them, may not be cosmically pervasive; but equally it is conceivable that the universe could 'develop' in a reversed sense from a reversed boundary condition![3] Of course new items of information are steadily coming forward, e. g. from fundamental particle physics where there have been clear indications of a breakdown of $T$ invariance. For all that, one has the unhappy impression that the guidance given by physics on the issue is likely to remain equivocal and uncertain.

*Three (or More) Views on Time*

Conflicting views about time have, of course, been a feature of Western philosophy ever since the ancient Greeks. "Archimedes," writes Whitrow [12] "is the prototype of those whose philosophy of physics presupposes the 'elimination' of time, that is, of those who believe that the temporal flux is not an in-

---

2 That is to say, it shows that there is an intrinsic distinction between time's *two* directions, although he denies that either one or the other direction should be picked out as *the* direction. As will be seen later, it is on this latter point that I shall adopt a contrary view.

3 The reader may wish to question, however, whether or not a reverse 'development', as here envisaged, is capable of being 'observed' by beings of our own sort, and therefore whether or not it is meaningful.

trinsic feature of the ultimate basis of things. Aristotle, on the other hand, is the forerunner of those who regard time as fundamental, since he insisted that there are real 'comings-into-being' and that the world has a basic temporal structure."

Present-day views about the time problem may be conveniently characterised, as has been proposed by R. M. Gale [3], by reference to McTaggart's terminology of the *A* and *B*-series of events. The *B*-series has as its generating relation *earlier* (or *later*) *than* (alternatively *before* or *after*) and it thus represents a serial ordering which runs from earlier to later according to relations of precedence and subsequence. The *A*-series, by contrast, runs from past through present to future, and is a series in which events are not so much ordered as *determined* – determined, that is to say, by non-permanent decisions concerning their pastness, presentness or futurity. Although the same events never change their B-ordering, they do change their A-determinations; they change from being future, to present to past, according to what we call 'happening', 'becoming', or 'coming into being'.

Concerning these two series, Mehlberg and those whose views are similar to his, may be said to maintain that no objective arrow can be attributed to the *B*-series and also that the *A*-series is entirely subjective. Grünbaum, on the other hand, together with other *B*-theorists, accepts that the *B*-series has an objective arrow whilst denying objectivity to any aspect of the *A*-series. The decision that an event is past, present or future he regards as being mind-dependent. Moreover these *A*-determinations are irrelevant to science which supposedly requires only the *B*-ordering.

The *A*-theorists, who form a third group, believe by contrast that the *B*-series is reducible to the *A* and that the latter is fundamental and cannot be dispensed with. Many *A*-theorists believe too that the use of tensed statements in temporal discourse is primitive and, generally speaking, they conceive time in a fully 'dynamic' sense by supposing that things actually 'come into being' along with our impression of their coming into being.

Since it is always of importance not to overlook the origins and presuppositions of one's thinking, I would like now to suggest that modern concepts of time, as well as having their counterparts in early Greek philosophy, may also have had certain infusions from theology. At a time when some of the basic notions of scientific theory were first being formulated, i.e. in the 16th and 17th centuries, theological doctrine was still, of course, a powerful influence and its own presuppositions may well have had a formative effect on the whole future development of science.

Concerning the concept of time it may be remarked that one important strand of Judaic and Christian theological doctrine favoured a view of the temporal sequence as being progressive and irreversible. Those sorts of events in which man himself was involved were regarded as genuinely 'happening' as in the time concept of the *A*-theorists. As Needham [7] puts it, Western theology presupposed "– – – a continuous redemptive time process, the plan of redemption, – – – a divine drama enacted on a single stage, with no repeat performances."

There was however a second strand of theological doctrine whose bearing on temporal concepts was entirely different and to this I wish to draw particular attention. This was the doctrine that the Deity is transcendent[4] and had created the world *as if from outside*. The world's material content was therefore regarded as if it were passive and inert and as having no intrinsic powers of creation or of self-creation. For apart from further acts of Divine intervention everything was already there in essence at the beginning and "– – – there is no new thing under the sun."

This view that God had imprinted certain henceforth changeless characteristics on the world at its creation has its clear counterpart in the mechanical notions of conservation, temporal invariance and determinism. Without necessarily claiming that these categories of scientific thought would not have originated but for the doctrine of transcendence, it may at least be suggested that this doctrine endowed them with a degree of authority, during the centuries of Newton, Laplace and Hamilton, which they might not otherwise have had. Thus it came about, I would suggest, that when the Second Law of Thermodynamics made its appearance, at a period quite late in relation to the development of mechanics, it was faced with an entrenched position in science to the effect that reversibility was to be expected as if it were the norm, whilst irreversibility was a merely tiresome exception. This corresponds to a static view of time, like Mehlberg's. Further support for this view came from the very successful interpretation of phenomena of change, on the basis of atomic and molecular theory, as being no more than alterations in the spatial distribution of preexisting and changeless entities. The "static" view of time was later further fortified, as we all know, by certain interpretations of relativity.

No doubt these two religious doctrines, so opposed in their influences on the theory of time, were by no means incompatible in their original context. In Christian theology man was set apart from nature and therefore the redemptive time process, in which man had a certain degree of freedom and had the power of acting on and changing his environment, was not inconsistent with the view that the rest of the natural order had been created once and for all, and subsequently remained essentially constant.

Obviously enough these theological doctrines no longer command their original degree of assent. Quite apart from the question of a transcendental Deity, a vast change has taken place in man's view of his own position in the world. He now regards himself as belonging to the natural order and as being understandable in naturalistic terms. If so, however, what applies to the world-as-a-whole must apply equally to man; and vice versa. Thus if the world is

---

[4] The opposing belief that the Deity is immanent in the world was, of course, an important Christian heresy.

fully determined so too is man; and conversely if man insists that he has free-will he must accredit all other natural entities with the same potentiality, to however small a degree. Quite apart from its humanistic and social implications, which way out of the dilemma we adopt will clearly have a profound influence on the metaphysical presuppositions with which we approach the problem of time.

*Before and After*

What we call "time" is pre-eminently a mental construct, one which gives unity to large areas of our experience, but which is not to be regarded as being an *existent* in the same sense as is "matter" or even "space". In short, we should not seek to reify time, as if there were a thing called "time" to which our theories have to adapt themselves. Therefore are we not wrong perhaps in supposing that questions such as: Does time have an arrow? or Is "becoming" objective? *could have* unique answers?

Be this as it may, it will be clear that the time concept as a whole is woven from many different strands. Some of these are metrical and "objective", but there are others, such as memory, anticipation and the instantaneous awareness of succession, which derive from consciousness. The existence of "thought trains" may well be the primary basis of the concept of temporal order. Therefore if we are to talk sense about time I believe we have to pay considerable attention to whatever bearing consciousness may have on the matter. It is from this point of view that I would like to discuss the *before-after* relation.

First I want to remark that the before-after relations of events are not given *directly*, any more than is the pastness, presentness or futurity of the same events. What natural processes actually display is a mutual consistency or parallelism (e.g. of entropy change, statistically if not absolutely) but they do not in themselves display the qualities of being *earlier than* or *later than*. These qualities have to be read into the physical situation by the observer.

What has just been said will no doubt be readily accepted but nevertheless it will be useful if the point is elaborated a little. For this purpose the thermodynamic group of processes will be used as an illustration along with a manner of looking at these processes due to Schrödinger [9]. They will be here considered phenomenologically and without taking account of the reversibility objection.

As is well-known, temporal presuppositions are inherent in the use of tensed verbs [2], [8], [10]. Moreover the restriction of language to the present tense alone may not always succeed in eliminating these presuppositions. Thus the present tense will frequently denote an action 'taking place' whose temporal direction is already taken for granted. For example when we speak of 'observing' some process, the *before-after* sequence is implicit because the 'unobserving' of the process, corresponding to a time reversal of the act of observation, is tacitly excluded. The same applies to 'preparatory' acts, such as

# In Defence of *the* Direction of Time

the 'putting together' of hot and cold bodies. For here it is assumed that the action is, in fact, a putting together and not a taking apart.

For the present purpose this difficulty can best be avoided by supposing that we have before us a filmstrip without knowing which of its ends corresponds to its 'beginning'. Each frame shows the dials of instruments which measure the temperature, pressure etc., of two virtually isolated systems, $A$ and $B$, whose entropy changes can thereby be calculated.

These entropy changes – leaving aside the statistical anti-Second Law anomalies – are found to occur in parallel (are co-directed) and it is *this* (and not anything immediately to do with *before* and *after*) which is the direct result of experiment. Thus if i and k refer to the numbering of any two frames (this numbering being made successively from one end to the other of the film strip), the following can be asserted quite independently of which of the two frames 'was taken' at what would be called the later instant:

$$\text{if } S_{Ai} \geq S_{Ak}, \quad \text{then } S_{Bi} \geq S_{Bk},$$
$$\text{or if } S_{Ai} \leq S_{Ak}, \quad \text{then } S_{Bi} \leq S_{Bk}.$$

Thus in either case we have

$$(S_{Ai} - S_{Ak})(S_{Bi} - S_{Bk}) \geq 0 \tag{1}$$

since the brackets have the same sign whether this be positive or negative. Relation (1), which holds for all *i* and *k*, asserts no more than the parallelism of the entropy changes of any two isolated systems over the whole duration of change.

With this direction-neutral statement the temporal consciousness of the 'observer' can now be linked up. Let this observer be *me* in the first place. What I 'observe' is that the higher entropy states of any isolated system are those which occur later in *my* consciousness. Thus if I had directly 'observed' system $A$ and its associated instruments at the instants at which frames *i* and *k* were 'exposed', I should have found:

$$\text{if } t_{Mi} > t_{Mk}, \quad \text{then } S_{Ai} \geq S_{Ak},$$
$$\text{or if } t_{Mi} < t_{Mk}, \quad \text{then } S_{Ai} \leq S_{Ak},$$

corresponding to the *i*'th frame being 'later' or 'earlier' than the *k*'th respectively. Here the symbol $M$ stands for *me* and the convention has been adopted that greater values of the temporal coordinate $t$ correspond to later times in consciousness.

Thus in either case[5]

$$(S_{Ai} - S_{Ak})(t_{Mi} - t_{Mk}) \geq 0, \qquad (2)$$

and of course the symbol $A$ in this relationship may be replaced by the symbols $B$, $C$ etc., referring to any other isolated physical systems.

A further relationship similar to (2) can be used to express the fact that any other 'observer' is aware of the same sequence of entropy states as I am myself. Thus

$$(S_{Ai} - S_{Ak})(t_{Yi} - t_{Yk}) \geq 0 \qquad (3)$$

where the symbol $Y$ stands for *you*. (1), (2) and (3) express a complete set of independent relationships in the scheme:

| M | A |
|---|---|
| Y | B |

where $M$ and $Y$ are *any* two persons, and $A$ und $B$ are *any* two isolated systems.[6]

The above formulation expresses, I believe, the whole of the empirical content of the sorts of experimental results on which the Second Law is based. It is the parallelism of the entropy changes, and *not* that higher entropy states occur later, which is the strictly objective conclusion. The criterion concerning the time sequence is offered, not by the physical events, but by the observer himself.

Of course once this criterion has been applied – i.e. the decision taken that higher entropies occur in the direction of *later* – the observer is entitled to stand back from the situation and to use *any one* physical process as a signpost for all the rest.[7] This does not reduce the Second Law to a tautology because of the mutual consistency expressed in relationship (1). For example the direction of 'decay' of a sample of radium might be used for defining the direction of *later than*, one such sample being used to overlap temporally with another as it 'approached' the end of its useful life. Such a sample – or a whole ensemble of them – might be called *the standard before-after system* – and be kept under lock and key in a Paris museum!

---

[5] Alternatively $\frac{S_{Ai} - S_{Ak}}{t_{Mi} - t_{Mk}} \geq 0$ or in the limit $dS/dt \geq 0$ which is the normally understood content of the Second Law.

[6] A fourth, but non-independent, relationship
$$(t_{Mi} - t_{Mk})(t_{Yi} - t_{Yk}) \geq 0$$
expresses the familiar fact that the time senses of $M$ and $Y$ are co-directed.

[7] Notice however that there would be no "law" of *increasing* entropy if the time senses of the various observers were not co-directed - i.e. if (2) and (3) did not apply in addition to (1).

However it is obvious enough that we have found it quite unnecessary to establish any such standard! And this for the simple reason that we have complete confidence in our own judgment concerning the sequence of *before* and *after*. (c.f. our fallible judgment concerning the metrical aspect of time when we prefer the evidence offered by a clock). Is it not the case therefore that the mental criterion of *before* and *after* has primacy over any criterion offered by science? This question will be returned to again later. For the moment it may be useful to remark with Wittgenstein [13]: "The description of the temporal sequence of events is only possible if we support ourselves on another process".

*An Absolute Irreversibility*

It will have become obvious by now what is my own bias concerning the time problem. One has to accept, I believe, what P. T. Geach [4] has called the grass roots character of much of our ordinary temporal discourse. Especially that part of it which concerns the use of *before* and *after*.

An important point is that mental processes, unlike physical, display an irreversibility of an absolute kind – i.e. in the sense that these processes never occur (in this case really never!) in their reverse temporal sequence.

Consider for example my seeing of a shooting star. First I have no knowledge of the event, then I see it as occurring and subsequently have knowledge that it has occurred. In a hypothetical reversal of these mental states I would first have knowledge of the event and this would precede my seeing of it. However the truly paradoxical character of the reversal lies not so much in this precognition as in its supposed temporal sequel: for this would consist in the elimination of my knowledge of the event *after* its occurrence. This knowledge would be instantly and finally deleted from my mind at the very instant of the shooting star being seen!

Obviously such a sequence is entirely contrary to the facts. Once we 'have seen' or 'known' something, we can never 'unsee' or 'unknow' it. The mind acts by the 'adding on' of new material, never by its 'subtraction'. To be sure we often 'forget' things. Yet this is not the reverse of cognition for what we 'do forget' will often 'come back'. Nor are the processes of 'carrying out' actions based on knowledge, or the 'giving' of this knowledge to others, examples of reversals. These are distinctive processes and are not the reversals of the primary processes in question. In short, the processes of perception and cognition, as these are known phenomenologically, are basically one-way in character and are thus unlike physical processes where reversals are always conceivable as possibilities.

The same point has been well made by Costa de Beauregard [1] when he writes: "Songeons par exemple à l'absurdité d'une supposition telle que celle-ci: si quelqu'un connaît une théorie scientifique ou philosophique, il pourra effacer cette connaissance qu'il a de la théorie en 'anti-lisant' de la dernière à la première ligne un ouvrage imprimé a elle consacré - - -".

An important conclusion to be drawn, I think, is this: that it now emerges as rational to speak of *the* direction of time. This point can best be understood by reference to Grünbaum. As has been seen, he believes, contrary to Mehlberg, that time *is* anisotropic, which is to say that time's *two* directions are intrinsically distinct. But he denies that either the one or the other direction can be singled out as being *the* direction. My own view is that time is indeed anisotropic but that we should also accept the guidance of consciousness as providing a criterion of *the* direction. By the latter I mean that we should dismiss all consideration of the opposed direction as having any bearing on the construction of reality.

Grünbaum's reason for taking the contrary view is that he regards the personal sense of *earlier* and *later* as being due to the 'laying down' of memory traces in the brain (the 'taking in' of information) and thus as being entirely physical in its origin. This may well be the case and I would agree that a "physicalistic" view of the mind-body relationship is likely to be more scientifically productive than any foreseeable alternative. Yet this does not impede me from believing that the criterion of the *before-after* relation offered by consciousness has to be regarded as primitive.

Let us envisage the possibility that a set of events, consciously experienced as being in the sequence $ABC$, is placed by a physical criterion (e.g. by use of the *standard before-after* system, as already discussed) in some quite different order such as $BAC$. In this case would we not find this criterion and this standard quite unacceptable? In other words, is it not the case that *what we really mean* by the *before-after* relation is the relation as obtained from consciousness?

Quite apart from the possibility of an "erring" physical criterion, let us suppose that we actually 'saw' an event 'taking place' in reverse – e.g. an apple 'rising' from the ground and 'attaching' itself to a tree. This might at first be regarded as due to trickery, but if this proved untenable we might well accept that we had suffered a visual delusion. While we would thus accept a delusion of the senses, I think what we would *not* accept would be an alternative proposal that we had suffered a reversal of our deeplying awareness of *before* and *after*. This awareness we always take as being absolute and entirely incontrovertible.

Therefore if we ask what *is* the temporal direction of events (i.e. in the sense of an identity) the answer must be that it *is* the direction of events based on the awareness of *before* and *after*.[8] Moreover this awareness must lie, I think, at a deeper level than the actual processes of perception or cognition, or even of the formation of memory traces. For the very fact that we can speak of 'seeing' and 'knowing', rather than of 'unseeing' and 'unknowing', shows that these

---

[8] This refers, of course, to events *as perceived* and is in no way contrary to the well known relativity of the *before-after* relation for a pair of distant events which are not causally connectible.

processes too are judged relative to a reference. And as regards the formation of memory traces it may be remarked that the intake of information from the external world is also appreciated as a 'taking in' (or as an 'adding on') and not as the reverse. Thus, contrary to what is claimed by Grünbaum, and also by Costa de Beauregard, the formation of memory traces is probably not the reference process which provides our primary awareness of *before* and *after*. Bearing in mind Wittgenstein's remark, this reference process must presumably be subconscious and introspectively unobservable. For to observe it would seem to require yet a further process as a reference, and so lead on to an infinite regress.

To argue as I have done is not, I think, to depart unduly far from "objectivity". One meaning of this term (although a relatively weak meaning) concerns that which is publicly agreed and we do, of course, all make the same judgments of the sequences of events. Using such judgments and recording them, man created a reliable and self-consistent system of historical ordering long before he began to think about entropy or about expanding wave fronts. The "objectivist" would surely be going altogether too far if he claimed that the statement: "The death of Queen Anne was later than the death of Elizabeth I", is not an objective item of knowledge because it depends on human testimony! A direction of time, as thus constructed, is also fully applicable to periods before life began, since inferential methods of extropolating the *before-after* series may be used, as is done in geology.

I thus regard the criterion of *before* and *after* which is offered by consciousness as having a primacy over any criterion offered by science. Perhaps this appears as a departure from the strict code of the physicalistic doctrine! Yet this is by no means to deny that man is a part of nature. My view is rather that consciousness is that particular feature of the natural order with which the time concept has the closest connection.

And in accordance with this view, and having the mental processes which we do have, I contend that we can best avoid paradoxes and absurdities by rejecting one of the two directions of time offered by physics and regarding the other as *the* direction. In so far as we are sentient beings talking about our world, we must (I believe) discuss its physical changes as taking place in the same temporal direction as that of the 'making' (rather than of the 'unmaking') of the perceptions and cognitions which we have of this world. We cannot conceive of a reversal of these acts of perception and cognition for that would be to make us 'unobservers' rather than 'observers'.

In conclusion two very brief remarks. The fact that cognition has a temporal direction implies that the cognitive conditions for prediction and retrodiction respectively are unequal. This kind of asymmetry has been discussed in a very interesting manner by Watanabe [11] but I am unable to follow him in his view that this asymmetry is all there is to the arrow of time. "--- the fundamental laws *per se*", he says, "are completely reversible, while the irreversible phenomena appear as the result of the particular nature of our human cognition". Such a view, it seems to me, is to set "mind" too far apart from the natural order. Also it seems entirely contrary to the massive irreversibility of large scale cosmic phenomena – e.g. the radiation of the sun – to which cognitive and preparative acts seem irrelevant.

My second remark is that, if the general purport of this paper be accepted, it would seem reasonable to suppose that improved consistency might be attained in science if we could introduce a thorough-going irreversibility into the physical laws themselves – e.g. by modifying the existing laws so that they become reversible only under idealised limiting conditions. Indeed a somewhat similar proposal has already been made by L. L. Whyte [14]. In so far as our theories are mental constructions it seems desirable that the "time" of the theories should correspond to the undoubtedly one-way temporal character of the whole of our mental activity.

*References*

1. Costa de Beauregard, O.: *Le Second Principe de la Science du Temps,* pg. 115. Paris: Editions du Seuil 1963.
2. Denbigh, K. G.: Brit. J. Phil. Sci. *4* (1953) 183.
3. Gale, R. M.: *The Philosophy of Time.* London: MacMillan 1968. *The Language of Time.* London: Routledge and Kegan Paul 1968.
4. Geach, P. T.: *Some Problems of Time.* Proc. Brit. Acad. *51* (1965) 321.
5. Grünbaum, A.: *Philosophical Problems of Space and Time.* New York: Knopf 1963.
6. Mehlberg, H.: In: *Current Issues in the Philosophy of Science* ed. Feigl, H., Maxwell, G. New York: Holt, Rinehart and Winston 1961.
7. Needham, J.: *Time and Eastern Man.* The Henry Myers Lecture, The Roy. Anthropol. Inst. (1964).
8. Prior, A. N.: *Past, Present and Future.* Oxford: Oxford University Press (1967).
9. Schrödinger, E.: Proc. Roy. Irish Acad. *53* (1950), (Sect. A) 189.
10. Sellars, W.: Minn. Stud. Phil. Sci. Vol. III (1962).
11. Watanabe, S.: *Revue de Métaphysique et Morale,* June 1951, 128.– *Louis de Broglie, Physicien et Penseur,* pg. 385, Paris: Michel 1953. – *The Voices of Time,* ed. J. T. Fraser, pg. 527. New York: Braziller 1966. – Phys. Rev. *84* (1951) 1008.
12. Whitrow, G. J.: *The Natural Philosophy of Time.* London: Nelson 1961.
13. Wittgenstein, L.: *Tractatus Logico-Philosophicus,* 6.3611. London: Routledge 1961.
14. Whyte, L. L.: Br. J. Phil. Sci. *6* (1955) 107.

# Creative Time

SATOSI WATANABE*

*Summary.* This paper deals with the basic philosophical issue with regard to the relationship between the past-future direction (direction of life) and the earlier-later direction (direction of entropy increase). Our freedom to create the future situation according to our value is guaranteed by the biunity of finality-causality. This implies that our creative freedom depends on science being capable of giving law-like predictive probabilities. Law-like prediction is possible, but prediction is factually impossible. All actual predictions depend on factual guesses. Law-like prediction is "successful" and "effective", because the entropy of our environment is increasing. This means that man can create order and organization in his world precisely because the entropy of his physical world is increasing, representing a tendency toward disorder and disorganization. The entropy-increase is neither law-like nor fact-like, but necessary. In an entropy-decreasing environment which may be partially realized in living matter, the situation will be the opposite. In such a special portion of the world, science has to be retrodictive, and under certain conditions teleological. This reflection reveals the profound reason why science has to be causal in our usual environment, a thesis taken for granted without explanation by philosophers. At the end, we shall discuss the problem of "becoming" from our point of view. A new geometrical conceptual aid is introduced to make it easier to visualize the true meaning of becoming which is lost in the usual four-dimensional space-time.

## 1. *The Role of Time in Life*

A scientific theory cannot be divorced, without becoming rootless and losing in its wholeness, from the intuitive picture of the world from which it has emerged ... a picture that cannot be adequately described in terms of definitions, assumptions, derivation or experimentation ... a picture that is neither true nor false ... a picture whose value cannot be judged without reference to one's imagination, desire and esthetic predilection. I would therefore first venture to convey my picture of the role of time in life, before presenting a more systematic thought about time. It is however quite possible that a reader who does not like my intuitive description may agree with my more rational explanation which comes later.

What is the most fundamental fact of life ... of my being alive? To me, the right answer at least in the context of the present discussion is that *I will*. In terms of more concrete manifestations, I desire, I want, I aim at. I hasten to add to this statement that I am often unconscious of my will and of what I will.

---

* Professor Satosi Watanabe, University of Hawaii, Dept. of Physics and Astronomy, Honolulu, Hawaii 96822, U.S.A.

Let us take an example. As one of the basic facts of life, I eat. My respectable scientific colleagues will tell me that as a causal effect of a certain state of internal body chemistry I feel hunger, which causes me to eat, and as a causal effect of the calory intake I survive. I eat, therefore, I survive. This is undeniable, yet what an empoverished picture of life! The picture will somewhat regain its natural balance, if I add to this causal description a finalistic description that my body will live therefore it eats. It is important to note that causality and finality are two aspects of the same fact.

What I have just described regarding nutrition intake applies also to lower animals without consciousness. What happens when consciousness enters the picture? What do I become primarily conscious of? When I am in a passive mood, I am conscious of what I have at hand, i.e., what I perceive. But, when my will starts to become active, I become conscious of what I want, overlapped on, but distinct from, what I have. What I want constitutes the future, what I have constitutes the present. My conscious will thus splits the present and creates the future. The future is not a later instant; it coexists with the present. My will wants the later instant to coincide with the future.

How do I go around to bring about such a coincidence? The only thing I can do is to act on my environment at present in a certain way. But, what kind of act should that be? To determine the act, I need the past. The past is not earlier time instants; it is an image of them at present. The necessity of choosing a right act at present in order to realize the future generates a structured memory. Thus, in our memory there is stored a vast network of association of events that have occurred at earlier instants. The event I want in the future (or a similar event) is found in this network of association. In particular, the antecedent events that had preceded this event emerge as having the power of producing the event in question at a later instant. They are conceived as a possible "cause" of the event. This causal association, scientific or prescientific, is the guidance of the action at present. We choose the present act so as to prepare the cause whose effect will be the desired event, i.e., the end, for which the present act plays the role of the means. Thus we see that causal knowledge is produced as a tool of will.

In summary, we may say: the awakening of consciousness implies the splitting of the present into future, present and past. Value, action and knowledge find their respective places in those three aspects of the temporal tri-unity of consciousness. Action, seen as produced by the end, is a means. Action, seen as generating an effect, is a cause. This bi-unity of finality and causality which is the very essence of the future-past relation explains also the basic role of science for man.[1]

---

1 This idea about the "origin of time" (time in the sense of future-past relation) was first published in 1947 in a magazine [12] and later included in 1948 in my book [13] entitled "Time" (in Japanese). The "origin" is not the genesis. The actual awakening of the sense of

It is a remarkable fact that the lower animals which apparently are conscious of neither end nor cause – hence aware of neither future nor past – behave nonetheless in a very similar fashion. This urges us to assume some underlying unity between developed animals and undeveloped animals. That is the will of life. When awakened, will implies creation. Creation is made possible only by the medium of time.

## 2. *Freedom, World, and Logic*

The biunity of causality-finality implies freedom of action. More precisely, freedom of choice of the cause. The reason for this is as follows. Causal knowledge tells you if you choose cause A you will get effect A', if you choose cause B you will get effect B', etc. Now, you have a certain goal, say, X' in mind. You consult your causal knowledge which may tell you: if you choose cause X you will get effect X' with good probability. Then, you decide to choose cause X as the means to realize your end X' as the effect. If you did not have the freedom of choice of the cause, you could not realize your goal. In the absence of such freedom, it would be meaningless to want anything. In short, will presupposes freedom. For these reasons, I take the following statement as a postulate which I use later for my more deductive thinking.

*Postulate of Freedom.* Within certain limits, we have the freedom to generate selectively a situation in our environments so that this situation as a cause will, according to our causal knowledge, give rise with large probability at a later time the effect which is desirable to us.

This postulate describing the temporal structure of our experience not only defines the role of science for man but also determines the nature of many basic concepts appearing in scientific thought. I will mention in this section the implications of the postulate with regard to the concept of the world and the nature of logic.

If the role of our knowledge is defined as in the postulate, the world about which our knowledge is concerned must be the environments in which we try to realize what we want by acting upon it in a certain fashion. The world of science is thus the object of our action. That is the world which we can observe, investigate, act upon, control, manipulate ... a world in which we can create. In contrast to this, there exists a notion of the world which includes everything, i. e., the object of our action as well as ourselves, the agents. That is a world that nobody can observe, study or control. It is a drama going on a stage without an audience. If the former is a scientific world, the latter is a theological world. If the former is the world to be acted upon, the latter is a world only to be "contemplated".

---

future and past in children may or may not coincide with the awakening of the sense of causality.

However, strangely enough, many scientists seem to have kept their nostalgia for theology and like to talk about the world as if it included everything. From the purely scientific point of view, such a world may be "thought of" as a kind of limiting concept, and may be even useful at times, but cannot be defined with rigor in a verifiable language. The world-to-be-acted-upon of a subject (one person or one group of persons in communication and acting together) and that of another subject are of course not the same. Let $W_a$ be the world of subject $A$, and let $W_b$ be the world of subject $B$. Suppose that subject $A$ recognizes subject $B$ as a portion $B_a$ of his world, then $W_a - B_a$ is the world for $A$ from which subject $B$ seen by $A$ is deleted. In the same way we may think of $W_b - A_b$ as the world for $B$ from which subject $A$ seen by $B$ is deleted. There is no guarantee that $W_a - B_a$ and $W_b - A_b$ should be identical. As a matter of fact, they are not. But, the resemblance encourages them to consider the world portion $W_b - A_b = W_a - B_a$ as "objective" and common to all people. Overlapping these world portions with different $A$ and $B$, in the manner of photomontage and joining in the manner of a jigsaw puzzle, they hope they can reach a unique and consistent picture of the entire world including all agents in it. But, such a world, which we denote hereafter by $W_o$ is a limiting concept at best and a fiction at worst and an unreal entity in any event. This does not mean that I do not use the notion of $W_o$ at all in this paper. In fact, we are so indoctrinated with the tenets of $W_o$ and we are so conditioned by the language based on $W_o$, that I myself cannot free myself from it entirely. We should only keep in mind that $W_o$ is an unreal entity sanctioned by "realist" philosophies.

Many sociological doctrines have capitalized on the common confusion between $W_o$ and $W_i$ (world for subject $i$). They say, for instance, the world (in the sense of $W_o$) will become so-and-so by its own causal development, therefore (!) you should work on the world (in the sense of $W_i$) to accelerate this process. The first part of this statement has no scientific content, because $W_o$ is not a scientific concept. Even if it had a scientific content, the second part of the statement is a non-sequitur, since, first, an "aught" does not follow from a "fact" and second the relationship between $W_o$ and $W_i$ is unexplained except by a mysterious dialectical logic.

Of course there is an excuse for $W_o$ of the following kind. $W_o$ has no direct scientific reality, but is an ideal costruct from which the observable worlds $W_i$'s can be theoretically derived, hence can be accorded some kind of indirect reality. We should know, however, that the so-called derivation of $W_i$ from $W_o$, is a purely metaphysical venture, quite different in nature from a derivation of a forecast from a scientific hypothesis.

Of cardinal importance is the fact that the concept of time in the sense of past-present-future relation belongs to the world view of $W_i$ and not to that of $W_o$, while the concept of time in the sense of a coordinate in the Minkowski space or a variable derivable from general relativistic space is, in its pure form, a concept belonging to $W_o$.

The most conspicuous feature of logical thinking is its incapability of coping with a real change in time. Logic is tenseless. The truth value of a proposition, $P(a)$, does not change with time. To conform with the true nature of logic, we should either consider object $a$ at $t$ and object $a$ at $t' \neq t$ as different objects or we should understand that at each $t$ there is a different predicate $P$. This makes as many propositions as there are values of $t$. And, worse still, there is no continuous connection among them. This punctualization of time which is probably more criticizable than its spatialization is an inevitable consequence of the traditional logic. The reason for this may be attributed to the postulate of truth set which underlies the usual logic. We may call this postulate Frege's postulate for convenience [5].[2] The postulate assumes that each predicate has a well-defined extension (a collection of all objects that affirm the predicate). By the virtue of this postulate, all the logical relations are reduced to the set-theoretical relations, and therefrom ensues the usual Boolean logic which is isomorphic with set-theory. If an object changes its property with time, we have to prepare a separate truth set at each time instant.

I do not pretend to offer a remedy to this fatal defect of the usual logic, but I should like to point out what seems to be a right direction in which to try to open up a new path for formalization of rational thinking. Our Postulate of Freedom makes it clear that the basic form of our knowledge should be of the type: If (cause $X$), then (effect $X'$). This suggests that we should restart all over again our logical inquiry on the basis of the if-then relation, or implication. This principle may be called Peirce's principle for convenience, because he expressed the idea that "illation" (implication) is the most basic logical relation [7], although admittedly his ground for stating this is entirely different from ours. I mentioned somewhere else several more reasons why I believe that Peirce's principle, instead of Frege's postulate, is the right starting point for formulating a logic of the future [24, 26]. The most conspicuous difference between the logic based on Frege's postulate and the logic based on Peirce's principle, is that in the latter the distributive law is not necessarily required, while in the former it is. This is extremely interesting because the so-called quantum logic is a special case of non-distributive logic. It is a conspicuous feature of the non-distributive logic that it can cope with situations where the act of observation changes the state of the object system [24]. Hence it can be used in the worldview of $W_i$ in which, in contrast to $W_o$, such an interference is inevitable in general. In this sense too, the Peircian logic has some basic affinity with our view expressed by the Postulate of Freedom.

It may be noted that quantum mechanics is so formulated that it cannot describe $W_o$, because the state-function (psi-function) has no meaning in the

---

[2] Frege was not only the first to emphasize the basic role of this postulate but also the first to cast some doubt over its validity. He said at one place, "Or must we suppose there are cases where an unexceptionable concept has no class answering to it as its extension." See p. 52 of Ref. [5].

absence of observation and to make observation possible, the observing system has to be excluded from the system described by the state-function. This fact is taken by many philosophers as one of the marks of dissatisfactory character of the modern physical science. Their argument is based usually on the blind faith in $W_0$, and betrays that they do not understand what science is all about.

## 3. *Factual Impossibility of Prediction*

The advice prescientific or scientific knowledge provides us with must primarily be of the type: If you do $A$ now, then you will have $B$ realized later. But, with the development of science, the statement of science has come to take the following "predictive" type: If the object-system is in initial state $S_I$ at the initial instant $t_I$, it will be in final state $S_F$ at the final instant $t_F$. This statement can play the required advisory role for our action, provided that we can, within certain limits, create a proper initial state in the object-system. But, in order that such a prediction may be successful, the object-system must be kept isolated from the exterior system during the period between $t_I$ and $t_F$ to prevent any unforeseen interference. If one wants to free oneself from this difficult-to-fulfill condition of isolation, one has to take the entire $W_0$ as the object-system. If $W_0$ is the object-system, where is the subject-agent, man? Is he in it, or outside? If he is in it, science is no longer science. If he is outside, the condition of isolation is not guaranteed to be fulfilled. As a good example of inevitable interaction between the predictor and the object-system between $t_I$ and $t_F$, we may mention the whole class of self-fulfilling and self-falsifying predictions [2].

Forgetting for the moment about these fundamental difficulties, let us try to defend predictability in principle by assuming an imaginary observer-predictor gifted with unlimited computing capability, i. e., Laplace's demon. His task is to predict the precise state of the world at a later instant. In order that his prediction may conern the same world that would exist and develop without his presence and predictive activities, he has to minimize his influence on the world, ... desirably he has to make it zero. I shall mention some of the basic difficulties that the poor demon will encounter.

In order for him to be able to carry out the computation (which consists in solving the dynamical equations with the given initial conditions), his brain or his computing machine may very well have to be as heavy as the Universe itself, giving rise to a disturbing gravitational field. Besides, where does he get his additional chunk of matter from? But let us assume that his alleged unlimited intelligence implies that he can calculate without a brain or computing machine.

Second, according to Brillouin, in order to obtain one bit of information, an observer has to spend $kT$ ergs of energy [1]. If, as in classical physics, the variables have a continuous range of variation, the number of bits required in observation is infinite. In practical human applications of physics, we need

Creative Time

no infinite precision, but for the special task of precision required of the demon cannot allow any error or approximation. Even if the number of bits required is enumerably infinite, or even finite, the energy the demon needs in collecting the necessary data about the initial state is incomparably larger than the entire energy available in the entire universe. Here again, we have to ask where he gets this energy from and how he can prevent the enormous amount of energy he expends from altering the state of the world.

The third and most crucial difficulty for the demon lies in a special space-time condition that hinders his collecting the necessary data about the present state, i. e., the state at $t_I$ of the universe. If the universe is infinitely large, it will take infinitely long time for the signal from the edge of the universe to reach the observer. This means that the state of the universe at time $t_I$ is knowable for the demon only at time $t_I$ plus infinity.[3] Therefore, even if the calculation itself does not take any time, the predication about the state at a finite future $t_F$ ($= t_I$ plus finite time) can be made only at an infinitely remote future.

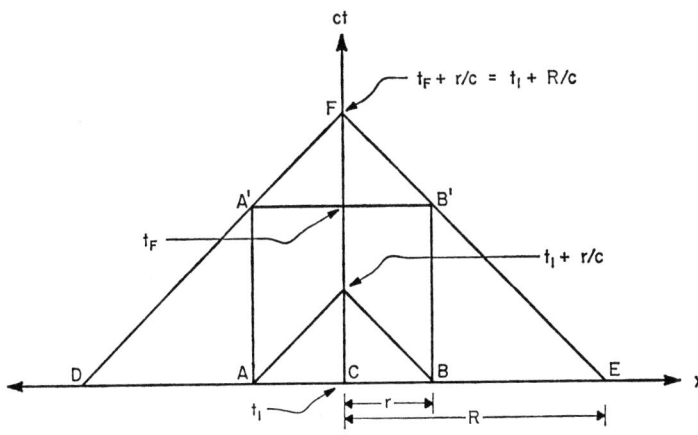

Fig. 1. The observer-predictor at $C$ wants to predict the state at $t_F$ of the finite world of radius $r$ (that is, the state on $A'B'$) basing himself on the state at $t_I$ of the same world (that is, the state on $AB$). The information about this state can reach him at $t_I + r/c$ which may be before $t_F$. In this sense, prediction is possible. However, in order for him to be sure that his world is isolated, he has to know the state at $t_I$ of the larger world of radius $R$ (that is the state on $DE$), where $R$ is just large enough so that any disturbance beyond $D$ or beyond $E$ cannot reach $AA'$ or $BB'$. The information about the state on $DE$ reaches the observer at best at point $F$, i.e., $t_I + R/c = t_F + r/c$. In this sense, he cannot factually PRE-dict with certainty. One might think that by shielding the system $AB$ by a specially designed wall, one may isolate $ABA'B'$ entirely from the rest of the world, but an atomic bomb starting somewhere in the triangle $BEB'$ can destroy the insulation wall

In a word, the demon is not only incapable of preventing his operation from altering the world thus vitiating his prediction, but also incapable of complet-

---

3 The reconstruction of the initial condition involves retrodiction, but we do not discuss this matter here.

ing his prediction before the predicted events happen. The last difficulty, lies in the very nature of the physical space-time and cannot be just ignored by simply assigning a special property to the demon.

This leaves us with a hope that we might be able to work out a meaningful prediction if we limit our object-system to a limited, isolated portion of the world. But, we can immediately see that the conditions of isolation cannot be guaranteed unless we know the present state of a larger portion of the world than the portion about which the prediction is made. And the size of this larger portion of the world is just so large that the collection of data about it can be completed precisely after the time $t_F$ to which the prediction refers.

Suppose that the observer-predictor intends to predict at $t_I$ the state at $t_F$ of a spherical portion of radius $r$ around himself. The state at $t_I$ of this spherical portion of the world is unknowable to him at $t_I$, but at $r/c$ seconds after $t_I$ he can know it where $c$ is the light velocity which is the fastest means of information transmission. If it is certain that no external influence penetrates the spherical surface from $t_I$ to $t_F$, then he can predict the state of the sphere at $t_F$ from the data (about the sphere at $t_I$) which he receives at $t_I + r/c$. Provided $t_I + r/c < t_F$, the prediction is made (assuming that computation requires no time) before the event. More accurately, he can check whether his prediction was correct only at $t_F + r/c$. As a result, provided $t_I + r/c < t_F + r/c$, i. e., $t_I < t_F$, his prediction has some meaning.

In this sense, as far as the natural *laws* are concerned, prediction of a finite isolated portion of the world is possible, but what is the status of the knowledge that the system is isolated. It has to be an observed *fact* or an inference with the help of a law from an observed fact, if his prediction is claimed to be scientifically well-founded. In order to be sure that no external disturbance arrives to the spherical surface of radius $r$ during the time period from $t_I$ to $t_F$, we have to know the state at $t_I$ of a larger sphere of radius $R$, where $R = r + c(t_F - t_I)$, because some disturbance from a source at this distance $R$ at $t_I$ can just reach the radius $r$ at $t_F$. Anything happening at $t_I$ beyond $R$ cannot affect our smaller sphere before $t_F$. Now, the information about the sphere $R$ at $t_I$ can reach the observer-predictor not earlier than $t_I + R/c$, which turns out to be equal to $t_F + r/c$, that is the time which we mentioned before as the time at which the observer can check the success of his prediction. Hence, from the point of view of fact, the prediction cannot be made with certainty before the predicted events happen. The predictor can gamble on the isolation and make a prediction, but its certainty is based on his belief on a certain fact, and not on an observed fact or a scientific knowledge.

This situation creates a curious asymmetry between prediction and explanation. In the case of prediction, isolation is a risky presumption, which may or may not prove to be the case later. In the case of explanation (made at $t_F + r/c$ in the preceding model), we can know whether or not the system has actually been isolated from outside. Explanation has more advantage than prediction.

Determinism is sometimes equated to predictability, but it is absolutely wrong. If determinism were tenable at all, it should be associates with "explanations" rather than prediction.

Another factual asymmetry between prediction and explanation, also overlooked usually by philosophers, is that a real-life prediction never uses the maximal description of the world at $t_I$ but uses only what seems to be more important, or more influential factors. But, an insignificant factor, which is usually ignorable, can at times affect the outcome (state at $t_F$) to a large extent, impairing the prediction based only on major factors. When we make explanations, we can reinstate these ignored factors and give a satisfactory explanation.

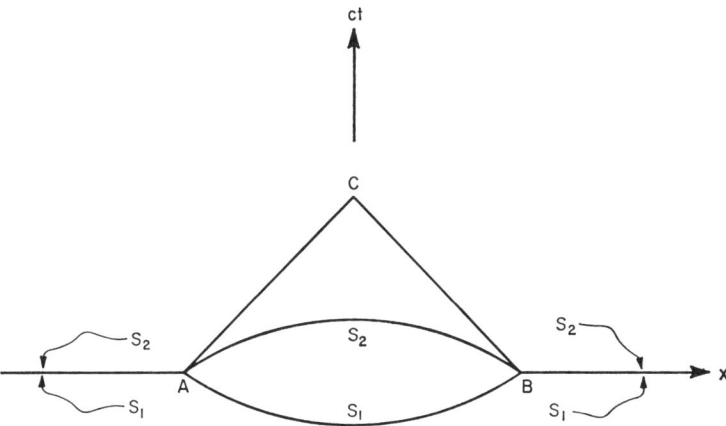

Fig. 2. Tomonaga's flying saucer. His theory allows to predict what happens on $S_2$ from the data on $S_1$ without bothering about the isolation of the boundaries. But, the data about the initial condition $S_1$ can be collected only at a time ($C$ in the figure) which is later than any point on $S_2$. $C$ is also the same time at which the final condition $S_2$ can be collected. Hence, what is done is not prediction. It is just confirmation of the theory with the experimental data and lacks any practical usefulness

Coming back to the problem of isolation of a finite system, it is interesting to note – although much practical usefulness cannot be expected of it – that Tomonaga's so-called "super-many-time" theory is free from the difficulties arising from the isolation condition [22]. See also related papers by the present author [11, 14]. Consider two 3-dimensional space-like surfaces $S_1$ and $S_2$ which are different from each other only in a limited spatial region and coincide entirely for the rest of the world. The condition that they are both space-like (i. e., their normal is time-like), make it possible to define which of the two, $S_1$ or $S_2$, is earlier than the other (according to some definition of the direction of time). The theory allows us to calculate the field quantities on, say, $S_2$ from the field quantities on $S_1$. But, this theory does not improve on the inherent trouble of the usual theory that the prediction can be made only after the predicted events have happened, and namely, prediction cannot be made earlier than the time at which the prediction can already be compared with what has actually happened. (See Fig. 2.)

The foregoing argument about the space-time condition concerns a portion of $W_o$, because it ignores the physical presence of the observer-predictor who observes $W_i$. In reality, observation itself alters the state of the observed world so that the actual prediction is not referring to the phenomena which would take place in the absence of the observer. Futhermore, isolation of the observed world from the observer-predictor can never be perfect after the initial observation. From this point of view, there is nothing paradoxical about the so-called reflexive prediction [2]. It is nothing but an interaction between the observer-predictor and the object-system. Confusion was created due to the lack of distinction between $W_o$ and $W_i$.

In summary, we may say that we are living in $W_i$, and there is no prediction possible in $W_i$ in the objectivistic sense of the word. Further, even in the idealized world of $W_o$, prediction is possible only in the sense that the natural laws are written in such a way that if we supply the initial and boundary conditions we can get the final condition. But, the fact is that we can collect the information about the initial and boundary conditions only after the time to which the prediction refers. The term "factual" is used in the title of this section to indicate that the impossibility does not stem from the dynamical laws themselves but from the auxiliary (fact like) conditions with which the dynamical equations are solved.

## 4. *Impossibility of Nomological Retrodiction*

All the natural laws we have are products of some kind of induction, and induction is a guess whose success is never fully guaranteed. In this sense already, any inference based on a natural law is in the nature of a belief. But apart from this inductive uncertainty about laws, prediction involves another basic uncertainty pertaining to the initial condition and/or boundary conditions. Prediction is possible only on a guess about the unknowable initial conditions which may or may not violate the presumption of isolation. Furthermore, in addition to these basic uncertainties, a practical prediction is possible only by omission of what is conjectured to be factors of minor importance. The last two of the three kinds of uncertainty can be avoided only in explanation, i. e., prediction in hindsight. Scientific prediction which is usually considered as a form of logical deduction, thus, in reality depends on a conjecture about facts. If so, the next question is: Does retrodiction fare better than prediction?

Retrodiction [16, 17] consists in inferring the unknown initial state at $t_I$ of an object-system on the basis of the known final state at $t_F > t_I$ with the help of natural laws. It is different from explanation which consists in deriving the known final state from the known initial state according to natural laws.

Retrodiction is in a certain sense the inverse of prediction which consists in inferring the unknown final state from the known initial state.

Let $\{I_i\}$ be the set of possible initial states at $t_I$, where the index $i$ specifying a particular member of the set may have a continuous or discrete range. In the latter case, $I_1, I_2, I_3, \ldots$, etc., are the possible initial states. Let similarly $\{F_j\}$ be the set of possible final states at $t_F$. The inference of $F_j$ on the basis of a given $I_i$ constitutes prediction, and involves the predictive conditional probability $p(F_j|I_i)$ [or $p(F_j|I_i)\,dj$ in the continuous case]. The inference of $I_i$ on the basis of a given $F_j$ constitutes retrodiction and involves the retrodictive conditional probability $p(I_i|F_j)$.

Predictive determinism means that for a given initial state $I_i$, there is only one final state $F_j$ allowed by the natural laws. This means, in the discrete case, that $p(F_j|E_i)$ is zero or one. In the continuous case, the probability density $p(F_j|E_i)$ is a delta-function in $j$. Similarly, retrodictive determinism means that for a given final state $F_j$, there is only one initial state $I_i$ conceivable according to the natural laws. Bilateral determinism means that both predictive determinism and retrodictive determinism are true. "Limping determinism", i. e., the case where only one of the two kinds of determinism is true, is conceivable. Classical physics, in its microscopic application, implies bilateral determinism. Macroscopic description in classical physics allows neither predictive determinism nor retrodictive determinism. It is essentially probabilistic. So is quantum mechanics. We shall come back to a delicate problem of the probabilistic analogue of limping determinism in a later section.

The central problem of this section lies in determining the epistemological status of the retrodictive probability $p(I_i|F_j)$. The first thing to do for this purpose is to note the mathematical symmetry that exists between $p(F_j|I_i)$ and $p(I_i|F_j)$. The conditional probability $p(F_j|I_i)$ is the ratio of the probability of both $F_j$ and $I_i$ being true to the probability of $I_i$ being true: $p(F_j|I_i) = p(F_j \cap I_i)/p(I_i)$. Similarly $p(I_i|F_j) = p(F_j \cap I_i)/p(F_j)$. Both together we can write in a symmetrical relation

$$p(I_i|F_j)\,p(F_j) = p(F_j \cap I_i) = p(F_j|I_i)\,p(I_i) \qquad (1)$$

From this we can derive two Bayes formulae

$$p(I_i|F_j) = \frac{p(F_j|I_i)\,p(I_i)}{\sum_k p(F_j|I_k)\,p(I_k)} \qquad (2)$$

$$p(F_j|I_i) = \frac{p(I_i|F_j)\,p(F_j)}{\sum_k p(I_i|F_k)\,p(F_k)} \qquad (3)$$

There is a widespread misinterpretation of the Bayes formula implying that it represents a certain mathematical asymmetry because the "inverse" conditional probability is given by a more complicated formula, for instance, $p(I_i|F_j)$ given in the right hand side of (2). The idea of using the Bayes formula in explaining the asymmetry of the H-Theorem is not new, and claim has often been made that the physical asymmetry is thereby reduced to a mathematical asymmetry. But, as the above derivation shows clearly that there is no asymmetry in the Bayes formula. If (2) is true, (3) is also true. The true asymmetry originates not from the mathematical nature of probability but from the empirical application of probability. To explain this, we have to introduce a simple mathematical theorem which the present author introduced and proved previously.[4]

*Theorem of Fixed Probabilities.* If the conditional probabilities $p(F_j|I_i)$ and $p(I_i|F_j)$ are both given for all $i$ and all $j$ in such a way that there is no mutual contradiction, then the unconditional probabilities $p(I_i)$, $p(F_j)$ and $p(I_i \cap F_j)$ are all determined for all $i$ and all $j$ by them except in the case of bilateral determinism. In the latter case, the conditional probabilities satisfy the relation $p(F_j|I_i) = p(I_i|F_j) = 0$ or $1$ and the unconditional probabilities satisfy $p(I_i) = p(F_j) = p(I_i \cap F_j)$ for such pairs $(i,j)$ that $p(F_j|I_i) = p(I_i|F_j) = 1$ and $p(I_i \cap F_j) = 0$ for other pairs. But, the values of the unconditional probabilities $p(I_i)$, $p(F_j)$ are not determined by the conditional probabilities. In the case where only $p(F_j|I_i)$ [$p(I_i|F_j)$] for all $i$ and all $j$ are given, $p(I_i)$ for all $i$ [$p(F_j)$ for all $j$] can be freely chosen, and $p(I_i|F_j)$ [$p(F_j|I_i)$] are thereby determined by the Bayes formula.

From this Theorem of Fixed Probabilities ensues the following theorem.

*Theorem of Free Marginal Probabilities.* Freedom of choice of $p(I_i)$ for all $i$ [$p(F_j)$ for all $j$] implies either that the case is that of bilateral determinism or that $p(F_j|I_i)$ [$p(I_i|F_j)$] for all $i$ and all $j$ are given and $p(I_i|F_j)$ [$p(F_j|I_i)$] are to be determined by the Bayes formula.[5]

All the statements up to this point are entirely symmetrical with respect to $\{I_i\}$ and $\{F_j\}$. Their validity is of mathematical origin and the assumption is only that these $p$-functions satisfy the probability axioms (being non-negative and summing up to unity) and all related by (1). Now we introduce for the first time an asymmetric element. Translated into the probabilistic language concerning the initial and final conditions, the Postulate of Freedom which defines the basic role of science for man implies:

*Law of Free Initial Condition.* Within certain limitations, we have the freedom of determining the initial probabilities $p(I_i)$.

This law combined with the Theorem of Free Marginal Probabilities implies:

---

[4] Theorem (3.1) in Ref. [19], Theorems 1 and 2 in Ref. [20], or Theorem A 3.11 in Ref. [25].

[5] The fundamental error committed by the necessary view of probability such as the view point of Carnap in his *book on probability* [3] lies in the contention that the value of $p(A|B)$ is determined by necessity by $A$ and $B$ and cannot be affected by human free choice. The present theorem contradicts the necessary view because $p(I_i|F_j)$ is affected by a free choice of $p(I_i)$.

*Law of Nomological Predictiveness of Scientific Laws.* Scientific laws must belong either to the case of bilateral determinism or to the case where the predictive conditional probabilities $p(F_j|I_i)$ are given by the law and the retrodictive conditional probabilities $p(I_i|F_j)$ are determined by the Bayes formula (2) in which the initial probabilities $p(I_i)$ are not determined by the scientific laws.

This law allow as legitimate scientific enterprises the following three classes of physical theories: (1) The classical physics (bilateral determinism)[6], (2) The classical statistical physics if formulatable so as to give the predictive probabilities and (3) Quantum physics if interpreted in a predictive sense [17]. To avoid misunderstanding, it should be emphasized that the foregoing derivation shows that the Law of Predictiveness is logically implied by the Postulate of Freedom, that is to say, the Law of Predictiveness is necessary in order to make the Postulate of Freedom possible. This, by no means, means that the Law of Predictiveness is somehow demonstrated from the point of view of basic physical laws. Of course, if we take the Postulate of Freedom as an empirical fact, then the Law of Predictiveness becomes an empirical law. However, physical science takes up as its burden the task of reducing all empirical laws eventually to the basic atomic laws. This, of course, is not done in this section, and we shall return to it later. The major claim of the present author in this area is that he related the basic asymmetry not just to the apparent asymmetry of the Bayes formula but to the basic role of science in life as formulated in the Postulate of Freedom. This was made possible by the small theorem regarding the conditional probabilities which he introduced and proved.

It is interesting to note that our law of predictiveness sheds new light on the problem of retarded and advanced potentials. In classical physics, if the actual initial, final and boundary conditions are all known for radiations and the motion of the source particles are given, the solution in terms of retarded potential and that in terms of advanced potential are equivalent, and there is no conflict between the two formulations. Usual textbooks are often misleading in this respect. But many problems are formulated in terms of the radiation generated by source particles assuming there is no external radiation impingent to be absorbed by the particles, hence the use of retarded potential alone is justified. In such a case, the motion of the particles is considered the "cause" and can be freely chosen in compliance with the Law of Predictiveness. In quantum physics too, the equivalence of retarded and advanced potentials is valid insofar as we are interested only in the temporal development of the state-function. But, if we interpret the formalism in terms of the actually observed initial condition and the actually observed final condition, this equivalence does not hold, because the actually observed final state is only one

---

[6] The reader will now see the profound implication of the fact that all the basic field equations are of the hyperbolic type.

of the (continuously) many possible final states to which the statefunction gives a probability distribution [10, 15]. For this reason, in quantum field theory, we have to stick to the initial condition problem, and to the retarded potential. This is, however, nothing peculiar to the advanced and retarded potential. Quantum physics in general, being essentially a probabilistic theory, has to be supplemented by an additional "direction for usage" regarding whether it should be used predictively or retrodictively, otherwise it does not give a unique answer [17]. Law of Predictiveness gives precisely the needed direction for usage. Quantum mechanics has to be used predictively.

The last part of the Law of Predictiveness can be paraphrased as

*Law of Nomological Impossibility of Retrodiction*. Except in the case of bilateral determinism, retrodiction is impossible by the scientific laws alone. It depends on the prior probability of the initial condition which depends on various kinds of „factual" information.

Some explanation about this law may be in order. The Bayes formula (2) means in words that the probability of an unobserved earlier event $A$ on the basis of the observed later event $B$ is proportional to the predictive probability of $B$ on the basis of $A$ times the prior probability of event $A$. The denominator in (2) is a constant insofar as the various possible earlier events are concerned and serves only the purpose of normalization, i.e., the purpose of making the probabilities of various possibilities sum up to unity. The predictive probability of $B$ on the basis of $A$ is the quantity which a good formulation of scientific theory must be capable of producing. The prior probability of $A$, on the other hand, has nothing to do with the nomological prediction and represents the degree of credibility or the subjective probability which the retrodictor would attach to event $A$ as what actually happened at the initial time independently of the knowledge of the given later event $B$. This subjective probability may depend on many factual information about the world. The predictive laws and the given later event $B$ alone thus cannot determine the retrodictive probability. Hence, from a nomological point of view retrodiction is impossible. In the case of prediction, the predictive laws and the given earlier event $A$ do determine the predictive probability. Hence from a nomological point of view, prediction is possible. The only difficulty lay in collecting information about $A$. For retrodiction, we need the extra-nomological evaluation of the plausibility of the retrodicted event $A$. For prediction, we do not need an extra-nomological evaluation of the predicted event $B$.

My assertion about nomological possibility of prediction and nomological impossibility of retrodiction may seem to contradict our daily belief that the past is certain and known whereas, the future is uncertain and unknown. We can easily clarify this apparent difficulty if we understand the role of law-like knowledge and that of fact-like knowledge. This is precisely the stumbling block which caused such an otherwise lucid thinker as Grünbaum to misunderstand the mechanism of retrodiction [6].

In order that the retrodictive probability of event $A$ may be large, it is necessary for both the nomological predictive probability $p(B|A)$ and the fact-like probability $p(A)$ to be large. For instance, if we see many footmarks on the beach ($B$) on Monday morning, we would give a large retrodictive probability to the event that there were many swimmers walking around on the beach ($A$) over the weekend. Many people will agree on the soundness of this inference for the conjunction of the following two reasons. First, the nomological predictive probability $p(B|A)$ is large, because if many people walked on the beach over the weekend it is very likely that many footmarks will still stay there on the Monday morning. This is a physical prediction. Second, the prior probability $p(A)$ is large because with or without footmarks it is plausible that there were many swimmers over the weekend. This is a conjecture based on our factual knowledge. The footmarks tend to increase the probability of $A$, just the same way as a positive evidence increases the credibility of a hypothesis. The rationale of our reconstruction of the past in our history-writing activity is essentially the same as the argument used in the above example. What we call a "record" of the past event plays the role of the footmarks in the above example and does not directly ascertain any particular past event. There could be noise (distortion) and our interpretation could also be wrong.

When we find a well-ordered stack of playing cards on a table we infer that there must have been somebody who handled it. The usual argument is that if there had not been anybody who had handled it the probability of such a well-ordered stack would be very small. But, this argument is incorrect. The special order of cards we call a good order is a particular permutation of 52 cards and its combinatorial probability is $1/(52!)$, just as is the probability of any other permutation. Random shuffling must result in one of the 52! possible permutations with equal probability. The correct argument is that if there was somebody who handled the cards (call this event $A$), then the conditional probability $p(B|A)$ of event $B$ of a well-ordered stack of cards lying on the table becomes large, and that, apart from fact $B$, there was a good chance of somebody passing by the table and handling the cards. The combination of the both leads legitimately to the above conclusion. Similarly, if we want to object to the usual evolution theory of chance mutation and natural selection, it is not sufficient to point out how improbable the present state of living organisms is. We would have to introduce a special agent or mechanism, whose existence is a priori probable, and whose influence would be such that the probability of the present state of living organisms emerging as the result becomes large. (See also another kind of argument on p. 179 of [17])

Before leaving this section, we should mention the relation between the consideration in terms of retrodictability versus irretrodictability and the consideration in terms of reversibility versus irreversibility. Except for the newest results which cast some doubt on its universal validity, reversibility of basic (microscopic) laws is usually accepted as one of the remarkable fundamental rules. Non-statistical classical physics is always retrodictable since there the initial and final states correspond one-to-one, but reversibility is an additional restriction. In quantum mechanics, reversibility means that probability of reaching $F_j$ at $t_F$ from $I_i$ at $t_I$ is the same as the probability of reaching $I_i{}^*$ at $t_F$ from $F_j{}^*$ at $t_I$, where the asterisk means the reversed state, i.e., the state in which the positions are the same and the momenta are reversed. Since quantum physics is essentially probabilistic, it is irretrodictable in general although it may be reversible in the foregoing sense. It has been my contention [17, 19] since many years that the entropy-increase law of thermodynamics has nothing to do with the conflict between reversibility and irreversibility but with the conflict between retrodictability and irretrodictability. Breakdown, even if true, of the microscopic reversibility would not affect our derivation, whereas one might have the impression in the old framework of thought that breakdown of microscopic reversibility might give some clue to the entropy-increase.

## 5. *Entropy-Increasing System*

We, as subjects of action, know and live the past-future relation, the future being something to be created. On the other hand, we as observers discern that our worlds $W_i$'s have a certain uniform direction of time. The radioactive substances are emitting and not absorbing their α-, β-, and γ-rays into the wide space. The rotation of moon is slowing down and not speeding up, inappreciable though the rate of deceleration may be. In a word, the entropy is increasing in one direction of time. This direction of time may be called early-late relation. The only fundamental problem of philosophical interest is the relation between the past-future direction and the early-late direction. Philosophers of science in the past cowardly avoided this fundamental issue and resorted to an "objective" problem in which there is no human being playing a role. What they (e.g., Reichenbach [9], Grünbaum [6]) considered as a theorem of profound philosophical significance seems to be a simple fact belonging to the pure domain of physics. The theorem may be summarized, in somewhat unrigorous wording, as follows: if a portion of the system branches off and reunites with the main system, the daughter system inherits the direction of entropy-change from the mother system. To me this theorem seems to be a simple corollary of another simple theorem:

*Theorem of Temporal Uniformity of Entropy Change.* If the entropy of an isolated system is increasing at an instant in one direction of time, it will continue with near-unity probability to do so in that direction of time until it reaches the maximum value (or a near-maximum value) under the existing constraints.

Applying this theorem to a splitting system, we can derive the branching theorem as follows: If the entropy is increasing just before the splitting (as is tacitly assumed) then the sum of the entropies of the two subsystems will continue to increase after the splitting toward the maximum value (under the splitting condition). If the daughter subsystem and the mother system change their entropies in the opposite directions, this trend toward maximization would not be possible. Note that due to the Theorem of Uniformity, the entropy of each system has also to change consistently in a fixed direction of time. The conclusion that both subsystems change their entropies in the same direction after the splitting will not be affected even if the boundary conditions are altered before the maximum value is actually reached. This completes the proof of the branching theorem. By the same argument, we can also conclude that systems which intermittently interact with one another must change their entropies in unison.

It should be noted that the Theorem of Temporal Uniformity does not give any attributive preference to one direction of time over the other. The validity of this Theorem can be easily understood in light of the following Theorem due to the Ehrenfests [4]. Suppose we find at a certain instant $t$ an isolated system with certain given constraints having a certain nonmaximum value of entropy. Assume further that we have no other knowledge about the system than this entropy value. (The wording "before-after" in the next sentence has no relevance to entropy.) Comparing the entropy at an instant just before $t$ and the entropy at an instant just after $t$ with the entropy at $t$, we can classify the possible situations at $t$ into four categories according as the entropy is (1) increasing (2) decreasing (3) at a local maximum or (4) at a local minimum. The point of inflection will be included in either (1) or (2) by taking larger intervals between observations.

*Ehrenfests' Theorem.* Under the conditions described above, the case (4) is overwhelmingly more probable than the cases (1) and (2) which are equally probable. The cases (1) and (2) are overwhelmingly more probable than case (3).

This theorem is also perfectly symmetrical with respect to both directions of time. If the entropy has been increasing in one direction of time, we have either case (1) or case (3). According to this Theorem, case (1) (increase) is overwhelmingly more probable than case (3) (local maximum). That is, the entropy will continue to increase with a probability overwhelming larger than suddenly switching to decrease. This is the content of the law of temporal uniformity.

Another important conclusion that can be drawn from Ehrenfests' Theorem is that if nothing else is known about the object system than the entropy at a certain instant $t$, the entropy value before and after $t$ should be expected to be larger. The reason is that the observed value is in all probability the local minimum, i.e., case (4). The second and much less probable group consists of (1) and (2), which cancel each other as far as the present question is concerned. Thus, the possibility of decrease comes only from the infinitesimal probability of case (3). This allows to predict an entropy increase. But this conclusion is perfectly symmetrical with respect to both directions of time, in apparent contradiction with the one-way increase of entropy.

This trouble could be avoided, if for one reason or another we were forbidden to use the foregoing inference to the past and allowed only to use it for the future. Now the reader will see that our Law of Nomological Predictiveness of Scientific Laws gives the clues to this asymmetry. Basic Predictiveness allows us to use science to forecast the future behavior of a physical system. The symmetrical entropy-increase is a conclusion based purely on the present state of the system developing under the dynamical laws, and it ceases to be valid for retrodiction in the case where we know something about the past of the system. In fact in the above derivation from Ehrenfests' Theorem, we used a premise that we know only the entropy value at $t$. But in reality we have various factual knowledge from which we can derive various other clues regarding the past. This violates the adopted premises. For the future, we have to rely on the present state of the (isolated) system and predictive laws, and hence we can decide that the adopted premises are more or less obeyed.

The above paragraph is the first place in this section in which a genuine asymmetry emerged, and it is highly significant that the asymmetry is somehow associated with "human" inference. As we shall explain presently, this does not imply that we are projecting our own asymmetry on something which does not have asymmetry. It does though imply that the asymmetry has a non-nomological origin, and our human mind is so marvelously adapted to this situation that it can work only in such an asymmetrical situation. We shall come back to this problem later. Predictiveness is the character we require of science and that character allows us to derive the entropy increase in the direction of our future.

The above mentioned paradox (entropy-increase in both directions) and its resolution can be explained also in connection with the H-Theorem. The proof of entropy-increase by the use of predictive probability $p(F_j|I_i)$ in the H-Theorem is very easy. The question is: why cannot we do the same thing with the retrodictive probability, $p(I_i|F_j)$, and infer a larger entropy in the past. To derive a symmetrical entropy increase in both prediction and retrodiction, we should proceed as follows. In physics there is such a thing as the "a priori" probability of each macroscopic state such as $I_i$ or $F_j$. (This a priori probability is proportional to the volume in the phase space or to the number of quantum states included.) The predictive probability $p(F_j|I_i)$ is large for a final macro-state $F_j$ which has a large a priori probability, and a large a priori probability means a large entropy value. That is a (perhaps overly) simplified explanation why the entropy of the final state is to be expected to be large.

How can we derive a retrodictive probability which does not depend on non-nomological (contingent) factors. This can be done by substituting the a priori probability of $I_i$ for $p(I_i)$ in formula (2). The retrodictive probability

$p(I_i|F_j)$ thus obtained from (2) may be called a blind retrodictive probability [17] because it gives essentially an equal prior probability to each unity of the phase space or to each quantum state. Such a retrodictive probability $p(I_i|F_j)$ will be large for an initial state $I_i$ whose a priori probability, hence also its entropy, is large. This will result in a large entropy assigned to the past. This will establish a symmetrical conclusion about the inferred entropy in prediction and retrodiction. (Note that I did not use the so-called reversibility argument here.) The reason why this conclusion is wrong in the retrodictive case is that the prior probability $p(I_i)$ does in reality depend on non-nomological factors as we explained already in Sect. 4 and cannot be equated to its a priori value.

We have thus succeeded to show human freedom of action, the essentially predictive character of science and the entropy increase in our environments are intimately bound up. But, we have not shown why predictive science, but not retrodictive science, can be formulated in the form of laws. We have indeed shown that the nomologically predictive science is the condition for both human freedom of action and the entropy increase, but we did not say how the macroscopic predictive science can be justified from the point of view of the fundamental microscopic science. Another question which we have not discussed is how "effective" such a macroscopic science can be. If a weather forecast says that it will either rain or not rain, the forecast may be successful, but it is not effective. If it gives 50% to rain and 50% to non-rain, it is not very effective either. If it gives 100% to either one, the rate of success may be suffer but its effectiveness is high if it succeeds. The question is now: why is predictive science in general effective as it is. We are going to discuss these two questions in the following two sections.

## 6. Why is Prediction Successful?

We have seen that our purposive action is made possible by the guidance of our scientific or prescientific causal knowledge, which may be quantitatively expressed by the macroscopic predictive probability $p(F_j|I_i)$. Now, the basic physical laws are formulated in terms of microscopic processes. Let us examine from such a microscopic point of view the conditions under which the law-like use[7] of a macroscopic predictive probability $p(F_j|I_i)$ but not a macroscopic retrodictive probability $p(I_i|F_j)$ is justifiable.

The initial state $I_i$ consists of many microscopic states $I_{ik}$, and the basic laws in principle can tell the probability $p(F_j|I_{ik})$ of the system in $I_{ik}$ at $t_I$ landing the final state $F_j$ (more precisely, any one of the microstates belonging to $F_j$) at $t_F$. The macroscopic predictive probability $p(F_j|I_i)$ has to be interpreted as an average of $p(F_j|I_{ik})$ with respect to the microstates $I_{ik}$ belonging to the macrostate $I_i$. But, this average value will depend on the weight one

---

7 We are therefore excluding the problem of knowability of the initial condition here.

places on each of the microstates within the same macrostate. The only way to obtain the average value that is independent of factors other than the given $I_i$ and $F_j$ is to use the socalled a priori probability $p_0(I_{ik})$ of microstates, which is constant per unity of the phase volume or per quantum state. The macroscopic predictive probability $p(F_j|E_i)$ has to be understood to be the average value of microscopic predictive probabilities in this sense.

Now suppose we have a system in the macrostate $I_i$ at $t_I$, but we do not know exactly in which microscopic state $I_{ik}$ this system actually is. In such a case we may think of a probability distribution $w(I_{ik})$ expressing our guess regarding the true microstate. The use of the average macroscopic predictive probability $p(F_j|E_i)$ would be justifiable in this case if the average of $p(F_j|I_{ik})$ based on our $w(I_{ik})$ and the average of $p(F_j|I_{ik})$ based on $p_0(I_{ik})$ are approximately equal. If this is the case, we shall say that the system we have at hand is a predictively quasi-random sample of $I_i$. In a very special case, where $w(I_{ik})$ and $p_0(I_{ik})$ are equal, we may say that the system is truly a fair random sample of $I_i$. The former case has to be qualified by the adverb "predictively" because the definition involves the predictive probabilities $p(F_j|I_{ik})$. The latter case does not need this qualification because $p_0(I_{ik})$ itself is a notion independent of the conditional probabilities $p(F_j|I_{ik})$.

A special retrodictive probability $p(I_i|F_j)$ analogous to the average predictive probability $p(F_j|I_i)$ based on the a priori weight $p_0(I_{ik})$ was called the blind retrodictive probability $p_0(I_i|F_j)$ [17, 19]. It can be obtained from formula (2) by inserting the a priori probability $p_0(I_i)$ for $p(I_i)$ in it. The $p_0(I_i)$ is the sum of $p_0(I_{ik})$ with respect to $k$ within $I_i$. The reason for this name is that due to the so-called Liouville Theorem, if we have the a priori probability distribution given by $p_0(I_{ik})$ at $t_I$ we shall have the a priori probability distribution $p_0(F_{jl})$ at $t_F$. As a result, if we assume that we have the weight distribution proportional to $p_0(I_i)$ for all $I_i$ at $t_I$ and if we pick up only those systems which land $F_j$ at $t_F$, we have an ensemble which consists of $F_{jl}$ with the a priori probability $p_0(F_{jl})$. This ensemble corresponds to the case where we know nothing but the final macroscopic state $F_j$, and we can consider the system as a fair random sample of $F_j$. We may also speak of a retrodictively quasi-random sample of $F_j$ if the estimate of the retrodictive probability $p(I_i|F_j)$ is the same as the corresponding "blind" value $p_0(I_i|F_j)$ although the probability distribution within $F_j$ is not quite $p_0(F_{jl})$.

It is easy to see that if we can prove an entropy increase in the positive direction of time with the use of the average predictive probability $p(F_j|I_i)$, then we can also prove an entropy increase towards the negative direction of time (i.e., a decrease in the positive direction) with the use of the blind retrodictive probability $p_b(I_i|F_j)$. (Note that we are not using the customary argument of the "Umkehreinwand" in terms of reversibility). To derive an asymmetrical conclusion from this perfectly symmetrical situation, we have to introduce an idea of uni-directional randomness. Indeed, if the state at hand at $t$ is a perfect random sample of the observed macroscopic state, then the entropy before and after must be larger than at $t$. According to the foregoing analysis, the only case where we can conclude the one-way change of entropy is such that the state is predictively quasi-random and not retrodictively quasi-random, or retrodictively quasi-random and not predictively quasi-random.

*Theorem of Quasi-Randomness.* Except for the maximum entropy state, the macro-state we have in our experience is not purely random but only quasi-random, either predictively or retrodictively.

This Theorem must be augmented by a theorem which determines in which direction the quasi-randomness is oriented.

*Theorem of Directed Randomness.* If the entropy of the macrostate at $t_2$ is larger than that at $t_1$ in an isolated system, the macrostate at $t_2$ is not truly random but quasi-random in the direction of $t_1 \to t_2$ and not quasi-random in the direction of $t_2 \to t_1$.

This Theorem is also formulated in such a way not to single out any privileged direction, but it allows one to derive a truly one-directional result. It eliminates case (4) of Ehrenfests' Theorem in our actual experience. In an actively prepared experiment, we have to generate the prescribed initial state. To do this, we usually generate first the corresponding boundary and start with some arbitrarily picked state. But, by the relaxation effect, after a very short time, the system reaches a "quasi-random" state under the given boundary condition. This quasi-random state is oriented only toward the future, because we know that at the beginning of the preparation time it started with a certain state whose entropy must have been smaller than at the end of the preparation period. The main part of the experiment starts after the end of the preparation of the initial state. The state we have at hand at the beginning of the main experiment can be treated as if it were a random sample of the initial state as far as the later development is concerned.

In a passive experiment, we have a chain of states, which are quasi-random states oriented all in the same direction. Suppose, for instance, we have two blocks of metal in diathermal contact, but thermally isolated from the outside. Assume also that the heat conductivity of the metal is sufficiently high and the heat conductivity of the diathermal wall is sufficiently low, so that each block has a uniform temperature at each instant. It may seem therefore that the microscopic state of the blocks can be considered as a true random state under the given temperature. But this cannot be true, because if we trace back the motion of each molecule in both blocks (and the separation wall) backward in time, the temperature difference will become larger and larger due to the bilateral determinism. This means that the state we have at each instant is a very special rare case as far as their past is concerned. But, on the forward direction of time the system behaves as if its state were random because it coincides with the prediction based on the random assumption. This situation continues to be true in agreement with the Theorem of Temporal Uniformity. This is a typical case of quasi-randomness in the positive direction of time. It is not quasi-random in the negative direction of time.

Of course, what is called positive and negative is arbitrary, as far as this system is isolated. But, the system must have been in interaction with the other part of the world and must have shared the same direction of entropy increase. For this reason, once we have fixed the positive direction of time by the entropy-increase of a partial system, the entropy of other particular system will be also increasing, i.e., the quasi-randomness will be oriented consistently in space and time in the same positive direction. It is also for the same reason, that we do not need to continue to observe the same system as the Theorem of Oriented Randomness might suggest in order to conclude which way the randomness is oriented. If we observe at a single time a system not in the maximum-entropy state, we know which way the entropy is increasing.

This consideration leads us to discover a fact of cardinal importance. We need predictive science in order to create future. Retrodiction cannot be done in the same simple way as prediction because we have a memory of the past. This epistemological argument causes us to justify the H-Theorem only for the future and not to the past, thus concluding the entropy-increase in the direction of our future. But this is a story told from our side of the world $W_i$. It is a separate question to ask, why we *succeed* with prediction and do not succeed with blind retrodiction when we describe the actual world $W_i$. The answer has been given in the course of this section. The reason for the success of prediction lies in the fact that the states of the physical world are not random states, but quasi-random states oriented to a particular direction of time which coincide with what we call future. The coincidence of the epistemological asymmetry between prediction and retrodiction with the orientation of the quasi-randomness may be characterized as another amazing case of harmony between life and its environments.

A natural conclusion seems to be that our life and mind have been created and adapted in such a way that our future coincide with the direction of the entropy increase, so that we can predict nomologically and therefore create future. In a lower living matter a similar coincidence must be going on unconsciously and making its life possible.

It is often stated that the entropy increase is not law-like but fact-like. It is agreed that it is not law-like (or non-nomological), but to state that it is fact-like or contingent is misleading. When something is fact-like, it means that as far as the laws are concerned it could be either yes or no (forward or backward, etc.), but it happened to be one way rather than the other way. In the case of entropy, this means that we have a certain direction of reference (such as our past-to-future direction) and the entropy increase could be either way but it happens to be in the positive direction (our past-to-future direction). But this is not at all the case. The entropy must be changing in one way or the other if it is not at its maximum. The animals can live in the world only in such way that their future coincides with the entropy increase. *The direction of entropy increase therefore is neither nomologically determined nor accidental. It is necessary, biologically, psychologically, epistemologically and metaphysically.*

## 7. Why is Prediction Effective?

Suppose you hold a book in a horizontal position and drop it from 6 inches above the center of the table. You know what will happen. It will stop at the center of the table and remain there. This will be the same if you drop the book from a height of 3 inches or one foot. You may even throw the book from the side and still get the book at the same position, if you aim right. Next, in another experiment, suppose you see the book lying at the center of the table. You ask yourself: where the book has come from? Was it dropped from 6 inches, 3 inches, 12 inches, or was it thrown sideways from somewhere else? You cannot tell. There are an infinity of possibilities and the probability is thinly distributed over all of them. Prediction is specific and almost deterministic, in the sense that the answer is almost unique, whereas retrodiction is utterly unspecific. Why? In other words, why is prediction more effective than retrodiction?

This question of effectiveness has to be distinguished from the question of success we discussed in the last section. A guess can be mathematically expressed by a probabilistic distribution, whereby the probability is one's subjective degree of expectation. This guess will be successful, if this probability distribution agrees with the actual relative frequency. But a successful guess may be worthless as a practical guide of conduct if the probability distribution is widespread over many alternatives. On the other hand, if the probability distribution is sharply concentrated on a few alternatives, the guess is useful because we can prepare ourselves effectively assuming that one of these few cases will really be the case. This effectiveness which depends on the sharpness of probability distribution, becomes essentially what was called "predictive power of a hypothesis" [25] when the notion is applied to a prediction derived from a hypothesis.

A superior effectiveness of prediction over retrodiction is characteristic of the system with increasing entropy. That the experiment described above involves an entropy increase can be understood easily when one thinks about the energy transfer that takes place when the book stops on the surface of the table. The mechanical energy is transformed mainly into heat energy. The oriented motion of the molecules of the falling book gives rise to random motions of the molecules of the table, the air, the book, etc.

The reason for the effectiveness of prediction in an entropy increasing system is as follows. Entropy increase implies that the final macroscopic state consists of many microscopic states, which in turn implies that no matter what initial macroscopic state one may start with, there is a large probability of landing this final state with many microscopic states. This means that the predictive probability distribution with respect to macroscopic states has a structure very much like many-to-one correspondence. This is evidently a consequence of the two facts, that the microscopic predictive probability has a structure like one-to-one correspondence (Liouville Theorem) and that there is a macroscopic state with overwhelmingly many microscopic states.

If the predictive probability distribution has a many-to-one correspondence leading to a large-entropy final state, the retrodiction starting backward from the large entropy state will become one-to-many and the retrodicted initial macro-states are widely spread. Hence retrodiction is ineffective. The situation is very close to what we called limping determinism in one of the earlier sections of this paper.

## 8. *Entropy Decreasing System and Counter-Causal Science*

Since the time the second law of Thermodynamics was formulated, there have been many scholars and thinkers who expressed some uneasy feeling or suspicion about the applicability of the law to living matters. Many of these opinions were ill-founded, but not all of them can be dismissed as nonsense. I think we can conveniently classify into three categories the various facts which seem to violate the entropy-increase law.

The first category pertains to the fact that living matter seems often to be provided with a tremendous amplifying mechanism in it, which can amplify a small fluctuation into a global change. If this is the case, a second law of Thermodynamics is bound to become a very "poor" statistical law, just the same way as it is a poor statistical law for a system consisting of a small number of molecules. The opinion that the concept of entropy cannot legitimately be applied to a living matter is, in reality, often intended to indicate such an amplification mechanism.

The second category is related to the fact that living matter by definition is an open system whereas the entropy-increase law is applicable only to an adiabatically isolated system. To live, the living matter has to exchange matter and energy (including heat). But this explanation is only negative in the sense that the entropy-decrease in an open system is *not forbidden* by the second law. Of course the existence of an open system whose entropy is decreasing is very easy to substantiate. A simplest example is a thermostat losing heat through the boundary, but such a system has no resemblance with a living organism. In this respect, Prigogine's recent paper [8] is more interesting, describing a chemical reaction which has some remote resemblance with production of living matter. We shall soon come back to a system whose physical entropy is decreasing, but before that, let me describe the third category of entropy-decreasing phenomena.

It is a common observation that living systems tend to generate heterogeneity, difference, structure, organization, order etc., whereas inert systems tend to generate homogeneity, equality, chaos, disorganization, disorder etc. Since the second law more or less represents the latter tendency, the former tendency seems to be a violation of the second law. But a closer examination reveals that what is usually considered as a case of the former tendency is not formulatable in terms of the physical entropy, but is often formulable in terms of an entropy defined in terms of non-physical probabilities [18, 21].

This last idea was first introduced by me in connection with inductive learning process, and the mathematical theorem obtained there was called "inverse H-Theorem". In fact, it is almost obvious that the entropy must decrease in the inductive process if it is defined by the probabilities (credibilities) placed on various competing hypotheses, because such a probability (credibility) distribution must become more and more concentrated on fewer and fewer hypotheses in the measure as our experience grows. An interesting thing is that such a tendency can be given a rigorous mathematical proof. If such an inverse H-Theorem is valid for inductive learning, then it is not farfetched to guess the same tendency to exist in other types of learning such as conditional reflex and skill development. These are verified as the behavioral inverse H-Theorem and its validity can be tested easily by defining the entropy by the response probabilities [18, 25].

Learning may not be the only field of animal behaviors in which an inverse H-Theorem is valid. The obvious structure-building character of animals may also be expressed as an entropy-decrease law provided the probabilities are suitably defined. What is of prime importance is that *animals can create structures suitable for their survival and self-development precisely because they have the freedom of initial condition which is a manifestation of the entropy-increasing (structure-destroying) tendency of their physical environments.*

Now let us go back to the second category of entropy-decreasing phenomena peculiar to some open systems (where the entropy is understood in the sense of physical entropy). We are not accustomed to physical systems of this kind, and as a result, it is difficult to imagine what kind of scientific theory will be valid for them. However, we can guess that the inverse of what has been said in Secs. 6 and 7 will become true. In particular, the macroscopic state at hand will be quasi-random, oriented to the backward direction of time, hence the retrodictive and not predictive probabilities can be nomologically determined. This will justify to apply a version of the H-Theorem toward the backward direction of time, explaining the entropy-decrease. Retrodiction will be more successful than prediction [23].

We can also conjecture that retrodiction will be more effective than prediction for a system whose physical entropy is decreasing. This must be so, because a passage to a state with a smaller entropy means a passage from one macroscopic state to one of the many possible macroscopic states, hence the number of probable final states is large, and as a consequence, prediction is not effective.

This peculiarity must have a deep connection with the often recognized fact that a "teleological" explanation is very easy in biology whereas causal explanation is possible only in a round-about fashion. Rigorously speaking, one should not call it a teleological explanation because it would suggest the existence of a consciously conceived end, which is not always the case. Beside this, even if the end is consciously conceived, the final state is only hoped to coincide with the end, and the coincidence is not a unfailing rule. What counts is the relation between the earlier and later event, and, in particular, an explanation (derivation) of the former from the latter. Of course, a living organism as a whole may or may not be entropy-decreasing. But, it may be safe to guess that certain parts of it are entropy-decreasing systems.

In any event, we may put forward a thesis that *legitimate science for an entropy-decreasing system must be retrodictive (counter-causal) just as much as legitimate science for an entropy-increasing system is predictive (causal). Philosophers and scientists of the past, and of the present, are completely prejudiced on this point in claiming without any foundation that prediction and causal argument are the only legitimate enterprise of science, or even of rational thinking in general.*

## 9. *Becoming and Creation*

Let us first point out some of the salient features of the concept of "becoming" in ordinary language and then try to explicate them in a more theoretical framework.

(1) Becoming involves a change or changes. What is permanent does not become. This is obvious and does not require further comments.

(2) Becoming is not a haphazard change. What you see through a train window is a series of changes, but can hardly be called becoming. The "six" after a "two" in die-throwing is not a becoming. This point is connected with point (4) below.

(3) Becoming, however, implies a certain selection out of many posibilities. If some event is conceptually the unique possibility, it cannot be called becoming, even if it is a change. This point is connected with point (5) below.

(4) Becoming has a connotation of design, fulfillment, achievement, and meaning. The passage of a bud into a flower is a becoming. Artistic creation is a becoming.

(5) Becoming is a change from the status of being undecided to the status of being decided.

(6) Becoming is a passage from the future to the past. It is located at the present.

Let us now try to translate these statements in the language of the objectivistic world $W_0$. A typical representation of this world is the Minkowski space in which every (four-dimensional) point can be assigned to some physical event. The physical existence of all human observers and their behaviors are supposed to be described in this four-dimensional space. It is not the essential point here whether the later events are deterministically determined by earlier events or are only probabilistically determined (according to some laws) by the latters. Even in the latter case only one physical event can happen at one point in space and at one instant in time, and that unique physical event is assigned to the four-dimensional point on the four-dimensional map.

It is obvious that there is no room for becoming in such a picture of the world, because there is nothing undecided, there is no "now", there is no room for choice or design. In an effort to reconcile the world view of $W_o$ with the idea of becoming, some scientists and philosophers resorted to a grotesque image of a conscious being living in this world $W_o$. They said that a conscious being travels along his world line in this Minkowski space. But, this is obviously self-contradictory. If they say that he (the conscious being) travels, it implies the elapse of time associated with this motion. But, what kind of time is this? Time is already described as a fixed coordinate in the space-time, hence one cannot introduce another time along which a point can move around in this four-dimensional space.

In an effort to avoid such an inconsistency, a more careful writer who nevertheless wants to retain the world view of $W_o$ formulates his thesis as follows: Becoming is a point or points in the above-mentioned map being noticed by consciousness, or something similar. But this reformulation does not really avoid the inconsistency it pretends to overcome. Insofar as the objectivistic world $W_o$ is used, the neuro-physiological-behavioral counterpart of any mental phenomenon is permanently registered on the map. The taking cognizance of any event is a mental phenomenon but its objectivistic record is already registered all along the world line of the observer. There is no now on this world line, hence there can be no true becoming. The mind cannot be detached from its bodily counterpart and fly around like a ghost; nor can it have a property to which the body does not have any counterpart. This failure is to be expected from the beginning because the objectivistic point of view underlying the world $W_o$ leaves no room for the subject or the observing or acting mind in true sense.

Let us now try the world view $W_t$. In this world, we have clear distinction between future and past: Future is what we can to some extent influence and past is what could have influenced us. We ourselves are seated at now which is the frontier of the two regions, where action and perception take place. The future event is only a potentiality because we can change it. But, precisely for that reason, it can be planned, designed and given meaning. Such a potentiality becomes actuality when it passes through the present. In our mind we have a series of events that have turned from potentiality into actuality. Hence, there are constant changes, which are neither haphazard nor deterministically determined, but which could be molded and which could convey meaning. These facts are precisely what I enumerated as features of becoming. Becoming has an adequate explication in the world view of $W_i$. The concept of time in this world view may be called "Creative Time", because only in this view of the world with its corollary concept of time we can think of creation of future under the direction of our Will.

The last question that we would like to ask is: Can we somehow geometrize the situation or at least graphically represent $W_i$ as we could in the world view of $W_o$? The answer is no, but we may suggest the following picture as a kind of help for visualization of the situation although it really lacks many of the essential features of the true situation. At each point of the world line (of a person), we can define a Minkowski coordinate system with its origin coinciding with that world point in such a way that the time axis is tangent to the world line. We can then pass from the coordinate system at one world point on the world line to the coordinate system at another by a Lorentz transformation. This is a coordinate system attached to, and traveling with the person, but only with a peculiarity that the time origin is shifted so that the time is counted from each world point. Note that each coordinate system can be identified by the proper time value of its coordinate origin measured from one fixed world point on the world line. Now suppose we overlap all these coordinate systems putting corresponding coordinate-axes together and putting the coordinate origins together. Then you get a four-dimensional coordinate system just like the usual Minkowski in appearance, but requiring a quite different interpretation. To have a name for this new space, let us call it a (temporally) "telescoped" Minkowski space.

In the ordinary four-dimensional Minkowski space there is no such thing as a motion of a point. But in the telescoped Minkowski space a point moves somewhat like in the usual three-dimensional space, the time parameter here being the relative local time as measured from the origin. A fixed world point (an event qua a geometrical point) in the ordinary Minkowski will move from inside the future light cone first into the region outside the both light cones and then finally into the past light cone. We shall call the locus of such a moving point an event-line to distinguish it from the world-line in the ordinary Minkowski.

The here-and-now of the person is located at the origin of this telescoped Minkowski, but we should not claim that we have produced a mathematical expression of the here-and-now from something which did not have. In fact, the origin of the telescoped Minkowski is a superposition of nows, but not "the" now. What happens at the origin at a particular time corresponds to "a" now. Besides, the geometrical construction we used can be applied to any world line whether or not a sentient being is located on it. In this way we can introduce a "pseudo-here-now" and corresponding "pseudo-future" and "pseudo-past" for any object or any point (moving with a velocity less than light).

Now going back for a while to the ordinary Minkowski space, we can say this. If a person is located at the origin of the space and he is excluded from the physical description of the world, all the world points in the future cone can be reached by his action, hence the physical content of such a world point is undecided by the physical world (minus him). From his point of view, it is an event that can be influenced or controlled. For instance, suppose he knows that there is a tree at $l$ meters which he wants to destroy. His cannonball will reach the tree at a time later than $l/c$ seconds after his decision but not earlier, where $c$ is the light velocity in meters per seconds. The physical content of the space-time point at which the tree is shot is under his control. The future cone is undecided, controllable and of course, unobservable.

On the other hand the physical content of the world point in the past cone can be observed and cannot be altered. Hence it is decided (up to retrodictive uncertainty). Besides, it is a world point from which an influence could have reached the person at the origin. The past cone is decided, controlling and observable.

The extra-conical zone is obviously unobservable, uncontrollable, and uncontrolling. Whether it should be called decided or undecided depends on one's metaphysics. A realistically inclined person would like to call it decided, but an idealistically inclined person may prefer to call it undecided. In any event, because the light velocity $c$ is very large in the human scale, the temporal thickness of the extra-conical zone is extremely small (at a reasonable spatial distance) so that the length of stay of a world point in this limbo is extremely "short". In this connection, it may be mentioned that some philosophers proposed to consider an event that is in the extra-conical zone as "real" on the grounds that there is a coordinate system such that it becomes simultaneous with the now of the person. But, from the present point of view, this idea is most preposterous. The points in the extra-conical zone are in fact the remotest points for the person because they are unobservable, uncontrollable, uncontrolling, and uncertain about their being decided. We shall explain shortly why this illusion has occurred.

Now going back to the temporally telescoped Minkowski space, we can say that the future events with their controllability and undecidedness are all mapped onto the future cone here regardless of which now they refer to. Similarly for the extra-conical zone and the past zone. The event qua an empty geometrical point gradually moves out of the future zone, enters the extra-conical zone and finally enters the past cone. All along this trip, the point keeps its identity as a receptacle for a physical event, but the status of the physical content with regard to decision, influence, and observation changes.

When no person is present at the origin of the telescoped Minkowski space, the characterization of zones in terms of being controllable and controlling, observable and non-observable disappears. The question of determination (status of being decided) is more subtle, but in my opinion, the characterizations of the zones in terms of determination loses its sense as soon as a sentient agent is absent. In spite of this, one can formally define the pseudo-future, pseudo-present and pseudo-past as soon as a world line is given even if no sentient agent is present at the origin. I think this is the reason why we can sometimes use the word "becoming" to an inanimate object considered by itself. Perhaps, we tacitly imagine some supernatural hand is intervening at the pseudo-present of the object.

An unanswerable question in this visualization of the world-to-be-acted-upon in terms of the telescoped space pertains to the alternatives: Is the agent part of the picture or not? This is supposed to be the world-to-be-acted-upon by him, hence he cannot be there in the picture. But on the other hand, we said that he is located at the origin of the space. This is an inevitable dilemma, because we tried to construct a picture of the world-to-be-acted-upon $W_i$ from the picture of the all-including world $W_o$, whereby $W_i$ and $W_o$ have a certain basic irreconcilable contradiction.

But, we can say this much. The bodily extension of a human (or any other) agent is finite. Its nervous system, too, occupies a certain finite domain. The transmission of nervous stimuli takes finite time. It is also psychologically confirmed that the psychological "present" has a finite duration of physical time. For this reason, it is more faithful to the fact to depict "here-and-now" as a finite domain around the origin of the telescoped space.

The temporal finiteness of the present gives birth to a peculiar consequence in the structure of our world of experience $W_i$. If $\tau$ is the order of magnitude of the temporal duration of the present and if $c$ is the velocity of a signal, then any signal coming from a distance less than $l = \tau c$ can reach the observer within the duration of the present. I stated in Sect. 3, and again in the present section, that the extra-conical zone is inobservable, that is, inobservable at present. But if the present has a finite duration, the region of radius $l$ becomes a presently observable region. If we put $\tau = 0.01$ seconds and $c = 3 \times 10^{10}$ cm., we get $l$ equal to three thousand kilometers which is indeed quite large.

This is the basic reason for which we developed a fantastic illusion about the meaning of simultaneity. (Of course, we are not talking about the old controversy with regard to simultaneity raised by relativity. We have here an entirely new problem.) We thus have an unalterable feeling that events happening elsewhere but at the present instant are something very real, more real than past or future. Not only common sense, but respectable philosophers could not free themselves from this illusion. For the same reason, people believed that prediction on the basis of the present data is a reliable enterprise. Nobody indeed seems to have noticed that the present data are actually unknowable, except in a fuzzy and limited sense mentioned above. The presently observable region $l$ will shrink to zero if we impose a sharply defined time instant for the initial condition.

I have certainly gone far beyond the time and space allowed to my presentation. Let me finish this paper by summarizing it by saying that

>Will implies Creation,
>Creation implies Time,
>Time is Life.

*References*

1. Brillouin, L.: *Vie, Matière et Observation*. Paris: Albin-Michel 1959.
2. Buck, R., Grünbaum, A.: Philosophy of Science, 30 (1963) 359 and 370.
3. Carnap, R.: *Logical Foundations of Probability*. Chicago: University of Chicago Press 1950.
4. Ehrenfest, P., Ehrenfest, T.: *Encyclopaedie der Mathematischen Wissenschaften*, Vol. IV-4 Leipzig: Teubner, 1909-1911.
5. Frege, G.: *Philosophical Writings* (translated by Geach and Black), Oxford: Blackwell 1952.
6. Grünbaum, A.: *Philosophical Problems of Space-Time*. New York: Knopf 1963.
7. Peirce, C. S.: *Collected Papers*, Vol. 3. Cambridge: Harvard University Press 1960.
8. Prigogine, I.: *Structure, Dissipation and Life*, presented at the International Conference on Theoretical Physics and Biology, Institut de la Vie, Versailles, France. 1967.
9. Reichenbach, H.: *The Direction of Time*. Berkley: University of California Press 1956.
10. Watanabe, S.: Scientific Papers of I.P.C.R., Japan. 31 (1937) 109.
11. — Kagaku (in Japanese), 14 (1944) 3; 14 (1944) 138.
12. — *On the Origin of Time* (in Japanese). Science of Thought, 1947.
13. — *Time* (in Japanese). Tokyo: Hakujitsu Publishing Co. 1948.
14. — Progress of Theoretical Physics, 2 (1947) 71; 3 (1948) 378; 4 (1948) 1.
15. — Physical Review, 84 (1951) 1008.
16. — Article in: Louis de Broglie, Physicien et Penseur, p. 385. Paris: Albin-Michel 1952.
17. — Reviews of Modern Physics, 27 (1955) 179.
18. — IRE Transactions Information Theory, IT-8 (1962) 246.
19. — Progress of Theoretical Physics Supplement, Extra No., (1965) 135.
20. — In: Fraser, J. T. (editor), *The Voices of Time*. New York: Braziller 1966.
21. — Progress in Biocybernetics, (editors: Wiener and Schade). 3 (1966) 152.
22. — Progress of Theoretical Physics Supplement, No. 37, 38, (1966) 350.
23. — Progress of Theoretical Physics, Supplement, Extra No. (1968) 495.
24. — Information and Control, 15 (1969) 1.
25. — *Knowing and Guessing*. New York: John Wiley 1969.
26. —*Logic of the Empirical World*. In: The Proceedings of the International Conference on Philosophical Problems in Psychology, 1968, Honolulu. To be published by the University of Hawaii Press.

# Temporal Order as the Origin of Spatial Order in Embryos

B. C. GOODWIN*

*Summary.* Embryological development takes place in four dimensions and requires the existence of time and space measuring processes within the embryo. It is suggested that space measurement (the establishment of embryological fields) results from periodic intracellular events which propagate at different rates from cell to cell in developing tissues. Some general consequences of this model are investigated.

## Introduction

It has long been recognized that the unfolding of the developmental process in embryos involves the intimate interplay of temporal and spatial factors. The emergence of spatial order and form in an organ such as the vertebrate limb or eye occurs in a well-defined temporal progression which, if seriously disturbed, results in a deficient structure. The question I would like to direct my attention to is the nature of the relationship between the temporal or sequential constraints which operate in embryonic cells, and the spatial constraints which result in the well-defined morphology of tissues and organs. My thesis will be that temporal order in cells gives rise to spatial order in cell aggregates or tissues.

One of the obvious temporal constraints which operate in embryos is that which arises from events in the biochemical or molecular substratum, the description of whose behaviour is the preoccupation of many biologists today. Enzymes catalyze reactions at certain rates and macromolecular syntheses occur within certain well-defined rate limits. The biochemical events which underly embryological processes impose certain velocity bounds on the overall process of development which clearly cannot be violated. This is the domain of biochemical time, the t which enters the rate equations for biochemical reactions. At the molecular level events are necessarily discrete or quantized, since an enzyme molecule, for example, undergoes a well-defined reaction or work cycle producing one molecule of product in each cycle. But above this level, biochemical time is usually regarded as structureless: averages of molecular events over time periods which are long (minutes or hours) compared with the molecular work cycles (milliseconds or seconds) are temporally homogeneous.

---

* Dr. B. C. Goodwin, School of Biological Sciences, The University of Sussex, Falmer, Brighton, Sussex, England.

Furthermore, spatial sampling from the biochemist's test tube also gives a uniform distribution of molecular species. We may say that the macroscopic world in which the biochemist usually works has neither space structure nor time structure. Clearly this is not the world of biological systems.

In recent years considerable attention has been directed towards the problem of understanding how structure can emerge from systems which, like the biochemical world described above, are initially uniform in their properties. One of the most important observations has been that described recently by Hess (1968). He and his colleagues have reconstructed, in the test tube, biochemical systems which have a well-defined time structure at the macroscopic level. This structure, the occurence of continuing periodicities in the concentrations of the molecular species involved in the reaction sequence, is one of the simplest that can be imagined. It is also one of the most important, having very far reaching consequences for the organization of biological systems. This elementary time structure gives to the system a macroscopic time-measuring capability; and with this capacity to order and measure events in time comes the possibility of ordering and measuring events in space, as will be discussed. Of course these ordering and measuring processes are not those of the physicist, used to measure physical events. They are essentially biological, and give to the organism possessing them its own coordinate system which may and may not correspond to that of the physical world. In certain instances it evidently is to the advantage of the organism to make such a correspondence, to have certain of its own internal time-measuring activities work in the units of solar time. We then observe the rather striking phenomenon of biological clocks, whose nature and properties have been the subject of fairly intense study in recent years. Such clocks are, however, no more than a particular specialization of a general time-measuring capacity of organisms which arises from a very basic feature of the biochemical and physiological organization of biological systems: the capacity to generate continuous, stable oscillations or rhythms. The biochemical oscillator described by Hess is probably the best understood and fully analyzed instance of this general property of controlled biological processes.

*Measurement, Temporal Anisotropy, and Memory*

We may ask at this point if there is any connection between the time structure which emerges with biochemical oscillations and the one-way flow of time which is certainly a characteristic of biological systems. It would appear, at first sight, that time invariance could well hold for biochemical reactions whether or not there are periodicities. All biochemical reactions are reversible, and at the molecular level where reactions are described by kinetic equations constructed in terms of collisions, transition states, activation energies, etc., time invariance holds just as it does for the equations of motion of gas molecules.

However, as Landsberg (1970) has shown so clearly in his contribution to this conference, the moment one introduces an averaging instrument to measure mean values of molecular concentrations or molecular collisions, one introduces irreversibility and strict time invariance is destroyed. Mathematically this is because the averager changes the nature of the transformation describing the trajectories in dynamical phase space from a group, having an inverse, into a semigroup, with no inverse. Landsberg has introduced an intermediate category of process between time invariant and oneway, which he calls weakly time invariant. The transformations describing such processes are contraction mappings. An example of such a process is diffusion, where time-reversal is not entirely excluded; it has a finite probability. Here again it is the averaging process which produces partial irreversibility.

I would like to suggest that strict time invariance is never satisfied by biological systems because the very basis of biological process depends upon the operation of an averaging device at the molecular level: the allosteric enzyme. The essence of the averaging or measuring activity with which I am concerned is not resident in the catalytic action of the enzyme, but in its response as a control device to the concentration of molecular species or ligands which are sterically quite unlike the molecules involved in the catalysed reaction, whose rate is controlled by the concentrations of the ligands. The enzyme which is the key to the oscillation described by Hess, phosphofructokinase, is an excellent example of such an allosteric catalyst. This enzyme has allosteric sites which "recognize" or bind ATP and FDP in amounts dependent upon their concentration. Such measurement involves an averaging process, the detection of the mean concentration of ATP and FDP in the vicinity of the enzyme molecule by the enzyme itself. The result is the setting of the rate of phosphorylation of FDP at a value which is a well-defined function of the ligand concentrations, ATP and FDP. Thus macromolecular averaging devices, the allosteric enzymes, introduce directional time-flow into the biological process at its most elementary level.

I believe it is useful and instructive to look upon such an averaging device as having a primitive sort of memory. In order to carry out a time average over elementary events such as collisions of molecules with a surface, it is necessary that the responding system have a relaxation time which is long compared with that of the elementary events. This is clearly true of the proverbial piston in the gas cylinder: its inertia is such that single molecular collisions have no observable effect, but the collective kinetic energy of the molecules exerts a certain pressure and holds the piston in a certain position. Similarly, the ligand molecules forming a cloud in the vicinity of the allosteric site of an enzyme are constantly exchanging with one another on the site, and the region of the catalytic site of the enzyme maintains a configuration which does not immediately flip during the interval between one ligand molecule leaving and another arriving to take its place. The relaxation time for configurational changes in a protein molecule is long compared with collision frequencies of ligand molecules with sites. Such averaging systems can in fact "remember" for a certain period what the concentration of ligands was at a slightly earlier time, in view of this slower relaxation time; and similarly for the piston in the gas cylinder. This kind of memory is a very primitive one, insofar as it involves no time ordering: the system cannot distinguish between the order in which events occur in time. For example, if X moles of ligand A are added to a solution of an enzyme and then Y moles of a second ligand B, then the final result in terms of reaction rate will be the same as if B was added first and then A. At least, this is the usual observation. If it were violated, then allosteric enzymes would have the properties of a genuine memory.

The point I am making here is that the distinction between systems with time invariance and those without is one which can be simply represented in terms of memory. In a memory-free universe, all processes will be time-invariant. This is the world of microscopic physics, of mechanics. The moment one enters the macroscopic domain, time invariance is violated because of the properties of the measuring devices which give the macroscopic averages: they destroy information about the dynamical trajectories. And biology enters this domain from its very inception, at the level of macromolecular control units, allosteric enzymes. Thus biology operates in a world with memory, and time invariance is always violated. This observation is related to but is not deduced from the second law of thermodynamics, which is usually regarded as the origin of time's arrow. Biological processes also put an arrow on time, but not because they undergo an entropy increase, which may not be true since the second law applies to biological process plus environment as a closed system, not to be biological process itself. The arrow in the biological process derives from the occurrence of averaging operations within the system itself.

*The Origin of Spatial Order in Embryos*

There are two fundamentally different ways in which ordered spatial patterns can arise in aggregates of embryonic cells. One involves the spatial analogue of the dynamical instability which give rise to oscillations in time. Recent studies by Gmitro and Scriven (1969) and by Lefever *et al.* (1967) have demonstrated the conditions under which spatial periodicities in concentrations can arise in initially uniform systems. Such spatial structure arises independently of any underlying time-measuring process, and it is not clear just how one could use this kind of structure to generate a space-time coordinate system such as is required by the embryo to organize itself in four dimensions. Within single cells, at the level of organelles and membranes, it seems likely that such "symmetry-breaking" processes will be found to be important as space-structuring forces. Above the level of the cell, however, in the organization of embryonic tissues, one of the simplest ways of generating a coordinate system is to use time structure to build space structure, thus coupling the dimensions into a complete coordinate system.

Spatial order can emerge in an aggregate of cells only if there is some kind of interaction between them. So besides postulating the occurence of temporal order in embryonic cells of the type described by Hess, and discussed somewhat more generally in my book (Goodwin, 1963), it is necessary to suppose that signals can pass from cell to cell in an aggregate. There is in fact good experimental evidence for the existence of communication channels between cells in embryonic systems (Potter, Furshpan, and Lennox, 1967). A simple consequence of these two postulates is that waves of occurrence of a periodic process in the cells can travel over the aggregate, as is shown in detail in a paper by Goodwin and Cohen (1969). These waves arise from the simple observation that a periodic event occurring in one cell can initiate the same event in an adjacent cell, and thus in the next cell, and so on. Under the natural assumption that there is an upper limit to the frequency of the periodic process under consideration, so that there is a refractory period of a particular duration after the initiation of one cycle before the next can commence, one gets directed wave propagation of the periodic process in an entrained or synchronized set of cells. If, furthermore, one assumes the existence of an initial frequency gradient with respect to the autonomous periodic event over the array of cells, then there will be a spatially localized centre from which the wave propagates. Those cells with the greatest autonomous frequency of the periodic event will be the centre or the pacemaker region of the tissue, and will drive the other cells at this frequency. This type of behaviour is familiar in the working of the heart, where the periodic event involves the visible contraction of the heart tissue; a wave of contraction spreads out from the pacemaker, which sets the frequency of the periodic contraction wave.

If there is a second event which follows the first, periodic event, and propagates from cell to cell with a velocity less than that of the first event, then we have the basis for a space-measuring axis in the cell aggregate. For in a cell at any given distance from the centre where the propagation of the two events originates, there will be a particular time interval between the occurrence of the two events. This time interval, which can be represented as a phase angle between 0 and $2\pi$ radians, where $2\pi$ radians is equal to T, the period of the oscillatory event, now becomes the space-measuring interval. It determines the biochemical state of a cell in any position in the aggregate, in view of the fact that the temporal ordering of metabolic processes in cells can determine their biochemical potential. Then spatial distance is measured by means of a time interval between two events in any cell, both of which have the same periodicity. The unit of linear measure is determined by the limit of the resolving power of the system in relation to time intervals between different biochemical events. For example, let us suppose that the period of the oscillating process in the pacemaker cells in 3 min. If the minimum time interval between the two periodic, propagating processes which produces a significant change in the biochemical state of the cell is 30 sec, then the maximum number of distinguishable units or regions along a linear axis would be $\frac{3 \times 60}{30} = 6$. If the axis consists of 120 cells in length, then each distinguishable region will consist of 20 cells. This would be the limit of resolution of such a system. Many embryological tissues show a much greater degree of resolution of the differentiation process than this. For example, the vertebrate retina shows a complex mosaic of rods and cones in which there can be a very fine differentiation pattern, adjacent cells being different. However, in this case the details of the mosaic are almost certainly determined by secondary, local interactions between cells, not by a spatial ordering at the global level over the whole retina. It is the latter process that is being considered here, the establishment of the early embryonic field which organizes developing tissues over domains which are large in relation to single cells.

If the length of an embryonic tissue along the axis of propagation of the waves is such that the time interval between the occurrence of the two waves reaches values greater than the period of the pacemaker oscillation, then the units of measurement along the axis repeat themselves and a spatial periodicity can occur. Thus measure along the axis is modulo $2\pi$ radians or T minutes. Spatial periodicities are frequently observed in embryonic patterns, so it is convenient to have a space-measuring process which gives rise easily to such repeating spatial cycles. The periodicity of the underlying temporal process in the model being considered results very naturally in such a result.

A further important property which a space-ordering process in embryonic systems must have is the capacity for regulation. There is a class of embryos in which a reduction in size of an embryo by accidental damage or deliberate surgery at an early developmental stage results in the production of a complete but small organism. Evidently the units involved in the specification of cell states along the developmental axis can be adjusted according to the size of the embryo. This requires a size-sensing mechanism of some kind. In terms of the theory under consideration, such a mechanism must adjust the slope of the phase gradient generated by the two propagating waves; i.e., the relative rates of propagation of the two events must be subject to adjustment according to the size of the embryo. A relatively simple way of achieving this result is suggested in the paper giving the detailed description of the model (Goodwin and Cohen, *op. cit.*). The proposed regulation process depends upon the phase angle of the two primary events, hence on the size of the aggregate. Once a particular linear dimension (and hence a particular phase angle) has been exceeded in an embryo or a tissue, it is postulated that a regulation event is initiated which adjusts the relative rates of wave propagation in the tissue until the 'correct' phase angles fit into the available tissue.

It is a simple matter to increase the number of coordinate axes in an embryonic tissue by postulating the occurrence of secondary, tertiary, and other centres from which waves of different kinds propagate. The tissue can thus be organized spatially in multiple dimensions. Furthermore, these coordinate axes can all be established in relation to one single clock, the primary pacemaker, all other propagating events originating in response to the wave generated from the pacemaker. The coordinate axes so obtained are not orthogonal axes, but one need not assume that the embryo is greatly concerned about orthogonality.

*Space-Time in the Organism*

The type of model under consideration, in which spatial structure arises from and depends upon temporal periodicities, gives rise to some fairly definite predictions regarding the type of experimental treatment which should affect the organization of the system. Clearly any interference with the underlying periodicity of the pacemaker should result in a perturbation of the field which it is organizing. One of the obvious ways in which to attempt to reorganize the coordinate axes in an embryonic system, and hence to alter normal development, is to introduce into the system at a particular location a periodic signal which mimics pacemaker activity. This type of experiment has actually been carried out with the fresh-water coelenterate, *Hydra littoralis*, which has been used for many years for the study of development because of its regenerative capacities. It was found that the introduction of an electrical periodicity into the proximal region of a regenerating section of the digestive zone resulted in the establishment of a partially-reversed axis of regeneration such that the animal developed two 'feet' instead of the normal single foot. The periodic signal apparently had the effect of initiating a second polarity along the developmental axis so that instead of there being a unique direction of measurement, there were two, originating from a medial point on the longitudinal axis. It is as if the origin of the coordinate had been shifted along the axis, and spatial ordering then proceeded in both directions from this new origin, resulting in an animal with a head (origin) in the middle and two feet, one at either end.

The significant properties of the signal producing such a reorganization of the regeneration field are its frequency and the location of its introduction into the tissue. Frequency and spatial location are precisely the factors which should be relevant in producing specific spatial order from localized developmental 'clocks' by means of a wave-propagation process of the type described. In the case of *Hydra*, the frequency which produced maximal response was 1 pulse every 2.5 min (the current was a DC pulse of 1.5 volts lasting for 150 msec, introduced through a $40\mu$ platinum wire). This, then, is the frequency of the developmental clock in the hypostomal region, if the above interpretation of the observations is correct. This frequency suggests a metabolic oscillator of the general type described by Hess.

The frequency spectrum available to cells is quite extensive, and different developmental systems could use different frequencies. There is one consideration, however, which indicates that there may be a relatively narrow frequency band which is used in developmental systems throughout the range of phyla, so that all developmental clocks may have periods between, say 1 and 20 min. The biochemical oscillator which is postulated to underly the time and space-ordering process in embryos must have certain properties if it is to function reliably for the developmental process. In particular, the propagation capacity depends upon the occurrence of some transmissible signal which is generated as part of the oscillation. This signal might be an ion such as a sodium or potassium or some other small molecule which can pass easily from cell to cell via a membrane structure such as the tight junction or desmosome. The oscillation must have certain other properties also which are discussed in detail in the original paper. It is of interest to observe that biochemical oscillations of the type under consideration here occur in single-celled organisms such as yeast, so that we may assume the pre-existence of such oscillations before the appearance of the metazoa and the evolutionary emergence of the developmental process. Once a basic space-ordering process such as that proposed here had emerged naturally and spontaneously from the behaviour of cell aggregates, it seems likely that it would be conserved as a basic developmental mechanism throughout the phyla. This would imply that the mechanism has universal properties, and may exist with only relatively small variations throughout the plant and animal kingdoms. If this were the case, then one would expect that the frequency range of the oscillation would be quite restricted, since the underlying biochemical process would be essentially the same throughout the phyla. This further suggests that the oscillation may well be closely connected with the basic energy-generating processes of the cell, as is the case with the oscillator described by Hess. In this case the frequency of the developmental clock in an organism would be intimately connected with the rate of energy production. The time scale of developmental processes would then be determined by the overall metabolic rate, a correlation which is known to be generally valid and must in any event obtain.

Universality of the biochemical mechanism whereby coordinate axes or embryonic fields are generated in no way precludes the emergence of the great variability in developmental detail which we see today among the species. This variability can arise from different "interpretations" of the information in the embryonic field, a point which has been made very clearly by Wolpert (1969) in his interesting paper on the subject of positional information in embryonic systems. Different genotypes would be expected to show differing responses to a phase gradient, or indeed to any gradient system.

Another aspect of the model under consideration which is of some interest in relation to evolution is the obvious similarities that exist between a system with pacemaker centres and wave propagation, and the behaviour of the central nervous system. The model we are considering is like a primitive neural network, wherein transmission of signals from cell to cell is achieved by an ion or small molecule acting in a manner rather similar to a neural transmitter substance. The coordinate system or information field which results is one which provides the embryo with a framework for self-reference, within which the process of individuation (Waddington, 1956) takes place according to self-organizing principles. Within the central nervous system there is a similar ongoing, self-organizing process. The nervous system is in fact that part of the adult which, in its plasticity and developmental capacity, resembles the embryological process. It seems reasonable to suppose that embryological-type properties are retained in the higher vertebrate nervous systems, giving them their remarkable capacity for restructuring their own information fields in response to 'inductive' stimuli from the sensory environment. The occurrence of clocks in the central nervous system such as the one demonstrated by Pöppel would then be an entirely natural expectation; and we might even suggest that their function may not be solely to keep time, but in fact to organize also the spatial distribution of information in the central nervous system, possibly connecting their activity with the organization of associative and distributed memory fields. Such processes would be dependent upon the occurrence of propagating waves of activity in neurones additional to the familiar nerve impulse, probably involving biochemical activities. Biochemical changes of some kind are required in any event for the changes of biochemical state which apparently occur in long-term memory. Interesting possibilities are evident here in relation to the temporal to spatial mappings which occur in the brain, and the inverse of these mappings, spatial to temporal, which underly behaviour conditioned by past experience. Such considerations arise naturally from the wave-propagation model described above for embryological phenomena, suggesting how the remarkable properties of the central nervous system may have arisen, and in what manner temporal order could be intimately involved in the processes of information encoding and decoding in neural networks. Biological time would then be inextricably woven into the fabric of biological organization throughout the levels of complexity that have arisen in the evolution of the metazoa.

*References*

Gmitro, J. I., Scriven, L. E.: *A Physicochemical Basis for Pattern and Rhythm.* In: Towards a Theoretical Biology 2 (1969) 184–203. Edinburgh: University Press.
Goodwin, B. C.: *Temporal Organization in Cells.* London: Academic Press 1963.
—, Cohen, M. H.: *A Phase-Shift Model for the Spatial and Temporal Organization of Developing Systems.* J. Theoret. Biol. 25 (1969) 49–107.

Hess, B.: *Biochemical Regulations*, in: *Systems Theory and Biology*, pp. 88–114 (ed. M. D. Mesarovic). New York: Springer 1968.

Landsberg, P. T.: *Time Concepts in Statistical Physics, Special Relativity, and Some Limiting Situations*. Studium Generale. (To be published).

Lefever, R., Nicolis, G., Prigogine, I.: *On the Occurrence of Oscillations Around the Steady State in Systems of Chemical Reactions far from Equilibrium*. J. Chem. Phys. 47 (1967) 1045–1051.

Pöppel, E.: *Oscillations as Possible Basis for Time Perception*. Studium Generale 24 (1971) 85–107.

Potter, D. D., Furschpan, E. J., Lennox, E. S.: *Connections Between Cells of the Developing Squid as Revealed by Electrophysiological Methods*. Proc. Natl. Acad. Sci. U.S. 55 (1966) 328—336.

Waddington, C. H.: *Principles of Embryology*. London: Allen and Unwin 1956.

Wolpert, L.: *Positional Information and the Spatial Pattern of Cellular Differentiation*. J. Theoret. Biol. 25 (1969) 1–47.

# Time in the Evolutionary Process

J. MAYNARD SMITH*

*Summary.* Two problems are discussed. The first is whether there has been time for evolution by natural selection to have occurred. The concept of a "protein space" is introduced, and it is shown that a fundamental inequality, concerning the proportion of all amino acid sequences which form functional proteins, must be satisfied if evolution is to occur. The second is whether there is any biological law (analogous to the second law) which enables us to put a time arrow on evolutionary processes. It is argued that Fisher's "fundamental theorem of natural selection" does not meet this need.

This paper will discuss two topics, related to one another only in that both have to do with time and evolution. The first is whether there has been enough time for existing organisms, with their fantastic complexity, to have evolved by a process as apparently inefficient as the natural selection of chance variations. The second is whether there is any biological law which might enable us to put an arrow on time in evolutionary processes, as the second law of thermodynamics enables us to put an arrow on physical processes.

When confronted by the richness of organic life, it is a common reaction, particularly among non-biologists, to argue that natural selection is an insufficient explanation. Is it really possible that an elephant can have arisen by selection acting on random variation? It is difficult to give a confident answer to this question because we do not know how improbable an elephant is, or, what amounts to the same thing, we do not know how much genetic information is required to control the development of an elephant. We shall not be able to answer this question until we know more about the process of development. However, for the time being, it is reasonable to assume that the genetic information in the DNA of the fertilised egg is sufficient to control development. In the case of mammals, this amounts to between $10^9$ and $10^{10}$ base pairs. There is no quantitative difficulty in seeing how such a length of DNA might have been programmed by selection since the origin of life some $2 \times 10^9$ years ago.

In recent years this argument has taken on an apparently quantitative form (e.g. Moorhead and Kaplan, 1967), as follows. A typical protein is 100 amino acid residues in length. There are 20 different amino acids used in making

---

* Professor John Maynard Smith, BA BSC, School of Biological Sciences, The University of Sussex, Falmer, Brighton, Sussex, England.

proteins. Hence if all sequences are permitted, the total number of possible proteins of that length is $20^{100}$. This is a very large number indeed, and much larger than the total number of proteins that have ever existed, or even than the total number which would have existed if the earth had been covered since the Cambrian by a layer of proteins several feet thick changing once a second.

It follows, or so it has been argued, that the chances of reaching the actual proteins in living organisms, which are beautifully adapted to their function, by a random walk is effectively zero. Natural selection does not help, because it can only ensure the survival of functional proteins if they arise; it cannot bring them into existence in the first place.

I have tried this argument out on a number of physicists during the past few years. Almost always their immediate reaction is to suggest that there must be some physical constraints on the mutation process whereby new proteins arise. Now it is almost certain that this is not the case. All the evidence suggests that any nucleotide sequence in DNA can exist and can arise by mutation; that if it does exist it will be translated by the cell into the corresponding protein; and that the protein will be made by the cell even if it is completely non-functional.

The way out of this dilemma can be best understood by analogy with a popular word game, in which it is required to pass from one word to another of the same length by changing one letter at a time, with the requirement that all the intermediate words are also meaningful in the given language. Thus WORD can be converted into GENE in the minimum number of steps, as follows:

WORD

WORE

GORE

GONE

GENE

This is an analogue of evolution, in which the words represent proteins; the letters represent amino acids; the alteration of a single letter corresponds to the simplest evolutionary step, the substitution of one amino acid for another; and the requirement of meaning to the requirement that each unit step in evolution should be from one functional protein to another. The reason for the last requirement is as follows: suppose that a protein A B C D .... exists, and that a protein a b C D .... would be favoured by selection if it arose. Suppose further that the intermediates a B C D .... and A b C D .... are non-functional. These forms would arise by mutation, but would usually be eliminated by selection before a second mutation could occur. Thus the double step from a b C D .... to A B C D would be very unlikely to occur. Such double steps with unfavourable intermediates may occasionally occur, but are probably too rare to be important in evolution.

It follows that if evolution by natural selection is to occur, functional proteins must form a continuous network, which can be traversed by unit mutational steps without passing through nonfunctional intermediates. In this respect, functional proteins resemble 4-letter words in the English language, rather than 8-letter words, since the latter form a series of small isolated "islands" in a sea of nonsense sequences. Of course this is not to deny the existence of isolated proteins, analogous to the 4-letter words ALSO and ALTO.

It is easy to state the condition which must be satisfied if meaningful proteins are to form a network. Let X be a meaningful protein. Let $N$ be the number of proteins which can be derived from X by a unit mutational step, and $f$ the fraction of these which are "meaningful", in the sense of being as good as or better than X in some environment. Then if $fN > 1$, meaningful proteins will form a network, and evolution by natural selection is possible. In estimating $N$ it is necessary to distinguish two classes of mutations:

(i) Substitutions of single amino acids, and additions or deletions of small numbers of amino acids, making only a small change to the protein, and

(ii) mutations producing a major change in amino acid sequence. (Examples are frame shifts and intra-molecular inversions; for non-biologists, the relevant point is that there are mutational changes in DNA which alter simultaneously all or most of the amino acids in a protein.)

Mutations of the former type are much more likely to give rise to meaningful proteins than the latter. In the same way, a single random letter substitution in a meaningful word is more likely to give rise to a meaningful word than the simultaneous alteration of all the letters. Although frame shift mutations are known to occur, it is not known whether they have ever been incorporated in evolution. Hence it is better to take $N$ as the number of possible substitutions of single amino acids. If all substitutions were possible in a single mutational step, $N$ for a protein of 100 amino acids would be 1900. In practice the genetic code limits $N$ to approximately $10^3$.

Hence $f$ must be greater than 1/1000. It does not follow that the fraction of all possible sequences which are meaningful need be as high as 1/1000. It is probably much lower. Almost certainly, there is a higher probability that a sequence will be meaningful if it is a neigbour of an existing functional protein than if it is selected at random.

No quantitative difficulty arises in explaining the evolution of proteins if $fN>1$. The argument of course says nothing about the origin of life, since it assumes that at least one functional protein exists as a starting point.

Before leaving this topic, it may be worth saying something about the geometry of the protein space; these ideas emerged during a conversation with Donald Glaser. We want space in which two proteins are neighbours if they can be converted into one another by a single mutation. For simplicity, I will assume that amino acid substitutions are the only possible mutations and, ignoring the code, that all substitutions are possible. I will start with the space representing all possible dipeptides. This will be represented by a 20 x 20 chess board, each of the 400 squares representing a different peptide. A single mutational step is equivalent to a rook's move. Any dipeptide can be converted into any other by two moves. However such a conversion may be impermissible in evolution if the intermediate is meaningless. Suppose two squares (dipeptides) are connected by a meaningful path, how long can the shortest meaningful path between them be? The answer is obviously the 38 rook moves required to travel from one corner to the opposite corner by moving one square alternately in a horizontal and vertical direction. This path would be shortened by the presence of any meaningful peptides other than those on the path.

Transferring these ideas to the space of all proteins 100 amino acids long, the space would be of 100 dimensions, and contain $20^{100}$ "squares". A mutation is still a rook's move, in any one of 100 directions. Any protein could be converted into any other by not more than 100 mutations. If two proteins are connected by a meaningful path, the shortest meaningful path between them cannot be greater than 1900 steps, and will usually be much shorter.[1] It follows that if, since the origin of life, protein X has evolved into protein Y, it could always have done so in less than 1900 steps. If it has in fact taken more, it has taken unnecessary detours. 1900 steps require approximately 1 step per $10^6$ years; most proteins have probably evolved more slowly (by factor of 10) than this.

As a convinced Darwinist, I published (Maynard Smith, 1961) the conclusion that $fN > 1$ before there was any direct evidence, since if it were not so evolution would not have happened. Since that time, it has turned out that many single amino acid substitutions can be made in proteins without seriously impairing their catalytic function, (for a review see King and Jukes, 1969).

The second problem I want to discuss is whether there is any law which plays the same role in biology as the second law of thermodynamics plays in physics. On a short time scale, measured in days, biological processes have an obvious direction. The cell cycle typically ends in the division of a single cell into two, and only rarely, in the sexual process, in the fusion of two cells to form one. This is a necessary feature of life. Life is most conveniently defined as consisting of entities with the properties which enable them to evolve by natural selection; i. e. the properties of multiplication, variation and heredity. The reason for choosing this definition is that the apparently purposive or adaptive features which characterise living as opposed to dead matter can evolve in entities with these properties but not in their absence.

---

[1] *Footnote added in proof.* This is wrong. The maximum "shortest meaningful path" could be longer than this by many orders of magnitude. Hence there may be large regions of the protein space as yet unexplored.

If this definition is accepted, then it is the property of multiplication which enables us to put a time arrow on biological processes. Heredity and variation are reversible; parents resemble (or differ from) their children as closely as children resemble their parents.

It is more difficult to say whether evolution as a whole has a direction. Thus suppose we are able to make observations on the members of a species at two points in time separated by millions of years, is there any way in which we could decide which set of observations was the earlier? At first sight it seems that we can do so by using Fisher's (1930) "fundamental theorem of natural selection", which states that for any population "the rate of increase of fitness of an organism is equal to the genetic variance of fitness". Since the variance cannot be negative, the law appears to state that the fitness of a population of organism must always increase. Thus, just as we can tell which of two states X and Y of a closed physical system is the later in time, by asking which has the greater entropy, so we should be able to tell which of two states X' and Y' of a population is the later in time by asking which has the greater fitness. Unhappily we cannot do anything of the kind.

The difficulty lies in the definition of the "fitness of an organism". The essential points can be understood by considering a parthenogenetic population consisting of two genetically distinct types, $A$ and $B$, in proportion $pA:qB$. Suppose that we count the $A$s at birth, and the number of offspring, also counted at birth, produced by these $A$s. Then the average number of offspring produced per $A$ is the fitness $w_A$ of $A$. The fitness $w_B$ of $B$ is similarly defined. Then the fitness $w$ of the population is defined as $w = pw_A + qw_B$. It is this fitness $w$ which, according to Fisher's theorem, necessarily increases.

There are three reasons why we cannot use this theorem to tell us which of two populations is the later in time:

(i) The fitnesses $w_A$ and $w_B$, and hence $w$, can only be defined for a particular environment. For example, a population whose life span is short compared to a year may evolve in one direction in summer and the other in winter. Even if the physical features of the environment remain constant, the biotic features will not. For example, $A$ may be rare but better at escaping a predator than $B$. If so $w_A > w_B$, and $A$ will increase in frequency. The predator may then evolve, or change in habits, so that it is better at catching $A$ *than B*. Then $w_B > w_A$, and evolution will reverse its direction. Such reversals may have been a common feature of evolution of defence against predators and disease.

(ii) The "fitness of an organism" $w$ is not in fact a measurable property either of a population or of an individual, but a function of the *relative* fitnesses of *individuals*. Thus suppose that a population consisting wholly of $A$s increases more rapidly than a population consisting wholly of $B$s. It does not follow that $w_A > w_B$, or that $A$s will replace $B$s in a mixed population. For example $B$s may be cannibals and $A$s not. If so, in a mixed population it may he that $w_B > w_A$, and $B$s will replace $A$s, and yet a population of *A*s might increase more rapidly than a population of *B*s. This objection is of more general application than the example of cannibalism might suggest. Characteristics which are favoured in intraspecific competition often do not increase the probability that a species as a whole will survive.

(iii) Even in a narrow sense, the theorem is not always true. For example, in a diploid sexually reproducing species, if at any locus the heterozygote is fitter than either homozygote, there will be an equilibrium at which there will be a genetic variance of fitness but no change in $w$ with time. This last difficulty can be overcome by referring to "additive genetic variance of fitness". However, it does bring out the point that Fisher's theorem has no empirical content other than the laws of heredity; if certain assumptions about heredity do not hold, then the theorem does not hold.

Thus Fisher's theorem cannot help us to put an arrow on evolutionary time. Yet it is in some sense true that evolution has led from the simple to the complex: prokaryotes precede eukaryotes, single-celled precede many-celled organisms, taxes and kineses precede complex instinctive or learnt acts. I do not think that biology has at present anything very profound to say about this. If there is a "law of increasing complexity", it refers not to single species, as does Fisher's theorem, but to the ecosystem as a whole. The complexity of the most complex species may increase, but not all species become more complex.

The obvious and uninteresting explanation of the evolution of increasing complexity is that the first organisms were necessarily simple, because the "origin of life" is the origin, without natural selection, of entities capable subsequently of evolving by natural selection, and without selection there is no mechanism for generating a high degree of improbability – i. e. complexity. And if the first organisms were simple, evolutionary change could only be in the direction of complexity.

Is there anything more interesting to be said? I do not know, but I have two comments to make. The first is that processes are known (e. g. duplication) whereby the genetic material of an individual can increase. Even if the additional material is redundant or nonsensical, it does provide raw material for the evolution of increasing complexity. It is less easy to imagine processes leading to a loss of genetic material, since most losses will involve losses of functions essential for survival. The only exception is in the evolution of organisms (e. g. viruses) living in an environment more complex than themselves, which may render previously essential functions unnecessary. It is significant that Spiegelman's (1968) "evolving" RNA molecules initially became simpler, and did so in an environment more complex than themselves.

The second comment is merely that we have at present no theory of evolving ecosystems, as opposed to the evolution of the species which compose them. In the absence of such a theory, it is hardly surprising that we can say little about the evolution of increasing complexity.

*References*

Fisher, R. A.: *The Genetical Theory of Natural Selection.* London: Oxford University Press 1930.

King, J. L., Jukes, T. H.: *Non-Darwinian Evolution.* Science *164* (1969) 788–798.

Maynard Smith, J.: *The Limitations of Molecular Evolution.* In: The Scientist Speculates, ed. I. J. Good London: Heinemann 1961.

Moorhead, P. S., Kaplan, M. M.: *Mathematical Challenges to the neo-Darwinian interpretation of evolution. Philadelphia:* Wistar Institute Press 1967.

Spiegelmann, S.: *The mechanism of RNA replication.* Cold Spring Harb. Symp. quant. Biol. *33* (1968) 101—124.

# The Measurement of Perceptual Durations

ROBERT EFRON*

*Summary.* It is often claimed that consciousness is unmeasurable because it has no attributes such as mass, extension, charge, etc., which can be quantified in physical units. For this reason some philosophers have referred to it, pejoratively, as "the ghost in the machine".

It will be argued that consciousness can be quantified with respect to time and that the "ghost in the machine" can, in fact, be measured. Recent experiments in which visual and auditory perceptions were measured accurately will be described. The results of these measurements and the theoretical framework used to account for them permit a novel interpretation of a number of time-dependent perceptual phenomena. One of these phenomena will be demonstrated.

## I. The "Ghost in the Machine"

I have recently concluded a series of experiments the purpose of which was to measure the duration of the perceptions of brief light flashes and tone bursts [7, 9, 10, 11, 12]. To claim to have measured the duration of a perception is equivalent to a claim to have measured the duration of a mental event. Yet we have been lectured by no less an authority than Gilbert Ryle [19] that the distinction between physical events and mental events is a spurious myth which has achieved the status of dogma. He asserted, with "deliberate abusiveness", that the mind-body distinction is a fallacy resulting from a Cartesian category-mistake and that those of us who continue to adhere to it are guilty of accepting what he called "the dogma of the ghost in the machine". In Ryle's terminology, it would seem that I have been engaged in the dubious pastime of measuring ghosts.

The concept of mind has been under increasing attack by many philosophers and scientists since the 19th Century. Few of these critics, however, have had Ryle's combination of imagination and sarcasm which allowed him to coin such a catchy and abusive slogan as "the ghost in the machine". What is the defect in the concept of mind which is presumed to be so cleverly exposed by this slogan?

Since the time of Galileo, Newton and Descartes, matter has been conceptualized as having mass and extension while mind, in contrast, is conceived of having neither of these attributes. Whatever mind is, all philosophers today agree that it is not a "thing in space". Since mind has no mass, extension, locus, charge or other attribute presently known to physicists, it is considered to be

---

* Robert Efron, M. D., Chief, Neurophysiology-Biophysics Research Unit, Veterans Administration Hospital, Martinez, California, 94553, USA.

entirely non-material. To certain philosophers, consciousness is no more real than a ghost. What more contemptuous jibe could be made by a mid-20th Century materialist than to accuse dualists[1] of believing in spooks?

To emphasize the ludicrous nature of the so-called "dogma of the ghost in the machine", the materialist frequently baits his opponent with an additional claim – one which he thinks is particularly devastating. He maintains that mind cannot be measured in terms of *any* of the units by which the physical scientist measures the attributes of material objects.[2] As a corollary of this proposition he also maintains that the so-called attributes of mind are all negatively defined, that mind is *without* extension, *without* mass, *without* charge, etc. If an existent is without *any* attribute, then it clearly does not exist.

The most absurd idea of all (at least in the minds (?) of the materialists) is the notion that mind has causal efficacy. How, they claim, could a non-material non-existent – a ghost – interact with a material existent? How could something mental, such as an intention, an emotion or a perception result in a physical action of the nervous system? The whole idea of causal efficacy of consciousness is dismissed as a universal delusion of mankind – a genetically determined madness from which we all suffer. The only exceptions to this universal malady appear to be those mutant philosophers who have been programmed (presumably by their genes) to "think"[3] that they are robots. There are other mutant philosophers who apparently have been compelled genetically to "think"[3] that other men have made a dreadful category-mistake.

All these attacks (and others not mentioned here) on the concept of mind by the materialists and their allies have long since been counterattacked by the dualists and their allies. It is not my intention today to provide either a detailed historical review or any last-minute battlefield reports on this interminable guerilla war in which neither side admits defeat. Specific battle reports with attached philosophical casualty lists can be found elsewhere [8, 6, 14, 13].

Instead, I propose a different line of approach: If it can be demonstrated that at least *one* attribute of mind *can* be measured in terms of the units used by physicists, then one of the more persistent attacks on the concept of mind can be definitively discredited.

---

1 In this context I use the term "dualist" in its broadest sense to refer to all those who accept a *metaphysical* distinction between mind and matter.

2 Some materialists, such as Ryle, frankly acknowledge that consciousness can never be "reduced to" the laws of physics and chemistry if consciousness cannot be measured in terms of any of the attributes of material objects. Other materialists (particularly the behaviorists), however, do not appreciate that it is contradictory to claim that consciousness is not measurable and also to claim that mind eventually will be reduced to the laws of physics. Some of the major contradictions of the behaviorist position have been discussed previously [8] and need not be elaborated upon here.

3 I use the conventional word "think" here since it is too cumbersome to say that they have been programmed to make laryngeal sounds (behaviors) as a function of reinforcement contingencies which in turn are no more than the operant responses of other organisms which in turn are consequences of their genetic and neurological structures.

Since mind is not a "thing in space" it could not have the measurable attributes of things in space. This does not necessarily mean, however, that it has *no* attributes which can be measured. If we conceive of mind as a *process* in time, then it should be measurable in the identical units that are used by scientists when they measure the temporal aspects of any material process. The specific goal of the experiments I will describe was to measure the durations of perceptions of brief flashes of light and tone bursts. Perceptual processes, as distinct from other mental processes such as emotions, thoughts, and dreams, were selected for special study since they result from physical stimuli which can be manipulated readily by the experimenter. However, the methods which can be used to measure the temporal aspects of perceptions are applicable, at least in principle, to other mental processes as well.

## II. The Experiments

It is not generally appreciated that the duration of visual perceptions were measured over two hundred years ago. So far as I have been able to determine [15] Seguer, in 1740, was the first to have performed the critical experiment. While a number of modifications of Seguer's experiment have been introduced in the last 230 years, the essence of his method has remained unaltered. A small light or ember is attached to a wheel which is then rotated at constant speed in a darkened room. At low speeds of rotation a comet-like object with a curved tail is seen. As the speed of rotation is increased, the tail becomes longer. By recording the length of the perceived tail and the speed of rotation, the duration of the perception of the light at each retinal point can be calculated. From experiments of this type, psychophysicists have concluded that visual persistence is of the order of 150–160 msec. That is to say, a momentary stimulation of a group of receptors at one locus on the retina gives rise to a process of awareness which lasts for approximately one-sixth of a second.

Seguer's method, even when modified by subsequent investigators, has one major limitation: It is applicable only to the visual system and an analogue cannot be elaborated readily to test the other perceptual systems such as audition and touch. Ideally, the psychophysicist would like a more general method of measuring the duration of perceptions which would be applicable to every perceptual modality.

The *most* general method of measuring the duration of any action, process or state is to make note of the time of its beginning and end. The difference between these two measurements is the duration. The measurement of duration is derived from *two* separate measurements of simultaneity: The simultaneity of the onset of the process with a particular clock reading and the simultaneity of the offset of the process with a second reading of the same clock.

The method for measuring the duration of perceptions which will be described in this paper relies on the subject's capacity to make judgements of the simultaneity or non-simultaneity of two physical events. The human being is capable of quite accurate judgements of this type, rarely making errors in excess of 10–15 milliseconds. In other words, if the stimuli are asynchronous by more than 15 msec, a subject rarely fails to detect this fact. An attentive observer may have appreciably smaller error [4, 5].

In the experiments which I have just completed the subject was asked to make two judgements of simultaneity. He was presented with *two* stimuli – one in the auditory modality, the other in the visual. One of these stimuli was very brief and will be referred to as the "index stimulus". The longer stimulus, presented in the second modality, produced the perception whose duration was to be measured. This stimulus will be referred to as the "control stimulus". In the experiments designed to measure the duration of visual perceptions, the index stimulus was a brief auditory click and the control stimulus was a longer flash of light. In the experiments in which the duration of auditory perceptions was measured, a brief flash was used as the index stimulus and a longer tone was the control stimulus. Since the two experiments are conceptually identical, I will restrict the illustration and the discussion to the experiment in which the durations of visual perceptions were measured.

The subject was first required to make a judgement of the simultaneity of the onset of the click (index stimulus) and the *onset* of the flash (control stimulus). He was then required to make a judgement of the simultaneity of the onset of the click and the *offset* of the flash. The difference between these two measurements is the duration of the perception of the flash.

The first simultaneity judgement is presented diagrammatically in Fig. 1 A, the second in Fig. 1 B. These two figures require some explanation. The small open rectangle denotes the index stimulus. The longer open rectangle denotes the control stimulus. These open rectangles are the *physical events* produced by the apparatus. The black bars above each stimulus denote the perceptions of each stimulus, i.e., the *mental events*. The intervals $X_i$ and $X_c$ represent the delay between the onset of the respective physical events and the onset of the perception of the physical event. These intervals, which at present are *unknown*, include all delays in the transmission of the neural information from the receptor organ to the brain, and any other delays which may occur before the subject is aware that the light or tone has been turned on. These delays will be referred to as the "perceptual onset delays". Similarly, $Y_c$ represents the delay between the offset of the control stimulus and the offset of the awareness of that stimulus. This will be referred to as the "perceptual offset delay". (The perceptual offset delay for the index stimulus is not included in Fig. 1 because it does not concern us in the present experiments).

In the sample experiment illustrated in Figs. 1 A and B, the control stimulus (the flash) was turned on exactly 300 msec after an electronic clock was started and was turned off 500 msec after the clock was started. Its physical duration was thus 200 msec. This 200 msec control stimulus gave rise to a perception

The Measurement of Perceptual Durations

denoted by the black bar – and it is the duration of this particular mental event which we are attempting to measure. It should be stressed that we cannot assume that the perception is also 200 msec in duration. This assumption would be valid only if the perceptual onset delay ($X_c$) and the perceptual offset delay ($Y_c$) are equal. The equality or non-equality of these delays will be ascertained by the experiments and no prior assumptions need be made as to their relative magnitudes.

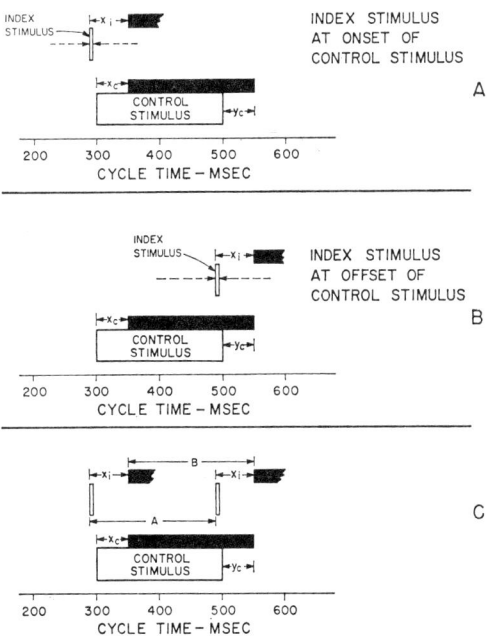

Fig. 1. Diagrammatic Representation of Experiment I

In the experiment illustrated in Fig. 1 A, the index stimulus was delivered at various times in reference to the *onset* of the control stimulus (indicated by the dashed arrows), and the subject was asked if its onset was or was not simultaneous with the onset of the control stimulus. By presenting the subject with thousands of trials, a reliable value was obtained for the temporal location of the index stimulus which the subject reported to be simultaneous with the *onset* of the control stimulus.

Upon completion of this experiment, the same subject was asked to make a judgement of the simultaneity of the onset of the index stimulus and the *offset* of the control stimulus. This is illustrated in Fig. 1B. Again, after thousands of trials, a reliable measure was obtained of the temporal location of the index stimulus which was reported to be simultaneous with the *offset* of the control stimulus.

The difference between the two temporal locations (See Fig. 1 C) of the index stimulus, obtained in Experiments 1 A and 1 B, is the measure of the duration of the perception of the control stimulus. In the example illustrated, the onset location was 295 msec after the clock was started (Fig. 1 A) and the offset location was 495 msec after the clock was started (Fig. 1 B). The difference between these two measures (200 msec) is the duration of the perception of the control stimulus (Fig. 1 C). In this particular example, the perceptual duration of the control stimulus (the inter-click interval) was equal to its physical duration, and we are on safe grounds to assume that in this particular case the perceptual onset delay ($X_c$) and the perceptual offset delay ($Y_c$) are equal. That this conclusion does not hold for *all* values of control stimulus duration can be seen from the results of a series of these experiments performed with different control stimulus durations.

It should be noted that the method used to calculate the duration of the perception of the control stimulus (Fig. 1 C) assumes that the perceptual onset delay of the *index* stimulus ($X_i$) remains of constant magnitude in Exps. 1 A and 1 B. This assumption is justified since the index stimulus is the *same* stimulus in both parts of the experiment. It is not necessary, however, that we have any knowledge of the absolute duration of $X_i$.

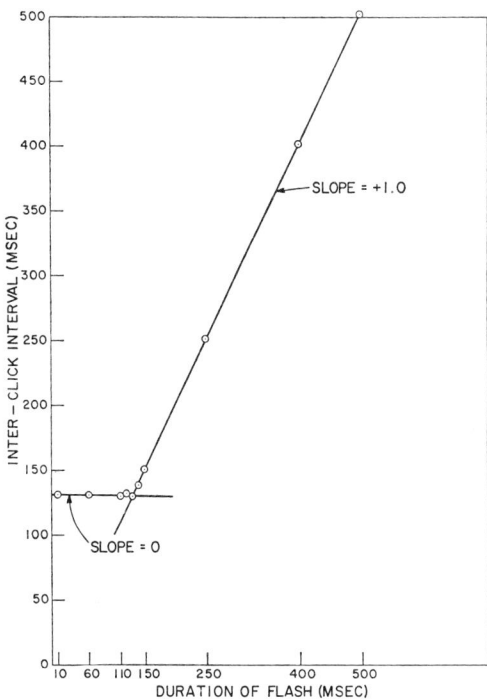

Fig. 2. The Durations of Visual Perceptions

Fig. 2 shows the results of measuring the durations of visual perceptions using an auditory click as the index stimulus. A number of different control flash durations were used, and for each of them the inter-click interval was measured experimentally. Two subjects were used for these experiments. Since their results were so similar, the data obtained from each was combined in Fig. 2. Fig. 3 shows the results of measuring, in the same two subjects, the durations of auditory perceptions using a brief flash as the index stimulus.

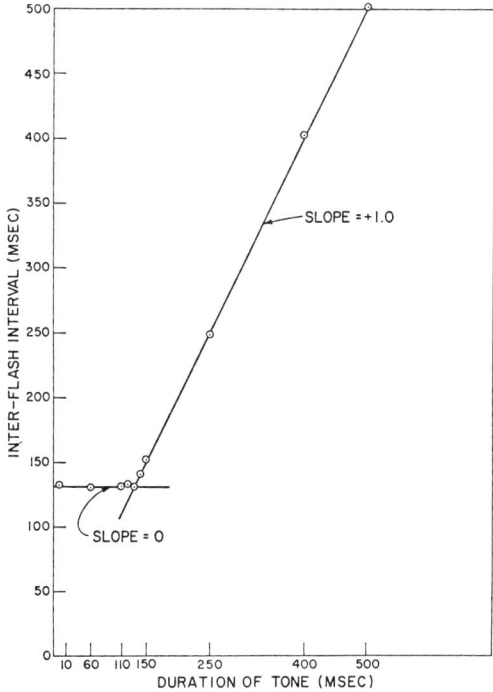

Fig. 3. The Durations of Auditory Perceptions

The conclusion drawn from these experiments can be expressed in two general statements:

(1) The duration of the perception of a stimulus is equal to the duration of the stimulus, provided the stimulus is longer than some critical value.

(2) For stimuli shorter in duration than the critical value, the perception is of constant duration.

A more detailed analysis of the data of Experiment 1 A (See Ref. [11]) revealed that the temporal location of the index stimulus was constant regardless of the duration of the control stimulus. Expressed differently, the duration of the perceptual onset delay of the control stimulus ($X_c$) was independent of the duration of the control stimulus. The detailed analysis of

Experiment 1B showed that the perceptual offset delays ($Y_c$) were equal to the perceptual onset delays ($X_c$) only when the stimuli were *longer* than the critical duration. As the control stimulus was made shorter than the critical duration, the perceptual offset delay became longer than the perceptual onset delay. This change in the duration of the perceptual offset delay is the explanation of the horizontal section of the curves in Figs. 2 and 3.

Again, it should be stressed that these experiments do not give rise to any knowledge about the absolute duration of $X_c$ and $Y_c$: They only permit a statement about their *relative* magnitudes under different experimental conditions.

The actual value of the duration of the shortest perceptions was 130 msec for the particular stimuli which were used. In further experiments of this type, which I will not report here, it was found that the shortest perceptual duration for brief visual stimuli was an inverse function of the stimulus luminance [11]. The value of 130 msec, therefore, does not represent some fixed biological constant, but is a dependent function of the intensity of the physical stimulus. (It also may be dependent on other stimulus qualities, and on levels of background intensity). The fact that the same value was obtained in Figs. 2 and 3 for the auditory and the visual perceptions was a fortuitous consequence of the particular values of stimulus intensity which were used in those experiments.

If the use of this method has provided us with a valid description of the temporal characteristics of the perceptual process, it might be expected that this process (and the neurophysiological mechanisms underlying it) operate not merely in the laboratory, but influence our daily experience of reality. The results suggest, for example, that two or more brief stimuli falling upon different retinal loci *within* a period of 130 msec might result in two or more perceptions which overlap temporally. That is to say, there would be some period of time during which the observer would be aware concurrently of two or more stimuli which were not *physically* concurrent.

This last implication of the results was studied jointly with David Lee of Harvard [12]. The experiment that was performed on a number of subjects is illustrated in Fig. 4. The subjects observed a white disc on which a single radial line was drawn. This disc was rotated at constant speed in a dark room and was illuminated with a periodically occurring brief flash. The duration of the flash was approximately one microsecond – well below the critical duration measured in the previous experiments. It was hypothesized that each brief flash would result in a perception of the line located at a particular place determined by the time of occurrence of the flash. This is illustrated as $S_1$, $S_2$, $S_3$, etc. in Fig. 4. We assumed that the duration of the *perception* resulting from each flash was $d$ secs. If the flashes are presented every $d/2$ sec, then the perception evoked by any one flash ($P_1$, $P_2$, $P_3$, etc.) should temporally overlap (by 50%) the perception evoked by the next flash. At any moment in time, the observer would be expected to see *two* radial lines and these two lines should appear to move in tandem through the circular path.

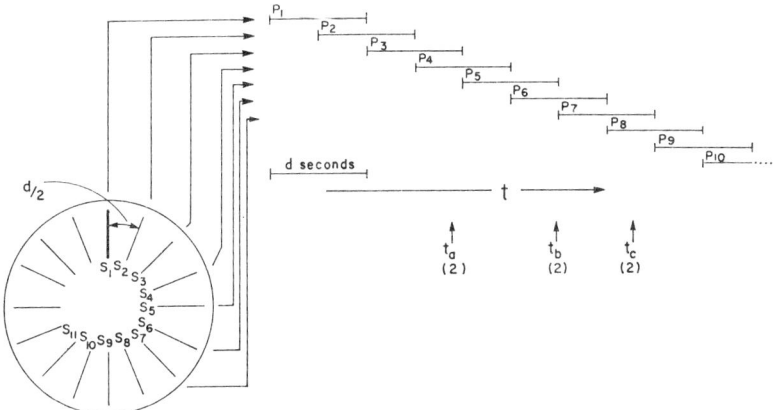

$S_1, S_2 \ldots\ldots S_n$ = Angular position of radius at time of flash ..... flash$_n$
$P_1, P_2 \ldots\ldots P_n$ = Duration of perception of radius due to flash ..... flash$_n$
$d$ = Hypothesized duration of perception
$d/2$ = Interval between flashes

Fig. 4. Diagrammatic Representation of Experiment II

Without going into the details of our argument which can be found elsewhere [12], it is possible to derive the prediction that the number of radial lines which will be seen should be equal to the minimum duration of the perception ($d$) multiplied by the frequency of the stroboscopic flash ($f$). If we have a report of the number of lines seen at each flash frequency, we should be able to infer from the slope of this function the value of the duration of the perception ($d$) at the luminance level used. This value should be identical to that discovered by the previous experiments for the same value of flash luminance.

The results of these experiments on four subjects are illustrated in Fig. 5 and reveal the predicted proportional relation between the number of radii seen and the strobe frequency. The discrepancy in the estimation of the duration of the perception ($d$) by the two methods was only 5 msec.

It will be recalled that in the experiment in which index stimuli were used as a measuring device, the minimum duration of a perception ($d$) was found to be an inverse function of stimulus intensity. If this conclusion is valid, the number of radial lines which appear to be moving as a group (for any particular rate of stroboscopic flashing) should increase as the intensity is decreased. This prediction was also confirmed (see Fig. 6).[4]

---

[4] The experiment just described was demonstrated to the participants at the conference.

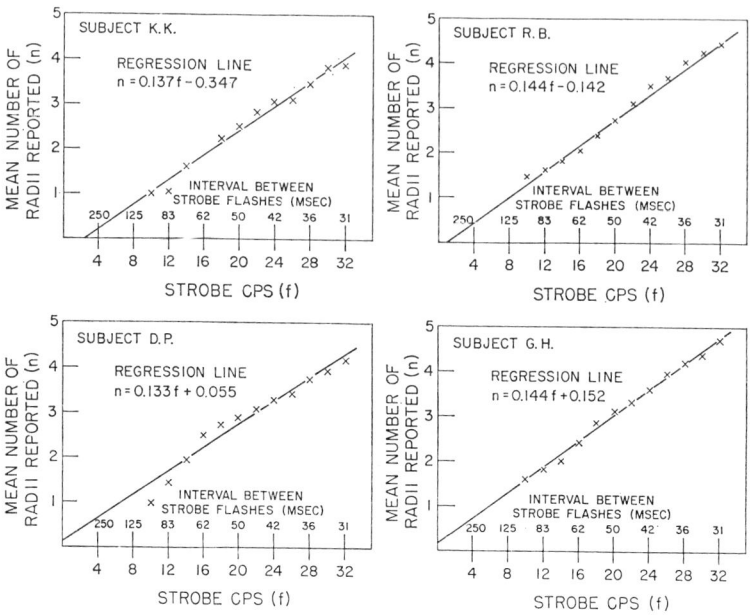

Fig. 5. Results of Experiment II in Four Subjects

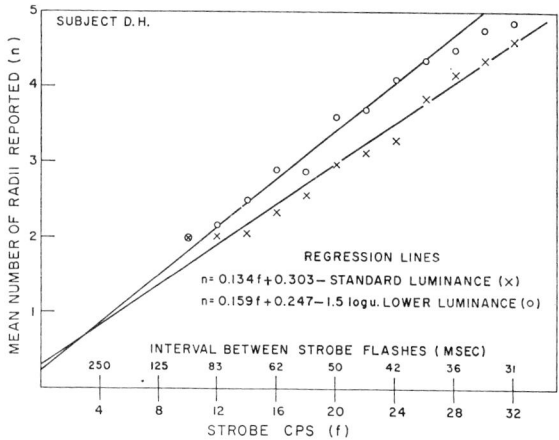

Fig. 6. Results of Experiment II at Two Luminance Levels

## III. Conclusion

The results of these experiments leave little room to doubt that the durations of perceptions can be measured in the same units that the physicist uses when he measures the duration of physical events. By the use of this method, we have confirmed Seguer's finding that the durations of perceptions may outlast the durations of brief stimuli. Further, we have shown that auditory as well as visual perceptions have the same general temporal properties. The method of using index stimuli is applicable, of course, to other perceptual modalities to which it will be applied in the future. Whether all perceptual modalities show the same temporal properties must await further research.

In this brief report of my own experiments I have not attempted to describe the many contributions of other investigators[5] in this general area of research or to relate their results and theoretical formulations to my own. To do this adequately would take much too long; to do it superficially would be discourteous. Instead, I will use my remaining time to focus on two of the philosophical implications which I see in these experiments.

You may recall from my introductory remarks that many philosophical materialists assert that the distinction between mental events and physical events is spurious, that mind cannot be measured, and that mind is defined exclusively in terms of the *absence* of attributes. It must be obvious by now that I disagree most heartily with these views. I have tried to show that the duration of a perception, a *mental* process, is measurable – and that it is measured in precisely the same way and in the same units as the duration of any *physical* process. The "ghost in the machine" has some fascinating properties in the temporal domain!

The second philosophical point relates to the often discussed distinction between the so-called "publicly" observable data of science and the so-called "private" domain of mind. Too much has been made of this issue. Perceptions, qua mental events, are, of course, "private" insofar as they cannot be experienced *directly* by another mind. To admit this does not mean that we must accept the false alternative that they are therefore entirely "beyond" the realm of objective science. In this paper I have tried to show one of the procedures by which we can gain access to this not-so-private realm. Although we cannot perceive each others perceptions or other mental processes, they are not so private that we are unable to study and measure them by the perfectly "public" procedures of science. Some of the methods which have been used to study perceptions (such as scaling procedures) do not provide measurements in the same units used by physicists. Other methods (such as those described in this paper) provide measurements identical to those used by physicists. Both categories provide objective and "public" measurements of various aspects of a "private" process. In sum, I do not see any *epistemological* difference in the methods used by the experimental psychologist when he studies the processes of mind and the methods of the experimental physicist when he studies the processes occurring in the atomic nucleus.

In addition to perception, there are other mental processes such as dreams, thoughts and emotions which are potentially measurable in the time domain.

---

5 Allport [1], Babington Smith [2], Crawford [3], Lichtenstein [16], Mayzner et al [17], Mollon [18], Sperling [20], and White [21] to name only a few.

Techniques for studying quantitatively the temporal characteristics of other mental processes have not yet been developed. They are less likely to be developed, however, if the philosophical Zeitgeist maintains that the measurement of mind is logically impossible.

## *References*

1. Allport, D. A.: *Phenomenal Simultaneity and the Perceptual Moment Hypothesis*. Brit. J. Psychol. *59* (1968) 395–406.
2. Babington Smith, B.: *On the Duration of the Moment of Perception*. Paper read to Brit. Psychol. Assn., Leicester, 1964. cf. Bull. B. Psych. Soc. *55* (1964) 27A.
3. Crawford, A.: *Measurement of the Duration of a Moment in Visual Perception*. Bull. Brit. Psych. Soc. *17* (1964) No. 54, 2A–3A.
4. Efron, R.: *The Effect of Handedness on the Perception of Simultaneity and Temporal Order*. Brain *86* (1963) 261–284.
5. — *The Effect of Stimulus Intensity on the Perception of Simultaneity in Right- and Left-Handed Subjects*. Brain *86* (1963) 285–294.
6. — *The Conditioned Reflex: A Meaningless Concept*. Perspectives in Biology & Medicine *9* (1966) 488–514.
7. — *The Duration of the Present*. N. Y. Acad. Sci. *138* (1967) 713–729.
8. — *Biology Without Consciousness – and its Consequences*. Perspectives in Biology and Medicine *11* (1967) 9–36.
9. — *The Relationship Between the Duration of a Stimulus and the Duration of a Perception*. Neuropsychologia *8* (1970) 37–55.
10. — *The Minimum Duration of a Perception*. Neuropsychologia *8* (1970) 57–63.
11. — *The Effect of Stimulus Duration on Perceptual Onset and Offset Latencies*. Perception and Psychophysics 7 (In Press).
12. — Lee, D.: *The Duration of Perceptions of Moving Stroboscopically Illuminated Targets* (1971) (To be Published in: American J. Psychol.)
13. Feigl, H.: *The "Mental" and the "Physical"*. Minneapolis: University of Minnesota Press 1967.
14. Koestler, A.: *The Ghost in the Machine*. New York: Macmillan 1967.
15. Le Grand, Y.: *Light, Colour and Vision*. London: Chapman and Hall 1957.
16. Lichtenstein, M.: *Phenomenal Simultaneity with Irregular Timing of Components of the Visual Stimulus*. Percept. Motor Skills, *12* (1961) 47–60.
17. Mayzner, M. S., Tresselt, M. E., Helfer, M. D.: *A Provisional Model of Visual Information Processing with Sequential Inputs*. Psychonomic Monog. Suppl. Vol. 2, No. 7 (1967) 91–108.
18. Mollon, J. D.: *Two Approaches to the Perceptual Moment Hypothesis*. Paper read to Exp. Psychol. Soc. Oxford 1969.
19. Ryle, G.: *The Concept of Mind*. London: Hutchinson 1949.
20. Sperling, G.: *Successive Approximations to a Model for Short Term Memory*. Acta Psychol. *27* (1967) 285–292.
21. White, C. T.: *Temporal Numerosity and the Psychological Unit of Duration*. Psychol. Monog. *77* No. 12 (1963).

# Oscillations as Possible Basis for Time Perception*

ERNST PÖPPEL**

*Summary.* A short review of hypotheses on human time perception emphasises the distinction between endogenous and exogenous concepts. Methods are described, and the related problem of stationarity is discussed. A hypothesis is formulated which assumes oscillatory processes as the basis for the perception of short temporal intervals. Several experiments are described which test this hypothesis. The results of the first experiment suggest a temporal constant in the range of the conscious present. Periodic components in subjective random series which are demonstrated in another experiment, indicate a sequence of temporal units. The influence of physiological factors on time perception is illustrated; diurnal variations are discussed as an example. However, it is shown that the observed variations in time perception are not dependent on body temperature alone. The importance of informational cues is noted and an experiment is described which indicates the role of individual differences.

## 1. Hypotheses on Endogenous and Exogenous Factors in Time Perception

The experimental study on human time perception began approximately one hundred years ago (Vierordt, 1868), but has, for a long time, not attracted as much attention as, for instance, space perception. Thus, some years ago Adams (1964) in a review wrote: "Time perception is a venerable, tired topic in psychology that interests very few active investigators any more, perhaps because no one bothered to explore the mechanisms of time perception and how it might enter into meaningful interaction with other mechanisms" (p. 197). In the last years, however, interest in the psychology of time has increased considerably and also the 'mechanisms of time perception' called forth active work (e.g. Creelman, 1962; Michon, 1967 and 1970; Ornstein, 1969; Treisman, 1963). Since 1960 approximately as many papers on time perception have been published as between 1860 and 1960. The new interest in problems of time perception is also reflected in the increase of monographical papers and books; (Bergius, 1969; Boring, 1933 and 1942; Cohen, 1967; Fraisse, 1964 and 1966; Gilliland *et al.*, 1946; Gooddy, 1969a; Heiss, 1961; James, 1890; Lévy, 1969; Nichols, 1891; Orme, 1969; Ornstein, 1969; Piéron, 1923; Quandt, 1906; Schaltenbrand, 1963; Wallace and Rabin, 1960; Weber, 1933; Woodrow, 1951). This increase within

---

\* This paper is dedicated to the late Ernesto Blohm.
\*\* Dr. Ernst Pöppel, Max-Planck-Institut für Verhaltensphysiologie, BRD-8131 Erling-Andechs, Germany.
The author acknowledges the assistance of Dr. Ronald Chase for his help in preparing the English version of this manuscript and Mrs. E. Borowietz for bibliographic assistance.

the last years might be an indication of the general scientific explosion, but it is possibly also dependent on some interesting developments in other areas which influenced experimental psychology like information theory and experimental biology where the "temporal organization of living systems" has been demonstrated (Aschoff, 1959; Pittendrigh, 1961).

If one tries to summarize some of the hypotheses on the fundamental mechanisms in time perception which have been discussed or which were implicitly assumed, one finds two basic concepts which have also played important roles in other areas: On the one hand it is thought that perception of time has an endogenous basis, on the other hand, exogenous cues are thought to be determinant for time perception.

Authors who stress the endogenous basis have discussed the cortical representation of time perception (e. g. Davis, 1956; Dimond, 1964; Hoff and Pötzl, 1934). They have also speculated about general physiological cues which somehow trigger time perception (Bell and Watts, 1966; Boring and Boring, 1917; Hawkes et al., 1962; Münsterberg, 1892; Piéron, 1923; Schaefer and Gilliland, 1938). In many cases "internal clocks" have been assumed (Gooddy, 1969b; Hoagland, 1933; Pöppel, 1968a, 1969 and 1970a; Stroud, 1955) which often are identified with the alpha rhythm of the electroencephalogram (EEG), (Anliker, 1963; Cahoon, 1967; Fraisse and Voillaume, 1969; Holubář and Machek, 1962; Legg, 1968; Werboff, 1962; Wiener, 1963).

Those authors who stress the exogenous basis of time perception base their concepts on information processing. One of their protagonists is the french philosopher Guyau (1890) who tried to refute the a priori approach of Kant (1781) and concluded "que le temps n'est pas une condition, mais un simple effet de la conscience" (p. 117). There are some studies which seem to be directly influenced by information theory and often use a computerlanguage (Bechinger et al., 1969; Lévy, 1969; Michon, 1965, 1967 and 1968; Murphy, 1966; Ornstein, 1969; Vroon, 1970), but there is also a great number of studies in which the information concept is only implicitly assumed; the mental content is looked on as the basic determinator for time perception (e. g. Deppe, 1969; Frankenhaeuser, 1959; Hoche, 1923; Sturt, 1925).

Within the exogenous approach not only the *processing* of information has been considered as a crucial parameter for time perception, but also the *processed* information which has generated a reference system. For example, this concept is illustrated in adaptation-level theory (Helson, 1964). It has been shown that pooling processes and anchor stimuli affect time perception (Adamson, 1967; Behar and Bevan, 1961; Eson and Kafka, 1952; Goldstone, 1967; Goldstone et al., 1957; Nelson et al., 1963; Postman and Miller, 1945; Sixtl, 1963; Zoltobrocki, 1965). These studies which are oriented towards a specific way of adjustment to temporal stimuli are paralleled by those which consider learning in general as a suitable interpretation for time perception (Aiken, 1965; Craik and Sarbin, 1963; Piaget, 1955).

The studies mentioned in these last three paragraphs are more or less oriented towards an explanation of time perception or reveal implicit hypotheses so that a categorization is possible. In those studies where *influences on* time perception are discussed, one can again separate the endogenous and exogenous approaches. The endogenous concept has inspired efforts to determine the manner in which physiological conditions influence time perception. This is done by either experimentally varying the physiological conditions or by using different naturally occuring physiological conditions. Body temperature has always been of special interest (Baddeley, 1966a; Bell, 1965 and 1966; Bell and Provins, 1963; Fox *et al.*, 1967; François, 1927 and 1928; Hoagland, 1933; Kleber *et al.*, 1963; Pfaff, 1968). Some authors claim that time perception is almost exclusively determined by body temperature. An other factor which has been studied is the influence of different pharmacological substances (Boardman *et al.*, 1957; von Bodman, 1969 and 1970; Goldstone *et al.*, 1958; Otto, 1963a; Rutschmann and Rubinstein, 1966; Steinberg, 1955; Sterzinger, 1935 and 1938) or hormones (Gardner, 1935; Kleber *et al.*, 1963; Stern, 1959).

A further influence which has been studied is the effect of time of day on time perception. It is probable that the daily variation of nearly all physiological and psychological functions are dependent on an endogenous circadian oscillation. In experiments with human subjects who lived for several weeks under constant conditions and without any time cues from the environment, it has been shown that representative physiological functions are still rhythmic and that they behave like selfsustained oscillations (Aschoff and Wever, 1962; Aschoff *et al.*, 1969; Pöppel, 1968b). If time of day has a systematic influence on time perception, it is therefore probably due to variations of physiological state and not to variations of external cues (Pöppel and Giedke, 1970). Time of day effects on time perception have been shown by several investigators (Halberg *et al.*, 1965; McLeod and Roff, 1936; Pfaff, 1968; Pöppel, 1969; Stephens and Halberg, 1965; Thor, 1962), but it is unknown which physiological parameters are responsible for the diurnal variation of time perception, although body temperature is considered important by many authors.

The exogenous approach is concerned with environmental cues which are affecting the behavioral conditions of the organism and thus influence time perception. In some classic studies of time perception attention was considered as an important factor (Benussi, 1907 and 1913; Katz, 1906; v. Kries, 1913). In more recent work the physiologically oriented concept of arousal has been discussed, and the influence of activating stimuli on time perception has been demonstrated (Aitken and Gedye, 1968; Bokander, 1965; Boulter and Appley, 1967; Cahoon, 1969; Falk and Bindra, 1954; Hirsh *et al.*, 1956; Langer *et al.*, 1961; Lockhart, 1967; Melges and Fougerousse, 1966; Otto, 1963b; Schönpflug, 1969; Smets, 1969; von Sturmer, 1966; von Sturmer *et al.*, 1968; Warm *et al.*, 1967). The subject's motivation is also important (Filer and Meals, 1949; Meade, 1959 and 1960; Warm *et al.*, 1964), and the influence of the experimental task which is designed to vary the mental content and perhaps changes the motivation of the subject, has an established influence on perceived duration (Axel, 1925; Dewolfe and Duncan, 1959; Gilliland and Martin, 1940; Gulliksen, 1927; Hall and Jastrow, 1886; Harton, 1938; Josenhans, 1959; Kafka, 1957; Loehlin, 1959; Müller, 1965; Ploeger, 1966; Spencer, 1921; Swift and McGeoch, 1925).

## 2. Methods in Time Perception Studies

Discussion about the methods used in the experimental analyses of time perception was intensified during recent years (Bindra and Waksberg, 1956; Björkman and Holmqvist, 1960; Carlson and Feinberg, 1968 and 1970; Fraisse et al., 1962; Hornstein and Rotter, 1969; Marum, 1968; McGrath and O'Hanlon, 1967; Ochberg et al., 1965; Richards and Livingstone, 1966; Ross and Katchmar, 1951; Stevens, 1957; Warm et al., 1963; Weber, 1965; Webster et al., 1962). Commonly, four different methods are used.

a) Pure estimation. The experimenter presents a stimulus of a defined duration and the subject responds reporting how long (in conventional time units) the interval appeared to be.

b) Comparison. The experimenter presents two stimuli in sequence, and the subject decides which one of the two is longer.

c) Production. The experimenter asks the subject to produce an interval of a certain duration specified in conventional time units.

d) Reproduction. The experimenter presents a stimulus and the subject reproduces it without necessarily being aware of its duration in conventional time units.

The methods of comparison and of reproduction do not need conventional time units, whereas those of verbal estimation and of production do require conventional time units. The last two methods, consequently, imply that the subject can use a reference system for time perception which he has acquired from his social environment. As the method of reproduction and of comparison are free from this implication, these methods are more useful for studies in which the main interest is the intrinsic temporal organization. In the terminology of Gooddy (1969b) the methods of reproduction and of comparison reflect "personal time", while the methods of verbal estimation and production reflect "government time".

Special methodological problems arise if measurements are repeated. It has been shown that there are systematic changes due to repetition alone (Carlson and Feinberg, 1970; Emley et al., 1968; Falk and Bindra, 1954). Even the learning of specific intervals with different feedback techniques (Denys and Richelle, 1965; Kelm, 1962; Schoeffler and Poole, 1967) does not eliminate systematic trends, as will be shown in this paper (Fig. 8).

There is no agreement as to whether a single common factor of time perception is studied by the four methods discussed above. Some results indicate only a small correlation between the methods (Clausen, 1950; DuPreez, 1963; Kruup, 1961; Siegman, 1962). The insignificant correlations have even been taken as an indication of fundamentally different mechanisms in time perception. Experiments in which factor analysis was used also indicate that there may be more than one "time perception factor" (Loehlin, 1959; Spreen, 1963). By using further assumptions, however, Fraisse *et al.* (1962) interpreted their results on the basis of a single mechanism.

## 3. Stationarity as an Implication of the Experimental Methods

If different methods of time perception are used without criticism, one important implication of these methods may be overlooked which seems to be crucial for experimental work on subjective time. These methods were originally developed for psychophysical studies not mainly concerned with time perception. They were then adapted for time perception studies under the assumption that there is no qualitative difference for instance between intensities and durations. Apparent time is considered as a typical example of a prothetic continuum (Stevens, 1957). This assumption implies that the perceptual process responsible for subjective duration must be stationary i. e. that an increase of physical duration is increasing linearly apparent duration. Thus, according to the hypothesis of stationarity the duration of the intervals is irrelevant for the process of perception. It is assumed that there is no qualitative difference in the experience of one second or one hour. The process responsible for time perception is considered not to change during the measurement (hypothesis of homogeneity) and not to change from measurement to measurement (hypothesis of invariance).

The hypothetical assumption of stationarity cannot be accepted without qualification. There are every-day observations and experimental evidences which suggest inhomogeneity of time perception, or whose interpretation demands the rejection of stationarity. The most obvious and trivial indication of non-stationarity is the fact that we loose conscious control over time during dreamless sleep. Furthermore, the existence of different mechanisms in time perception was stressed long ago by James (1890) who thought that intervals up to approximately five seconds can be perceived as a unit. Longer intervals are coded symbolically and can only be estimated, not perceived. A similar distinction between the perception of long and short intervals has also been proposed by Edgell and Waller (1903). It can be postulated that the perception of short intervals is limited by the duration of the subjective present (Ahrens, 1963; Boring, 1933; Quasebarth, 1924; Sixtl, 1962; Stern, 1897; Wundt, 1911). If there is indeed a real difference in the perception of long and short intervals and if it can be shown that the "subjective present" is a real temporal unit, then the hypothesis of stationarity in time perception is not valid. In the case of non-stationarity results dealing with different temporal ranges (or gained with different methods) should not be compared without critical discussion.

*4. Hypothesis: A Periodic Process as a Reference System for Short Time Perception*

The question arises what mechanism could be responsible for non-stationarity of time perception. Periodic processes represent one obvious case of non-stationarity. The different phases within one period are of course not equivalent; only identical phases of succeeding periods are comparable. Identical procedures in an experiment may give rise to different or even contradictory results depending on phase of an underlying periodic process.

Periodic processes which might be the basis for human time perception have occasionally been discussed, but have never been explicitly studied. One of the earliest remarks came from Mach (1885): "Es wird hiermit die Vermutung nahe gelegt, daß die Empfindung der Zeit mit periodisch oder rhythmisch sich wiederholenden Prozessen in nahem Zusammenhange steht" (p. 212). In an experimental study Estel (1884) showed a rhythmical variation in time perception with a period of 0.75 seconds; however, this result could not be replicated (Mehner, 1885; Stevens, 1886). James (1890) and Wundt (1911) also favored the concept of periodic processes as basis for time perception. Recently there have been experimental results by Richards (1964) and Pöppel (1968a, 1969 and 1970a) which suggest a rhythmical organization in time perception, and the data of Bechinger *et al.* (1969) support these studies. Gooddy (1969b) concurs on theoretical grounds.

The relevance of periodic processes for human time perception will be demonstrated in the following sections. It seems possible that the previously discussed exogenous and endogenous interpretations can be reconciled under the assumption of periodic processes (probably of the type of relaxation oscillations).

In the next chapter (4.1) it is shown that there exists a special time interval up to 10 seconds which is experienced differently than longer intervals. In chapter 4.2 some evidence is given that there is a periodic sequence of intervals within this range. In chapter 5 the dependence of time perception on physiological conditions is demonstrated. In chapter 6 the importance of information processing is mentioned. In chapter 7 individual reference systems are shown and finally, in chapter 8, the experiments are briefly discussed.

4.1. Inter-Individual Stability of the Indifference Interval

The so-called indifference interval was described in one of the earliest works on time perception (Vierordt, 1868). Interpretative discussion of this phenomenon still continues (e.g. DuPreez, 1967; Hörmann, 1964; Sixtl, 1963 and 1964). If the method of reproduction is used, the indifference interval is that duration which is reproduced correctly. In general, stimuli shorter than the indifference interval are reproduced longer and stimuli longer than the indifference interval are reproduced shorter. Several investigators have reported indifference intervals, but there is no agreement about the absolute duration (Anderson, 1936; Brown and Hitchcock, 1965; Carlson and Feinberg, 1968; Fox, 1952; Woodrow, 1934 and 1935).

There have been several attempts to interpret the indifference interval on the basis of the fading trace theory (Köhler, 1923). The attraction of this physiological interpretation is diminished, however, if Helson's (1964) arguments are accepted. He interprets the indifference interval as an artefact generated by the experimental conditions. According to his interpretation, a central tendency in judgment (Hollingworth, 1913) leads to the establishment of a reference system, and the indifference interval is viewed as the main reference point.

Fig. 1 shows the results of an experiment which gives some support to such an interpretation. A subject had to reproduce time intervals between 0.5 and 7.0 seconds. Intervals in 0.5 steps within this range were each presented optically five times in a random sequence. Each stimulus was followed by a pause of 1.0 seconds before the reproduction began. As can be seen in Fig. 1, the subject reproduced short stimuli longer and long stimuli shorter than the actual stimulus duration. The line $S = R$ indicates perfect accuracy. Between 3.0 and 3.5 seconds the reproduction line crosses the line of perfect accuracy. At the crossing point subjective equality corresponds to objective equality, thus defining the indifference interval for this subject.

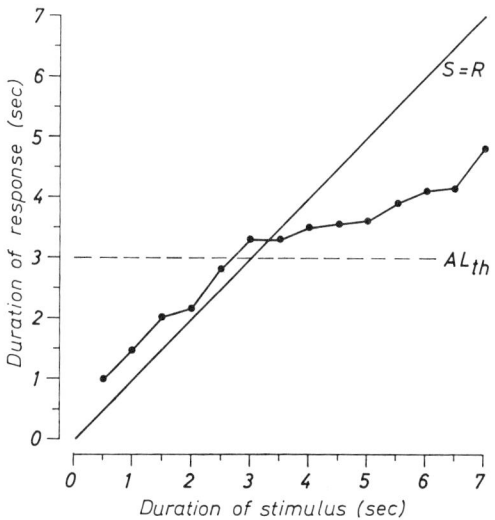

Fig. 1. Reproduction of temporal stimuli between 0.5 and 7.0 seconds duration. Results from one subject; each point is the mean of 5 trials. The 70 stimuli (14 × 5) were presented in random order. Light was used as a stimulus. At S = R the duration of the stimulus equals that of the reproduction. $AL_{th}$ is the geometric mean of all stimulus durations. Between 3.0 and 3.5 seconds a change from positive to negative errors is indicated, thus giving the indifference interval for this subject

On the basis of adaptation-level theory one would interpret this effect in the following manner. Because the subject has to reproduce many stimuli in the course of this experiment, one can assume that he establishes a time reference system. Within this established reference system he is able to locate the different durations. The mean point of this reference system is given by the geometric mean of all stimulus durations. In Fig. 1 this point is indicated by $AL_{th}$ (theoretical adaptation-level). As can be seen, this theoretical point roughly coincides with the subjective indifference interval.

Fig. 2 shows the pooled results of 11 subjects from the same experiment. The mean indifference interval of this group (between 2.0 and 2.5 seconds) no longer corresponds with the $AL_{th}$. This difference between the mean indifference interval and the $AL_{th}$ is not in favor of, but also not by itself, a strong argument against an interpretation on the basis of the adaptation-level theory, for particular experimental conditions, conscious or unconscious anchor effects, or strong residual effects may be responsible for such a shift.

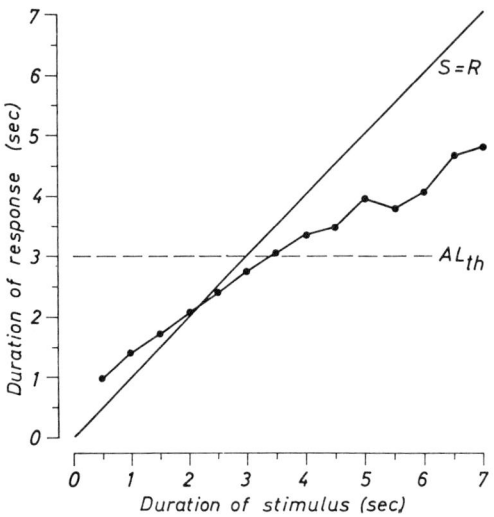

Fig. 2. Reproduction of temporal stimuli between 0.5 and 7.0 seconds duration. From 11 individual curves obtained as in Fig. 1 the mean values are given. A mean indifference interval between 2.0 and 2.5 seconds is indicated

The appropriateness of the adaptation-level theory for an interpretation of the indifference interval can be tested experimentally. On the basis of this theory a specific prediction can be made: An indifference interval should only be established when a subject is required to make a series of reproductions of different lengths. No such interval would be expected when each subject in a group reproduces only one interval, although taken collectively the group reproduces intervals over a given range. If in such a situation an indifference interval is observed – in this case an inter-individual indifference interval – this observation would be incompatible with the adaptation-level theory, since no individual reference system exists in this situation.

According to the adaption-level theory, indifference intervals should be observed in any temporal range. On the other hand, if the indifference interval is a constant value then, obviously, to reveal this interval one must work in an appropriate temporal range. As the subjective present has been associated with the indifference interval (Boring, 1942) it would be reasonable to look for the indifference interval within the range of the subjective present. This was done in the experiment described below.

The range of reproduced intervals for the experimental group was between 0.5 and 5.0 seconds, for the control group between 10.5 and 15.0 seconds. There were 38 and 28 subjects in each experiment respectively. Each subject had to reproduce only one stimulus duration, and this only once. The results of both experiments are presented in Fig. 3. As can be seen on the left side of Fig. 3 (experimental group) there is again an indifference interval. The crossing over

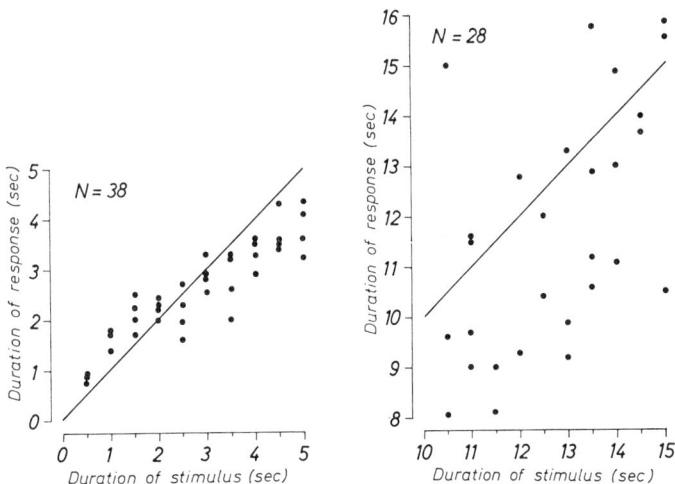

Fig. 3. Reproduction of stimuli from two different temporal ranges. Each subject responded to only one interval. Left side: 38 subjects reproduced stimuli between 0.5 and 5.0 seconds. Right side: 28 subjects reproduced stimuli between 10.5 and 15.0 seconds. Only within the smaller range a change from positive to negative time error is indicated (between 2.0 and 2.5 seconds), thus giving an inter-individual indifference interval

from longer to shorter reproductions is observed between 2.0 and 2.5 seconds. This inter-individual indifference interval corresponds to the mean indifference interval shown in Fig. 2. In the control experiment there was no indication of an indifference interval as is shown on the right side of Fig. 3.

The interpretation of the indifference interval on the basis of the adaptation-level theory is therefore not valid. This is concluded on two grounds: 1. Although the subjects had no opportunity to develop a reference system, nonetheless, an indifference interval was observed within the smaller range, thus indicating an inter-individual constant. 2. This effect could be observed only in one temporal range.

## 4.2. Temporal Grouping in Subjective Random Series

It must be asked whether the organization of temporal behavior is influenced by the temporal constant which has just been described. The preceding experiment does not decide whether this constant should be understood as a continuous function or whether it indicates discrete time. If this constant is a continuous function, it defines an interval of equal duration for each point in the temporal continuum, i.e. for each temporal point there is a past interval of constant length. Alternatively, if this constant indicates discrete temporal units, subjective time is quantized in successive intervals of the same length. (This distinction relates to the discussion of stationarity and non-stationarity above.) For both models this interval possibly reflects the conscious present. Subjective experience does not provide an obvious indication which of these models holds true. However, experimental methods allow to decide between the alternative models.

Grouping effects which support the hypothesis of time quanta have been reported. It has been demonstrated that regularly occuring stimuli, such as the beating of a metronome, are subjectively grouped into units. (Bolton, 1893; Dietze, 1885; Koffka, 1909; Temperley, 1963; Woodrow, 1909.) The rhythmical grouping has a temporal limit, but there is no agreement as to magnitude (Boring, 1933).

A more objective test was designed in order to test experimentally the hypothesis of temporal grouping. Such a test is provided by subjective random series. In such an experiment the subject is requested to produce a random sequence of arbitrary symbols. If in such a situation grouping effects occur, they are especially significant, because they appear despite the subjects intention to randomize.

Many investigators have shown that subjects are not capable of producing truly random series (Baddeley, 1966b; Bakan, 1960; Chapanis, 1953; Day, 1956; Guttmann, 1966; Melendez, 1966; Mittenecker, 1958; Ramsay and Broadhurst, 1968; Rath, 1966; Schmitt, 1964; Tune, 1964). Furthermore, in a previous paper periodic components within subjective random series have been reported (Pöppel, 1967), which indicate grouping effects. However, the experimental design which was used in that experiment did not permit an answer whether temporal factors were involved in the grouping.

The experiment which will now be discussed was designed to decide between two possible explanations for grouping. The first holds that grouping is mainly dependent on the number of alternative symbols available; the second says that temporal factors are responsible.

In this experiment 17 subjects verbally produced random series with the numbers 1, 2, 3, 4, and 5. The random series were produced at two different rates: 1.0 and 0.9 seconds per digit. The time-giver for generating the random series was a light flash; at each flash the subject had to give a response. Five hundred consecutive calls were made at each rate. In the first part of the experiment, 9 subjects responded at a rate of 1.0 seconds and 8 subjects at 0.9 seconds; the sequence of responding rates was reversed for the second part.

This design provides the following prediction. If a temporal factor determines the period, the length of the period (measured in numbers of symbols) should increase when the responding rate is increased. If no temporal factor is involved, the period should remain the same when the rate is changed.

The subjective random series were analyzed by power spectra analyses (Blackman and Tukey, 1958) and nonparametric techniques (Pöppel, 1970b). Two examples of power spectra are given in Fig. 4. The responses of a subject under the two conditions have been analyzed. The peaks of these spectra are indicated by an arrow and represent the prominent periodic components

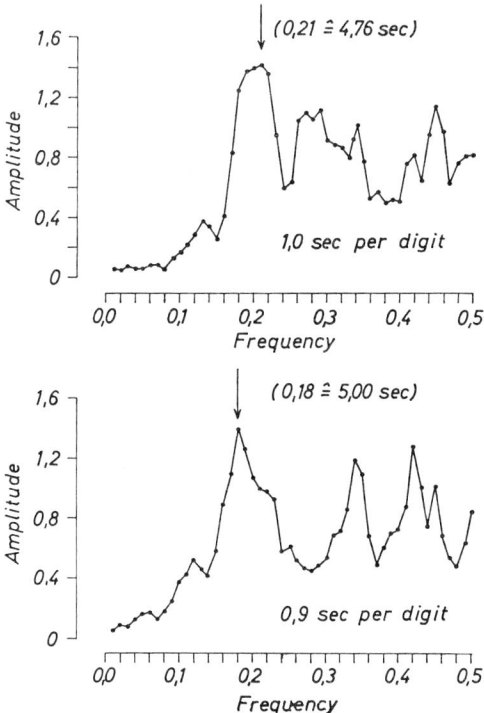

Fig. 4. Two power-spectra showing the periodic components in subjective random series. The series were produced by one subject under two conditions. Top: Speed for the production of the random series was 1.0 seconds per digit. Bottom: Speed was 0.9 seconds per digit. The subjects' task was to produce the numbers 1, 2, 3, 4 and 5 in a random sequence. The most prominent periodic components within the random series are indicated by an arrow; the period was measured in numbers of symbols. The corresponding times for these peak periods are given in parentheses

within the subjective random series. As can be seen, an increase in responding rate results in a decrease of the frequency, i.e. the periodic component is lengthened under the higher rate.

The Table summarizes results of the power spectra analyses. As can be seen the mean periods under the two conditions are significantly different ($t$-test). But the difference between the corresponding times for these periods is insignificant. It can therefore be concluded that in both conditions a temporal factor and not the number of alternatives, determined the length of the periods.

Table. *Summary of the power-spectral analyses of the subjective random series of 17 subjects. The subjects produced random series under two different conditions (1.0 and 0.9 seconds per digit)*

|  | Mean period and S. D. (Number of symbols) | Mean time and S. D. (seconds) |
|---|---|---|
| 1.0 sec per digit | M = 5.62<br>s = 1.86 | M = 5.75<br>s = 1.97 |
| 0.9 sec per digit | M = 6.44<br>s = 1.75 | M = 5.86<br>s = 1.67 |
|  | t = 2.36<br>p < 0.05 | t = 0.37<br>p ≈ 0.70 |

The temporal factor probably reflects a mechanism which is organizing subjective time by producing discrete consecutive time units. This experiment does not necessarily indicate the absolute length of the temporal constant, for the observed periodicities can be interpreted to reflect either multiples or fractions of oscillations with other periods. However, the next experiment suggests a periodic process with a period between 4 and 7 seconds (c.f. chapter 5 and Fig. 5).

A choice can now be made between alternative concepts of the temporal constant which were discussed at the beginning of this chapter. It appears that the temporal constant is not a stationary value but that it varies according to the phase of an endogenous oscillation. Only identical phases can be viewed formally as equivalent. It is suggested that the continuum of physical time is quantitized. This interpretation implies that there is no passive representation of time such that at *each* temporal point a just-past interval of constant duration is experienced as present. More likely, this duration may vary systematically according to the phase of the endogenous oscillation.

## 5. The "Temporal Constant" and Its Dependence on Physiological Conditions

The next three chapters will discuss briefly some influences on time perception which were mentioned in the introductory chapter.

In an experiment with six subjects, subjective random series were measured at different times of day. The subjects were requested to write two numbers (0 and 1) in random order. A metronome beating every second was used as a time-giver for the generation of the random series.

The results of this experiment are summarized in Fig. 5. At the top of the figure an individual example is given. It is obvious from these raw data that this series is not random in a statistical sense although the subject attempted to produce a random series. Alternating series of Zeroes and Ones yield a periodicity.

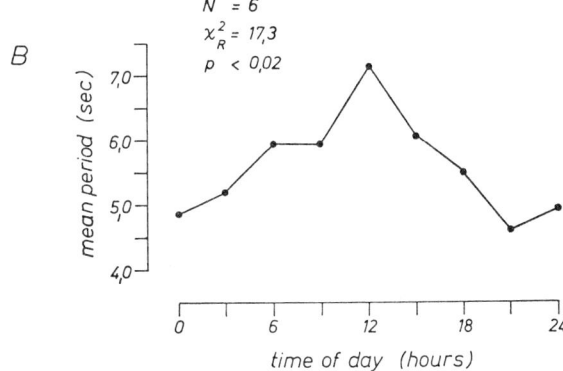

Fig. 5. Preliminary results on diurnal variation of the periodic components in subjective random series. The 6 subjects produced random series of two symbols which were transcribed into the numbers 0 and 1 for the computer analysis. A: An example of a sequence of 60 "calls" at 9 p.m. of subject C. B: Summarized data for the 6 subjects. The means of the prominent periodic components are shown at each time of day. The $p$-value was obtained with the Friedman two-way analysis of variance

For statistical analysis, the period length was determined for each subject at each time of day. The daily variation of these values was evaluated with the Friedman two-way analysis of variance (Siegel, 1956) and indicated a significant variation. The longest periods were observed at noon, the shortest periods at 9 p.m. and at midnight. These values vary approximately between 4.5 and 7 seconds. Although the number of alternatives in this experiment (2) were different from the number in the preceding experiment (5), the periodic lengths were in the same range. In the experiment just discussed the diurnal variation was used as a tool to further support the hypothesis on the temporal constant. It is on the other hand worthwhile to discuss briefly the physiological basis for this rhythm.

It has been suggested that diurnal variations of body temperature are responsible for diurnal variations of time perception (c.f. Pfaff, 1968). If there is a causal relationship between body temperature and time perception a minimal requirement would be that maxima and minima for both variables coincide in time – or that the phase difference between both oscillations is at least very small –, and that an experimentally manipulated change in body temperature be accompanied by a corresponding change in time perception.

Fig. 6 illustrates the results of an experiment that tested these expectations. Measurements of time perception were taken every 3 hours. Body temperature was registered continuously with a rectal thermometer. Subjects ($n = 12$) were

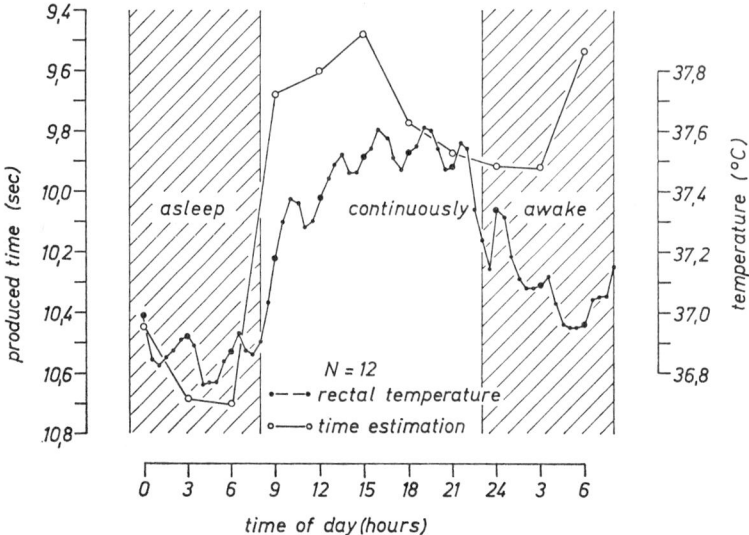

Fig. 6. Summarized data of an experiment on 10-sec production. 12 subjects produced acoustic intervals. The measurements were done in 3 hour intervals. During the first experimental night the sujects were awakened at midnight, 3 a.m. and 6 a.m. During the second night the subjects were required to stay awake. Body temperature was measured continuously

asked to produce intervals of 10 seconds. During the first experimental night the subjects were awakened for measurements at midnight, 3 a.m. and 6 a.m. During the second night the subjects had to stay awake. As can be seen both body temperature and time perception show clear daily variations. However, two results suggest that the variation in time perception is not determined exclusively, if at all, by body temperature: 1. From 3 to 6 p.m. there is a drop in time perception, but body temperature is still increasing. 2. During the second experimental night time perception remains relatively high as compared to the first experimental night but body temperature again falls appreciably. Obviously other factors besides body temperature must be responsible for the variation in time perception. Although these factors are still unknown, there are probably many involved (Pöppel, 1968c; Pöppel and Giedke, 1970; Pöppel et al., 1970).

## 6. Dependence of Time Perception on Knowledge of Temporal Intervals

The influence of informative stimuli on the reproduction of temporal intervals has been tested using the same experimental design as has been described in chapter 4.1. There were two conditions. In the first condition the subjects reproduced intervals between 1.0 and 9.0 seconds, without knowing the duration of the stimulus in seconds. In the second condition the subjects were told the absolute duration of the stimulus which was to be reproduced; the range of the stimuli was the same as in the first condition. In both conditions the presentation of stimuli was randomized.

The results of this experiment are illustrated in Fig. 7. Under both conditions indifference intervals were observed. The indifference interval in the condition where stimulus length was unknown lies between 2.5 and 3.0 seconds (only slightly different from the indifference intervals reported in the previous

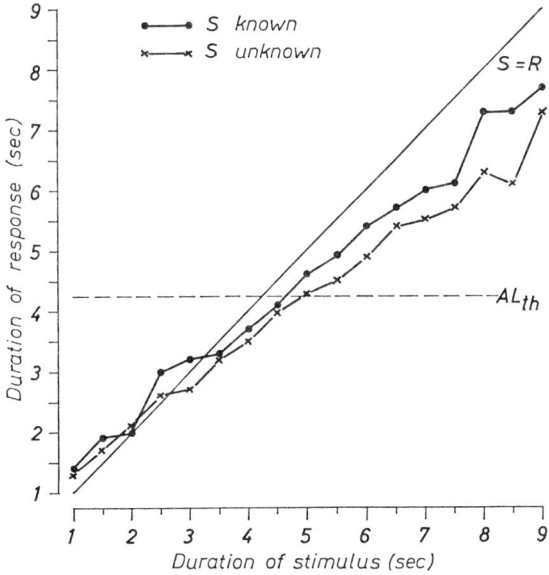

Fig. 7. Reproduction of temporal stimuli between 1.0 and 9.0 seconds. Summarized results for 18 subjects. The experiment was done under two conditions. For one part of the experiment conditions were identical to those described in Figs. 1 and 2. The second half was undertaken 24 hours later. Before each stimulus presentation the subjects were told the absolute length of the interval which was to be reproduced. With one exception (2.0 seconds) knowledge of interval length led to longer reproductions. The indifference interval for the unknown stimuli corresponds to the indifference intervals in the other experiments

experiments). The indifference interval under the condition of the known stimulus was longer, namely between 3.0 and 3.5 seconds. Not only the indifference intervals varied, but also all known stimuli were reproduced longer than the unknown stimuli (with one exception).

Since supplementary information did not increase accuracy of reproduction, the subjects probably did not rely on a reference system based on conventional time units. An alternative interpretation might be that information itself led to a lengthening of the indifference interval.

## 7. Dependence of Time Perception on Individual Reference Systems

In this last experiment it will be shown that there may be considerable differences in individual reference systems. Furthermore, these differences seem to be independent of prior learning. Fig. 8 illustrates this phenomenom.

In this experiment, feedback was used initially in order that subjects could learn to produce accurate 10 second intervals. As can be seen on the left side of Fig. 8, the subjects adjusted rather quickly to the set point. In this figure only the first 10 of 30 learning trials are shown. After the learning phase subjects lived alone for several weeks in an underground bunker were feedback was no longer provided (Pöppel and Giedke, 1970, Experiment 6; Wever, 1969). During this experiment they had to produce 10 second intervals several

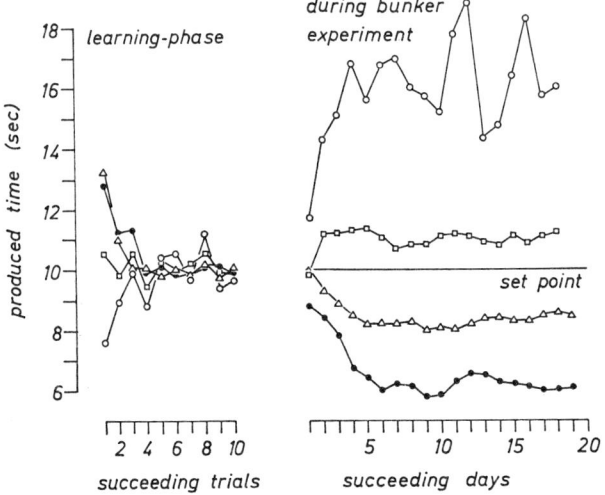

Fig. 8. Example of 10 second production of 4 subjects under two conditions. Left side: During a learning phase subjects were given feedback. The set point was reached already after a few trials. Right side: Subsequent bunker experiment (isolation from any time cues of the environment) with no feedback. The subjects reached individual levels different from the set point in general after several days. Each point is the mean of about 7 productions made throughout one circadian period

times a day; on an average there were seven productions each day. The productions of each day were averaged, and these mean values are shown in Fig. 8 on the right side. The subjects' productions drift from the learned set point towards a new subjective set point. This adaption to individual set points may take several days. The considerable differences between individual set points can be seen by comparing highest and lowest individual set points in Fig. 8. The shift to individual set points cannot be interpreted as decrease of accuracy, because then no set points should be reached. A decrease of accuracy or the dominance of the learned set point would support an exogenous basis of time perception. The individual set points, however, favor the hypothesis of endogenous reference systems for time perception.

*8. Conclusion*

The results presented in this paper suggest that the temporal continuum is subjectively quantitized into discrete units, which successively follow each other. The duration of these units is indicated by the grouping effects in subjective random series and is found to be roughly between 4 and 7 seconds. It is thus approximately equivalent to earlier estimates of the conscious present. The quantization can formally be described as an oscillation.

The duration of "quantum-period" seems to be influenced by several physiological conditions. The dependence on time of day has been discussed here. However, the physiological mechanism underlying this oscillatory process is unknown. A psychological influence on the length of the quantum-period can be assumed since informational cues have been shown to affect the temporal constant. Individual differences can also be considerable.

The biological purpose of this oscillatory process might possibly be to organize temporally the generation of subjective information as for instance language. The basic structure of a sentence (Chomsky, 1965) has to be anticipated before it can be uttered. Because of the limited capacity of the short term memory this anticipation cannot be extended into an infinite future, but has to be restricted. The quantum-period discussed here possibly has an intimate relation to this temporal organization of language.

*Bibliography*

Adams, J. A.: *Motor skills.* Ann. rev. psychol. 15 (1964) 181–202.
Adamson, R.: *Anchor effect limits.* Psychon. Sci. 9 (1967) 179–180.
Ahrens, R.: *Störungen von Sukzessivgestalten und Zeiterleben bei Aphasie.* G. Schaltenbrand (ed.): Zeit in nervenärztlicher Sicht. 100–103 Stuttgart: Enke 1963.
Aiken, L. R.: *Learning and retention in the estimation of short time intervals: a circuit and a study.* Perc. and motor skills 20 (1965) 509–517.
Aitken, R. C. B., Gedye, J. L.: *A study of two factors which affect arousal level and the apparent duration of a ten-minute interval.* Brit. journ. psychol. 59 (1968) 253–263.
Anderson, S. F.: *The absolute impression of temporal intervals.* Psychol. bull. 33 (1936) 794–795.
Anliker, J.: *Variations in alpha voltage of the electro-encephalogram and time perception.* Science 140 (1963) 1307–1309.
Aschoff, J.: *Zeitliche Strukturen biologischer Vorgänge.* Nova Acta Leopoldina N. F. Bd. 21, Nr. 143 (1959) 147–177.
— Pöppel, E., Wever, R.: *Circadiane Periodik des Menschen unter dem Einfluß von Licht-Dunkel-Wechseln unterschiedlicher Periode.* Pflügers Arch. 306 (1969) 58–70.
— Wever, R.: *Spontanperiodik des Menschen bei Ausschluß aller Zeitgeber.* Naturwissenschaften 49 (1962) 337–342.
Axel, R.: *Estimation of time.* Archives of psychology 12 (1925) 1–77.
Baddeley, A. D.: *Time-estimation at reduced body-temperature.* Amer. journ. psychol. 79 (1966a) 475–479.

— *The capacity for generating information by randomization.* Quart. journ. exp. psychol. 18 (1966b) 119–129.
Bakan, P.: *Response-tendencies in attempts to generate random binary series.* Amer. journ. psychol. 73 (1960) 127–131.
Bechinger, D., Kongehl, G., Kornhuber, H. H.: *Natural 2-second cycle in time perception and human information transmission.* Naturwissenschaften 56 (1969) 419.
Behar, I., Bevan, W.: *The perceived duration of auditory and visual intervals: cross-modal comparison and interaction.* Amer. journ. psychol. 74 (1961) 17–26.
Bell, C. R.: *Time estimation and increases in body temperature.* Journ. exp. psychol. 70 (1965) 232–234.
— *Control of time estimation by a chemical clock.* Nature 210 (1966) 1189–1190.
— Provins, K. A.: *The relation between physiological responses to environmental heat and time judgments.* Journ. exp. psychol. 66 (1963) 572–579.
— Watts, N.: *Personality and judgments of temporal intervals.* Brit. journ. psychol. 57 (1966) 155–159.
Benussi, V.: *Zur experimentellen Analyse des Zeitvergleichs.* Arch. ges. Psychol. 9 (1907) 366—449.
— *Psychologie der Zeitauffassung.* Heidelberg: C. Winter 1913.
Bergius, R.: *Vom Zeitsinn zum Verhaltensparameter Zeit.* M. Irle (Hrsg.) Ber. 26. Kongr. DGPS. Göttingen: Hogrefe, 1–21 (1969).
Bindra, D., Waksberg, H.: *Methods and terminology in studies of time estimation.* Psychol. bull. 53 (1956) 155–159.
Björkman, M., Holmkvist, O.: *The time-order error in the construction of a subjective time scale.* Scand. journ. psychol. 1 (1960) 7–13.
Blackman, R. B., Tukey, J. W.: *The measurement of power spectra.* New York: Dover 1958.
Boardman, W. K., Goldstone, S., Lhamon, W. T.: *Effects of lysergic acid diethylamide (LSD) on the time sense of normals.* A. M. A. Arch. neurol. psychiat. 78 (1957) 321–324.
von Bodman, K.: *Pharmakawirkung auf Kurzzeitschätzungen bei Ratten.* Naturwissenschaften 56 (1969) 335.
— *Die Wirkung von Pharmaka auf Kurzzeitschätzung und circadiane Periodik von Säugetieren.* Zschrft. vergl. Physiol. 68 (1970) 276–292.
Bokander, I.: *Time estimation as an indicator of attention–arousal when perceiving complex and meaningful stimulus material.* Perc. and motor skills 21 (1965) 323–328.
Bolton, T. L.: *Rhythm.* Amer. journ. psychol. 6 (1893) 145–238.
Boring, E. G.: *The physical dimensions of consciousness.* New York: Appleton-Century-Crofts 1933.
— *Sensation and perception in the history of experimental psychology.* New York: Appleton-Century Crofts 1942.
Boring, L. D., Boring, E. G.: *Temporal judgments after sleep.* Studies in Psychology. Titchener Commemorative Vol., 255–279 (1917).
Boulter, L. R., Appley, M. H.: *Time and effort as determiners of time-production error.* Journ. exp. psychol. 75 (1967) 447–452.
Brown, D. R., Hitchcock, L.: *Time estimation: dependence and independence of modality-specific effects.* Perc. and motor skills 21 (1965) 727–734.
Cahoon, R. L.: *Effect of acute exposure to altitude on time estimation.* Journ. psychol. 66 (1967) 321–324.
— *Physiological arousal and time estimation.* Perc. and motor skills 28 (1969) 259–268.
Carlson, V. R., Feinberg, I.: *Individual variations in time judgment and the concept of an internal clock.* Journ. exp. psychol. 77 (1968) 631–640.
— — *Time judgment as a function of method, practice, and sex.* Journ. exp. psychol. 85 (1970) 171–180.
Chapanis, A.: *Random-number guessing behavior.* Amer. psychol. 8 (1953) 332.
Chomsky, N.: *Aspects of the theory of syntax.* Cambridge: The Massachusetts Institute of Technology Press 1965.
Clausen, J.: *An evaluation of experimental methods of time estimation.* Journ. exp. psychol. 40 (1950) 756–761.
Cohen, J.: *Psychological time in health and disease.* Springfield, Ill.: Charles C. Thomas 1967.

Craik, K. H., Sarbin, T. R.: *Effect of covert alterations of clock rate upon time estimations and personal tempo.* Perc. and motor skills 16 (1963) 597–610.

Creelman, D. C.: *Human discrimination of auditory duration.* Journ. acoust. soc. Amer. 34 (1962) 582–593.

Davis, H.: *Space and time in the central nervous system.* EEG clin. neurophysiol. 8 (1956) 185 bis 191.

Day, W. F.: *Serial non-randomness in auditory differential-thresholds as a function of interstimulus interval.* Amer. journ. psychol. 69 (1956) 387–394.

Denys, W., Richelle, M.: *Régulations temporelles simples chez des malades mentaux.* Schweiz. Ztschr. Psychol. u. Anwend. 24 (1965) 263–267.

Deppe, A. H.: *Overload and sensory deprivation: time estimation in novice divers.* Perc. and motor skills 29 (1969) 481–482.

Dewolfe, R. K. S., Duncan, C. P.: *Time estimation as a function of level of behavior of successive tasks.* Journ. exp. psychol. 58 (1959) 153–158.

Dietze, G.: *Untersuchungen über den Umfang des Bewußtseins bei regelmäßig aufeinanderfolgenden Schalleindrücken.* Phil. Studien 2 (1885) 362–393.

Dimond, S. J.: *The structural basis of timing.* Psychol. bull. 62 (1964) 348–350.

Dupreez, P. D.: *Relation between verbal estimation and reproduction of a short time interval: preliminary study.* Perc. and motor skills. 17 (1963) 45–46.

— *Reproduction of time intervals after short periods of delay.* Journ. general psychol. 76 (1967) 59–71.

Edgell, B., Waller, A. W.: *On time judgments.* Amer. journ. psychol. 14 (1903) 154–174.

Emley, G. S., Schuster, C. R., Lucchesi, B. R.: *Trends observed in the time estimation of three stimulus intervals within and across sessions.* Perc. and motor skills 26 (1968) 391–398.

Eson, M. E., Kafka, J. S.: *Diagnostic implications of a study in time perception.* Journ. general psychol. 46 (1952) 169–183.

Estel, V.: *Neue Versuche über den Zeitsinn.* Phil. Studien 2 (1884) 37–65.

Falk, J. L., Bindra, D.: *Judgment of time as function of serial position and stress.* Journ. exp. psychol. 47 (1954) 279–282.

Filer, R. J., Meals, D. W.: *The effect of motivating conditions on the estimation of time.* Journ. exp. psychol. 39 (1949) 327–331.

Fox, G. J.: *The effect of interval and duration of visual stimuli upon the time error.* Psychol. Nwsltr. 45 (1952) 1–17.

Fox, R. H., Bradbury, P. A., Hampton, I. F. G., Legg, C. F.: *Time judgment and body temperature.* Journ. exp. psychol. 75 (1967) 88–96.

Fraisse, P.: *The psychology of time.* London: Eyre and Spottiswoode 1964.

— *Zeitwahrnehmung und Zeitschätzung.* W. Metzger (Hrsg.) Handbuch der Psychologie: Allgemeine Psychologie, Band I, 1, (1966) 656–690, Göttingen: Hogrefe.

— Bonnet, C., Gelly, N., Michau, G.: *Vergleich der Zeitschätzungsmethoden.* Ztschr. Psychol. 167 (1962) 268–277.

— Voillaume, C.: *Conditionnement temporel du rythme alpha et estimation du temps.* L'année psychol. 69 (1969) 7–15.

François, M.: *Contribution à l'étude du sens du temps. La température interne comme facteur de variation de l'appréciation subjective des durées.* L'année psychol. 28 (1927) 186–204.

— *Influence de la température interne sur notre appréciation du temps.* C. R. soc. biol. 98 (1928) 201–203.

Frankenhaeuser, M.: *Estimation of time: an experimental study.* Stockholm: Almqvist and Wiksell 1959.

Gardner, W. A.: *Influence of the thyroid gland on the consciousness of time.* Amer. journ. psychol. 47 (1935) 698–701.

Gilliland, A. R., Hofeld, J., Eckstrand, G.: *Studies in time perception.* Psychol. bull. 43 (1946) 162–176.

— Martin, R.: *Some factors in estimating short time intervals.* Journ. exp. psychol. 27 (1940) 243–255.

Goldstone, S.: *The human clock: a framework for the study of healthy and deviant time perception.* Annals New York Acad. Sci. 138,2 (1967) 767–783.

— Boardman, W. K., Lhamon, W. T.: *Effect of quinal barbitone, dextro-amphetamine, and placebo on apparent time.* Brit. journ. psychol. 49 (1958) 324–328.
— Lhamon, W. T., Boardman, W. K.: *The time sense: Anchor effects and apparent duration.* Journ. psychol. 44 (1957) 145–153.
Gooddy, W.: *Disorders of the time sense.* P. J. Vinken and G. W. Bruyn (eds.) Handbook of clinical neurology vol. 3. Amsterdam: North-Holland Publ., 229–250 (1969a).
— *Outside time and inside time.* Persp. biol. medicine 12 (1969b) 239–253.
Gulliksen, H.: *The influence of occupation upon the perception of time.* Journ. exp. psychol. 10 (1927) 52–59.
Guttmann, G.: *Komplexe Ordnungstendenzen in Verhaltensabläufen und ihre differentialdiagnostische Bedeutung.* Z. exp. angew. Psychol. 13 (1966) 19–30.
Guyau, M.: *La genèse de l'idée de temps.* Paris: Alcan 1890.
Halberg, F., Siffre, M., Engeli, M., Hillman, D., Reinberg, A.: *Etude en libre-cours des rythmes circadiens du pouls, de l'alternance veille-sommeil et de l'estimation du temps pendant des deux mois de séjour souterrain d'un homme adulte jeune.* C. r. acad. sci. Paris 260 (1965) 1259–1262.
Hall, G. S., Jastrow, J.: *Studies of rhythm.* Mind 11 (1886) 55–62.
Harton, J. J.: *The influence of the difficulty of activity on the estimation of time.* Journ. exp. psychol. 23 (1938), 270–287, 428–433.
Hawkes, G. R., Joy, R. J. T., Evans, W. O.: *Autonomic effects on estimates of time: evidence for a physiological correlate of temporal experience.* Journ. psychol. 53 (1962) 183–191.
Heiss, R.: *Die Zeit des Menschen und die Zeit überhaupt.* Freiburger Dies Universitatis (1960/61) 117–131.
Helson, H.: *Adaptation-level theory.* New York: Harper and Row 1964.
Hirsh, I. J., Bilger, R. C., Deatherage, B. H.: *The effect of auditory and visual background on apparent duration.* Amer. journ. psychol. 69 (1956) 561–574.
Hoagland, H.: *The physiological control of judgments of duration: evidence for a chemical clock.* Journ. general psychol. 9 (1933) 267–287.
Hoche, A.: *Langeweile.* Psychol. Forschung 3 (1923) 258–271.
Hoff, H., Pötzl, O.: *Über eine Zeitrafferwirkung bei homonymer linksseitiger Hemianopsie.* Zschrft. ges. Neurol. Psychiatrie 151 (1934) 599–641.
Hollingworth, H. L.: *The central tendency of judgment.* Arch. psychol. 4 (1913) 44–52.
Holubář, J., Machek, J.: *Time sense and epileptic EEG activity.* Epilepsia 3 (1962) 323–328.
Hörmann, H.: *Kritische Bemerkungen zu einer Untersuchung von F. Sixtl.* Zschrft. exp. angew. Psychol. 11 (1964) 667–670.
Hornstein, A. D., Rotter, G. S.: *Research methodology in temporal perception.* Journ. exp. psychol. 79 (1969) 561–564.
James, W.: *The principles of psychology.* Vol. I. New York: Henry Holt 1890.
Josenhans, W.: *Versuch einer Objektivierung von Änderungen im Zeiterleben und Zeitempfinden.* Naturwissenschaften 46 (1959) 538–539.
Kafka, J. S.: *A method for studying the organization of time experience.* Amer. journ. psychiatry 114 (1957) 546–553.
Kant, I.: *Kritik der reinen Vernunft.* Hamburg: Felix Meiner 1956; Orig. 1781/1787.
Katz, D.: *Experimentelle Beiträge zur Psychologie des Vergleichs im Gebiete des Zeitsinns.* Zschrft. Psychol. Physiol. d. Sinnesorgane 42 (1906) 302–340, 424–450.
Kelm, H.: *Consistency of successive time estimates during positive feed-back.* Perc. and motor skills 15 (1962) 216.
Kleber, R. J., Lhamon, W. T., Goldstone, S.: *Hyperthermia, hyperthyroidism, and time judgment.* Journ. comp. physiol. psychol. 56 (1963) 362–365.
Koffka, K.: *Experimental-Untersuchungen zur Lehre des Rhythmus.* Zschrft. Psychol. 52 (1909) 1–109.
Köhler, W.: *Zur Theorie des Sukzessivvergleichs und der Zeitfehler.* Psychol. Forschung 4 (1923) 115–175.
von Kries, J.: *Über die Bedeutung des Aufmerksamkeitssprunges für den Zeitsinn.* Dt. Zschrft. Nervenheilkunde 47, 48 (1913) 352–370.
Kruup, K.: *Influence of method on time judgments.* Austr. journ. psychol. 13 (1961) 44–53.
Langer, J., Wapner, S., Werner, H.: *The effect of danger upon the experience of time.* Amer. journ. psychol. 74 (1961) 94–97.

Legg, C. F.: *Alpha rhythm and time judgments.* Journ. exp. psychol. 78 (1968) 46–49.
Lévy, J. C.: *Le temps psychologique.* Paris: Dunod 1969.
Lockhart, J. M.: *Ambient temperature and time estimation.* Journ. exp. psychol. 73 (1967) 286–291.
Loehlin, J. C.: *The influence of different activities on the apparent length of time.* Psychol. Monogr. Nr. 474 (1959) Vol. 73, 4.
Mach, E.: *Die Analyse der Empfindungen und das Verhältnis des Physischen zum Psychischen.* Jena: G. Fischer 1919 – 8 (1. Aufl. 1885).
MacLeod, R. B., Roff, M. F.: *An experiment in temporal disorientation.* Acta psychol. 1 (1923) 381–423.
Marum, K. D.: *Reproduction and ratio-production of brief duration under conditions of sensory isolation.* Amer. journ. psychol. 91 (1968) 21–26.
McGrath, J. J., O'Hanlon, J. F.: *Method for measuring the rate of subjective time.* Perc. and motor skills 24 (1967) 1235–1240.
Meade, R. D.: *Time estimates as affected by motivational level, goal distance, and rate of progress.* Journ. exp. psychol. 58 (1959) 275–279.
— *Time estimates as affected by need tension and rate of progress.* Journ. psychol. 50 (1960) 173 bis 177.
Mehner, M.: *Zur Lehre vom Zeitsinn.* Philos. Studien 2 (1885) 546–602.
Melendez, F.: *Number preference as a variable influencing human S's inability to generate a truly random series of numbers.* Perc. and motor skills 23 (1966) 1201–1202.
Melges, F. T., Fougerousse, C. E.: *Time sense, emotions, and acute mental illness.* Journ. psychiatric res. 4 (1966) 127–140.
Michon, J. A.: *Studies on subjective duration II. Subjective time measurement during tasks with different information content.* Acta psychol. 24 (1965) 205–219.
— *Timing in temporal tracking.* Soesterberg Institute for Perception RVO-TNO 1967.
— *A model of some temporal relations in human behavior.* Psychol. Forschung 31 (1968) 287–298.
— *Processing of temporal information and the cognitive theory of time experience.* Studium Generale 23 (1970) 249–265.
Mittenecker, E.: *Die Analyse „zufälliger" Reaktionsfolgen.* Zschrft. exp. angew. Psychol. 5 (1958) 45–60.
Müller, K.: *Die phänomenale Dauer visueller Sukzessionen. Experimentelle Untersuchungen zum Problem der Abhängigkeit der Zeitwahrnehmung von der Geschehensstruktur.* Zschrft. exp. angew. Psychol. 12 (1965) 98–123.
Münsterberg, H.: *Beiträge zur experimentellen Psychologie.* Freiburg 1892. Heft 4.
Murphy, L. E.: *Absolute judgment of duration.* Journ. exp. psychol. 71 (1966) 260–263.
Nelson, T. M., Bartley, S. H., Jordan, J. F.: *Experimental evidence for the involvement of a neurophysiological mechanism in the discrimination of duration.* Journ. psychol. 55 (1963) 371–385.
Nichols, H.: *The psychology of time.* Amer. journ. psychol. 3 (1891) 453–530.
Ochberg, F. M., Pollack, I. W., Meyer, E.: *Reproduction and estimation methods of time judgments.* Perc. and motor skills 20 (1965) 653–656.
Orme, J. E.: *Time, experience and behaviour.* London: Iliffe. New York: Elsevier 1969.
Ornstein, R. E.: *On the experience of time.* Harmondsworth: Penguin 1969.
Otto, E.: *Unterschiedsempfindlichkeit für Zeitintervalle und automatisierte Motorik nach Injektion von Amytal-Na.* Ztschrft. Psychol. 168 (1963a) 176–194.
— *Modifizierende und konstituierende Faktoren der Zeitwahrnehmung beim Menschen.* Biol. Rdsch. 1 (1963b) 2–18.
Pfaff, D.: *Effects of temperature and time of day on time judgments.* Journ. exp. psychol. 76 (1968) 419–422.
Piaget, J.: *Die Bildung des Zeitbegriffs beim Kinde.* Zürich: Rascher 1955.
Piéron, H.: *Les problèmes psycho-physiologiques de la perception du temps.* Année psychol. 24 (1923) 1—25.
Pittendrigh, C. S.: *On temporal organization in living systems.* The Harvey Lectures. New York: Acad. Press, Ser. 56 (1961) 93–125.
Ploeger, A.: *Zeiterleben in einer Extremsituation.* Ztschrft. Psychotherapie u. med. Psychol. 16 (1966) 13–20.

Pöppel, E.: *Signifikanz-Artefakte in der experimentellen Parapsychologie.* Ztschrft. Parapsychol. Grenzgeb. Psychol. 10 (1967) 63–72.
— *Oszillatorische Komponenten in Reaktionszeiten.* Naturwissenschaften 55 (1968a) 449–450.
— *Desynchronisationen circadianer Rhythmen innerhalb einer isolierten Gruppe.* Pflügers Arch. 299 (1968b) 364–370.
— *Tagesperiodische Veränderungen der akustischen Adaptation und des psychomotorischen Tempos mit und ohne Nachtruhe.* Pflügers Arch. 300 (1968c) R 11–R 12.
— *Oszillatorische Vorgänge bei der menschlichen Zeitwahrnehmung.* M. Irle (Hrsg.) Ber. 26. Kongr. DGPs. Göttingen: Hogrefe, 388–398 (1969).
— *Excitability cycles in central intermittency.* Psychol. Forschung 34 (1970a) 1–9.
— *Frequency measurement in time series data.* Life sciences and space research 8 (1970b) 234.
— Aschoff, J. C., Giedke, H.: *Tagesperiodische Veränderungen der Reaktionszeit bei Wahlreaktionen.* Zschrft. exp. angew. Psychol. 17 (1970) 537–552.
— Giedke, H.: *Diurnal variation of time perception.* Psychol. Forschung 34 (1970) 182–198.
Postman, L., Miller, G. A.: *Anchoring in temporal judgments.* Amer. journ. psychol. 58 (1945) 43–53.
Quandt, J.: *Das Problem des Zeitbewußtseins.* Arch. ges. Psychol. 8 (1906) 143–189.
Quasebarth, K.: *Zeitschätzung und Zeitauffassung optisch und akustisch ausgefüllter Intervalle.* Arch. ges. Psychol. 49 (1924) 379–432.
Ramsay, R. W., Broadhurst, A.: *The non-randomness of attempts at random responses: relationships with personality variables and psychiatric disorders.* Brit. journ. psychol. 59 (1968) 299–304.
Rath, G. J.: *Randomization by humans.* Amer. journ. psychol. 79 (1966) 97–103.
Richards, W.: *Time estimates measured by reproduction.* Perc. and motor skills 18 (1964) 929–943.
Richards, W. J., Livingston, P. V.: *Method, standard duration, and inter-stimulus delay as influences upon judgment of time.* Amer. journ. psychol. 79 (1966) 560–567.
Ross, S., Katchmar, L.: *The construction of a magnitude function for short time-intervals.* Amer. journ. psychol. 64 (1951) 397–401.
Rutschmann, J., Rubinstein, L.: *Time estimation, knowledge of results and drug effects.* Journ. psychiatric res. 4 (1966) 107–114.
Schaefer, V. G., Gilliland, A. R.: *The relation of time estimation to certain physiological changes.* Journ. exp. psychol. 23 (1938) 545–552.
Schaltenbrand, G.: *Zeit in nervenärztlicher Sicht.* Stuttgart: Enke 1963.
Schmitt, E.: *Untersuchungen an Binärprädiktoren, insbesondere bezüglich ihrer Anpassungsfähigkeit und ihrer Vorhersageleistung gegenüber Versuchspersonen.* Kybernetik 2 (1964) 93–102.
Schoeffler, M. S., Poole, D. M.: *Accuracy and variability in the production of short durations.* Psychon. Sci. 7 (1967) 423–424.
Schönpflug, W.: *Bezugssystem, Urteilsakzentuierung und Aktivierung.* M. Irle (Hrsg.) Ber. 26. Kongr. DGPs. Göttingen: Hogrefe, 405–413 (1969).
Siegel, S.: *Nonparametric statistics for the behavioral sciences.* New York: McGraw-Hill 1956.
Siegman, A. W.: *Intercorrelation of some measures of time estimation.* Perc. and motor skills 14 (1962) 381–382.
Sixtl, F.: *Die Erfassung von Sukzessionen bei Ausschaltung der aktiven Vergegenwärtigung.* Arch. ges. Psychol. 114 (1962) 337–377.
— *Der Zeitfehler (time-order error) beim Schätzen der Reizzeit und als Funktion der Reizlänge, der Intervallzeit und der Versuchswiederholung.* Ztschrft. exp. angew. Psychol. 10 (1963) 209–225.
— *Erwiderung und Ergänzung zu Hörmanns Diskussion der Fading-trace-Theorie.* Ztschrft. exp. angew. Psychol. 11 (1964) 671–678.
Smets, G.: *Time expression of red and blue.* Perc. and motor skills 29 (1969) 511–514.
Spencer, L. T.: *An experiment in time estimation using different interpolations.* Amer. journ. psychol. 32 (1921) 557–562.
Spreen, O.: *The position of time estimation in a factor analysis and its relation to some personality variables.* Psychol. records 13 (1963) 455–464.
Steinberg, H.: *Changes in time perception induced by an anaesthetic drug.* Brit. journ. psychol. 46 (1955) 273–278.
Stephens, G. J., Halberg, F.: *Human time estimation.* Nursing Res. 14 (1965) 310–317.

Stern, M. H.: *Thyroid function and activity, speed, and timing aspects of behaviour*. Canad. journ. psychol. 13 (1959) 43–48.
Stern, W.: *Psychische Präsenzzeit*. Ztschrft. Psychol. 13 (1897) 325–349.
Sterzinger, O.: *Chemopsychologische Untersuchungen über den Zeitsinn*. Ztschrft. Psychol. Physiol. Sinnesorgane 134 (1935) 100–131.
— *Neue chemopsychologische Untersuchungen über den menschlichen Zeitsinn (das Problem der 5-Minutenzeitstrecke)*. Ztschrft. Psychol. Physiol. Sinnesorgane 143, (1938) 391–406.
Stevens, L. T.: *On the time-sense*. Mind 11 (1886) 393–404.
Stevens, S. S.: *On the psychophysical law*. Psychol. Rev. 64 (1957) 153–181.
Stroud, J. M.: *The fine structure of psychological time*. H. Quastler (ed.) Information theory in psychology. Glencoe, Ill.: Free Press, 174–205 (1955).
Sturmer, G. von: *Stimulus variation and sequential judgments of duration*. Quart. journ. exp. psychol. 18 (1966) 354–357.
— Wong, T., Coltheart, M.: *Distraction and time estimation*. Quart. journ. exp. psychol. 20 (1968) 380–384.
Sturt, M.: *The psychology of time*. London: Kegan Paul 1925.
Swift, E. J., McGeoch, J. A.: *An experimental study of the perception of filled and empty time*. Journ. exp. psychol. 8 (1925) 240–249.
Temperley, N. M.: *Personal tempo and subjective accentuation*. Journ. general psychol. 68 (1963) 267–287.
Thor, D. H.: *Diurnal variability in time estimation*. Perc. and motor skills 15 (1962) 451–454.
Treisman, M.: *Temporal discrimination and the indifference interval: implications for a model of the "internal clock"*. Psychol. Monogr. Nr. 576, (1963) Vol. 77, 13.
Tune, G. S.: *A brief survey of variables that influence random-generation*. Perc. and motor skills 18 (1964) 705–710.
Vierordt, K.: *Der Zeitsinn nach Versuchen*. Tübingen: Laupp 1868.
Vroon, P. A.: *Effects of presented and processed information on duration experience*. Acta psychol. 34 (1970) 115–121.
Wallace, M., Rabin, A. I.: *Temporal experience*. Psychol. bull. 57 (1960) 213–236.
Warm, J. S., Greenberg, L. F., Dube II, C. S.: *Stimulus and motivational determinants in temporal perception*. Journ. psychol. 58 (1964) 243–248.
— Morris, J. R., Kew, J. K.: *Temporal judgment as a function of nosological classification and experimental method*. Journ. psychol. 55 (1963) 287–297.
— Smith, R. P., Caldwell, L. S.: *Effects of induced muscle tension on judgment of time*. Perc. and motor skills 25 (1967) 153–160.
Weber, A. O.: *Estimation of time*. Psychol. bull. 30 (1933) 233–252.
Weber, D. S.: *A time perception task*. Perc. and motor skills 21 (1965) 863–866.
Webster, F. R., Goldstone, S., Webb, W. W.: *Time judgment and schizophrenia: psychophysical method as a relevant contextual factor*. Journ. psychol. 54 (1962) 159–164.
Werboff, J.: *Time judgment as a function of electroencephalographic activity*. Exp. Neurology 6 (1962) 152–160.
Wever, R.: *Untersuchungen zur circadianen Periodik des Menschen mit besonderer Berücksichtigung des Einflusses schwacher elektrischer Wechselfelder*. Bundesministerium für wissenschaftliche Forschung, Forschungsbericht W 69-31 (1969).
Wiener, N.: *Kybernetik, Regelung und Nachrichtenübertragung im Lebewesen und in der Maschine*. Düsseldorf: Econ 1963.
Woodrow, H.: *A quantitative study of rhythm*. Arch. psychol. New York 18 (1909) Nr. 1.
— *The temporal indifference interval, determined by the method of mean error*. Journ. exp. psychol. 17 (1934) 167–188.
— *The effect of practice upon time-order errors in the comparison of temporal intervals*. Psychol. review 42 (1935) 127–152.
— *Time perception*. S. S. Stevens (ed.) Handbook of experimental psychology. New York: Wiley 1224–1236 (1951).
Wundt, W.: *Einführung in die Psychologie*. Leipzig: Voigtländer 1911.
Zoltobrocki, J.: *Über systematische Fehler bei wiederholter Schätzung von Zeitintervallen*. E. Rausch (Hrsg.): Psychologische Arbeiten. Frankfurt/Main: Kramer 1965.

# Processing of Temporal Information and the Cognitive Theory of Time Experience

JOHN A. MICHON*

*Summary.* For man as an information processing system, time is one of the experiential dimensions of information, and it should be considered equivalent to other, non-temporal, aspects of this information, such as intensity, size, etc. Since as a processor man has a limited capacity there will be necessarily a trade-off between temporal and non-temporal information, which is open to quantification. Research in this area is reviewed. Most contemporary models of time evaluation incorporate a-specific "pulse counter" mechanisms to account for the internal clock by which time is measured subjectively. The rate of this internal clock is thought to be influenced by the information processed by the subject. In this paper an alternative formulation is defended: time evaluation is a cognitive reconstruction of contents of the interval. The latter formulation avoids the unnecessary assumption of the former. It explains the same phenomena equally well, while moreover it can handle various matters that offer difficulties to models stated in terms of clock mechanisms.

## Time Constants and Time Experience

The temporal structure of human behavior currently attracts considerable interest. Time has become a parameter in psychological models in a different sense than being only an ordering relation on stimulus or response sequences. To a large extent this is a consequence of the penetration of experimental psychology by cybernetics and information theory which introduced such concepts as feedback and channel capacity. If, for instance, a subject is given feedback about his performance, it is important when precisely this feedback arrives. Dramatic consequences may derive from a delay in the auditory feedback a speaker obtains about his own voice. When his voice reaches his ears through earphones with a delay of approximately 0.2 sec, the speaker will prolong vowels and duplicate or triplicate syllables (see Smith, 1966). Also, when the delay is in the order of 1 to 2 sec, whole words may be repeated one or more times (Berko, 1965).

Classical psychological theories were essentially Platonic, timeless structures, but as soon as one starts considering the human organism as an active information processing system, the question of its time constants becomes acute. The

---

* Prof. Dr. John A. Michon, Institute for Perception RVO–TNO, Kampweg 5, Soesterberg, Netherlands, and State University Groningen, Netherlands.

amount of information transmitted by a subject cannot be defined as long as we do not specify how much in what interval.

These time constants of human behavior may refer to purely physiological processes or describe hypothetical psychological mechanisms. On one hand a refractory period, i.e. a period of diminished functional sensitivity immediately following action, has been measured in the single cells of the central nervous system. On the other hand a psychological refractory period also has been observed. This phenomenon is based on the observation that when two independent stimuli arrive in rapid succession – say 100 or 200 msec, – the response to the second stimulus will be delayed. Neurons being vastly simpler than man, their refractory period has been investigated with accordingly greater precision than the psychological refractory period.

It is somewhat surprising – and for a devoted time psychologist also somewhat discouraging – to find that in all relevant research the time constants are treated as if they apply to a physical system. Subjects are considered as systems without temporal awareness, having a strict physical time keeper.

We should know better by now: people simply have no constant rate clocks. Moreover, they are consciously aware of the order of events in time and of the duration of these events. Thus we may expect that in any situation in which time is a factor of concern, variable "clock rate" and awareness of order and duration will exert their influence on behavior, and therefore should be incorporated in models of such behavior. This is not an idle claim. The skilled execution of many perceptual-motor tasks depends on evaluation – be it conscious or implicit – of the duration of events.

For example, in a study of the skill of Air Traffic Controllers (ATCs) Michon, Wagenaar, Lazet and Koutstaal (1965) found that the subjective estimates of the ATC about the mental load imposed by his task is dependent on the average time he needs for reaching a decision. When we tried to introduce an improved information updating and display system, the ATCs felt as if they could handle only half as much information as they had processed in the conventional system. Only when we succeeded, by training them, in reducing their average decision time below the original average, their subjective mental load fell below that of the old procedure.

Similar findings for somewhat longer intervals were obtained by McGrath and his coworkers (McGrath and O'Hanlon, 1967). They studied the effect of average interval between stimuli in a vigilance task. Some relevant studies dealing with much shorter intervals were reviewed in a recent paper by Schmidt (1968), in which he discussed the problem of anticipation in skilled behavior, such as sport and games. There, time estimation is of great and continuous importance. When I want to kick a soccer ball for example, I even have to make two estimates at the same time: I not only need to determine when the ball will arrive at the contact point, but also when my foot will be there (cf. Schmidt 1968, p. 636).

In summary, any theory of human performance and in particular any theory that deals with the interaction between man and machine, must incorporate the fact that man is aware of order and duration but has no regular physical internal clock to evaluate it.

Time is not just a homogeneous, continuous parameter of input – output relations. It also means *information* to man.

This point of view will be worked out in more detail in the present paper. We shall consider the role of temporal information in skilled performance, and see how the gap that exists between contemporary time psychology and performance theory is being bridged. This task has some likelihood of success, because the "Zeitgeist" appears to be favorable for various reasons.

Already did I mention the rise of interest in temporal factors in general.

Furthermore, some recent, quantitative studies have equipped "time psychology" with a better image than it has had for a long time. One particular aspect of this trend is the insight that measures of central tendency are not necessarily the most interesting data about time experience. Instead the variance and the distribution of time data appear to be much more revealing. This establishes a strong tie with modern psychophysics, which also leans heavily on statistical properties of its data other than means or medians.

Finally, the strong tendency to treat man as an information processing system is a facilitating factor. This trend in psychological theory is exemplified by the "cognitive psychology" advocated, among others, by Miller, Galanter and Pribram (1960) and Neisser (1967). Time psychology has a strong tradition in cognitive or "cue" theory (Guyau, 1890; Janet, 1928; Woodrow, 1951).

Altogether the tide seems running high to make time psychology the respectable branch of experimental psychology again that it was in the days when Titchener considered it to be a "microcosm perfect to the last detail" (Titchener, 1905).

*Current Theories of Time Experience*

Let me start by sketching the main trends in psychological theorizing about the experience of duration. Although I recognize the "dating" or "event labeling" aspect of human time experience I shall only occasionally touch upon it.

Broadly we can distinguish two main formulations, which are known as the *pulse counter* or internal clock theory, and the *information storage* or cognitive theory. The first has the elegance of simplicity, the second the elegance of flexibility. For the rest they appear, at first sight, to be alternative formulations for one and the same problem. Let me elaborate on this.

For a long time the psychology of time perception has attempted to specify the physiological basis underlying time experience. In another place I have exposed this endeavor in more detail, and shown that essentially the same formal pulse counter model underlies all of the various physio-mythological explanations offered this last century (Michon, 1965 a). The counter model has as its central concept the internal clock. This is a hypothetical mechanism which is driven by the state of specific activation of the subject. When the level of specific activation is high the clock rate will be fast, and when the level is low the rate will slow down. The word "specific activation" was devised by Treisman (1963) to indicate that it is not necessarily general arousal that is referred to, but only the driving of the internal clock. There is a marked tendency in the literature though, to equate the two. General arousal can be determined from the EEG pattern, from the activity or irritability of the subject and from various autonomic responses, such as the galvanic skin response. Specific arousal has no observational basis.

In the first place external stimulus conditions will influence the state of specific activation – leading to such well known results as the dependence of stimulus intensity on perceived duration. On the other hand body temperature, emotional states, and drugs are known to alter the state of arousal, and consequently are thought of as changing the rate of the internal clock. The "pulses" generated by this "clock" are fed into a counting mechanism, which counts the number of pulses produced during an interval of a certain length. This number will be small in a given interval if the clock rate is low, thus leading to a subjectively fast passage of time. If the arousal state is high, the number of pulses in the same interval will be high, and physical time will seem to pass slowly. The clock rate is not strictly constant, and this will cause successive judgments to vary. Models of this kind were proposed most recently by Creelman (1962), and by M. Treisman (1963). Especially Creelman's paper is, in my opinion, an extremely important contribution. One reason is purely historical. It was the first completely quantitative account of the properties of an internal clock mechanism, and of the errors of judgment that may derive from incorrect "readings" of this clock (Creelman 1962, Michon 1965 a, 1967).

The alternative theoretical position is based on insights about human information processing that have reached prominence through the development of information theory, cybernetics and computer science. It has also earlier forms, for instance in the work of Guyau (1890) and Woodrow (1951).

Essentially the pulse generating and counting mechanism is replaced here by a judgmental process that bases its estimates about duration on the number and nature of the events that occur during the interval. The difference between the two formulations seems to be formal rather than essential, at least at first sight.

We shall see however that some experimental results can be explained in terms of a pulse counter only at the cost of additional assumptions, while some other findings clearly do not fit this theory at all. The degree of equivalence between the two theories has hardly been a matter of concern. Frankenhaeuser for instance, who stated her work on memory for intervals originally in terms of information processing and memory (Frankenhaeuser, 1959), now maintains that she is a representative of the arousal theory and has always been so (Berglund, Berglund, Ekman and Frankenhaeuser, 1969). However, a recent book by Ornstein (1969) presents many arguments that enhance the distinction between what the author calls the "sensory process metaphor" and the "storage size metaphor".

Ornstein's arguments are very illuminating, although his experiments offer only indirect evidence for the cognitive theory. They present no direct evidence since the subjects report in terms of relative lengths of lines rather than by reproducing the intervals. Like verbal estimation this method explicitly calls for a symbolic representation of the interval and not for a "temporal trace", as in the reproduction method. There is indirect support though for a cognitive mechanism, which requires too much "editing" if we are to explain it in terms of a counter. The essence of Ornsteins approach is that he presents subjects with a task for a given length of time. Afterwards he manipulates the memory of what went on during the interval. This can be achieved in various ways, for example, by providing the subject with a recoding principle, by which he may reorganize what he has seen or done. Subsequently the subject is asked to estimate the period during which he worked originally. The result is that time estimation is affected by the reorganization of the original material.

In one of his experiments Ornstein used paired associate learning of word-sound pairs. It was established that words paired with harsh sounds tend to drop from memory, while word-mellow sound pairs were retained more easily. It was found that the estimated duration of the learning period decreased in proportion to the number of word-harsh sound pairs in the list. At first sight this may not seem to be convincing evidence: the fact that a memory storage is affected need not imply that the original estimate was not a pulse count. However, when we realize that the estimation afterwards must have been based on a memory image in which the original information was still present – albeit influenced by the properties of the stimulus pairs of the list – then the arousal hypothesis becomes less attractive.

A quite different line of argument comes from animal conditioning experiments, where it was found that rats that are conditioned to intervals of a certain length start to display what is called "collateral" behavior. The result is an improved response rate. Preventing the animal from displaying this behavior results in a marked decrement of performance (Laties, Weiss and Weiss, 1969).

Warm, Smith and Caldwell (1967) showed that induced muscle tension (known to increase arousal) does not affect estimates of short intervals, although for intervals of 24 and 48 seconds estimates did decrease. This is in the predicted direction, but it seems strange that only when the muscle strain becomes an attention attracting factor do we find an effect, and not where "counts" would be most precise but the muscle tension is not yet a disturbing (cognitive) factor.

Finally, Aitken and Gedye (1968) compared task-filled and idle intervals in which there either was or was not distraction stress. They used manual tracking as the task filled condition. They found that changes in activation level were very small compared with the changes in time estimates. More important however is their finding that both the tracking task and the distraction stress raised the level of arousal, although the first induced shorter and the second longer time estimates.

This paradigm deserves careful attention, and requires replications with various measures of arousal. In principle the evidence is very strong indeed.

Although we could use some extra data to substantiate this conclusion, there seems to be plausible evidence that correct estimation of duration depends on storage of detailed information about the interval (or the means of regenerating it), rather than on just an arousal driven pulse count. If in recall only temporal information is required of the subject, he may be able to regenerate a counter-like state as "partial image". From this the "counter metaphor" may have originated introspectively.

*The Equivalence Postulate*

How should we place the theoretical position that we took with respect to time experience within the framework of human information processing? This question may be restated as a postulate expressing the equivalence of temporal and non-temporal information.

Duration appears under a dual aspect. On the one hand it is a physical construct. On the other it appears as a "property" of patterns of information in the real world, which we call objects and events. There is no useful way for distinguishing between this latter aspect and the other properties of information, such as size, intensity, or locus.

This postulate – or rather hypothesis – requires some explanation. First, let me be somewhat more explicit about the cognitive theory of human information processing. One of the basic aspects of this theory is the assumption that human behavior – in its widest sense such as to include perception – is guided by internal "models" of the real world. These "models" or "representations" may be considered as cognitive structures that encompass a subject's explicit or implicit knowledge of (part of) the real world. They are not just rigid replicas of reality but are abstractions of the concrete situations in which the information pertinent to the representations was acquired.

When I say "I know this city", I mean that I have an internal representation of it that I can regenerate in bits and pieces when somebody wants me to point out to him the shortest route from A to B. I usually will not recall under what specific circumstances first learned this route (Neisser, 1967, Ch 11); I even may never actually have travelled this particular route at all (Michon, 1968).

Whatever the properties of the internal model may be, it has become clear that the information processing system cannot be entirely verbal as was maintained by the behaviorists. We need to assume the existence of an "imaging" as a mediative process.

This is *a fortiori* true if we wish to include temporal features in our concept of "internal model", as we would in the case of anticipatory behavior. Anticipatory tracking of continuously moving target and manœuvring of vehicles has been studied quite extensively and various mathematical models have been developed to account for the behavior of the subjects in these studies. Only recently the subjects' expectations and decisions about the future course of target or vehicle have been taken into consideration.

As a result these new theories necessarily incorporate as a cognitive element a spatio-temporal "model" of the possible courses of behavior of the system. A recent study by Young (1969) offers an excellent exposition of the internal model as a basis for action in tracking and control situations. A more general discussion is to be found in Miller, Galanter and Pribram (1960), who defend the information processing approach to, what they call, "images". It should be stressed at this point that the "internal representation", or "model" or image is not necessarily a visual image appearing before the mind's eye. In fact we may determine the properties of models and the relations with overt behavior, without reference to the modality in which they "appear".

Activities are based on decisions taken by extrapolating from the internal model. If the act (perception or motor) does not confirm the predictions that were made, the "model" will be updated, and behavior strategies redefined. The contact with the external world is maintained through a complex monitoring system, which scans the available input and output channels, and continuously evaluates the priority of the information in each channel. If a high priority signal is detected the activity that is in progress may be interrupted and another routine may take over. This switching of attention has been studied in great detail (Egeth, 1967; A. M. Treisman, 1969). Less is known about the way in which reponse patterns are organized in memory. It is unlikely though that all alternative patterns of behavior are stored in memory. Rather one should think in terms of sequences of operations from which the act is reconstructed. These sequences may be hierarchically organized and allow for geat flexibility. They are the opposite of the classical rigid reflex arc (Miller, Galanter and Pribram 1960).

Research has led to a generally accepted functional distinction between "iconic" memory, short term memory (to be split up, according to some investigators, in very short term memory and short term memory proper), and long term memory, in which information may be stored quasi-indefinitely (Norman, 1969; Shiffrin and Atkinson, 1969). It has been shown that there is a very short storage of visual (and auditory) information which is available for about 1 sec. and can be selectively attended to. An array of $3 \times 4$ letters will normally yield a recall of 4 or 5 letters when it is presented in a very brief flash. If the subject is asked – after the presentation – to selectively report a particular row, his results show that his retention is much better (Sperling, 1960; Averbach and Coriell, 1961). During this second, information is transferred into a short term storage mechanism, which has a limited capacity of about 7 "chunks" of information, and probably operates in a strictly serial fashion.

In this short term memory, which enables us to look up to a number in a telephone directory and subsequently dial the number without looking again, rehearsal may take place to reinforce the fading contents. From this short term memory, information will gradually pass on to a more permanent storage system, one in which it will remain as long as is functionally required, or really permanently stored. The last distinction is probably one of degree depending on deliberate strategies of remembering, so called mnemonics (Miller, Galanter, Pribram 1960; Luria 1969; Norman 1969). The time limits imposed on the three memory systems are respectively of the order of 1 sec, 5–20 sec and a lifetime. It seems very plausible that the well known distinction, between temporal events, of *continuity, interval,* and *succession* perception, stressed by Fraisse (1956) is related to this functional organization of memory. Therefore we should also expect durations in each of these three categories to be evaluated by the cognitive processes which characterize the specific storage. That is, perception of long intervals must be based on transfer of shorter segments of information into long term memory, and make extensive use of mnemonics. Short intervals on the other hand will fall within one single time "chunk" of short term memory, while very brief intervals will be judged on the basis of a momentary "iconic image" where we may expect interactions, not only with the information content of the intervals, but also with the energy distributions of the stimuli. Temporal integration is demonstrably not just a matter of physiological time constants of the system such as in Bloch's law which states that for light flashes the product of stimulus intensity and duration is constant for intervals shorter than 0.1 sec. Ekman and his coworkers have shown a logarithmic relation to exist over much longer intervals, of the order of 0.5 sec at least (Ekman, Frankenhaeuser, Berglund und Waszak, 1969; Berglund, Berglund, Ekman and Frankenhaeuser, 1969).

There is more to be argued in favour of the equivalence postulate. Not only is there a time-energy trade off at very short intervals. It appears that a flow of non-temporal information through the system must alternate with temporal information when this is asked for in the response, in much the same way as it must alternate with other non-temporal information (A. M. Treisman, 1969). In other words, selective attention may be directed to temporal inputs. This becomes evident in tasks in which the subject is pressing buttons in response to one of a set of lights, while at the same time he is serially producing intervals.

Michon (1967) found that only if these two tasks are highly correlated, that is when the responses to the lights are identical to the interval production response, the two tasks are performed well.

If no such connection is possible, the reaction task, and to a much greater extent the time production task are disturbed. This interference – result of attention switching – is found in many other "dual task" situations (Woodworth and Schlosberg, 1954; Brown, 1964; Michon, 1966).

In particular some results obtained by Noble and Trumbo are quite revealing in this light (Noble, Trumbo and Fowler, 1967; Trumbo, Noble and Swink, 1967). They found that when their subjects tried to follow a moving spot on an oscilloscope screen with a marker, while at the same time performing a second task, it was the timing of the response and not the shape of the tracking response that was affected. Michon (1967) demonstrated the exact equivalent of this result in a temporal tracking task which consisted of key tapping in synchrony with an auditorily presented click pattern. It was found that if the response curves are corrected for errors in timing (onset) of the response, the shape of the average response is identical with the response that is found when no secondary task is performed simultaneously. My last example of the close similarity between temporal information and non-temporal information as they are processed by man, concerns the important question whether dual task interference originates from the input, "digestion" or output channels. This problem was also tackled by Trumbo and Noble with respect to non-temporal tasks, and by Michon (1965b) for time-evaluation.

Trumbo and Noble (Trumbo, 1969) studied the influence, on tracking, of tasks in which they had subjects push buttons in response to auditorily presented numbers (1 to 5). When there was choice reaction required, tracking performance was not affected. On the other hand, when the subjects were presented with a series of clicks, and were required to respond randomly with one of the response buttons, they in fact generated information that was not present in the original stimulus. This condition greatly affected tracking performance. On the basis of this and additional experiments they reached the conclusion that response selection rather than input or execution is the interfering factor. An essentially equivalent conclusion was reached by Michon (1965b) in a study of serial production of 2 sec. intervals. Here the subject had to respond to one of 6 stimuli by pressing the appropriate button. The requirement was, to do so exactly every 2 sec after the presentation of a new random stimulus.

The number of stimuli was 1, 2, 4 or 6, the number of responses likewise 1, 2, 4 or 6, and all possible input-output combinations were used.

The results indicated that input information by itself had no effect on the average 2 sec estimate. Increasing transmitted and output information on the other hand produced a shortening of the 2 sec estimate. The function relating the two variables appeared to be negatively accelerated.

These examples may be sufficient to illustrate the point that we can not only remain close to the mainland of psychological theory when discussing time estimation, but also that the hypothesis stating the equivalence between temporal and non-temporal information is supported by empirical findings.

## Current Trends in Research

Now that we have stressed the importance of the relations between time experience and information processing in general, let us see to what extent recent years have produced research that is relevant in this wider context. Let me start with a few negative remarks. I feel that too much time and effort is spent in unimportant research which just drags on and adds no new information to what was known in essence a long time ago. We know that there are sex differences, that filled intervals seem to pass more quickly than so called empty intervals and we know that emotion affects time estimation in a predictable way. If we overlook the fact that many of the effects are marginal and suffer from inadequate control of stimulus- and response variables, there still is the fact that they do not surpass the stage of the "existence proof". Instead the investigator should try to establish a truly quantitative, i.e. parametric, relation between the experience of duration, and the conditions he created. Or he should attempt to verify specific hypotheses about functional relations between the constituents of his theory or model. Only then he may expect to *add* to a growing *body* of consistent knowledge. It is not enough any more to conclude that a particular result "is in the direction predicted by a hypothetical internal pace-maker", or a "time quantum".

A similar argument applies to the studies of response control, in which the merits and limitations of the basic experimental paradigms – production, reproduction, verbal estimation and comparison – are compared.

Again, in too many research papers, there is exclusive concern for the fact that there are discrepancies between these methods, rather than for the questions which mechanisms are responsible for them and how large the discrepancies are as a function of duration or condition. There is however an increasing awareness of the importance of this problem, and recent years show this by a number of relevant research papers (e.g. Treisman, 1963; Brand and Holborn, 1966; Schmidt, 1969). Methodologically it is time to make up the balance of a century of experimental work.

After this rather negative view of the persistent "classical" trend in time psychology, I now want to present some thoughts on what I consider the more promising directions for research. I want to do this in two short sections: the trends imposed by mathematical models as they are used in current psychological theories, and the parametric approach by which the functional relations between duration and information processing are studied.

## Formal Models

Psychology has gradually drifted away from its highly complex, verbal, theories which lacked logical consistency and predictive power. Instead, many of the current theoretical issues are stated in terms of purely formal models, which only get their psychological significance through the data to which they are applied. An example of such a theory is Signal Detection Theory (SDT). Its point of departure is the statistical probability that an observer will be able to distinguish signals from non-signals, when the two are not too different, and moreover are fluctuating in a random fashion. The theory predicts optimal performance for any given signal – non-signal difference, amount of fluctuation and cost of making errors of judgment. In psychology this model has been applied with great success, and SDT has largely replaced the old psychophysical theory of sensory threshold mechanisms. Creelman (1962) applied SDT to the discrimination of short time intervals. By considering intervals that differ by a small amount $\Delta T$, he established the SDT formulation of threshold for duration. Important is that SDT allowed him to specify in detail the parameters of all functional units in the pulse-counting model, which he adopted as his theoretical framework. In doing so, he in fact requires from all future investigators that they account for his and their own results in a better way – quantitatively – or that they falsify his assumptions quite specifically. A new model should deserve consideration only if it leaves less variance unexplained. A similar approach to the same goal of being specific and consistent, was taken by the present author in his study of synchronization tapping (Michon, 1967). Synchronization with a varying rate series of clicks is conceptually similar to tracking a moving dot on oscilloscope screen. Quite powerful tools to analyze and describe tracking behavior of humans – derived from the theory of control systems, – have been used since the late forties (Licklider, 1960; Poulton, 1966). Michon applied some of these tools to describe "temporal tracking" or synchronization. Description here also means prediction, since the model will predict what a subject will do next when it is in a given state.

In this case too, a better description of synchronization behavior may well be found soon, which will make it necessary for me to throw my model in the wastebin. It should however, explain more of the variance in the data, do so in a set of simpler assumptions, or explain more features of the data on an equal number of assumptions. Thus the importance and strength of mathematical models – and by this I also refer to computer simulation programs – is not only their elegant summarizing, but also their vulnerability to very specific tests. It should be added that psychology has not yet succeeded in creating a very large body of very general models of human behavior. This should not stop our efforts in trying to formulate exact and consistent models however.

*Parametric Studies Relating Temporal and Non-Temporal Factors*

The experimental work that has been done on the interrelations between time estimation mechanisms and other psychological mechanisms deserves our special attention. As late as 1964 Adams pointed out the fact that time perception was so remote from other branches of psychology, because nobody had cared to study the interrelations between the various mechanisms (Adams, 1964).

This is no longer true. I have already mentioned several cases of studies in which the interaction between temporal and non-temporal information was investigated. There are numerous studies which illuminate other connections. Let me discuss a few examples. Again, they serve the purpose of illustration and are not intended to be exhaustive.

Recency

In the analysis of the working of memory, the time during which an item has been stored in memory until recall, is named recency. Recency is known to influence the correctness of recall quite considerably. In an attempt to find out if the subjectively experienced recency of an item is determined by the strength of the memory trace (or the activity of the memory process to use a more neutral terminology), Yntema and Trask (1963) and later Peterson (1966) studied judgment of recency of an item in a list of auditorily presented words in relation to the chance of correct reproduction of these words. If the subjective recency estimate – which in fact is a judgment of the interval elapsed since the representation – is a consequence of the strength of the memory trace, one would expect the judgment to covary with the probability of correct recall. This turned out not to be the case. The temporal cue derived from the number of intervening items in the presented list of words is largely independent of the strength of memory for that item. This finding was confirmed by other experimenters. Wolff (1966) for instance, manipulated the association value of the items in the list, and also found practically no effect.

Later studies by Brelsford, Freund and Runders (1967) and especially by Hinrichs and Buschke (1968), have shown that the recency phenomenon does indeed follow the rules of subjective time, and that time perception therefore is a powerful aid in the study of memory.

Pay-Off

It has been demonstrated that signal detection, psychophysical judgment, estimates of the subjective probability of events, and other measures of behaviour can be influenced by pay-off schemes known to the subject. By penalizing one type of response and rewarding another, judgments can be shifted in a particular direction without conscious control of the subject.

The effect of pay-off has also been studied in the realm of time evaluation. At this point I should refer to a large number of studies on animals. Innumerable cats, rats and other laboratory animals have been put on reinforcement or shock schedules and pay-offs (for instance, shock if pressing a lever too early, but not, if too late) have been manipulated. I want to restrict this exposition to human performance, however. In a study by Kornblum (unpublished), subjects pressed a button some seconds after the onset of a burst of white noise in a headphone. The subject was facing a display showing a line with vertical marks extending to the left and right of a zeropoint. After pressing the button, this display would show the deviation of the response. When no pay-off scheme was given, the subject produced an error distribution which was essentially normally distributed, as has been found by other authors. (Ehrlich, 1957; Michon, 1967; Allen, 1967). If however, an asymmetrical pay-off was given, the response distribution would become skewed, and follow most exactly, the pay-off distribution. Although this result needs a follow up with various pay-off schemes, it appears that this paradigm may shed some light on a specific feature of the memory mechanisms involved in time estimation.

The problem is this: if the subject is tapping a series of intervals, trying to copy the previous interval, he has to store the "image" of this previous interval while the time of the second interval is passing. It is well known, that during this period the "image" is subject to decay. In the literature there is much confusion about the nature of this decay: both negative time-order errors and positive time-order errors are reported. I have tried to cope with the two possibilities in a single model (Michon, 1967). At the end of an interval to be remembered, it is stored and from that moment on, in an arbitrarily small elementary unit of time, a "unit of remembered temporal information", may be added *or* subtracted from what is currently in store. If the probability of gaining a unit equals that of losing one we expect the mean to remain constant and the variance of reproduced intervals to be normally distributed. What if the probabilities do not match however? Then the second interval will be either shorter or longer, and the distributions of response will be skewed. Thus far it seemed difficult to manipulate the probabilities of loss and gain, but now appears it to be possible to achieve this by means of the pay-off paradigm. In this way we may be able to establish a connection between research on time and subjective expected utility.

Response Strategies

Closely related to the previous example is a topic which figures in skilled behavior: the response strategy. From the cognitive point of view a response strategy is a routine or program by which the subject is able to act in a given situation. As I mentioned before, the response strategy is associated with an expectation, not unlike the expected utility, which may change with experience. If in a given situation the expected result of an action is not achieved, the system will switch to another strategy, and so on, until it finds one that meets the requirements of the situation (this may include retreat, tantrums or even suicide). Response strategies in dealing with time estimation are well known – but have been suppressed by almost every investigator of time as being artifacts.

Respiration cycles, subvocal counting, etc., in fact should be restored to a relevant category of response variables. Even animals make use of such "collateral" behavior to provide themselves with temporal cues, as we saw before (Laties, Weiss and Weiss, 1969).

Response strategies were clearly present in the author's experiments on synchronization behavior (Michon, 1967). In one experiment the subject was confronted with a series of clicks, randomly distributed in time. Since the series was unpredictable no strategy would reduce uncertainty completely. How will the subject behave? Apart from the normal strategy, in which the subject bases his performance on a weighted score of some 5 or 6 previous intervals, there are other strategies – known from tracking and guessing experiments – which he might chose in trying to cope with his impossible task of synchronizing with a random event. Such alternatives are: constant rate producing, matching of the statistical distribution of the input, reacting to each click, or copying the last previous interval. It was established that naive subjects more or less try to match the properties of the input, while experienced subjects drop the weighted average strategy, and will predict on the basis of foregoing events: they just copy the last previous interval.

This finding gives us some insight in the adaptation to changed temporal conditions: when a highly specific program does not work anymore, will a subject try next more general program in his response hierarchy?

The Influence of Subjective Time on Task Performance

Instead of asking how certain conditions affect time evaluation, we may induce a distorted time estimate and see how it affects performance. The usual ways of achieving this have been change of body temperature and drug use. In a series of interesting experiments McGrath and his coworkers have explored this alternative (McGrath and O'Hanlon, 1967; O'Hanlon and McGrath, 1967). They achieved a change in the subjective passage of time by changing the rate of a wall clock which was clearly visible in the room where the subjects were working. McGrath was primarily interested in studying the relation between "temporal orientation" and vigilance performance (McGrath and O'Hanlon, 1967). In a viligance task the subject is required to pay prolonged attention to a display which presents stimuli irregularly at a very low rate. The probability of detection of a stimulus has been shown to be highly related to the time distribution of the stimuli. One particular theory of vigilance expectance theory, explicitly employs the concept of time evaluation. The gist of this theory is that subjects build up a response set which has its peak at the expected moment of arrival of a stimulus. If the stimulus arrives too early or too late expectancy will be lower, and the response will be lower or even not occur at all (missed signal). Baker (1963) suggested that subjects may average previous interstimulus intervals and base their expectancy on his average. A corollary of his hypothesis is that those subjects who customarily underestimate the period gone by, will perform better in a vigilance task. They will be better prepared, since their expectancy is still "building up" when – on the average – the signal arrives.

By changing the pace of a physical reference clock, subjects could be manipulated into behaviour that was consistent with the prediction from expectancy theory (O'Hanlon and McGrath, 1967). What is most surprising in this kind of experiment, is perhaps the extent to which people can be fooled with fast or slow mowing clocks (Rotter, 1965). Obviously estimates in real time can easily be affected a factor 2 or more up and down. This is consistent with the usual finding of large intra-individual variances in time experiments.

These few examples, in no way intended to be exhaustive, from such divergent areas as memory, subjective expected value, response theory and vigilance, may demonstrate to what extent ideas about time experience and human performance interact in present day research.

*Conclusion*

In this paper I took up an argument where I left it in my study on timing in Temporal Tracking. The epilogue concluded thus:

> ... "we had to refrain from connecting our inquiry to the extensive work on topics which are in a sense related to the problem of timing, like immediate memory, response latencies, the psychological refractory period, etc." ...
> "We are convinced that temporal relations – with which man is confronted in several ways – provide him with information which must be processed in a way that is not essentially different from the way any other kind of sensory, symbolic or proprioceptive information is handled. In this respect the psychology of time does not deal in any sense with a "microcosm" but on the contrary with a very essential part of all human behavior" (Michon, 1967, p. 109).

I have traced some of these threads and connections by reviewing some recent research in the realm of human time experience. Let me summarize briefly the main interrelations that emerged.

First I outlined the interdependence of the fields of time psychology and temporal factors in behavior, and we saw how much the two need each other.

Second, I sketched the relation between the two major theoretical positions regarding time experience, the internal clock theory and the cognitive theory. The available evidence appears to favor the latter approach.

Third, I stressed the "equivalence postulate", in which the close analogy between temporal and non-temporal information is stated explicitly.

Fourth, I pointed out that theorists of time psychology should stay close to the trend toward the formulation of functional, quantitative models. In particular the areas of memory and skilled behavior, both searching for quantitative models with well defined temporal constants, offer many relevant ties.

Research should stress these relationships. It should be directed at parametric investigations into the properties of the mechanism of time evaluation as they emerge from our analyses.

I am greatly indebted to Mr. A. van der Heyden, University of Leiden, who read an earlier draft of this paper, and suggested several improvements.

## References

Adams, J. A.: *Motor Skills.* Ann. Rev. Psychol. *15* (1964) 181–202.
Aitken, R. C., Gedye, J. L.: *A study of two factors which affect arousal level and the apparent duration of a ten-minute interval.* Brit. J. Psychol. *59* (1968) 253–263.
Allen, G. D.: *Two behavioral experiments on the location of the syllable beat in conversational American-English.* Unpubl. Doctoral Dissert. Univ. Michigan, 1967.
Averbach, E., Coriell, A. S.: *Short term memory in vision.* Bell Systems Tech. J. *40* (1961) 309–328.
Baker, C. H.: *Further toward a theory of vigilance.* In: D. N. Buckner and J. J. McGrath (eds.), Vigilance: a symposium. New York: McGraw-Hill 1963.
Berglund, B., Berglund, U., Ekman, G., Frankenhaeuser, M.: *The influence of auditory stimulus intensity on apparent duration.* Scand. J. Psychol. *10* (1969) 21–26.
Berko, M. J.: *Amelioration of athetoid speech by manipulation of auditory feedback.* Unpubl. Ph. D. Diss. Cornell Univ., 1965.
Brand, W. G., Holborn, S. W.: *Temporal context effects with two judgmental languages.* Psychon. Sci. *6* (1966) 151–152.
Brelsford, J., jr., Freund, R., Runders, D.: *Recency judgments in a short-term memory task.* Psychon. Sci. *8* (1967) 247–248.
Brown, I. D.: *A comparison of two subsidiary tasks used to measure fatigue in car drivers.* Ergonomics *8* (1965) 467–474.
Creelman, C. D.: *Human discrimination of auditory duration.* J. Acoust. Soc. Am. *34* (1962) 582–593.
Egeth, H.: *Selective attention.* Psychol. Bull. *67* (1967) 41–57.
Ehrlich, S.: *Le mécanisme de la synchronisation sensorimotrice; étude expérimentale.* Année Psychol. *58* (1958) 7–23.
Ekman, G., Frankenhaeuser, M., Berglund, B., Waszak, M.: *Apparent duration as a function of intensity of vibrotactile stimulation.* Percept. Motor Skills *28* (1969) 151–156.
Fraisse, P.: *Les structures rythmiques.* Louvain: Studia Psychologica 1956.
Frankenhaeuser, M.: *Estimation of time: an experimental study.* Stockholm: Almqvist and Wiksell 1959.
Guyau, M.: *La genèse de l'idée de temps.* Paris: Alcan 1890 (2nd edition 1902).
Hinrichs, J. V., Buschke, H.: *Judgment of recency under steady state conditions.* J. Exp. Psychol. *78* (1968) 574–579.
Janet, P.: *L'évolution de la mémoire et de la notion de temps.* Paris: Chahine 1928.
Laties, V. G., Weiss, G., Weiss, A. B.: *Further observations on overt "mediating" behavior and the discrimination of time.* J. Exp. Anal. Behav. *12* (1969) 43–57.
Licklider, J. R. C.: *Quasi-linear operator models in the study of manual tracking.* In: R. D. Luce (ed.), Developments in mathematical psychology. Pp. 171–280. Glencoe: Free Press 1960.
Luria, A. R.: *The mind of a mnemonist.* London: Cape 1969.
McGrath, J. J., O'Hanlon, J.: *Temporal orientation and vigilance performance.* In: A. F. Sanders (ed.), Attention and Performance. Pp. 410–419. Amsterdam: North-Holland 1967.

Michon, J. A.: *De perceptie van duur.* Ned. Tijdschr. Psychol. 20 (1965) 391–418 (a).
— *Studies on subjective duration II: subjective time measurement during tasks with different information content.* Acta Psychol. 24 (1965) 205–219 (b).
— *Tapping regularity as a measure of perceptual motor load.* Ergonomics 9 (1966) 401–412.
— *Timing in temporal tracking.* Soesterberg: Institute for Perception RVO-TNO 1967.
— *On the internal representation of associative data networks.* Nederlands Tijdschrift voor de Psychologie 23 (1968) 428-457.
—, Wagenaar, W. A., Lazet, A., Koutstaal, G. A.: *Over de operationele bruikbaarheid van de automatische afleesborden van SATCO fase II.* Institute for Perception RVO–TNO, Tech. Rep. 1965–C5.
Miller, G. A., Galanter, E., Pribram, K. H.: *Plans and the structure of behavior.* New York: Holt 1960.
Neisser, U.: *Cognitive Psychology.* New York: Appleton-Century-Crofts 1967.
Noble, M. E., Trumbo, D., Fowler, F.: *Further evidence on secondary task interference in tracking.* J. Exp. Psychol. 73 (1967) 146–149.
Norman, D. A.: *Memory and Attention; an introduction to human information processing.* New York: Wiley 1969.
O'Hanlon, J., McGrath, J. J.: *Temporal patterns of signals and vigilance performance.* Hum. Factors Res. Tech. Rep. (1967) 719–3.
Ornstein, R. E.: *On the experience of time.* London: Penguin 1969.
Peterson, L. R.: *Search and judgment in memory.* In: B. Kleinmuntz (ed.), Concepts and the structure of memory. New York: Wiley 1967.
Poulton, E. C.: *Tracking behavior.* In: E. A. Bilodeau (ed.), Acquisition of Skill. Pp. 361–410. New York: Academic Press 1966.
Rotter, G. S.: *Time rate as an independent variable in research.* Proc. 73th Ann. Convention Amer. Psychol. Ass. 1965, 51–52.
Schmidt, R. A.: *Anticipation and timing in human motor performance.* Psychol. Bull. 70 (1968) 631–646.
— *Movement time as a determiner of timing accuracy.* J. Exp. Psychol. 79 (1969) 43–47.
Shiffrin, R. M., Atkinson, R. C.: *Storage and retrieval processes in long-term memory.* Psychol. Rev. 76 (1969) 179–193.
Smith, K. U.: *Cybernetic theory and analysis of learning.* In: E. A. Bilodeau (ed.), Acquisition of skill. Pp. 425–482. New York: Academic Press 1966.
Sperling, G.: *The information available in brief visual presentations.* Psychol. Monogr. 74 (1960), whole nr. 498.
Titchener, E. B.: *Experimental psychology,* Vol. II, Part 2. London: MacMillan 1905.
Treisman, A. M.: *Strategies and models of selective attention.* Psychol. Rev. 76 (1969) 282–299.
Treisman, M.: *Temporal discrimination and the indifference interval: implications for a model of the internal clock.* Psychol. Monogr. 77 (1963), whole nr. 576.
Trumbo, D.: *Some response strategies in skilled tasks.* Paper to the 19th International Congress of Psychology. London, 1969.
—, Noble, M. E., Swink, J.: *Secondary task interference in the performance of tracking tasks.* J. Exp. Psychol. 73 (1967) 232–240.
Warm, J. S., Smith, R. P., Caldwell, L. S.: *Effects of induced muscle tension on judgment of time.* Percept. Motor Skills 25 (1967) 153–160.
Wolff, P.: *Trace quality in the temporal ordering of events.* Percept. Motor Skills 22 (1966) 283–286.
Woodrow, H.: *Time perception:* In: S. S. Stevens (ed.) Handbook of experimental psychology, Pp. 1224—1236. New York: Wiley 1951.
Woodworth, R. S., Schlosberg, H.: *Experimental psychology.* London: Methuen 1954, (3rd ed.).
Yntema, D. B., Trask, F. B.: *Recall as a search process.* J. Verb. Learning Verb. Behav. 2 (1963) 65–74.
Young, L. R.: *On adaptive manual control.* Ergonomics 12 (1969) 635–674.

# The Psychophysical Structure of Temporal Information

PATRICK MEREDITH*

*Abstract.* As a concept for ordering and analysing real events with variable amounts of information "time" is much more complex than a simple clock-measure. Psychophysics has traditionally dealt with one-way processes from stimulus to sensation, creating information. In human action the reverse process occurs. The information in a plan of action anticipates the outcome. The latter is subject to uncertainties and the planned timing must be elastic, requiring a topological calculus for the relative timing of planned processes. Human actions can now have consequences of planetary orders of magnitude, giving unpredictable quantities in astronomical space-time, and local thawing in the frozen framework of The Minkowski continuum.
Quotations are from "The Principle of Relativity" by Lorentz, Einstein, Minkowski and Weyl (1923 translation, Dover Edition).

## *The Concept of Time*

We obtain temporal information from natural objects, such as the cross-section of a tree-trunk; from graphic diagrams such as the trajectory of a rocket; from mathematical expressions, such as Clerk Maxwell's Equations; and from verbal expressions, for every sentence has a temporal aspect. Our concept of time is shaped by all these four types of information and is therefore very rich and varied. It cannot be summed up in a single definition. In particular it cannot be limited to the supposedly strict metric expression of time derived from mechanical clocks. For these, in fact, are not the invariant embodiments of ideal uniform Newtonian time which mechanical mythology would have us believe. They are man-made devices, wound up by man, set by man, speeded or slowed by man, varied and stopped by natural forces, and can even be reversed.

## *Quantification of Time*

In order to express temporal information we must, in some way, *quantify* time, and this means dividing intervals into parts. But we need to distinguish between divisibility, continuity and uniformity. Man cannot stop the Universe,

---

* G. Patrick Meredith, Emeritus Professor of Psychophysics, University of Leeds, Department of Psychology, Epistemic Communication Research Unit, Leeds 2, Great Britain.

and, in this sense, he cannot divide the continuity of natural time. What he can do is to divide material space. In so far as events in time have left their marks in space, in some kind of order, these divisions of space yield temporal information. An ordinary clock has a circular dial which is a spatial object. It is not strictly a *time-keeper*, for it keeps no record of the rotation of the hands. It is the reader of the clock who marks the coincidence of the hand with a figure on the dial, usually at irregular intervals in his memory. He quantifies these memories on the assumption of a uniform clock-rate.

*Variability of Standards*

For many purposes of daily life the assumption of a uniform clock-rate is irrelevant to the actual use to which we put temporal information. Outside the laboratory we do not often use clocks for measuring the velocity of natural events, and in the laboratory the search for precision only serves to underline the variability of clocks. Even atomic clocks must vary with the gravitational field, and when light is being used, not only as a measure of time but also as a medium for transmitting it, the question of an ultimate uniform standard is beset with ambiguities – especially if the information is transmitted across space covered by varying gravitational fields. But whereas the metric duration of events is subject to error their mutual *ordering*, given certain conditions, is an absolute.

*Topological Structure*

Given that temporal information is expressed in space, and that intervals between events are subject to variability in the recording, we can apply the concept of *elasticity* to the metric aspects of time whilst making valid use of digital arithmetic to describe the mutual ordering relations between events. And given this elasticity in the measurement of durations we can begin to envisage a *topology of temporal information*, in which we have rubber strings or nets or sheets on which events are recorded as knots or other singularities, expressing certain necessary relations between events, relations which hold good even though the rhythms of all the clocks concerned are variable. This is, in fact, the situation which holds in most living organisms. Thus the structure of biological time needs to be conceived in topological terms.

*The Spacing of Time*

When an event leaves a mark on a recording surface we may say that "time has been spaced". Since the mark records an event which, after the recording, is in the past, and hence unalterable, the causal relation disappears from the record. In this sense Einstein eliminated force by placing temporal information in a spatial framework. Whatever causal inferences we draw from a spatial distribution of information these cannot be expressed dynamically since the coordinate space is static. Corresponding with the concept of an effect in time being *produced* by a cause we now have a structure in space being *constituted* by component parts. Thus the part-whole relation symbolizes the cause-effect relation. But whereas any given set of forces combine to produce a determinate effect, a given whole structure can be partitioned in many different ways into constituents.

*Direction and Sense*

Thus geometrical analysis of spatial records cannot alone yield an unequivocal set of propositions concerning past causal relations. In probability theory we do not even attempt this. A given shuffle of cards cannot be unshuffled in practice, even though the number of possible permutations is finite and could, in theory, be examined one by one. But many records of scientific experiments show a complexity of data at least equal to that of a shuffle of cards and yet, by reason of a knowledge of certain limiting factors, we can often eliminate the majority of the partitions because most of them would creat logical contradictions. These arise not from geometry alone but from a geometry whose lines have *sense* as well as direction, and whose surfaces have determinate *orientations*. To construct this geometry we must scrutinize Euclid's assumptions.

*Kinematic Implications of Geometry*

In two main particulars Euclid was inadequately explicit. In allowing the use of straight-edge and compass he overlooked the fact that their use involves not only physical materials but also a temporal order of operations. Secondly in prescribing the condition for congruence, which is the fundamental relation in establishing most geometric propositions, the power to translate and rotate triangles as solid objects is a requirement going beyond the axioms. The fact that there is a certain kinematic assumption in geometry does not of itself make geometry a branch of mechanics, for no assumptions are made concerning mass and acceleration, but it does implicate movement, and hence *time* in the geometrical analysis of space. But it need not be metric time. The rate at which a geometric transformation proceeds is of no relevance to the structure.

*Chronotopical Similarity*

Our approach here may be described as a "pragmatic" one, not in the everyday sense of being content with compromises but in the strict sense of analysing the possibilities and impossibilities of *action*. Most physical actions, whether those of an experimenter in a laboratory, a ballet-dancer on a stage, a surgeon in an operating theatre, an astronaut in space, or even of a mathematician using chalk on a board, implicitly involve a considerable complexity of temporal judgments and decisions. Often the metric aspects of the events are quite variable, as between one performance and another, but what makes two actions "similar" in what we may call a "chronotopical" sense, as distinct from chronometrically equal ratios, is the distribution of *coherent timings*. It is this concept which underlies every pragmatic skill and which we here seek to formalize.

## Limitation of Traditional Psychophysics

The formalization of a concept is essentially a constructive activity in which an author works as an artist, manipulating an expressive medium – ink, pigment, clay etc., under the drive of forces in his brain which he may variously describe as "intuitions", "images", "ideas". This pragmatic drive towards the physical embodiment of a psychological state has been curiously neglected by traditional psychophysics, although Fechner, the founder of the science, obviously could never have founded it without this demonic drive to impose his mind on matter. The science itself has, for over a century, concentrated narrowly on the reverse relation, which should logically be called "physico-psychical" i.e. the imposition of physical stimuli on the psychic recipient.

## The Inner-Outer Field of Psychophysics

Since nothing could be said or written, let alone constructed or acted, without processes in which the temporal order is psycho → physical, the science of psychophysics cannot continue indefinitely to hop along on one foot. But it must be admitted that the word "psychic" has acquired so many strange semantic overtones that its use as a scientific term is hopelessly compromised, and even the term "physical" is by no means free from ambiguity. Every human being is both a person and a thing. As a person he has inner impulses, images, ideas, impressions and a continuing identity (all words which, in English, conveniently begin with the letter "i"). As a thing, a body, he is an obtrusive, observable, operative, organized object (all words conveniently beginning with "o"). Both the "i" set and the "o" set are manifested by connected sequences of events whose relations form the field of psychophysics.

## Variables and Constants

Since the range of "i" covers the whole human race considered as personal individuals, and the range of "o" covers all observable objects whether human bodies or other material forms, these two letters, as symbols, must be treated as variables. In any formal notation we need constants as well as variables but it is not easy to define the term "constant" itself without begging questions. We may *call* any letter of the alphabet a "constant" but unless we make it of platinum and lodge it in a Bureau de Poids et Mesures, its actual process of existence cannot even simulate constancy. And even Paris moves with the earth and is exposed to cosmic radiation. We might regard "constancy" as representing man's dream, never to be realized, of escaping from the ravages of time. In a notation, however, it has a finite prosaic utility.

*Context Determination*

If we use the capital letter I to stand for a single person in any defined context we can say that "I" is a *constant* for all references within that context. Similarly if O stands for a single object this is likewise constant in the context. This constancy guarantees semantic consistency, without which our language breaks down. Further guarantees are then required as we move from one context to another. To speak of persons and objects without specifying the context we use the variables "i" and "o". The fact that I and O are constant in any given context C does not mean an escape from time, for C itself is moving on. But *within the framework* of C the positions and relations of I and O are fixed and space-like. This fixity is not an assumption but a material requirement.

*Rigidity as a Psychophysical Concept*

Mathematicians are often regrettably vague about the requirements for securing the constancy of their coordinate framework. It is not enough to specify that, for physical measures, a framework must be "inertial", though this at least requires it to have mass and hence a material basis. But a cloud of gas or a drop of water would be inertial frameworks if this were all. To serve the purposes of coordination there must be an internal constancy of spatial relations among the constituents of the framework. This condition is satisfied only by the solid state (and even then it must strictly be an ideally rigid solid). We treat a framework as rigid if there are no noticeable vibrations or distortions. Since, in terms of pure physics, this is impossible, "rigidity" is essentially a psychophysical concept.

*Asymmetry of Experience*

It is, perhaps, surprising to find a psychophysical requirement built right in to the axiomatic foundations of coordinate geometry. It is for this reason that we place our "I–O" relation at the centre of our notation. Geometrical and arithmetical statements are *propositions*. We are so used to handling propositions impersonally that we forget that no proposition would come into existence without a *proposer*, a "first person singular", an "i" using his "o" (i. e. his body as an operator) on a second "o" (i. e. some objective material) to convey information about other "o's" to some fourth "o" who, as an interpreter is, internally, an "i". There are significant asymmetries both between the two "i's" and between "i" and "o". In a perfectly isotropic space all relations are symmetrical. We can never experience such a space. For our experience itself introduces asymmetry.

*Asymmetry in Time and Space*

Experience implies an asymmetry of time in that the "i" is different *after* from what it was *before* the experience. When it becomes a memory we cannot undo it. A memory may be destroyed – we may forget an experience but we cannot *remember* its opposite. Thus psychophysical time is irreversible. Purely "psychic" time, i. e. the mutual order of imagined or even remembered events, can be traversed internally in any order we please, as demonstrated by the fact that we can tell the end of a story before its beginning. The question of the reversibility of purely "physical" time is a matter of current controversy, with paradoxical semantic overtones, suggesting that the protagonists are in the realm of metaphysics rather than pure physics. But what is a matter of observable fact is that in *space* we have the phenomenon of "laterality" or "parity".

*Psychophysical Consequences of Asymmetry*

The human body is, superficially, bilaterally symmetrical. The left side *appears*, more or less, to be a mirror image of the right side, but a cast of one hand would not fit into a mould of the other. Internally, however, the heart is not in the centre and there are many other asymmetries. The head in no way resembles the feet. The front differs markedly from the back. Psychophysics has paid far too little attention to the systematic consequences of these asymmetries on our sensory experience, our locomotion, our manipulation, our linguistic behaviour, our memory and our concept-formation. This system of essentially topological relations dominates, by a set of plus and minus signs, all the arithmetic of measured behaviour. A single linear arithmetic cannot represent the mathematics of behaviour.

*The Kronecker Delta*

A similar system of pluses and minuses also enters into the Tensor Calculus and is handled by a device known as the "Kronecker Delta". If we can penetrate the significance of this device, going beyond its purely formal function in calculating the outcome of a series of reversals, asking *what* is being reversed, and *where*, and *when*, we begin to see the Kronecker Delta as an itinerary for marking out a determinate course of physical values in space-time. This is not easy to visualize so long as we are tied to a gratuitous assumption that empirical space is inherently "three-dimensional", and that multi-dimensional space is to be conceived as an unimaginable extended analogue of this. We are the cultural victims of a good deal of misleading late 19th Century science fiction concerning "the 4th dimension".

*The "Fourth Dimension"*

Although epistemological problems in physics became explicit, early this century, in the discussions on the implications of Relativity and Quantum Theory, it is curious that so little critical thought was given to the specific epistemic problem raised by Minkowski. His 4-dimensional space-time continuum actually merged space and time, in theory, but in the formalism the time-coordinate was explicitly treated as *the* 4th dimension. When the framework rotates all four axes appear as homogenous representations of space-time variables, the distinction between space and time now being lost. But what of the rotation itself? Does not this introduce a *fifth* dimension? And if the empirical space of our perception is three-dimensional what is the status of the two extra dimensions? Indeed, what *is* a dimension?

*Epistemic Status of Coordinates*

Some of these problems underlay the development of Hamilton's Quaternions in the 19th century and the (at first) highly controversial "operators" of Heaviside's calculus towards the end of the century. They are not problems for mathematicians alone, nor even for mathematics and physics alone. Nor can they be handled by pure logic which must accept atomic propositions as *known* to be true or false without asking *how* the truth is known. If we cannot say how the truth of a proposition is known not only can we never demonstrate it – we cannot even understand it. And since the formalism of physics is expressed in a coordinate language there are propositions concerning coordinate frameworks whose epistemic status must be established if physics is to be meaningful.

*The Minkowski Framework*

It was Minkowski himself who implicitly recognized the necessity for a *psychophysical* postulate to give meaning to his framework: "Everywhere and everywhen there is something perceptible". Also, although his framework was mathematical his views of space and time had "sprung from the soil of experimental physics, and therein lies their strength". It is worth quoting his vivid statement of the nature of a "world-line" as showing that, in origin, this framework was by no means a piece of abstract formalism, whatever it became subsequently, but a model derived from a genuine attempt to portray the realities of experiment which include not only physical substances but percipients. And not only percipients but writers and chalk. The allusion to the fact that the latter is also carried along by the earth is particularly significant.

### The Introduction of Space-Time (Minkowski 1908)

"A point of space at a point of time, that is, a system of values $x$, $y$, $z$, $t$, I will call a world-point. The multiplicity of all thinkable $x$, $y$, $z$, $t$ systems of values we will christen the world. With this most valiant piece of chalk I might project upon the blackboard four world-axes. Since merely one chalky axis, as it is, consists of molecules all a-thrill, and moreover is taking part in the earth's travels in the universe, it already affords us ample scope for abstraction; the somewhat greater abstraction associated with the number four is for the mathematician no infliction. Not to leave a yawning void anywhere, we will imagine that everywhere and everywhen there is something perceptible. To avoid saying "matter" or "electricity" I will use for this something the word "substance". We fix our attention on the substantial point which is at the world-point $x$, $y$, $z$, $t$, and imagine that we are able to recognize this substantial point at any other time. Let the variations $dx$, $dy$, $dz$ of the space co-ordinates of this substantial point correspond to a time element $dt$. Then we obtain, as an image so to speak, of the everlasting career of the substantial point, a curve in the world, a worldline, the points of which can be referred unequivocally to the parameter $t$ from $-\infty$ to $+\infty$."

### Epistemological Criterion

Einstein, in his epoch-making 1916 paper on "The Foundation of the General Theory of Relativity" was quite explicit concerning the *epistemological* dilemma of Newtonian Mechanics, and first identified by Ernst Mach, viz. the exceptional character of relative rotation, as compared with linear translation. "No answer can be admitted as epistemologically satisfactory unless the reason given is an *observable fact of experience*." If we compare this with Minkowski's "everywhere and everywhen something perceptible" we see that both were agreed on the essential criterion for the admittance of an "answer" (i. e. a theoretical proposition on space-time coordinates) to the category of "epistemologically satisfactory" propositions. It is a psychophysical criterion of observability. And this requires a relation between two events "o" and "i" resulting in a deposit of information.

### Psychological Requirement

Now Minkowski decided to use the word "substance" to specify the *physical* requirement of this criterion. Neither he nor Einstein wished to commit themselves to specifying the *psychological* requirement, and this reluctance has left an *hiatus* in the epistemology of Relativity ever since. It is time to seize this bull by the horns, and for us to have, to-day, the courage of their convictions of yesterday. Einstein gave a clue, by implication, in his 1916 paper. The ultimate implication of the ubiquitous fact of unceasing motion, wherever we turn, in a universe containing observers who seek to coordinate their experiences in spite of all their relative motions, is to eliminate *privileged frameworks*: "The general laws of nature are to be expressed by equations which hold good for all systems of coordinates, that is, are co-variant with respect to any substitutions whatever".

## *Identification of Persons and Things*

Einstein goes on "This requirement ... takes away from space and time the last remnant of physical objectivity". Taken literally this would appear to mean that spatial information and temporal information have no objective reference. How then can the laws of physics be derived from clocks and measuring-rods? How indeed, except by *physicists* who are *persons* who perceive, and remember, and calculate, and travel and communicate? Two astronauts, mutually acquainted, meet on the Moon, having traveled there by different trajectories. Each can identify the Moon, the Earth and the other astronaut without any calculation whatever. And this capacity for the ready identification of unique persons and things (as well as the generic identification of *types* of persons and things) is the psychophysical requirement implicit in Minkowski's "something perceptible".

## *Abolition of Privilege*

The "physical objectivity" eliminated from space and time by the requirement of co-variance was previously a Newtonian myth associated with a fictitions privileged framework. The relativistic requirement of co-variance is "with respect to any substitutions whatever". In other words we have here a declaration of epistemic democracy which may be stated thus: – "Since your motion is no more privileged than mine, my measurements are as valid as yours. Since they disagree in what they say about physical space and time each set of measures must refer to our own individual space and time. But we ourselves are moving in the universe and our *accelerations* are *not* independent either of one another or of the rest of the universe. It is the mutual dependencies of second differentials which show the objective laws of the universe".

## *The Third Differential Coefficient*

By providing a calculus of substitutions Einstein was able to "assign to Caesar the things which are Caesar's and to God the things which are God's". If only economists could perform the same feat with our currencies we might eliminate financial crises – but only if we also eliminated human greed. An unnoticed assumption was smuggled into Einstein's doctrine of co-variance viz. that there are no independent *third* differentials. In other words the motion of every body is universally and exactly controlled by the gravitational field. It is a universe of passive objects devoid of any *auto-kinetic* capacity. No spontaneous locomotion, whether by muscle or motor, can occur in such a system. The astronaut's trajectories, whilst *using* the laws of Relativity, introduce accelerations beyond relativistic prediction.

*Impact of Human Action*

The humanly engineered fact of making *impacts* on the Moon has, in however small a degree, altered the Moon's orbit and made all existing lunar tables obsolete. In principle it makes the future orbits in the whole Solar System unpredictable, for we have no calculus of NASA intentions or of the evolution of technological capabilities. This is a very dramatic fact and it is necessary to stress it because Einstein's elimination of *physical* objectivity from space and time is all too easily misinterpreted (and *was* misinterpreted by Eddington). Its implication is *not* that space and time are "subjective", but that coordination and substitution are *objective human activities* which can have objective physical consequences.

*The Psychophysical Range*

The "everywhere and everywhen" of Minkowski's frozen 4-dimensional space-time continuum does not qualify for the criterion of what is "epistemologically satisfactory". For επιστημη means *knowledge* and this is a state of a human being. Human beings cannot penetrate everywhere and everywhen, and, even where they can penetrate, the range of their perceptual capacities is strictly limited. This range has been vastly extended by physical instrumentation and it has now given us, essentially, five different cross-sectional views of the universe viz. an ultra-spectral view down to the limits of Cosmic rays and nuclear particles; an optical view within the visible spectrum; an infraspectral view (the radio universe); a mechanical view (of tangible motions) and a chemical view.

*Information and Technology*

When psychophysics learns to walk on two legs (the I → O leg of pragmatic technology as well as the O – I leg of theoretical physics) we can begin to frame mathematical answers to the question "What is Man to the Universe?" as well as to the traditional question "What is the Universe to Man?" It will be a different kind of mathematics. Traditional physics derives information from facts. Technology creates new facts from information. This, of course, could still be a determinate process if the processing of information is itself determinate. But there are two important indeterminacies in physics itself which exclude total predictability. One is the timing of individual breakdowns in radio-active atoms, and, given that they are spinning, the directions of propagation of the products are also indeterminate. The other is the fact that the direction from which radiation is absorbed by an atom cannot affect the direction of subsequent emission.

## Internal Structural Autonomy

We seem to have here an important fact about time itself or rather (since it is dubious if time can be said to have a "self") about the structure of temporal information. A pattern of waves impinges on an object, a pattern describable in terms of spatial directions in some arbitrary coordinate system. To preserve the conservation of energy we assume that whatever subsequent pattern issues from the object is *quantitatively* determined by the energy of the incident pattern. But internally the pattern, as a structure, dissolves amidst the internal processes of the object and no physical law can predict the structure of the outcoming pattern. Thus it matters not whether we call the interior processes "subjective" or merely "internal", they are autonomous as regards structure. Determinism, as demonstrated by experiment, is quantitative not structural.

## The Thawing of Space-Time

Thus the frozen space-time 4 D continuum, a purely structural concept, is forever thawing locally "everywhere and everywhen" at which emissive matter is to be found. It is only *between* these "material" nodes of knotted fields that structural propagation, free from spontaneous reshuffling, can be conceived as proceeding with strict determinacy. In the freedom of space we find the determinacy of light and gravitation. In the interior of matter, quantitatively limited and encapsulated by finite, closed, potential barrier surfaces, the only *directional* determinacy is derived from the internal history of the object. And here a new quantitative factor arises. If the object is an atom the duration of its internal history is no more than the interval between an absorption and an emission.

## History and Prediction

But in a rigidly distributed distribution of atoms, such as a crystal, there is a history *external* to the atoms, but determined by any linear motions they undergo (as distinct from dynamically stable vibrations which preserve the crystal structure). This history is *internal* to the crystal. It may be determined partly by random slipping of atoms but also by external forces, from whatever source.

When the source is a living agent we have a technological change. We have the internal history of the agent changing the course of internal history in the crystal. The latter may, as regards its *past*, be statistically analysable, which would make its future statistically predictable if there is no interference. But man's interference is *not* predictable, even statistically. And the local thaws can produce avalanches of physical events which no physicist could predict.

*Rigidity and Indeterminacy*

Thus the pursuit of Einstein's epistemological questions to their psycho-physical conclusions leads to a thaw in the space-time framework. So far we are directly aware only of the local thaw produced by Man's technology but we do not know what other technologies may be at work in remote planetary systems. The intriguing fact is that it is the existence of the rigid solid state (which made machines possible) which allows Man to escape from the total determinacy of radiation. But in between these two we have states of matter expressed in such properties as plasticity, elasticity, divisibility, fluidity, expansiveness and ionization. It is in terms of this information, intimately concerned with temporal changes of structure, that Man designs his technological adventures in changing the course of history. And these require a very flexible mathematics.

*The Timing of Innovation*

We can look at technological processes at two markedly different stages in their history. When an invention has been adopted it tends to become a routine, and many technological systems proceed as impersonally and predictably as the Solar System itself (at any rate between strikes, bankruptcies and other "acts of God"). Here conventional mathematics is usually adequate to describe the system. But when we examine the creative stages of technology, the processes of innovation, we have an interaction between unpredictable acts of imagination and sets of calculable *possibilities*, determined by the known properties of matter. We may know everything possible about how matter can behave in prescribed conditions but it is *the timing of the establishment of those conditions* which decides what will actually occur.

*The Planning of Timings*

We can conclude only with a sample of the mathematics required for what is the crux of all technology viz. the *planning of timing*. Given all that is known about physical possibilities, and given that every partial change in a system has a certain duration, the fact that no matter is totally rigid, and no propagations are instantaneous, requires us to introduce a certain limited variability into the separate timings of the whole complex of release-mechanism which control the actual onset of physical processes in a technological system. Absolute simultaneity is unattainable. We therefore make such specifications as "A must finish before B starts" or "A must start after B, but must finish before B finishes" or "A and B must each occur within a given interval but their order is immaterial" etc. etc.

## *Modular Time and Elastic Logic*

Given that clocks are variable any realistic planning of operations must allow for margins of tolerance in the mutual phasing of related events. Also, given the irreversibility of time and the finitude of energy available for implementing any plan, all margins of tolerance have their limits. The limits are rigid and metrical. The tolerances are elastic and topological. These two characteristics are not opposed but complementary. In the process by which the human voice is imposed on an electromagnetic carrier-wave through a microphone the rhythms of the voice are imposed on the rhythms of sound which, in turn, are imposed on the rhythms of electro-magnetic force. This exemplifies the process of "modulation" and serves as a model for the application of the elastic psychophysical I → O relation in a pragmatic context.

## *Limits of Elastic Timing*

A vocal expression is a linear sequence of varied phonemes. But when it carries prescriptive propositions, i. e. commands for action, this represents a requirement for various events in space to be mutually timed in such a way as to secure various causal sequences. From the fact that utterance precedes response all language necessarily has a future implication. In that future there are many uncertainties and, in the phrasing of commands, the temporal implications of words have to be combined in such a way that a predetermined path through the minefield of uncertainties is secured in the wording. Mostly we contrive these verbal structures intuitively, but in complex or dangerous operations our language must make *certain* that the *uncertainties* are circumvented. Thus every elasticity must have a strict limit.

## *Convergence to an Instant*

The concept of an "instant" of time, like the concept of a geometrical "point" has been translated in modern times into the concept of a sequence of values converging to a limit. A "point" is therefore not a single entity but a set. Since it is an infinite set its members can never be enumerated but any one member of the set can serve as a representative. It might be a circle or a triangle or any other form. The sequence of similar forms, diminishing in magnitude, defines the "point". A point thus has a formal character and there can be an infinity of different types of points. But we must also consider the convergence-process. If the point is the vertex of a cone the convergence is uniform but this need not always be the case. The rate of convergence may be expressed by some function giving the "cone" a surface with a curved longitudinal cross-section.

*Function of Convergence-Cone*

Any one circle or polygon (for the "cone" might be a pyramid) could be a member of any number of different convergence-sets. Thus it does not uniquely represent any particular set and the "point" is insufficiently defined. By specifying the convergence-function, as well as the representative element, we completely define the "point" subject to one condition viz. that the axis of the "cone" of convergence is straight. But there is no reason why it, too, should not be curved. The triangular longitudinal section of the cone then bends round in accordance with a second rate-function. Thus a third item of information is required to give a value which can represent the "point". The curved axis of the "cone" now defines a plane at right angles to the first section. This plane, in turn, may be curved, requiring a third rate-function and defining another plane, which in turn may be curved, and so on.

*Clocks and Dimensions*

However complex these convergence-functions may be, they can, according to Fourier's Theorem, be expressed as sums of circular function. A circular function can be represented by a clock. To correspond with the repeated curving of the axis of the cone we must allow the dial of each clock to be set at a different angle in space. It is still ordinary stereoscopic space, but the infinity of possible directions for the axes of the clocks allows for an infinity of "time-dimensions". Any physical experiment is necessarily finite and any one reading is performed by an observer whose psycho-physical limitations define a "just-noticeable difference" which pragmatically defines the representative element of the sequence of converging moments to stand for the "instant" of time. Thus every measured duration is necessarily finite.

*The Planning of Timings*

Now in any one plan of operations, whether it is a physical experiment requiring instruments to be switched on or off at various times, readings to be taken at various times etc., or some administrative, military, political, economic or educational performance, in all cases the plan includes a set of decisions on the *relative timing* of constituent events. However elastic the different rates of the various processes may be to allow for inertia, weather-conditions, human

# The Psychophysical Structure of Temporal Information

factors etc. the success of the whole performance depends on a rigorously defined set of decisions concerning the *order* of the events. These decisions are determined by an analysis of the available information concerning both the rates, and variations of rate, of the different processes, and also of the rates of the clocks used and of the *rate of transmission-time for temporal information*.

## Simple Mutual Timing Relations

This analysis has here been expressed in very general terms but the essential point is that any "instant" of time must be conceived as a multi-dimensional function capable of representation by a finite number of clocks and readers. The accompanying chart displays one possible plan of analysis in which two constituent processes, whose relative timing is variable, and whose respective

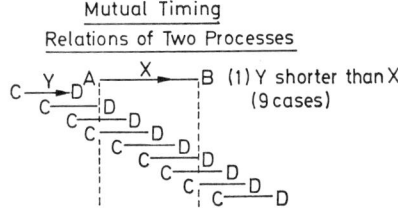

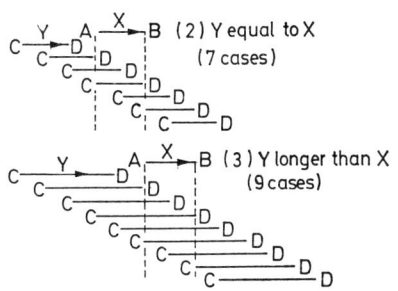

Fig. 1

Process X begins at time A and ends at B.
Process Y begins at time C and ends at D.
The interval AB is held constant.
The interval CD can vary so long as it conforms to the limits specified.
The uni-directional character of time restricts the range of possible relations of A with C and D, and B with C and D to $9+7+9=25$ cases in all.

durations may be equal or unequal, but whose occurrence can be perceptually identified, and timed by a clock, are shown in all possible mutual timing relations. The logic of these relations in this simple case points the way to a Boolean calculus for more complex cases in which processes may be divided into phases, and more than two processes may have to be mutually timed, as in many real-life situations.

# The Dimensions of the Sensible Present

H. A. C. DOBBS †

*I. Introduction*

In the past two years there has been a revival of interest, on the part of some psychologists, in the structure of the so-called 'sensible present'. This term – and its equivalent 'specious present' – is used to describe experiences we are all familiar with: the fact that we can hear very short bursts of sound, in such rapid succession that, although we are aware of successiveness, we cannot give the items in a *reliable* order. It is as if the whole situation was, in one sense, enjoyed together; while in another (equally obvious) sense it involves different phases, which are successive in time. It is known by experiment that, where the time-separation between items of such an experience falls below a certain minimum extent – about fifty milliseconds of physical time – it is impossible to place the items in a reliable simple-linear order. If, for example, the experience is of two sounds lasting less than 10 milliseconds played on a tape, then a replay of the tape will often reverse the time order, as subjectively heard, when the interval between the onset of the sounds is less than about 50 milliseconds.

For this revival of interest James Gibson, University Professor of Psychology at Cornell U.S.A., is largely responsible. In a paper entitled '*The problem of temporal order in stimulation and perception*' he says [5]:

> 'The perception of time has been recognised in psychology, or by some psychologists, but they have never faced up to the contradiction between this kind of time (i.e. *perceived* time lapse) and memory time (i.e. *remembered* time lapse).'

In the same paper he says:

> 'One reason for the muddle of memory, I suspect, is our failure to consider the problem of temporal order objectively. We have habitually conceived time as consisting of past, present and future, instead of a dimension defined by the relation of prior to subsequent ... An exact division between the present and the past has never been possible; and we, therefore, do not know when perception leaves off and memory begins. The travelling moment of present time is not a razor's edge, as James observed, for that would cease

to exist; there is a 'sensible present', although no one has been able to measure its duration. I would add that we cannot be sure when perception stops and expectation begins, either, for a recent experiment at Cornell with an expanding silhouette in the field of view strongly suggests that the *perception* of approach and the *expectation* of collision are not separate ... A change of visual pattern in time ... is perceived not remembered ... Perception is an activity, not an instantaneous event shrunken to a razor's edge ... We have attempted to keep separate the problem of detecting patterns (objects) and that of detecting sequences (events). And hence the equivalence of pattern and sequence, of space and time, has seemed to be a puzzle which had better be swept under the rug than confronted ... Stimulation normally consists of *successivities* as well as *adjacencies* and either will excite a receptive system. This says that there are two kinds of order in stimulation, adjacent and successive, the latter being on the same footing as the former.'

In this paper I propose to confront you with 'the problem of detecting patterns' and 'sequences' in one of its simplest forms, which you can hardly sweep under the carpet. I shall explore for a short way some of the implications, for time-theory, of the alternations in perspective depth, seen in the case of figures presented in ambiguous perspective, on a plane diagram in ordinary three-dimensional space. I shall discuss certain simple empirical psychological phenomena, precisely on the basis suggested by Gibson in the passage quoted: of the 'equivalence of pattern and sequence, of space and time.'

I shall be concerned, that is to say, with the dimensional structure of the perceived space and perceived time of the physical world as *experienced*, rather than with the dimensional structure of the abstract world of physics. Since, however, the word 'dimension' can itself be ambiguous, I must make clear that I am using it in the straightforward topological sense, in which 'space' is commonly held to have the *three* (and only the three) 'dimensions' of height, width and length, i.e. extent in a direction outwards towards the horizon. Similarly, time is generally held to have one and only *one* dimension. This commonsense notion of dimension can in fact be made precise, and consistent with the topological definition; the latter being the one according to which a manifold is said to be of '$N$' dimensions if, and only if, that manifold can be rendered discontinuous by an '$N-1$' dimensional entity which divides it. For example a line, which is a one-dimensional thing, is separated into two disconnected parts by the removal of just one point, a point being something of zero dimensions having position but no size.

The particular instance of a figure showing perspective reversal which I shall discuss is the well-known Necker Cube, depicted on the first of the figures accompanying this paper. I shall argue that the visual experience of alternation of perspective which Fig. 1 provokes, when stared at, is *logically impossible* in any manifold having less than *four* dimensions in the topological sense. These alternations are distinguished for separate scrutiny in Fig. 2A and B and Fig. 3A and 3B. All the figures actually seen are perceived as three-dimensional spatial structures, having visible *height*, *width* and *extent* in the third dimension. These alternations in seen perspective occur because the perspective cues seen in the Necker Cube in Fig. 1 are ambiguous, in that at some times the square face $ABCD$ seems to be out in front, whereas at other times the face $EFGH$ is so seen.

My contention is that, while the perspective is seen to change, there are manifestly never more than the usual *three spatial* dimensions of height, width and depth to be seen in any of these alternating structures. Therefore the additional degree of freedom, required by the topological argument indicated in Sect. III of this paper, in order to make logically possible the transformation actually seen to occur, is an imaginary *time* dimension, to be regarded precisely as Gibson suggests as 'a dimension defined by the relation of prior to subsequent'. That a time dimension of this kind has the same logical properties as a *space* dimension is a suggestion already put forward by that most suggestive genius of applied mathematics, the late Professor A. S. Eddington [4] in his *Fundamental Theory*.

*II. Logical Equivalence of an Imaginary Time and a Real Space Dimension*

I want to devote most of this paper to a discussion of the empirical facts of perspective reversal in a perceptual situation, rather than to a formal discussion of the logical and mathematical concepts involved. Therefore I shall indicate as briefly as possible the logical basis for the identification of an imaginary time dimension with a real space dimension. This identification turns on the mathematical fact that they share a common ordering relation which is three-termed and often called 'betweenness'. This three-termed relation is itself derivable from a two-termed relation of 'partial precedence', or 'quasi-precedence' which seems to be precisely the one that Gibson had in mind in the passage I have quoted, in which he speaks of the relation of 'prior to subsequent'.

It may seem paradoxical to call the reflexive relation that I shall define in the next following paragraph 'precedence' (rather than some other term such as 'quasi-precedence') if we are accustomed to thinking solely in terms of a mathematically real time dimension which has intrinsically the same kind of non-reflexive serial order as the set of mathematically real numbers. In a set having serial order of that kind as Bertrand Russell [7] remarked in his *Introduction to Mathematical Philosophy* 'no term must precede itself'. But Russell admits in the same passage that this kind of simple linear order does not apply either to the set of mathematically complex numbers or to the set of events in time. Anyway the sense of 'precedence' I have in mind is not one of simple linear order, but is a relation which generates the partial order of events in such experiences as hearing a very rapid succession of sounds. For example, two sounds of brief duration, such as a hiss and a click, can be recorded on tape so close together in physical time that, when the tape is played back, the two sounds are heard as successive; yet as falling within a single temporal 'Gestalt', in such close temporal proximity that they cannot be placed reliably and stably in a particular sequence (Broadbent [2]). When the same tape is replayed a number of times each will be heard to come first on different occasions. The abstract mathematical theory of such partially ordered sets has been worked out in detail by a number of mathematicians. I will only mention here Leonard Blumenthal who has given a definition of a relation he calls 'precedence', which I reproduce below (1). This relation has the appropriate characteristics for describing the order between phases of events (such as sounds in rapid sequence) where we experience successiveness – so that in one way the phases of the event seem to have the relation of 'prior to subsequent' in Gibson's language – and yet in another way they are all com-present together in a single simultaneous chunk of time. For it is characteristic of this relation that a term can *both precede* another *and be simultaneous* with it, as is evident from condition (2) in the next paragraph.

'Precedence' in the foregoing sense can be defined as follows: a set, of either spatial or temporal elements, is said to be partially ordered by the relation of 'precedence' symbolised by '$\prec$', when:

$$\text{for each element } x \text{ of the set, } x \prec x; \text{ and,} \qquad (1)$$
$$\text{if } x \prec y, \text{ and } y \prec x, \text{ then } x = y; \text{ and,} \qquad (2)$$
$$\text{if } x \prec y, \text{ and } y \prec z, \text{ then } x \prec z. \qquad (3)$$

Now if we take three particular elements $a, b, c$ of the set to which the postulates (1), (2) and (3) above apply, we may define a triadic relation of 'partial order-betweenness' $B(a, c, b)$: to mean '$c$ is between $a$ and $b$' if we have the relation '$a \prec c . c \prec b$'. Such a relation of order betweenness has been taken as the basis, both of the set of points along an undirected line of space and of the order characterising the set of durations along an undirected line of time. The precise logical form of the relation can be different in different geometries. In general such partial order betweenness is a *necessary*, (but not a *sufficient*) condition for the kind of *metrical* betweenness characteristic of the order of points on a line of space in Euclidean geometry, and on a line of time in Minkowski's four dimensional space-time geometry. For it is convenient, in a *purely metrical* geometry, to require that a point '$c$' which is between two points '$a$', '$b$' does not coincide with either '$a$' or '$b$': whereas in partial order-betweenness clearly any two of the three points may coincide, or all three of them may do so.

However there are features common both to the partial-order-betweenness and to metrical-betweenness which are as follows, where the expression '$B(acb)$' means 'the point $c$ is between $a$ and $b$' in both the sense of 'partial order betweenness', and of 'metrical order betweenness':

(a) Symmetry between the outer points: so that if $B(acb)$ then also $B(bca)$.

(b) Speciality of the inner point: so that if both $B(acb)$ and $B(abc)$ then $b = c$; and similarly, if both $B(acb)$ and $B(cab)$ then $a = c$; but we do not have identity between the outer points $a = b$ unless all three points coincide – which is impossible in *metrical* order betweenness, but permissible in *partial* order betweenness.

The feature of symmetry, between the outer points of a betweenness relation, common to both non-metrical partial order and to metrical (simple linear) order, entails complete reversibility rather than the kind of *serial* order characteristic of the field of mathematically real numbers, ranging from minus infinity to plus infinity, and of the succession of cause and effect in time. In a set of elements with a *serial* order there is an intrinsic sense of direction, holding between each element and others belonging to the set, which is definite and irreversible. The relation generating this order can be expressed by the symbol $\langle$ defined as follows:

$$a \langle b, \text{ if and only if, } a \mathrel{-\!\langle} b, \text{ not } b \mathrel{-\!\langle} a.$$

This proposition states that $a$ irreversibly precedes $b$.

The lack of *serial* order is characteristic of a time dimension assimilated to a space dimension, as in the time measure of everyday life – the 'mean solar second' – which is defined in terms of the diurnal rotation of the earth with respect to the notional motion of the so-called 'mean sun', such motion involving 'corrections' to take into account irregularities of motion. Exactly as many 'mean solar seconds' would be held to have elapsed if, instead of revolving in one directional sense about its axis, it suddenly reversed its rotational motion, with respect to some externally defined direction. This argument is not affected by the introduction in 1964 of Atomic Time which is equally devoid of serial order.

The characteristic feature of 'imaginary time', contrasted with real time, is that it has *reversible* simple linear order, like that of the points along an undirected straight line in Euclidean space, but it does *not* have *serial* order like that of a *directed* line. I have some more to say on this subject of reversibility later on.

By adjoining an imaginary time variable to a real time variable we get a complex time variable, in the mathematical sense. It is characteristic of a complex variable that there is neither serial order nor simple linear order between its values, which are complex numbers in the mathematician's sense.

These form partially ordered sets, in the sense of partial order discussed earlier; although the so-called 'real' and 'imaginary' parts do form sets having simple linear order. Moreover, the *absolute* values (moduli) of complex numbers do have linear order. (Indeed it can be argued that, since they are always either zero or are *positive* they have *serial* order.) In this paper I claim that the partial order of complex time is indeed manifested by the subjective contents of the seen reversals of perspective experience, obtained by fixation upon the so-called Necker Cube.

But of course a series of happenings in a *real* (as distinct from a complex or imaginary) time order has both simple linear order and an inherent sense of direction, (defined objectively by the relation of cause and effect), that is irreversible. This irreversible sense of direction, inherent in the real time serial order of actual happenings, is the feature which the poet Edward Fitzgerald captured metaphorically in his famous simile of the 'moving finger' that 'writes' and cannot be 'lured back'. In other words, the difference between space and time dimensions only emerges fully when one takes into account the irreducibly transitory aspect of any actual process, in which it is logically impossible for the effect to precede irreversibly the cause.

Such an irreversible transitory aspect does not come into the equations of motion in mathematical physics which, alike in relativity and quantum and Newtonian mechanics neglect dissipative forces, and all expressions valid equally for $+t$ and for $-t$. It is however, an essential feature in theories of physical prediction, where the time point $t = 0$, which divides positive and negative values of the time variable, has a direct physical significance: as in extrapolation methods, applicable only to the *past* of stationary time series. In this area of physics we encounter mathematical operations which are significant when applied to negative time values (the past) but are meaningless when applied to positive values (the future). A full discussion is to be found in Norbert Wiener's *Interpolation, Extrapolation and Smoothing of Stationary Time Series with Engineering Applications* [8].

Wiener demonstrates that no realisable predictor can operate on the future of a time series, but only upon its past. In this respect therefore the transitory aspect of temporal relations does come into a part of physics. Indeed it does so in every case where we are essentially concerned to deal with events in 'real time' as the engineers say (that is with events as they actually happen) as opposed to 'dead time' (the time relations between events as stored or recorded e.g. as sounds on magnetic tape). The latter – dead or storage time – reduces very obviously to the *spatial* relations between different parts of the record: for example before recording begins one end of the tape is wound on first and is thus nearer the capstan than the other, and will therefore be heard last when played back after winding back.

We can sum up this part of my discussion by saying that the required additional dimension of 'imaginary time' is in part 'dead' time – the time of some record storage system no matter what precise form that storage system may take. But it also contains 'expectation time'. It is in fact what has been called 'static' time by Colin Cherry [3]: i.e. it is time bereft of becoming or happening. A series of events related by imaginary time relations is then equivalent to a set of elements (points) in space ordered by the relation of simple order. Thus in its mathematical structure an imaginary time dimension is equivalent to a real space dimension. The relevance of this equivalence will be clear after I have analysed a typical experience of the oscillation of depth perspective, by a characteristic trembling or twinkling pulsation, elicited by the well-known diagram often called the *Necker Cube*. I shall argue that such perceived reversal requires us to adjoin the dimension of 'storage', 'dead', or mathematically imaginary time to the usual real time dimension, in such perceptual situations, thus arriving at a partially ordered complex time with mathematically real and imaginary parts. This would be analogous to the introduction of a mathematically imaginary time into the quantum mechanical equation for the motion of the electron by Bunge, in order to cope with the paradoxical aspects of the Zitterbewegung introduced by Dirac's theory. Of course there is a vast difference in the extent of the time period involved in the two cases. In Bunge's treatment of the electron the extent of the mathematically imaginary time period, which he associates with the spin of the electron, is of the order of the period of its Compton wave-length (viz. $h/4\pi m c^2$). Whereas, in the case of the perceived oscillation of the Necker Cube, the period of the spin of consciousness is several seconds. But this difference in scale is not necessarily a barrier to the usefulness of the analogy. It may simply be a reflection of the vast difference in scale between the electron and the human brain, which must be evident in any theory.

## III. Perception of a 4-Dimensional Transformation in the Perspective Alteration of the Necker Cube

The Necker Cube – so called after its inventor Julius Necker – is perhaps the simplest example of a figure drawn on a sheet of paper which exhibits reversal of extent in the third spatial dimension (the dimension orthogonal to the plane of the paper). This reversal can be seen in Fig. 1 which shows a skeletal cube drawn so as to exhibit alternately the face $ABCD$ in front of the face $EFGH$, and *vice versa*. It is this alternation or reversal of visible extent in the third spatial dimension, which everyone gets sooner or later on fixing their attention upon the Necker Cube, that I want to discuss in the remainder of this paper.

To experience the phenomenon of reversal of perspective it is advisable to keep the eyes focussed on the middle of Fig. 1. It will then be perceived as the skeleton of a cube: as the kind of three dimensional structure which children put together in play sets in the game called *Kugeli*. This skeletal three dimensional structure is seen as having a definite extent in each of the three spatial dimensions of visible length, breadth and height. It will be seen as having two square faces, which are defined by the lines $ABCD$ and by the lines $EFGH$ respectively. One or other of these is seen initially in front in the sense of seeming to be nearer the observer than the other. But, if the gaze is fixed upon Fig. 1, before long a reversal of this perspective will occur in such a way that if, for example, the face $ABCD$ is seen initially as out in front of $EFGH$, then, shortly afterwards, it will be seen to move behind the face $EFGH$ which protrudes. The face $EFGH$ will then stay out in front for a few moments before it spontaneously retires behind the face $ABCD$. Then, if fixation upon the figure is maintained, the cycle of alternations of seen perspective will continue indefinitely.

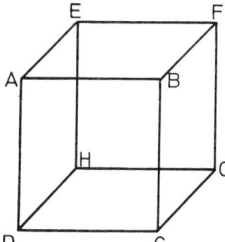

Fig. 1. A Necker Cube

In order to follow the rest of my argument it is important to get clear the precise significance of the phenomena I have just described, which everyone can observe for himself by fixing his vision upon the Fig. 1. What is directly perceived in this (and similar perceptual situations) is NOT just the seeing of one *two-dimensional plane* figure followed by another *plane figure*. It is true that the pattern of retinal stimulation, which gives rise to visual perception, is physically practically a two-dimensional structure. But nevertheless the object directly perceived – in this case a skeleton cube – is a 3-dimensional entity, having manifest length, breadth and height.

In the case of the Fig. 1 a three-dimensional entity – a skeletal cube – is seen in which either the face $ABCD$ is projecting out of the paper, or the face $EFGH$ is so seen. Precisely which of these two *alternative* (and mutually exclusive) 3-dimensional configurations is seen at any particular moment, by a particular observer, cannot be predicted in advance. But with prolonged fixation, a reversal of the depth perspective of the initially seen 3-dimensional figure will certainly occur: so that if one starts by seeing the face $ABCD$ as jutting outwards, this experience will be replaced by one in which the face $EFGH$ protrudes and $ABCD$ retires to the background plane. The first of these two (different) transformations is depicted in the Fig. 2A and 2B; and the reverse transformation is illustrated by the Fig. 3A and 3B.

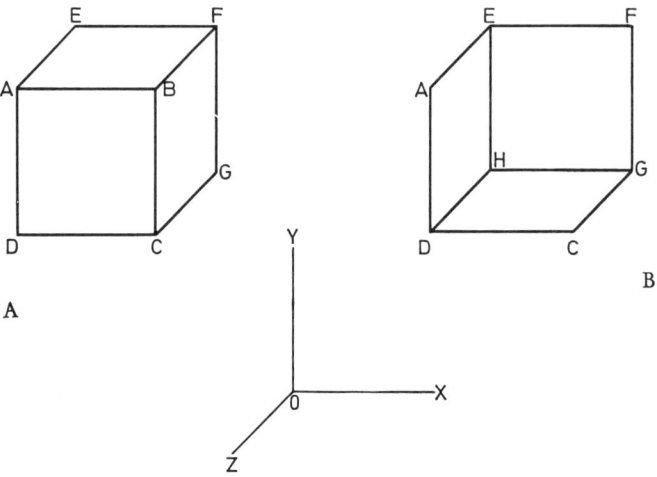

Fig. 2A and B. Two different aspects of perspective transformation induced by fixation of Figure 1, involving a 4 – Dimensional Rotation about a Plane

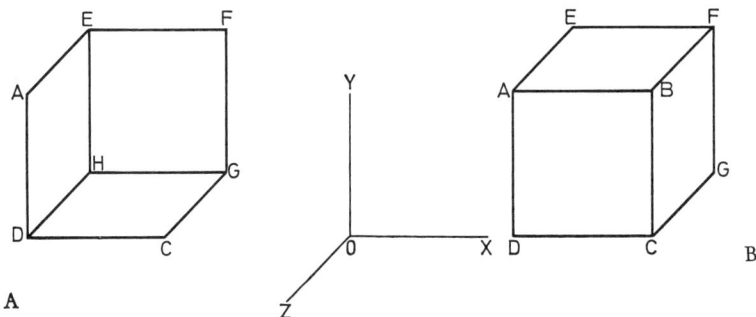

Fig. 3A and B. The inverse of the transformation shown in Figures 2A and 2B, involving a 4 – Dimensional Rotation about a Plane

The casual observer of such reversals of perspective might suppose that the phenomenon simply consists in the transformation of one apparent 3-dimensional spatial configuration, perceived at one point of time, into another precisely similar and congruent 3-dimensional structure, perceived at a slightly later point of time. If this were all, the phenomenon would be simply equivalent to a rotation of the apparent 3-dimensional configuration (the skeletal cube) about the vertical axis $OY$. This is indeed how Fig. 4 A is transformed into Fig. 4 B.

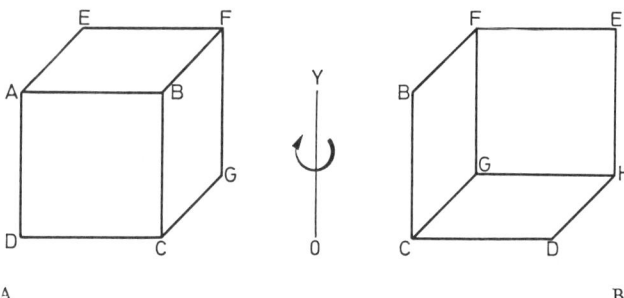

A                                                                                               B

Fig. 4A and 4B. Two different aspects of ordinary 3 – Dimensional Rotation of a Cube about a Line

In the perspective reversal, seen after prolonged fixation of Fig. 1, whereby Fig. 2A is transformed into Fig. 2B, and B drops back as A comes forward, there is rotation about the *plane* YOZ.
Here the reversal of perspective is in the opposite direction to that of Fig. 2A and 2B. Fig. 3A is transformed into Fig. 3B by a rotation in the *opposite* sense, about the *plane* YOZ. In this transformation, whereby Fig. 4A is transformed into Fig. 4B, there is a rotation, through a little less than 2 right angles, of the skeletal cube about the vertical *line* axis OY. Such a rotation is of course a common event in 3-dimensional space; but it is different from those in Fig. 2 and 3 which are NOT possible in 3-space.

But it is clear that this particular transformation, from 4A to 4B, is NOT what we see happening in Fig. 1. For, when the perspective is reversed in Fig. 1, we get the transformation depicted in Fig. 2A and 2B. In such a rotation we get exactly *the opposite* arrangement of the corners of the cube: the front face in Fig. 2B reads $EFGH$ instead of $FEHG$ as in Fig. 4B.

This fact means that the two structures in Figs. 2B and 4B, seen as skeletal cubes, are two basically different structures, related to each other as mirror images; or, as the mathematicians say, as 'enantiomorphic forms', structures which are counterparts but which cannot be put into congruence in a 3-dimensional space. Hence they have been called 'incongruous counterparts'. Perhaps the most familiar examples of incongruous counterparts are the left and right hands of an ordinary human being, each of which has five fingers, a palm and a wrist, but which are so arranged with respect to each other as to be mirror images. They *cannot* be converted into each other by a rotation about a *line* in three dimensional space.

We find, then, that the 3-dimensional structures which are perceived by looking at Fig. 1 alternate, sometimes the plane $ABCD$ being seen in front of the plane $EFHG$, and at other times *vice versa*, these alternate transformations being depicted in Figs. 2 and 3. Two questions then arise: (1) how can one see the kind of transformation between incongruous counterparts exhibited in Figs. 2 and 3? (2) What are the implications of the seen occurrence of such transformations?

There are several features to notice in connection with the transformations between alternate perspectives, which occur spontaneously when we look at Fig. 1, and which are separately depicted in Figs. 2 and 3.

(1) The transformation as seen is an instance of a *continuous transformation*. For the transformation manifestly does not involve any *break* in the continuity of the lines joining the corner points of the skeletal cube.

(2) The transformation seen, whereby one of the two faces $ABCD$ or $EFGH$ advances towards the observer, while the other recedes, seems to involve a reorganisation within the manifold by a kind of *twinkling pulsation*.

(3) Not only does this transformation seem to occur within a sensible present or 'mental now'; but it is seen to happen in just the same direct kind of way as the motion of the second hand of a watch. One *can actually see* the skeletal cube swing some 30 degrees of arc from right to left (and *vice versa*) in azimuth; and also tilt by about the same angle in the vertical plane.

(4) This transformation, which I have called a twinkling ('Augenblick') pulsation is *between two 3-dimensional incongruous counterparts*.

It is only right to mention here that there is a difference in the subjective reports of observers as to the structure of the experience, which came out in the course of discussion. Some agreed with the writer that they observed a continuous movement, of the nature of an 'Augenblick puls'. Others seemed to experience a sort of discontinuous quantum jump between the incongruous counterparts. A similar difference in experience was found by Sir Cyril Burt in the course of experiments with a set of subjects. (Personal communication.)

I am not at the moment clear as to the precise significance of this difference in perceptual response. It might, for example, be merely a matter of the precise manner of presentation which could be eliminated by a more sophisticated technique such as the one suggested by Robert A. Knapp in the course of discussion. (This suggestion, which it is hoped to pursue, was to create the visual appearance of the Necker Cube by means of suitably spaced lights which, with properly phased switching would give rise to the phi-phenomenon). Alternatively, the difference might be found to be due to some perceptual deprivation, comparable to colour blindness affecting some subjects.

The 19th century geometer Möbius suggested [6] that continuous transformations between incongruous 3-dimensional counterparts are mathematically impossible, within a 3-dimensional manifold. For such transformations involve rotation about a whole *plane*, and not merely about a single line within a plane; and this kind of transformation requires at least a 4-dimensional manifold for achievement. Indeed it can be shown that rotations within an N-dimensional manifold can take place about, at most, an $(N-2)$-dimensional entity. For example, within a line, which is a 1-dimensional manifold, rotation can take place at most about a 1–2 dimensional entity which, having the dimensions —1 is the null-set with no members: so that rotation cannot occur at all within a line. In a plane, which is a 2-dimensional manifold, rotation can occur at most about a $2-2=0$ dimensional entity, which is a point. In a 3-dimensional manifold rotation can occur at most about a $3-2=1$ dimensional entity which is a line.

We must conclude that any continuous transformation between 3-dimensional incongruous counterparts – such as those which are seen to occur in the course of fixation upon the Fig. 1, and which are separately shown in Figs. 2 and 3 – can only take place in a manifold which is at least 4-dimensional. It might, however, be objected that the transformation which is seen to occur after looking long enough at Fig. 1, (shown separately in Figs. 2 and 3) is not in fact a 'genuine' rotation about a plane. Because it does not involve 'genuine' 3-dimensional entities (viz physical skeletal cubes) but only pseudo-cubes; the third dimension of 'length' being only a quasi-dimension in the *mind* of the observer. On this line of argument the only 'genuine' entities involved would be certain *plane* figures, drawn on the 3-dimensional surface of a sheet of paper, the appearance of extent in the third dimension, which they convey, being deceptive.

But such a line of argument is quite untenable. If it was sound it would serve to demonstrate that *all* perceptions of the third spatial dimension in a visual field are deceptive. For it is a physical fact that the pattern of *retinal* stimulation at the surface of the human eye is always a 2-dimensional pattern, no matter whether we are looking at sheets of paper, or at more complex stimulus fields. In *every* case, the structure of the array of photons momentarily impinging on the retinal surface at any moment is necessarily a 2-dimensional structure. From this essentially 2-dimensional pattern of stimulation, produced by the impact of light quanta upon the sheet of retinal receptors, followed by its relay to the central nervous system, the brain derives the ordinary 3-dimensional 'solid' perception of objects with which we are familiar in everyday life.

It follows that the 2-dimensional character of the merely *physical* configurations in Fig. 1, which serve as stimuli for the perception of the oscillation of perspective, is irrelevant to the question of the number of dimensions to be ascribed to the final product of such stimulation. The 'apparent' 3-dimensional structures which are seen, although they are the end-products of the stimulation of the 2-dimensional surfaces of the retina of the human eye have, as regards the topology of visual space, just as much right to be treated as 'genuine' 3-dimensional entities, as do the perceptual appearances of any of the ordinary 3-dimensional physical objects, encountered in the course of everyday perception of the physical world.

Now, if it is logically necessary for the field of perception to be at least a 4-dimensional manifold, at each moment of a perceptual situation in which reversal of perspective occurs, presumably the momentary structure of *all* human visual perceptual situations is 4-dimensional. For there is no topological difference between perceptual situations displaying alternation of perspective and other perceptual situations, in which the third dimensional perspective remains stable. But it has, I think, always been held that in all ordinary perceptions there are only three *spatial* dimensions – those of visible breadth, height and length. Certainly those visual fields which manifest alternation of perspective do not seem to involve any new and independent order of spatial arrangement, over and above the familiar three-fold ambience of breadth, height and length. It is apparent, however, that we cannot be said to 'see' in the figures depicted in this paper, the additional dimension of the manifold, required to make continuous reversals of depth logically possible, in any sense equivalent to that in which we do 'see' the three *spatial* dimensions of breadth, height and length. What, then, is this extra dimension, and how shall we identify it?

The answer to this question is, I believe, simple and inescapable. The fourth dimension of any visual field, directly perceived as an Augenblick in the course of a sensible present, is what I have called in section II of this paper the 'imaginary time' dimension. The significance of this phrase 'imaginary time' needs a little more elaboration at this stage of the argument. The word 'imaginary' is not in this context intended primarily to have a subjective or psychological meaning, but rather a strictly mathematical sense. In other words this 'imaginary' time is to be conceived as having the same mathematical relationship to the usual time dimension (as ordinarily conceived in connection with psychological and physical events), as the so-called 'imaginary' part of a complex number to the so-called 'real' part, in the mathematical theory of complex variables. We should again recall here Eddington's remark that 'formally an imaginary time' is 'equivalently a real space dimension'.

If we admit an imaginary time as equivalent topologically to a real space dimension it will follow that in a momentary act of perception, falling within the so-called 'mental' or 'sensible present' of a percipient, the content perceived is a four-fold, comprising a short, but finite, extent of time. This finite extent of time will constitute a *duration*, comprising a temporal sequence of different instantaneous configurations of three-dimensional space. For example the perception, within a single sensible present, of the movement of an apparent solid 3-dimensional spatial structure, from one place to another in a visual field (e.g. the seen flight of a tennis ball over the net), involves a momentary (or temporally punctiform) synthesis of a temporal sequence of different 3-dimensional spatial configurations, each of which is incompatible with its predecessor. Each of these incompatible 3-dimensional configurations is an instantaneous spatial location of the ball in a different place within the visual field; it being logically impossible for the single seen ball to be both at place 1 and also at place 2 at the same point of time, if time is a 1-dimensional manifold.

However, the *compresent* presentation (within a single Augenblick), of a *succession* of such incompatible spatial configurations in visual fields as being that of a *single* visible object, e.g. the tennis ball is a *necessary* condition for the direct intuitive experience of the movement of any object in a 3-dimensional space. Without it, we should only be aware of the *present* appearance of the object in a particular place in the visual field – call it place '$n$' – together with the *recall in memory* of it having been, just previously, in the adjacent places '1' '2' ... '$(n-1)$'. This is a different kind of experience, which we do have from time to time. But it is not the same experience as the *seeing* of the object in movement as it actually takes place in real time. It is rather what one might call a cognitive *reconstruction* of a movement in memory after the event, rather than seeing it happening in real time. Many people do indeed have this latter experience, when they try to reconstruct, in their mind's eye, a movement which has already taken place. And it is also the nearest equivalent to seeing a continuous movement which patients suffering from the pathological condition known as 'polyopia' are able to enjoy. In such conditions, instead of continuous movement, a disconnected set of 'stills' are seen, apprehended together with memories of earlier seen 'stills'.

The essential point is that such experiences are quite different to these we describe as the normal seeing of a continuous movement, of which the experience of seeing the alternations of perspective in Fig. 1, are instances. Such coherent succession (and integration) of disparate and incompatible 3-dimensional spatial configurations, seen both in the alternation of perspective in Fig. 1, and in the ordinary perception of movement in general, clearly requires at least a 4-dimensional manifold. But when one takes into account *that within a single 'Augenblick puls'* there must be a 4-dimensional manifold – to accommodate the kind of geometrical transformation required to provide for reversal of perspective – then one is driven to conclude that the *history* in time of a sequence of such experiences – each of which is itself 4-dimensional in content – must be a 5-dimensional entity.

The view I am now suggesting, to account logically for the perception of movement (and change in a visual field generally), is that such perception involves, at any moment, the coherent conjuction, or integration, of a set of disparate and incompatible spatial configurations – for example the set of configurations comprising the same entity at two or more different places in the visual field. But such a coherent conjunction or integration of disparate d-dimensional manifolds, requires the co-existence of at least a 4-dimensional manifold: just as the coherent conjunction or adjunction of disparate (and therefore incompatible) locations of a *single entity* on a 1-dimensional manifold (a line in space), requires at least a 2-dimensional manifold.

It is logically impossible to have a single thing in two different places of a three-dimensional space at the same time. If one tries to persuade oneself that such an event can occur, what must in fact be happening is the momentary coherent integration over a finite time interval of a succession of *incompatible* spatial configurations. We know that the orderly combination of a set of different lines, each of which is in itself only a 1-dimensional entity, yields a 2-dimensional entity – viz a ruled surface. In the same way the regular combination or momentary integration of a set of different, and incompatible, 3-dimensional spaces yields a 4-dimensional configuration in space-time. But, as I have said earlier, when we perceive a 4-dimensional manifold of this sort, in the momentary experience during a single sensible present, of the reversal of spatial perspective, we are not acquainted with any *additional space* dimension. We can still only perceive the location of objects, with respect to the usual three space dimensions of breadth, height and length. What we do have is, at a moment of time, a coherent integration over time, of a *succession in time* of *different* 3-dimensional spatial configurations, each comprising various locations of the moving object, in respect of the three dimensions of height, breadth and length. It is clear that this account of the perception of motion requires two distinct dimensions of time: (1) the 'time *over* which' the integration is taken; (2) the 'time *at* which' the integration occurs. Therefore we conclude that the additional (fourth) dimension, found in the perception of movement within a single sensible present, is a time-dimension; and that its adjunction to the usually recognised time dimension yields a complex time variable. Furthermore, we find that this additional time dimension has a reversible order, its reversibility being clearly visible in the alternation of perspective seen in Fig. 1. For we can regard this visible spatial reversal either as a reversal of the third spatial direction, or as a reversal of an arbitrary and extrinsic sense imposed upon the order of imaginary time succession.

It is a characteristic feature of complex numbers, derived by the association of imaginary numbers with real numbers (in the mathematical sense), to be without directed serial order. The field of complex numbers – in contrast to the field of real numbers – has no simple linear order; nor is there any meaning in saying of any complex number that it is either positive or negative, or that it is larger or smaller than any other. However, an arbitrary extrinsic simple linear order can be associated with a complex variable, by considering its absolute magnitude or modulus, as distinct from the twofold class of its real and imaginary parts.

For, as I have said earlier, the modulus of a complex number is, by definition, a real number: being the *positive* square root of the sum of the squares of its real and imaginary parts. One way, therefore, in which we can eliminate the twofold character of a complex time variable is by considering only its so-called absolute value or modulus. For example, reverting to Bunge's use of a complex time variable '$t \pm iT$' where $i = \sqrt{-1}$, there is no simple linear order between the two conjugate complex time points $t+iT$ and $t-iT$. In fact both these time points have the same modulus viz the *positive* real number $+\sqrt{(t^2+T^2)}$.

If we now confine our attention to the modulus, and regard it as a measure of the time extent of the happening of events, then we can say that the becoming of time has a simple linear order. Indeed it has a *serial* order with a sense of positive direction – the order of increasing positive real numbers – always provided that the real part of the complex time is itself a monotonically increasing variable.

By associating a set of imaginary times with the set of real times in this way we get a complex time variable, in which the modulus – a real number – plays the role that we usually associate with the *becoming present* of events, as contrasted with the mere measure of already elapsed durations: that is to say the temporal role we associate with *happening* as opposed to *history*. The happenings of events have indeed as Eddington remarked 'distinctive directions between past and future' (*Fundamental Theory*, p. 124). This is a basic fact which every practical man such as an engineer knows to be true. But equally there is another 'time', which engineers often call 'dead time' or 'elapsed time' or 'recorded time', which is merely a variable used for the description of the extents of a fourth co-ordinate, required to specify one aspect of events that have happened in the past and are no longer happening. This 'dead' or 'elapsed' time of *recording*, (rather than *happening*) is exactly analogous to a one-dimensional space component, as we can see very simply by noting that it is conveniently measured in terms of so many feet of film or inches of magnetic tape. It is this kind of 'dead' time which I am calling 'imaginary time'. I believe it is necessary to adjoin such a time to the 'real' time to get an adequate time analysis of the sensible present.

This brings me back to Gibson, and to my starting point. In my view he is correct in saying that we observe 'successivities' in time, as simply and directly as spatial 'adjacencies': provided we interpret the relation of succession as a 'partial order' relation, not to be confused with the order generated by the relation symbolised by '$<$' or '$>$' associated with what Eddington called 'absolute distinction of sign'. This partial order relation is indeed to be identified, as Gibson suggests, with that of 'prior to subsequent'. So we then treat the different successive phases, or configurations, within a 'sensible present' as related by partial order relation of mathematically *imaginary* time. The *real*-time relation, between different sensible presents, has the absolute distinction of sign, characteristic of the real time order of happening, as distinct from the imaginary time order of anticipating and remembering (recording). The time order of sensory inputs to a perceiving organism is an irreversibly directed linear order (the direction being defined by the relation of cause to effect), which we associate with a mathematically real time order. But, as Gibson has told us – and nearly every other modern psychologist from James onwards – we cannot make a valid sharp division between the experienced contents due solely to present *sensory* input, and the accompanying immediately evoked intuitive memories and anticipations. These latter components serve to interpret the input data, filling them out and giving them the elaborated content of a typical perception. (One experimental psychologist – Richard Gregory, now Professor at Edinburgh – has referred to the sensory input as being a mere 'cartoon' of the perceived situation, which is immediately and unconsciously filled out by the subjective contributions derived from past experience.)

Thus we have to recognise that the purely sensory input, ordered serially and irreversibly in the real time dimension, is immediately and unconsciously associated with, and adjoined by intuitive memory and anticipation contents evoked from past experience, ordered reversibly by simple linear order in the imaginary time dimension. The imaginary time order involved in intuitive memory and anticipation adds the extra degree of freedom, required to elucidate the experience of reversal of perspective by a continuous transformation, as seen in the case of reversals in a sensibly present perception of the Necker Cube.

The 'intuitive', or 'interpretive' memory to which I refer here is, of course, quite different from the 'memory of recall' that I referred to earlier as a 'cognitive reconstruction'. The term 'intuitive memory', in the present context, refers to any stored contents of previous experience (attributable to sensory inputs, due to causes occurring in some interval of real time before the present), which are implicitly evoked by those inputs as an act of *interpretation NOT* as an act of conscious *recall*. They must, consequently, be distinguished both from:

(a) contents just experienced in the immediate past, and *held unchanged*, by persistence ('immediate memory', as some psychologists use that term). And from

(b) contents which are *explicitly recalled*, by conscious decision, as *cognitive reconstructions* of past experienced contents.

So we finally conclude that events fall into two time-order classes:

(1) those forming a series the terms of which are connected by relations of objective cause and effect. This class of events manifests a time with an irreversible serial order (real time), represented by the mathematically real numbers taken in the sense of monotone increasing magnitude, ordered by the relation $<$ defined earlier. This class includes all physical events which are *causes* of perceptions, and all psychological events which are *stimuli* for responses in experiments.

(2) Experienced durations in perceptions containing a set of elements, describable in terms of a complex time variable, between which a partial order relation obtains. The contents of such events include those of ordinary perceptual situations, as well as those of specially arranged situations: such as the perception of the Necker Cube. In so far as circumstances allow us to separate the (objective) sensory-input component from the (subjective) memory-anticipation-interpretation, in any given experiential situation, the former component is serially ordered in mathematically *real* time, and the latter component is simply linearly ordered in mathematically *imaginary* time.

*Summary and Conclusion*

(1) The ideas put forward in this paper may be summarised as follows:

(a) The transformation manifest in the seen perspectual reversal of the Necker Cube is logically impossible in 3-space.

(b) No *spatial* degree of freedom is manifest in these transformations beyond the usual triad of length, breadth and height. The extra dimension must be a mathematically imaginary time dimension.

(c) When we adjoin an imaginary time dimension to the usually assumed real time dimension we get a complex time variable. The order between elements of a set of complex numbers is a *partial* order not a *simple linear* order. This is analogous to the situation obtaining in the case of those events which are experienced durations having contents with uncertain temporal order, e.g. sounds heard in very rapid succession.

(d) The time order holding between the objective happenings which are causes of perceptions or stimuli for responses is, on the other hand, an irreversible serial order, appropriately described by the directed order relations between the real numbers taken with respect to increasing magnitude. It is always significant to say with respect to two real numbers $a$ and $b$ that $a < b$ or $b < a$. But it is meaningless to say of two complex numbers $\alpha$ and $\beta$ either that $\alpha < \beta$ or that $\beta < \alpha$; or that $\alpha$ or $\beta$ is either positive or negative.

(e) In the perceptual situations where we gaze fixedly at the Necker Cube we have instances of the reversal of the sense arbitrarily imposed on the second time dimension, by a kind of Zitterbewegungen, with a characteristic frequency of oscillation or alternation. Such alternations are not observed in ordinary perceptual situations, being inhibited by the subjectively supplied interpretation of the perception as it occurs, the perspectival meaning incompletely defined in the case of the Necker Cube.

(2) Tests should be conducted to establish the frequency in time of this alternation of perspective and the other significant parameters of the phenomena. The test should be directed at establishing the statistics of perspective reversal, not only in respect of a diversity of geometrical objects, but also in respect of a diverse range of personality types. Different displays should also be used such as drawings, illuminated wire models, and presentation by means of the phi-phenomenon.

*References*

1. Blumenthal, L.: *Theory and Applications of Distance Geometry* p. 4. Oxford 1954.
2. Broadbent, D. E., Ladefoged, P.: *Auditory Perception of Temporal Order*. J. Acous. Soc. Am. 31 (1959) 1539.
3. Cherry, C.: *On Human Communication*. 139. Cambridge: M. I. T. 1968.
4. Eddington, A. S.: *Fundamental Theory*, p. 125. Cambridge 1946.

5. Gibson, J. J.: *The problem of temporal order in stimulation and perception*. Am. Psychol. 62 (1966) 141–149.
6. Möbius, G.: *Der Bayrcentrische Calcul* p. 184. Leipzic 1827.
7. Russell, B. A. W.: *Introduction to Mathematical Philosophy*. London 1917.
8. Wiener, N.: *Interpolation, Extrapolation and Smoothing of Stationary Time Series with Engineering Applications*. New York 1949.

# Time, Time Stance, and Existence*

BERNARD S. AARONSON**

*Summary.* Time is analyzed as being those processes by which a system notes the processes which comprise its own existence. The directionality of time is given by the concepts, past, present, and future. To understand the meaning of these concepts, a set of experiments was carried out with four male subjects in which areas of time were expanded or ablated by means of post-hypnotic suggestions. These operations were carried out singly or in combination. The data suggest that the present is primarily associated with stimulus input, the past with the criteria for defining that input as well as the development of response inhibitions and the future with the directionality of that input. Marked personality changes which range from the occurence of catatonic-like states to obsessive types of reaction and manic-like reactions are associated with different combinations of time alterations. Some of the changes are in accord with the serialist position on time of J. W. Dunne. The concept of death is very important both with regard to one's view of time and how one lives. While it is important to live in the present, the present has no meaning without at least a past or a future. Implications of these findings for psychopathology in general and schizophrenia in particular are considered.

The world is filled with propaganda alleging the existence of time. From childhood on, our lives are filled with instruments for measuring time, and these become, for most of us, even more important as we grow older. It is a significant milestone in our growth when we can finally "tell time". We are constantly urged to be on time, to take enough time, and not to squander it or lose it. In school, we learn sets of equations in calculus in which time is a variable. We also learn its importance in those major scientific theories whose results have not only benefited us materially, but have cast the world into an increasingly comprehensible cognitive space.

Those of us who are concerned with religious matters learn that time is at least an attribute of deity. Jehovah is "the Rock of Ages." In the *Bhagavad Gita,* Krishna reveals himself in his ultimate divine form with the words, "I am Time." We know that many wise and famous men spent their lives dealing with the question of time, and the fact that their writings on the

---

\* This research has been supported in part by grants from The Parapsychology Foundation, The Ittleson Family Fund, The Erickson Educational Foundation, and U.S. Public Health Service Grant No. 1–SO1–05262–01.
\*\* Bernard S. Aaronson, Ph.D., Bureau of Research in Neurology and Psychiatry, Box 1000 Princeton, New Jersey 08540, U.S.A.

subject are at least often difficult to comprehend makes time seem even more real and important. Books are written on it and learned societies debate it. Surely something that doesn't exist would not arouse so much concern or activity, or become so important in our lives. Or would it?

The fundamental operation for the perception of time is discrimination. In order for any succession of events to be enumerated, in order for any "before" or "after" relationship to obtain, there must be some way of differentiating events. Long confinement in a monotonous environment which provides few such cues, as shown by the experience of Michel Siffre when confined in a cave alone for 63 days (Siffre, 1965), leads to underestimation of how much time has passed. It has been argued (Cohen, 1967) that he may have moved into a state approximating hibernation, but this is only to shift the onus for his perception of 63 days as having been an interval of 36 days from the set of events external to him to those internal. It is of interest to note in this regard that when metabolism is speeded up, as in fever (Hoagland, 1933; Bell, 1965) or under great stress (Grebe, 1962), so that more events occur per unit of external time, personal time is experienced as speeded up, environmental time is overestimated.

Those external markers which we create to measure time are similarly devices which produce discriminable events, whether these be the amount of sand in a glass, the movement of a hand around a dial face, or the peak-trough characteristics of a particular wave pattern. Mere succession, however, is insufficient to produce the sense of time. Not only must there be some capacity to delimit events, there must also be some capacity to recognize similarities between events. A counter may be set to be activated by the peak of a waveform, a conventional clock marks off points separated by equal intervals on the circumference of a circle, an hour-glass is turned over whenever the upper chamber is empty. Organisms similarly delimit events, even though, as Barker and Wright (1951) found when trying to count the acts of a young boy in the course of a day, the events may overlap when approached with any limited set of criteria. I have been informed by a number of individuals who have experimented with marijuana that among the most frequently occurring changes in time perception under the influence of this drug is the experience of time as discontinuous, as divided into segments, as a function of changes in concomitant events.

The ability to perceive differences between successive events implies also the ability to perceive non-differences, or similarities. When dealing with discriminations among successive events, both imply the existence of memory and some mechanism for calling up and comparing the present stimulus with those that preceded it. If there were no way of telling that a similar stimulus had recurred, there would be no basis for judging duration. In light shows, in which all of the senses are being bombarded by random bursts of stimulation without any noticiable recurrence of previous stimulus situations, the sense of time is apt to be rapidly disorganized.

While similarity is necessary to an appreciation of time, it is not sufficient. If two sets of similar events are separated by unequal intervals, the longer interval can be discriminated from the shorter. Moreover, studies with the *tau* and *kappa* effects (Cohen, 1967) show an interdependence between time and distance such that the longer it takes to traverse a given space, the greater the distance will be adjudged, and the faster a given space is traversed, the more its length is underestimated. These data can be explained if it is assumed that the amount of stimulation which the organism receives is greater as a result of paying attention for a longer interval and so receiving more stimulation. An analogous situation occurs whenever we take a trip for the first time to a place where we have never been, down unfamiliar roads. As we continue to go to that place, and the roads become more familiar, we need to pay less attention to the route and the experienced time for the trip seems to be less.

For time to be judged properly, some constant source and pattern of stimuli is required which is relatively invariant. Such a source is provided by the stimuli which we call the physiological and metabolic rhythms of the body as they change in interaction with the inner and outer environments of the body. These are sufficiently repetitive, similar, and occur with sufficient frequency to provide a basis for all other time judgments. In this connection, Efron (1964) has shown a relationship between length of evoked potentials and the reports of subjects on the duration of a light flash, and Thor and Hoats (1968) have shown a circadian periodicity in the rate with which a rat will press a bar for light when all other external cues to the passage of time have been controlled. When these stimuli increase in frequency, a situation occurs analogous to that which is created in other sensory domains when judgments of increased intensity are reported, and the subject reports an experience of fast time. When they decrease in frequency, a situation analogous to a decrease in intensity takes place, and a situation of slowed time is reported.

When perceptions take place in terms of other sensory modalities, the pattern of stimulation is simultaneous. Recognition may involve referral to memory, but memory is not crucial to perception as is the case with respect to duration. Perception in terms of any other sensory modality is limited to patterns in that modality, even though the patterns themselves may be wondrously variegated. When I look at a scene from my window, I see shape, color, texture, motion, but I do not normally see sound, hear color, or smell touch, unless I am fortunate enough to have some synesthesia. Duration and the passage of time can be estimated in terms of any sensory modality, and it is as meaningful to ask, "How soon after you see the lightning, do you hear the thunder?" as it is to ask someone to estimate the interval between two bolts of lightning or the duration of a peal of thunder.

From the standpoint of any device for measuring time, what is being measured are those events comprising its own existence Priestley (1964, pp. 42–43) cites Hood as stating that the quartz clocks at the Greenwich observatory keep better time than the earth itself. Are they measuring the earth's time in any other sense than by social convention, or are they recording the events of their own existence? When we ourselves make judgments of time or of duration, what are we doing other than marking the correlation between the passage of external events and the passage of our own existence? Merleau-Ponty (1962, p. 414) says that "consciousness unfolds or constitutes time." Gooddy (1966) notes that we each have our individual personal time, although because of the similarity in our structures, the scope of variation is comparatively small as compared to the range of living creatures.

Time is a useful social variable which not only regulates the flow of events in society, but is also useful in developing explanatory constructs for the way things function in the universe. When we consider ourselves as clocks, however, we are merely ticking off our own existences. We abstract the concept of time from that existence and reify it so that we seem to be passing our lives in it, without being able to say what it is we are passing our lives in. The concept of time makes sense only when considered as the measure of changes taking place in a larger system of which the system to which we are paying attention is a component. But even though organisms may not exist in time, they have a past, a present, and a future, and a variable rate at which their own existence passes.

It may well be that we structure the world in terms of our own body image. We are three-dimensional organisms characterized by a front-back, right-left, and top-bottom, and we structure space in terms of height, width, and depth. How might we structure space if we had four dimensions, or if we were spherical? In a study carried out in my laboratory (Aaronson and Mundschenk, 1968) on the spatial stereotypes of time, we found that in a sample of 226 college students, 67.7% chose depth as their first choice for being most like time. Width was the most popular second choice. When either first or second choices is considered, depth is the choice of 84% of the sample. The students overwhelmingly placed the past to the bottom, left, and back and the future to the top, right, and front, with the present running through them. These attitudes were so deeply ingrained that most of them could not say why they made their choices. A small but significant trend was observed for left-handed people to place future on the left and past on the right, although because of the small number of left-handers in the sample, this experiment should be repeated.

We tend to structure events in terms of the ways in which we experience them. In a previous experiment (Aaronson, 1968b), I hypnotized some subjects and altered the scale of the environment so that when they were awakened, they found themselves in a miniature world like that of Lilliput in *Gulliver's Travels,* or in a giant world like that of Brobdignag. Most subjects eventually solved their dilemma by growing or shrinking to the scale of the world in which they found themselves. It was distinctly easier for them to grow than to shrink. Most of us have had the experience of growing, few of us have had the experience of shrinking very much, and certainly not during young adulthood. In the same way, the experience we have of time determines the way we conceive of it. Keller (1969) has derived the Western view of time as a consequence of the subject-object distinction in its phenomenal world view. Most of us experience our existence unitarily and linearly, and as a result build a construct of time which is unitary and linear, even though a multidimensional nonlinear concept might raise a number of interesting questions.

As organisms, our existence involves the reception, processing, and response to stimuli which come to us from without or within. We call these activities, *the present*. We receive some of these stimuli from our memory banks and constantly refer new stimuli to be stored there and to be compared with what has been already stored there to ascertain meanings and to help determine appropriate responses. We call what is stored there, *the past*. Although we tend to think of the past as fixed, there is evidence that memories do change (McGaugh, 1966) and that depending on what the present is like, we may select different sets of comparison experiences from the store. When we are happy, we tend to refer to happy experiences, when we are unhappy, or angry, we are likely to choose appropriate experiences to buttress these moods. On the basis of the present and the past, we set up goals, deadlines and expectations about positive and negative events which may take place, and may even program sets of possible responses for these eventualities. We call these likelihoods, *the future*. The future is set up on the basis of what is going on in the present and what has gone on in the past, but once a concept of future has been set up, it may mold the present and even influence the kinds of memories chosen from the past. In its turn, the future can be changed by events in the present or even memories from the past. All three areas are in constant dynamic interaction, for "time is the one single movement appropriate to itself in all its parts" (Merleau-Ponty, 1962, p. 419).

The influence of the past and present on one another is easily shown in a pilot study carried out in my laboratory. We routinely collect the three earliest memories of all of our subjects. In this study, the subject was hypnotized and regressed in time to that of an early memory in which he was rejected. The subject was then led to relive the memory, but the end was changed so that he was accepted. He was then brought back to the present, and awakened. Originally diffident, insecure, and determined to please, he now became aggressive and self-confident. He had an acute sense of his own worth and his own rights. His performance on a battery of tests which he took before and after the reliving of the early memory was totally different. When asked to write down his three earliest memories, he produced three different memories than his initial set. The original set had expressed themes of loss and ambivalence. The new set expressed themes of positive self-regard and positive object relationships, and a sense that people went out of their way to express positive feelings toward him. He resisted mightily when the time came to change his memory back to its original form.

The influence of the future on the past is shown in any writing of a biography, for biography is not a chronicle of incidents. Dewey (1938) has pointed out that when the biography of a great man is written, the biographer seeks to derive, in the interplay of successive incidents how his greatness emerged. It is important to know that Lincoln was born in a long cabin only because he was a great Civil War president. In fact, we use the future to influence the past whenever we review our past experience for justification for future action or whenever, indeed, we appeal to precedent and call up those memories that make what we want to do, right.

Von Schelling (1964) has attempted to apply to space perception and time awareness the kind of joining that Einstein's special theory of relativity has performed for space and time in physics. He concludes that a double cone irradiates out from the present to the "active future and the passive past." Without agreeing with his concept of the past as "passive," it follows from his model that living takes place in the present, and that as one moves away from the present to the far reaches of the past and the future, influences and events become more vague and amorphous. Whatever the influences may be, however, different individuals and even different cultures stress different members of this past-present-future trio differently in the way in which they handle their existence. In a study of the way in which different groups orient themselves in time, for instance, Kluckhohn and Strodtbeck (1961) found that Spanish-Americans preferred to live in the present more than in the future, and in the future more than in the past. Anglo-Americans preferred the present and future about equally, and did not like the past. Zuni Indians preferred the present and past about equally over the future, while Navajo Indians liked the present alone. The consequences for behavior are not set forth, although the achievement orientation of the Anglos and the conservatism of the Zuni instantly suggest themselves. The question of how a focus on any of these three areas might affect behavior remains unanswered.

*Method*

In order to ascertain the behavioral consequences of different time stances, four male college students, ranging in age from 21 to 24, were hypnotized and given post-hypnotic suggestions expanding or eliminating the past, present, and future, alone or in combination. Three were trained hypnotic subjects, and one was a simulator, chosen for this task after extensive attempts to hypnotize him had failed to produce anything deeper than a light trance. The special task of the simulator was to act out the suggestions given as if he had been hypnotized. Only the experimenter knew that he was a simulator and not a true hypnotic subject. As will be seen, the role-playing itself put him into some sort of altered state of consciousness.

The data obtained with three of these subjects have been reported elsewhere (Aaronson, 1965, 1966a, 1967, 1968a), the fourth subject is new. Because of the catastrophic nature of the responses of one of the subjects, he was not used in all of these conditions.

The conditions included (1) *no past*, (2) *no present*, (3) *no future*, (4) *no past and present*, (5) *no past and future*, (6) *no present and future*, and (7) *control*. The contrasting set of conditions included (1) *expanded past*, (2) *expanded present*, (3) *expanded future*, (4) *expanded past and present*, (5) *expanded past and future*, (6) *expanded present and future*, and (7) *control*. The general hypnotic instruction for the ablation condition was, "Do you know how we divide time into the three categories of past, present, and future? (Hypnotist waits for subject to respond affirmatively.) When I wake you, the *(name or names of category)* will be gone. There will be no *(name or names of category)*." The general hypnotic instruction for the expansion conditions was, "Do you know how we divide time into the three categories of past, present, and future? (Hypnotist waits for subject to respond affirmatively.) When I wake you, the *(name or names of category)* will be expanded. The *(name or names of category)* will be expanded." In the *control* conditions the subjects were hypnotized, but no post-hypnotic suggestions of perceptual change were made. In presenting the conditions to the subjects, both ablation and expansion series were intermixed. No subject received any of the conditions in the same order.

Subjects were first given a battery of perceptual tests and completed a Q-sort of emotion adjectives based on Plutchik's theory of emotions (1962). They were then hypnotized and the post-hypnotic instructions were imposed. They were interviewed by the experimenter and then allowed two hours of free time, except for two subjects who were painters and were asked to paint a standard scene $1^1/_2$ hours afterward. The scene was the view from one of the windows of the experimental room. The subjects were then taken for a ride over a standard course and, on their return, asked to write an account of how their day had been. They were then interviewed by an outside observer, a psychologist or psychiatrist who was ignorant of the condition imposed. The subjects then took the Minnesota Multiphasic Personality Inventory (MMPI). One subject also completed figure drawings. The perceptual battery and Q-sort were readministered, the subject was interviewed by the hypnotist, and the suggestion was removed. The subject was interviewed again, and in the event that any negative state remained from his experiences was worked with until he felt in good shape. Three of the subjects, including the simulator, kept a diary of the events of the day and their responses to them. The simulator's diary stated what he had tried to do with the conditions and what had really happened to him.

*Results*

The conditions are presented here in purely arbitrary order for purposes of exposition. No subject was exposed to the conditions in the same order as any other. The results for each of the conditions are as follows:

*No Past.* The first subject became confused and disoriented. He maintained the memory of important persons, such as his wife, but could recall little of the specific events of his past. He seemed disinterested, bored, irritable, and given to mild verbal acting out. He seemed to lose inhibition and differentiation of meaning.

The second subject also lost the memory of many of the specific events of the past, while retaining the memory of important people, such as his wife. He became confused, unable to handle even simple questions, and spent as much time as possible asleep.

Fig. 1. Retrospective painting of experimental room as seen by third subject during the no past condition

The third subject lost all language and understanding. When I came near him, he seized me, buried his head, and mouthed like a small infant. He later said that he did not know who I was, but that he felt warmth emanating from me. He had no consciousness of self and all that he experienced was raw sensation.

He said he especially enjoyed smell, warmth, and touch. He liked the sound of my voice, but was frightened by the outside observer. The only color seemed to be golden-brown, except that at one time one of my eyes seemed to be green and yellow, with flecks of red, blue, and brown. (The experimental room is painted spice brown and my eyes are hazel.) He also said that everything looked strange in a way that he could not describe. He was asked to try to paint what he had seen, and produced a painting that suggested that he had been seeing contour lines (Fig. 1). He was questioned about this, and apparently did not know that studies on the development of visual perception in individuals previously congenitally blind (Senden, 1932) showed that it starts with contour lines from which more complex forms are subsequently differentiated.

The simulator felt that he could respond to this instruction by blocking past memories or by liberating himself from his past. He chose the latter, and responded with a feeling of rebirth. He responded to things as they were, not as previous experience had made them. He became oriented to action and involvement with others.

*Expanded Past.* The first subject became happy, but was experienced by all as more difficult to relate to. The accomodation that makes social intercourse possible was considerably lessened. As long as one wished to deal with what he happened to be interested in, everything went all right. If one attempted to deflect the conversation in any other direction, he ceased relating and went about his own affairs.

The second subject responded happily and spent the day reminiscing about his past. It was possible to turn the conversation to other things, but he quickly turned it back to his main area of preoccupation. As is true of most of us, his past was not entirely positive and he tried to avoid the sad memories by increasing his level of activity when these occurred.

The third subject reported feeling more relaxed and comfortable. He was mildly preoccupied with his past. His movements seemed slowed and he seemed rather introverted and taciturn. He did not seem to want to relate with just anybody, but with somebody who was interested in what he was interested in.

The simulator spent the day examining the way he now was as opposed to the way he used to be in the past as a result of bad conditionings imposed by neurotic authority figures. He, too, related well if one fell in with what he was interested in, but poorly if one wished to relate in terms of any other topics than those in which he was interested.

*No Present.* The first subject immediately became immobile and responded neither to his name nor any other stimulation. At the beginning, his eyes showed a slight tendency to flutter when his name was called, but this passed. An original mild waxy flexibility quickly became a catatonic-like rigidity. When I opened his eyelids manually, they remained in the position in which they were set. When he was stood up, he seemed a dead weight. He was balanced on his feet only with difficulty and began to fall over as soon as a breeze entered the room. When awakened, he described his experience as "a state of unbeing, like death." He had been aware of what was going on in the same way that a tape recorder might. He said that he had not even been frightened by the experience until he was brought out of it.

The second subject felt lonely and still, but not in a terribly bad mood when he could get himself to wake up enough. He spent most of the day asleep. He reported without emotion a constant vision of graveyards. In the post-session interview, he stated that he had allowed his sense of the present to lessen rather than let it go, because of a presentiment that something very bad would happen if the present disappeared.

The third subject didn't want to talk, but did so when pressured. He reported that he felt caught in a vise and that he had no relationship with objects around him or with people, who had become objects. He lost the sense of depth, and everything looked flattened and two-dimensional. Colors were pallid and uninteresting. His sense of his body boundaries was considerably weakened, and at one time he asked to have the door to the experimental room closed because he felt he was leaking away. He seemed withdrawn, depressed, childish, and hostile. He was extremely uncooperative, but on the drive in the car tried to rest his head on my shoulder, an uncharacteristic gesture for him. He tried to sleep but found that he could not do so, because of the repetitive thoughts that were going through his head.

The simulator reacted with good-natured aggressiveness. He felt that in the clash of hostility the masks and falseness that obscure being would be torn away. He felt that one must take violent action to affirm life in a world that denied it.

*Expanded Present.* The first subject became fascinated by lines and textures. He experienced a mood of great luminosity which he described as "mystical." He seemed very happy. He noted diminished concern about some problems that he was having outside of the laboratory.

The second subject became happy and relaxed. He felt much involved with the clarity of objects and sounds around him. On the car ride, he made me stop when we sighted a herd of cows in a field, and we spent a long time watching the cows chewing.

The third subject developed a sense of enhanced depth and a sense of seeing things unusually clearly and distinctly. He reported that he could hear colors and be aware of the shape of smells. He was in a rather exuberant mood and a bit disgruntled with the other people in the laboratory who were working and not enjoying what he felt was the most beautiful day of the Fall. He reported that time had slowed down and that he had a great deal of time to do whatever he had to do.

The simulator immersed himself totally in the experiences of the moment. Eventually the situation became more that he could bear, but he found himself unable to withdraw from it. At the end of the day he was still happy and active, but very tired.

*No Future.* The first subject entered a euphoric, semi-mystical state. Everything seemed to be occurring in a boundless, immanent present. He seemed totally free of anxiety and spent his day savoring the experiences of the present. He seemed interested in colors and textures, and involved himself in each interaction to its fullest. He seemed unusually easy to relate to.

The second subject became alert and interested in all that was going on. He found himself less anxious, as well as less motivated. He took things as they came and enjoyed himself.

The third subject felt that he had lost a whole portion of his identity which was based on what he was going to do and become in the future. He became tired, discouraged, depressed, and began to develop physical complaints and some minor distortions of body image. He felt that the physical world around him was far more real and solid than usual, but he and it had stopped growing. He also felt more in touch with his past. All of his ambitions had come to an end, and he found it extremely difficult to exist in a world where he had no ambitions.

The simulator became depressed at first, but this gave way to a mood of stoicism. He became rather philosophical and reported that he felt no anxiety, but also no anticipation of pleasures to come. He showed a loss of drive and seemed content just to sit.

*Expanded Future.* The first subject became happy, as he felt that for once there was ample time to meet all the environmental demands upon him. Deadlines became unimportant and he reported that, in particular, death became merely the end of life rather than an event to be feared. The realization of this elevated his mood even more.

The second subject looked forward to his life with happiness and anticipation. He felt that he could concentrate better and that he was more full and rich as a person.

The third subject was happy to begin with, and became happier. He described the experience as a bit like the opening of a good LSD trip. He felt fulfilled, calm, and with a sense that he had all the time he needed without pressure. He became more extroverted and saw everyone and everything around him literally bathed in a rosy light.

The simulator became expansive and introspective. He felt self-confident and able to meet any vicissitudes that might befall him. His life to this point did not seem to be the measure of what his life would be like. He felt there was ample time to take all experiences in stride.

*No Past, No Present.* Because of his response to the *no present* condition, this condition was not administered to the first subject.

The second subject became very optimistic and totally involved in future projects. In the interview following the end of the condition, the subject admitted that while he had let the past go to some degree, he had not let the present get away from him.

The third subject seemed comatose. He seemed relaxed but unresponsive, although there was a slight resistance to having his eyes opened manually. He himself described his state as being like a deep, dreamless sleep, with more complete muscular relaxation than was usual in sleep. Occasionally, he said, he would become aware of sounds which he experienced as mild pressures on his body.

The simulator felt that what was and what had been no longer had any relevance for him. He became totally involved in future plans and would deal and talk about nothing else. He fretted and fumed and paced restlessly, waiting for the time when he could put them into effect.

*Expanded Past, Expanded Present.* The first subject became concerned with the origin of things. Whenever he would look at anything, he would see in his mind's eye how it had looked at various stages in its history while he simultaneously thought how it had come about. He thought much about the superimposition of the transient on the permanent. He became joyful as he thought how at each moment he was "riding on the crest of history."

The second subject spent the day reminiscing nostalgically about his early life. This carried him forward into a concern with self-analysis and the gaining of understanding into himself.

The third subject was at first impressed by the beauty of things and by his sensory images. Gradually he seemed to become more withdrawn and lethargic. Finally, he just lay around compulsively craving companionship to lighten a dreary time.

The simulator became depressed to the point of seeming suicidal to the outside observer. He seemed withdrawn, self-pitying, and hypochondriacal. He felt the present was an extension of a past that he hated, totally determined by that past, and that he was trapped in it with no escape.

*No Past, No Future.* The first subject became rigid, as he had been during the *no present* condition. Subsequently he developed some mild waxy flexibility. His eyes never fluttered when his name was called. Apart from these two minor differences, his responses were identical with those obtained under the *no present* condition, down to his subjective report, and the unanimous description of outside observers that this was a state like catatonia. After the condition had been terminated and the situation worked through, he was asked to paint a picture showing what the state had been like. He produced a gloomy, mechanistic abstraction (Fig. 2) in answer to his own question, "How do you paint a picture of nothingness?"

The second subject again retained part of the time categories that had been presumably obliterated by the post-hypnotic instructions. He became bored, disinterested, seemed to lose memory and developed an almost fanatical

Fig. 2. Retrospective painting of state brought about by no past-no future condition in first subject

dedication to sleeping. There seemed to be as little inner life going on as there was any outer manifestation of life. He had no motivation for anything, although he seemed to be able to perk up briefly when his wife came in to see him.

The third subject became confused, paranoid, cold, and hostile. His speech became stilted, pompous, and circumlocutious. His affect seemed flattened and, at times, inappropriate. His sentences seemed to ramble and show looseness of associations. He seemed forgetful and not to know quite what was being asked of him.

The simulator became unaccountably sleepy, although he reported that he had slept sufficiently the night before. He napped for a while, but couldn't shake off the drowsiness. He seemed lethargic, uninterested, and uninvolved with his environment. He felt that everything tasted flat and he had no wish for even coffee or cigarettes. He seemed dependent, depressed, and passively hostile.

*Expanded Past, Expanded Future.* The first subject began ruminating extensively on the topic, whither have I been, whither am I drifting. He became very self-involved and spent the day reviewing his life. He became lost in reflection and obsessive thinking. At the end of the day, he remarked that he had been as close to being a philosopher as he could ever be.

The second subject showed little change. He reported that he had some difficulty concentrating because his thoughts were swallowed up by the past and future, but the only thing anybody else could see was that he seemed a little more passive and daydreaming.

The third subject became happy, warm, and relaxed. He felt that he was the center of a nexus of time. At first he had a sense of objects giving off energies and interpenetrating, but later on he denied any perceptual change. He showed some tendency to become hyperideational. At one point in the afternoon, he became very angry over a chance remark that he had overheard, which was not typical of him.

The simulator became very abstracted and felt himself swallowed up in an immensity of time and space, in which he was but an insignificant part. His ruminations became so profound and personal that it was hard for him to communicate. He said he felt suspended in a river, surrounded by mist, unable to see the shores or how it was flowing, and not caring.

*No Present, No Future.* Because of his response to the *no present* condition, this was not run with the first subject.

The second subject retained enough of the present and future to evade the full implications of this condition. On the other hand, he became trapped in an incident that had occurred when he was about seven, and wept bitterly and inconsolably about it for several hours. In manner and even physical appearance, he seemed like a young child at this time.

The third subject also resisted the suggestion and became rather childlike and reacted to everything like an early adolescent. His own feeling was that he had been behaving very much the way he used to when he was fourteen.

The simulator became sleepy and unhappy, for he felt hopelessly trapped in the past. All of his neurotic characteristics and defenses appeared in full strength and increased all day long.

*Expanded Present, Expanded Future.* The first subject described himself as being in a happy, mystical state. He felt he had triumphed over death by incorporating the future into the present. Behaviorally, after a period of his usual euphoria, he began to show pressure of speech, moved more rapidly and restlessly, and finally began to pin obscene slogans to the backsides of people in the laboratory. After he had gone off to relieve himself, he returned from the lavatory trailing ribbons of toilet paper with which he proceeded to festoon the doors of the laboratory. Subsequently it turned out that he had drawn a cartoon of a face on the mirror over the washbowl. When questioned about this behavior after the session, he reported that he had felt a gradual build up of tension all day which became considerably increased after the outside observer had led him to think about how he was feeling in the course of the interview.

The second subject had a happy, contemplative day, full of plans for the future. He described his experience as being similar to "a pot high." He was very much aware of sensory stimulation and enjoyed it through all the modalities to its fullest.

The third subject also experienced the day as like a psychedelic "high," although he also showed some pressure of speech. All sounds and colors were enhanced. He began to have experiences of synesthesia, in which every sound was seen in color, and every touch as shapes of light. He showed much stimulus hunger, and went from place to place in and around the laboratory to obtain new experiences.

The simulator displayed a mood of great optimism. He felt he could think unusually clearly and well and wanted to concern himself only with significant things and ideas. He reported a great deal of energy and his manner seemed rather grandiose. He tended to be rather aggressive in interpersonal contacts because of impatience with the trivia of such relationships.

*Controls.* No abnormality or alteration of behavior was shown by any subject on any control condition, except that the first subject became unaccountably vapid and withdrawn during his first control session. Neither of us could account for this behavior, except that it was there. It was not repeated by any other subject, nor by him during the subsequent control session.

*Discussion*

The data show the strong identification of the present with being alive. For all four subjects, the *no present* condition raised the issue of life and death. For two of the hypnotic subjects, major disorders of behavior appeared, the third withdrew into sleep. Expanding the present produced positive states in all four subjects, characterized by greater attention to stimuli. The often repeated injunction to "live in the present" becomes, in terms of these data, simply an admonition to be alive. The fact that all subjects paid more attention to ongoing stimulation under the *expanded present* instructions underscores the fact that as organisms we are primarily stimulus processors and that the sense of being alive is very much related to the kinds of stimulation that we receive. It should be noted that all the subjects sought out positive stimulation. Whether "aliveness" would also be involved in an affectively negative present is a moot point.

The schizophreniclike behavior and the withdrawal shown by the hypnotic subjects raises the question of whether schizophrenia may not be a psychic analogue to dying. It will be recalled that in the study on the spatial stereotypes of time, cited earlier (Aaronson and Mundschenk, *op. cit.*), the present was almost universally seen as running through the subjects. Fuse (1966) has postulated a breakdown in the connections between past, present, and future in catatonia, similar to what has been observed here, and a loss of time experience as related to the present in hebephrenia. Melges and Fougerousse (1966) also have observed fragmentation of temporal perspective in schizophrenic and severely delusional patients, with a greater focus on the present to the relative exclusion of past and future frames of reference on a personality questionnaire. Yaker and Franzblau (in press) in a study of verb tense usage by schizophrenics and normals, conclude that schizophrenics avoid using the present tense. The Yaker and Franzblau study is based on current behavior, that by Melges and Fougerousse on responses to a questionnaire. As the data in this study suggests, the present requires a past and a future interconnected with it to be meaningful, just as the past and future require a present in order to manifest themselves.

The past gives to the present meaning and inhibition. When the past was removed, the hypnotic subjects became confused and regressed in their behavior to a point, in the case of one, where he was close to babyhood in his responses. At its extreme, loss of the past involves a loss of sense of self and going back into the past, a movement back to an earlier self. This is clearly exemplified in the case of the two subjects who resisted the *no present, no future* instructions and found themselves caught at earlier ages in their life spans. The strong association of past with sense of self is also indicated by the narrowed span of egocentrically determined interests shown by all the subjects under *expanded past* instructions. As a source of definitions and a source of "what is me" and "what is not me," it follows that as the past lengthens, people should get more set in their ways. As it is also a source of increasing inhibition, it also follows that there should, in general, be an increasingly narrowed range of interests with age. Cumming and Henry (1961) have documented the increased self-involvement and narrowed interests with aging. The increased inhibition and rigidity has also been shown to be an important component of the popular stereotypes of personality change with aging (Aaronson, 1966b).

Because the past is the record of what has been experienced when past and present have been expanded together, the past has been seen as controlling. The orientation tended to be deterministic in those subjects who had happy childhoods, and fatalistic in those who thought their childhoods were unhappy. The bringing of the past into the present and seeing it as controlling the responses of the present is an operational definition of the psychoanalytic concept of transference. When the simulator responded to the *no past* by casting off the influence of the past on his present, he was casting off transference and responded with a sense of rebirth. Freud felt that the fundamental problem of psychoanalysis was to analyze away the transference which means that its fundamental problem is not to find some causal factor in the past for present behavior, but to eliminate the direct influence of that factor in the present. The behavior of the subjects here suggests that proper functioning is best served when the past exists in a consultative capacity to the present. When it dominates the present, the result is apt to be personality disorder.

The future is the source of ambitions, goals, and anxiety. Elimination of the future in these experiments was positive, insofar as it eliminated anxiety, negative insofar as it eliminated motivation. The future represents our estimate of what is likely to happen to us on the basis of what is happening in the present and what has happened in the past. If there is no future, many present activities become meaningless because they have been entered into for the promise of future reward or to avoid future punishment. It has been pointed out that individuals with a high achievement drive have a strong sense of the future (McClelland, 1951), but this is only to say that they are well-motivated. Melges and Bowlby (1969) have attributed sociopathy to the abandonment of the future and a seeking of goals in the present. Siegman (1961) has also shown lowered future time perspective among delinquents and correspondingly less impulse control in those with lower future time perspectives.

In order to produce the kinds of effects in the future that one desires, present behavior must be geared to produce those effects. Krauss (1967), who has made a strong case for the future direction of anxiety, has also shown with his collaborators (Krauss, *et al.*, 1967) that in normals anxiety produces greater attention to the present as a coping device, and that among patients moderate levels of anxiety are associated with greater attention to the present and future, while higher levels are associated with avoidance of the present and future (Ruiz and Krauss, 1968). The positive effects from expanding the future in these studies may derive from the general perception of the future as demanding rapid adjustment to the demands arising from the future, which was eliminated by the suggestion. Moreover, in all the conditions in which attention was directed to the future, there was also a concomitant increase in stimulus seeking and response output, as shown by the responses of the subjects in the *expanded present, expanded future* conditions especially.

That a present without a past and future is as inconceivable as a past and future without a present is strongly suggested by the great similarity of response to these two conditions. Both produce a preoccupation with death, both produce schizophreniform behavior. It will be noted that the studies cited earlier of changes in time perspective associated with schizophrenia apply equally as well to the *no past, no future* condition as they did to the *no present*. The implication may be that life, at least as it is lived on a human level, must carry with it some sense of direction, either from past to present, or present to future. The loss of a sense of being on the part of one subject and the grappling with situations involving death by all subjects during the conditions in which the present alone and the past and future together were eliminated supports the contentions of J. W. Dunne (1938, 1939) that when time is conceived in terms of the concepts of past, present, and future, it partakes of the character of an infinite series. To eliminate a term like the present, or terms like the future and past simultaneously, so that the remaining terms stand alone, robs those terms of all meaning.

The expansion of past and future produced some quality of hyperideation in all subjects and obsessiveness in two. The concern with past and future experience without the mediation of the present seems associated with increased cognition, but decreased motor response which might implement one's preoccupations. This represents the classic dilemma of the obsessive, who knows what he should do, but becomes so lost in means and ends that he does very little. Cottle (1967) has pointed out that a greater emphasis on past and future is characteristic of males rather than females. This is interesting in view of the observation that on the Minnesota Multiphasic Personality Inventory, Scale 7, or Psychasthenia, the scale that would be most typically elevated in obsessive-compulsive syndromes, is also one of the scales most typically elevated in males (Aaronson, 1958). Brown (1966) identifies the direct effect of the past on the future with the concept of Karma, the Eastern view that what happens to us now is the result of what past experiences have made of us. One traditional way of overcoming Karma is simply to stop doing those things that attract the influences you don't want, which means to act in the present. Melges and Bowlby (*op. cit.*) feel that depression arises as a direct effect of the past on the future in intensely future-oriented people. Here, too, the individual can remain caught in his conflict unless other behaviors in the present help create a new orientation.

If time, as was argued earlier, is existence, every mode of orienting to the positions of time has its existential consequences. Some of these have been set forth here. There are obviously others that have not been touched upon. There are other ways of conceptualizing time than those employed here. For instance, what if time were conceived as a set instead of a line? In any event, a full description of the psychological consequences of time requires a complete theory of behavior. And a complete theory of behavior requires a variable like – – – time?

The writer would like to thank Drs. A. Moneim El-Meligi, Frank Haronian, Richard I. Jontry, Harriet Mann, Humphry Osmond, Stanley R. Platman, Hubert Stolberg, and A. Arthur Sugerman for their assistance in this study.

## *References*

Aaronson, B. S.: *Age and sex influences on MMPI profile peak distributions in an abnormal population.* J. consult. Psychol. 22 (1958) 203–206.
— *Hypnosis, being, and the conceptual categories of time.* Presented at meetings of the New Jersey Psychological Association, Princeton, N.J. 1965.
— *Behavior and the place names of time.* Am. J. Clin. Hypn. 9 (1966a) 1–17.
— *Personality stereotypes of aging.* J. Gerontol. 21 (1966b) 458–462.
— *Hypnosis, responsibility, and the boundaries of self.* Am. J. Clin. Hypn. 9 (1967) 229–246.
— *Hypnotic alterations of space and time.* Int. J. Parapsychol. 10 (1968a) 5–36.
— *Lilliput and Brobdignag – self and world.* Am. J. Clin. Hypn. 10 (1968b) 160–166.
—, Mundschenk, P.: *Some spatial stereotypes of time.* Eastern Psychological Association meetings, Washington, D.C., 1968.

Barker, H. G., Wright, H. F.: *One boy's day: A speciment of behavior.* Hamden, Conn.: Shoe String Press 1951.
Bell, C. R.: *Time estimation and increases in body temperature.* J. exp. Psychol. 70 (1965) 232–234.
Brown, N. O.: *Love's body.* New York: Random House 1966.
Cohen, J.: *Psychological time in health and disease.* Springfield, Ill.: Charles C. Thomas 1967.
Cottle, T. J.: *The circles test: An investigation of perception of temporal relatedness and dominance.* J. proj. Tech. and Pers. Assess. 31 (1967) 58–71.
Cumming, E., Henry, W. E.: *Growing old.* New York: Basic Books 1961.
Dewey, J.: *Time and individuality.* In: *Time and its mysteries.* New York: Collier 1962.
Dunne, J. W.: *This serial universe.* London: MacMillan 1938.
— *An experiment with time.* 3d Ed. London: Faber and Faber 1939.
Efron, R.: *Artificial synthesis of evoked responses to light flash.* Ann. N.Y. Acad. Sci. 112 (1964) 292–304.
Fuse, K.: *An approach to schizophrenics. III. Time experiences of schizophrenics.* Hirosaki Med. J. 18 (1966) 75–86.
Gooddy, W.: *Disorders of orientation in space-time.* Brit. J. Psychiat. 112 (1966) 661–670.
Grebe, J. J.: *Time: Its breadth and depth in biological rhythms.* Ann. N.Y. Acad. Sci. 112 (1962) 1206–1210.
Hoagland, H.: *The physiological control of judgments of duration: Evidence for a chemical clock.* J. gen. Psychol. 9 (1933) 267–287.
Keller, I. J.: *Self and world. The one and the many.* Unpublished bachelor's thesis. Princeton University, Princeton, N.J. 1969.
Kluckhohn, F., Strodbeck, F. L.: *Variations in value orientations.* Evanston, Ill.: Row, Peterson 1961.
Krauss, H. H.: *Anxiety: The dread of a future event.* J. individ. Psychol. 23 (1967) 88–93.
—, Ruiz, R. A., Mozkierz, G. J., Button, J.: *Anxiety and temporal perspective among normals in a stressful life situation.* Psychol. Rep. 21 (1967) 721–724.
McClelland, D. C.: *Personality.* New York: William Sloane Associates 1951.
McGaugh, J. L.: *Time-dependent processes in memory storage.* Science 153 (1966) 1351–1358.
Melges, F. T., Bowlby, J.: *Types of hopelessness in psychopathological process.* Arch. gen. Psychiat. 20 (1969) 690–699.
—, Fougerousse, Jr., C. E.: *Time sense, emotions, and acute mental illness.* J. Psychiat. Res. 4 (1966) 127–140.
Merleau-Ponty, M.: *Phenomenology of perception.* London: Routledge and Kegan Paul 1962. (Translated by C. Smith.)
Plutchik, R. E.: *The emotions: Facts, theories, and a new model.* New York: Random House 1962.
Priestley, J. B.: *Man and time.* Garden City, N.Y.: Doubleday 1964.
Ruiz, R. A., Krauss, H. H.: *Anxiety, temporal perspective and item content of the incomplete thoughts test (ITT).* J. clin. Psychol. 24 (1968) 70–72.
Schelling, H. von: *Experienced space and time.* In Schaeffer, K. E. (Ed.) *Bioastronautics.* New York: Macmillan 1964.
Senden, M. von: *Raum- und Gestaltauffassung bei operierten Blindgeborenen vor und nach der Operation.* Leipzig: Barth 1932.
Siegman, A. W.: *The relationship between future time perspective, time estimation, and impulse control in a group of young offenders and in a control group.* J. consult. Psychol. 25 (1961) 470–475.
Siffre, M.: *Beyond time.* London: Chatto & Windus 1965.
Thor, D. H., Hoats, D. L.: *A circadian variable in self-exposure to light by the rat.* Psychonom. Sci. 12 (1968) 1–2.
Yaker, H. M., Franzblau, R.: *The perception of time and disturbed behavior.* In Yaker, H. M., Osmond, H. and Cheek, F. E. (Eds.) *Man's place in time.* New York: Doubleday. In press.

# Personality and the Psychology of Time

R. H. Knapp *

The purpose of this paper is to discuss several methods which we have evolved for the study of individual differences in viewing and accommodating oneself to time. As such, this paper ought to be considered an exploration of individual personality differences. But before reporting on these studies, I should like to comment on the problem of time itself as a phenomenological dimension of man's experience.

With our physical and mathematical scientists, I am quite obliged to recognize that time may not be considered independently of space. Indeed, since the formulation of the ancient paradoxes of Zeno, I think philosophers have well recognized this fact. But the final precise relation of these two "extensities" by which man gauges and fixes his experience has not, to my best knowledge, been definitely and universally agreed upon, and the paradoxes of Zeno have commanded reexamination in our own time for the puzzles they contain.[1] Be this as it may, we must recognize that for most of mankind, the two "extensities" of time and space are separately viewed. Each extensity has its own terminology and vocabulary, each has its own units of measurement, and each confronts man with seemingly different perplexities.

Space, for instance, has been commonly viewed as defined by three orthogonal parameters, where time has but one. Space is probably, in a sense, a more benign medium. One can seemingly (but, of course, not really) move, stop, rest, or return in space, but time offers no such convenience. One may flee his position in space or remain standing, but time moves inexorably and in only one direction. Space has its geometry, time none, as Piaget oberves.[2] Thus I think we may say that time is the more demanding and vexatious of the two two primary extensities of our environment.

Philogenetically, too, I suspect that the management of time has come to living organisms more recently than spatial orientation, organic schedules in lower life forms notwithstanding. Thus the capacity for delayed responses is

---

\* Professor Robert H. Knapp, Ph. D., Wesleyan University, Psychological Laboratory, Middletown, Connecticut 06457, U.S.A.

1 Whitehead, A. N., Russell, B.: *Principia Mathematica*. Cambridge, England: Cambridge Univ. Press 1925.

2 Piaget, J.: *Le développement de la notion de temps chez l'enfant*. Paris: Presses Universitaire de France 1946.

sharply limited in infrahuman species as compared with man, while spatial orientation and manipulation seem well and even fantastically developed at the infrahuman level, as in the flight of birds and even insects. As Piaget has observed, the sense of time emerges but slowly in childhood[3] in consequence of successive and laborious abstraction. And, of course, we have no organ specifically for sensing time. So it would seem that man's orientation to time is a peculiarly recent and troublesome aspect of his worldly adjustment.

If, in our reality-oriented life, time is the most vexing and demanding of our two phenomenological extensities, then, when outer controls begin to fail, time may lose its tyranny. Freud has insisted that the world of dreams knows not time, and M. Bonaparte[4] has likewise insisted that time becomes ever more plastic and obedient to the pleasure principle as we regress from full ego control. In these circumstances of dreams, fantasy, hallucination, and the arts, time indeed may stand still or even reverse itself as though it then obeys our deep wish that it assume the flexibility of spatial extensity and yield to the relative mastery by which we command space.

I have undertaken a few half-whimsical examinations of the usages of the words "time" and "space" which may amuse you somewhat. First of all, I have examined the periodic literature in Psychology to see which has commanded the most interest from psychologists.[5] In point of fact, the *Cumulated Subject Index to Psychological Abstracts* lists an almost equal number under the two classifications if we admit "depth perception" to constitute a study of space. The incidence of the two words found in Lorge-Thorndike lists,[6] shows "time" among the first five hundred and "space" among the first thousand of the words used in common English prose. We cannot, from this date, set a precise ratio for the two, save to observe that both are in very common usage and that "time" is not overwhelmingly favored.

Our third excursion in this inquiry lay in the examination of well-established anthologies of famous quotations and aphorisms. In Bartlett's,[7] there are 392 quotations ascribed to "time" as against only 28 for "space", a ratio of about 14 to 1. Stevenson[8] yields an even more disparate result of 450 to 9, or a ratio of about 50 to 1. *The Oxford Dictionary of Quotations*[9] reveals 296 to 12 or a ratio of 24 to 1. Other less well known sources yield similar or more extreme ratios. The point seems established that, at least in our Western culture, the dimension of time is a singular focus of fascination, vexation, and curiosity, which

---

3 *Ibid.*
4 Bonaparte, M.: *L'inconscient et le temps.* Rev. Franc. Psychanal., (1939) 11 61–105.
5 *Cumulated Subject Index to Psychological Abstracts.* Vols. 1 and 2 (1927–1960), 1966, and 1st Supplement (1961–1965) 1968, published Boston: G. K. Hall and Co.
6 Thorndike, E. L., Lorge, I.: *The Teacher's Word Book of 30,000 Words.* N. Y. Teachers College: Columbia Univ. 1944.
7 *Bartlett's Familiar Quotations.* 14th Ed., Boston: Little, Brown 1968.
8 *Stevenson's Home Book of Quotations.* 8th Ed. New York: Dodd, Mead 1956.
9 *Oxford Dictionary of Quotations.* 2nd Ed. New York: Oxford University Press 1953.

draws ever the attention of our most sensitive poets, philosophers, and aphorists. We have not come, it would seem, to some settled and comfortable peace with this provocative aspect of our existence.

If this be so, and I believe that it is, then men and cultures should still be striving for some final adjustment to the experience of time, and differences, both cultural and individual, should readily be found. Few would doubt the existence of cultural differences in attitudes toward time. Spengler[10] devotes, for example, a large and bountifully illustrated discussion of the different time orientation of our own and classical civilization. Apparently, the day was divided only into hours in the age of Pericles; sun dials were not introduced until middle and late Hellenistic times. He contrasts this with the western time sense which almost from the first, established itself in the life of the Western monastery. By the seventeenth century, many thousand of clocks were ticking forth their temporal messages; bells, unknown to the ancients, tolled forth their hours and even quarter hours, and the stage was set for our time-obsessed scientific and industrial revolution. Spengler, rightly, I believe, sees in these two contrasting time attitudes, a quintessential difference in the *psychology* of the two civilizations.

Anthropologists, to be sure, have not neglected the time orientation of cultures and, indeed, have considered it one of the prime parameters upon which cultural evaluations may be reckoned. Beginning with the work of Florence Kluckhohn[11], time, as a basic dimension of culture, is directly related to social status, familial structure, economic forms, etc. Green's paper presented in the same series (cf. Studium Generale 23 (1970) 571—586) is essentially a cultural study of the time sense of black Africa, with its very significant message concerning the entire structure of the African life style.

But our first interest here is in the individual differences in "time" awareness and the manner in which measurements of such differences relate to each other and to interdependent aspects of personality and temperament. Let me first describe the measures we have evolved or employed.

I. *The Metaphor Scales:* Here we assembled a number of metaphors that might be taken as representative of time in poetic or other metaphorical allusion. They were then rated by a substantial number of persons according to the individual's sense of their appropriateness. From this data we identified, mathematically, three primary clusters; first, the *vectorial*, or images involving rapid motion in one direction, i.e., a fleeing thief, a racing locomotive; second, the *oceanic*, i.e., a vast expanse of sky, a quiet ocean; and third, the *humanistic*, involving figures of human persons or artifacts, i.e., a staircase, a revolving wheel, or an old woman spinning.

---

10 Spengler, O.: *The Decline of the West*. New York: Knopf 1927—28.
11 Kluckhohn, R.: *Dominant and Variant Value Orientations*. Chapter 21 in Kluckhohn, Clyde, Murray, H. A., Schneider, D. M.: *Personality in Nature, Society, and Culture*. New York: Knopf 1956 (2nd Ed.).

II. *The Anticipator Test:* This test was applied in two forms, first with a clock face and again with a tube in which the fluid was slowly elevated. In any event, the test consisted of permitting the subject to observe the constant movement of the hand (or rising water) and then asking him to estimate when the moving element would reach a given mark after it had passed behind an obscuring shield. It was scored by observing the accuracy of the judgments made and the subject's tendency to fall short of, or pass, the target.

III. *The Time Questionnaire Instrument:* This consists of a series of questions referring to personal attitudes and practices respecting time and was administered to a substantial sample of students. It was factor analysed to yield two primary factors. The first expressed a feeling of harassment with the management of time, together with an effort to control it period. The second factor expressed attitude of "time efficiency vs. time obliviousness".

IV. *The Metronome Test:* Here the subject was given an electrical metronome and asked, alternately beginning at a very rapid tempo, and then a very slow tempo, to adjust it until the frequency was most pleasing or "comfortable." The procedure was repeated several times and a score obtained showing the average tempo preferred by the subject as well as the "anchorage" effect of starting at a higher or lower frequency.

V. *The Musical Minutes Test:* In this particular test, the subject was given an illustrative example of a 20-second interval. Then he was instructed to mark off with the electrical key, such intervals for several minutes while listening to music. He was scored by the accuracy of his judgments and his tendency (was in most cases) to increase the interval with time or decrease it.

Let us start by reviewing those studies employing the time metaphor scales. The prototype of these was done by Garbutt and myself in which we employed not metaphors, but paired antithetical adjectives.[12] We required that each of our 75 student subjects rate the concept of time on each of 12 such scales. Moreover, we have previously determined, by McClelland's Method, the level of each subject's achievement motivation. We next established a correlation matrix relating our 12 pairs of adjectives to the need-achievement measure and to each other. Upon examining the results, we find that need achievement was found to be positively associated with descriptions of time as "active, fast, tense, hot, happy," etc., and negatively to such qualities as "passive, dull, slow, relaxed," etc. It is, in short, clear that with subjects possessing high achievement motivation, time is identified as a vigorous and dynamic quality imbued with energy, and not as an inert, passive conception.

These results led us immediately to develop the metaphor scale previously described,[13] which was reduced to 25 items breaking out factorially into our

---

12 Knapp, R. H., Garbutt, J. T.: *Variation in Time Descriptions and Need Achievement.* Journal of Soc. Psych. 67 (1965) 269–272.
13 Knapp, R. H.: *A Study of the Metaphor.* Journal of Projective Techniques 24 (1960) 389–395.

vectorial, oceanic, and humanistic clusters. This study involved the measurement of metaphor preference in relation to achievement motivation as described by McClelland from the Thematic Apperception Test. Here again we confirmed our essential finding, that preference for vectorial images correlates positively with achievement motivation, while preference for oceanic and humanistic metaphors yields negative correlations.

We may pause here to observe that vectorial images of time seem to be particularly characteristic of our Western civilization in its more recent centuries. Elizabethan literature, and particularly Shakespeare, is shot through with images of time as a thief or usurper, and it is surely not a coincidence that Newtonian physics conceives of time vectorially and unidirectively. Indeed, this entire attitude toward time seems more or less a precondition of the development of Western European physics and its immediate child, the Industrial Revolution and modern science. The oceanic image of time, on the other hand, seems to be identified with Asian religions and, above all, with Buddhism, whose stress on timeless meditation is distinctive and characteristic. The third imagery complex, namely, the Humanistic, is less definable in terms of philosophic origin, but one might suggest that it be identified with classical Mediterranean thought and perhaps, more specifically, with the concept by Protagoras, that "man is the measure of all things."

A further study involving the Time Metaphor Scale was conducted with Paul Lapuc.[14] In this study we again presented the 25 time metaphors and required our subjects to rate them for their appropriateness. Next, we administered the Meyers-Briggs Type Indicator. We also requested that they produce a protocol describing how they would conduct themselves in solitary confinement, in order to maintain psychological integrity, emotional stability, and physical personal well-being. A large number of our subjects outlined various physical exercises and regimes that would insure their well-being, but a substantial number of them spoke of intellectual exercises or moral disciplines designed to occupy their mental life or to prevent despair and personal collapse. We were able to show that persons preferring oceanic images of time did, indeed, score highly on the Meyers-Briggs introversion scale, and it was this group, too, that was notable for its refuge in moral and intellectual defenses in the circumstances described.

Studies involving the Anticipator Test are two in number: the first was conducted by Helen B. Green and myself at Wesleyan and involved a tube apparatus described earlier.[15] As independent variables, we employed two additional tests, namely, the Tartan test of aesthetic preferences, and an Events

---

14 Knapp, R. H., Lapuc, P. S.: *Time imagery, introversion and fantasied preoccupation in simulated isolation.* Perceptual and Motor Skills 20 (1965) 327–330.

15 Green, H. B., Knapp, R. H.: *Time judgment, aesthetic preference, and need for achievement.* Journal of Abnormal and Social Psychology 58 (1959) 140–142.

test, in which the subject was to estimate the proximity of past events to the present, i.e., the outbreak of the Korean war, the health of Stalin, etc., as well as the likelihood that future attainments would soon be realized. We were able here to demonstrate that the tendency to underestimate time for the rising water in the tube to reach an established mark was clearly correlated with high achievement motivation. Moreover, the tendency to recall past events as more recent and future events as more imminent was also positively correlated with achievement motivation and preference for somber, blue-green tartans. Nor should we neglect to note in passing, that preference for such tartan designs had been shown earlier to be related to high achievement motivation. In our conclusion, we observed, "Such persons (scoring high in these three variables) wish to create in the psychological present a sort of event density." We speculate that the future is already upon them while the past has not yet dropped away, and that the universe confronting them is teeming with opportunities for manipulation and achievement. We may tentatively propose a dynamic complex relating parsimonious time attitudes, asceticism of aesthetic taste, achievement motivation, and the perception of "event density." This syndrome may be identified with the character qualities associated with Weber's Protestant **Ethic** and the rise of European industrialism and entrepreneurship.

The next inquiry to be reported[16] relates the first factor identified on our time questionnaire, **as reported earlier,** to three independent variables, namely, aesthetic preference on the Tartan test, "science interest" and the underestimation of time intervals on our anticipator test, as inferred from performance on our Tartan instrument. The correlations between performances on our first factor which define "time harassment versus Olympian indifference" are all positive to a highly secure degree with the three independent measures, need achievement, "science interest," and underestimation of time as the anticipator. The second **factor,** describing more strictly actual practices rather than sentiments about **time,** does not seem significantly involved with the independent variables reported here.

The *musical minutes test* described earlier[17] was applied to 77 student subjects who had already been measured on achievement motivation by the protective device employed by McClelland and his associates. Helen B. Green and I divided our subjects into a high, middle, and low group, and then required them to mark off successive intervals of 20 seconds after an example had been given them. To provide distraction, loud music was played during the period of their performance.

---

16 Knapp, R. H.: *Attitudes toward time and aesthetic choice.* Journal of Social Psychology 56 (1962) 79—87.
17 Knapp, R. H., Green, H. B.: *The judgment of music-filled intervals and $n$ achievement.* Journal of Social Psychology 54 (1961) 263—267.

Those high in achievement showed a significantly more faithful retention of the 20-second interval under these conditions of distraction, while those in the low-achievement group appeared to succumb to the musical distraction so that the length of their estimates progressively increased, a difference widely exceeding the 1 percent level of confidence.

**The Metronome Test has already been** simply described and has been employed in our research to a very limited degree. Its simplicity, at least, commends it, and would appear to determine the preferred rhythmatic tempo for different temperaments.[18] One of my students has shown that aesthetic preference for slow tempos is related to preference for bland foods, low impulsivity and tenacity of purpose, but such results may be regarded as only tentative. That the andante-allegro preference between cultures may be of real significance I can hardly doubt. I am immediately mindful of the andante tempo of the Puritan hymn in contrast to the vivacity of our Southern music in all that it conveys of the Weltanschauung of these two cultures.

In only one study have we made an effort to give to a group of subjects several of our time tests simultaneously. In this case we administered the metaphor test, the metronome test, and musical minutes test and the anticipator to 45 mature scientists being tested at the Institute for Personality Assessment in Berkeley.[19] The selection of subjects was, for general **purposes,** psychological, perhaps unfortunate because they constituted a highly talented and homogeneous body of engineers. Still, it proved possible to compare for the same subjects their performance on these four tests, and further to examine the relationship of these performances to scores on the California Personality Inventory for which each subject had been scored. It will come as no surprise to learn that preference for vectorial images was generally conspicuous in this group. Still, this preference correlated with the tendency for premature halting of the anticipator, and stability of time judgment on the musical minutes tests. Preference for oceanic time images are associated with overshooting the target on the anticipator and increase of judged intervals while listening to music. But an examination of these correlations and a review of some comparable studies in the recent literature **suggest** that attitudes toward time and per-**formance on time tests** cannot be explained by ordering on some simple continuum.

The foregoing has presented a variety of small and diverse studies, all related, to be sure, to personality attributes and time awareness, but quite unsystematically designed **and administered.** It is clear that a bold and coherent study is **in order, with** a large and well-controlled body of subjects relating such measures to stable personality and temperament dimensions. Such has

---

18 Unpublished study done at the Institute of Personality Assessment and Research, Univ. of California (Berkeley).

19 Unpublished study done at the Institute of Personality Assessment and Research, Univ. of California (Berkeley).

not yet been accomplished. But the results I have presented should, I believe, assure you of both the worth and feasibility of such a study.

May I conclude with a final thought on the problem of time in relation to personality structure. We have, in the past half century or more, fallen very heavily under the influence of a climate of thought which makes the interpersonal exchanges of infancy and childhood primary determinants of the emerging adult. I doubt not the truth of much of this, for man is surely a social animal. But man is also a metaphysical animal harboring basic and important assumptions concerning the nature of time, space, causality and choice. These are of profound significance in the evolution of self-awareness and man's self-image. Piaget alone has fully realized this. His thought has come to us as a sagacious corrective. But his main effort has not been directed to the study of individual differences in personality organization to which we are especially committed.

I would hope that a new thrust in personality research might be directed, especially in the light of new existential developments, toward the examination of the individual's orientation to time and space, those primary extensities in which the individual finds his metaphysical orientation. Both of these take on vital, yet highly varied interpretations whose influence on the personality structure is profound. Such a development in our sciences would also restore to philosophy and metaphysics their importance for the interpretation of personal individuality, largely neglected, I believe, in the last century of personality research.

# The Notion of the Present

A. N. Prior †

Before directly discussing the notion of the present, I want to discuss the notion of the real. These two concepts are closely connected; indeed on my view they are one and the same concept, and the present simply *is* the real considered in relation to two particular species of unreality, namely the past and the future. So let's begin with the real in general.

Philosophers often speak as if the real world were just one of a number of different big boxes in which various things go on, the other boxes having such labels as "the mind" or "the world of Greek mythology". For example, centaurs exist in the world of Greek mythology but not in the real world, aeroplanes exist in the real world but not in the world of Greek mythology, and horses and men exist both in the real world and in the world of Greek mythology. Again, Anselm addresses himself to people who held that God does not exist in the real world but only in the mind, and claimed to have a proof that if God exists in the mind he must exist in the real world too. Leibniz contrasted the real or actual world with an infinity of merely possible worlds in which various things happen which do not happen in the actual world. All these ways of talking suggest that the real world or the actual world is just a *region* of some larger universe which contains other regions as well – possible worlds, imaginary worlds, and so on.

I want to suggest – I don't of course claim that there's anything original in this suggestion – that this way of conceiving the relation between the real and the unreal is profoundly mistaken and misleading. The most important way in which it is misleading is that it minimises, or makes a purely arbitrary matter, the vast and stark *difference* that there is between the real and every from of unreality. For talking of the real as one "region" among others immediately suggests the question, "In that case, what is so special about the real world in contrast with all other regions? – is it not a kind of narrow-mindedness and parochialism to think that it has anything special about it that none of the others have?" One philosopher, Meinong, has indeed said precisely that it *is* just narrow mindedness and parochialism to single out the real world as a region of special interest; the "prejudice in favour of the actual", he called it. Well, I want to argue that this is *not* just narrow-mindedness and parochialism, and that it becomes obvious enough what is so special about the real world as soon as we drop this metaphor of boxes or regions and become a little more literal.

To say that there are centaurs in the world of Greek mythology is surely *not* to say that there are centaurs in some remote and peculiar region, but just to say that *Greek myth-makers have said that* there are centaurs. Similarly, to say that there are centaurs in some person's mind is to say that *that person thinks or imagines that* there are centaurs. And to say that there are possible worlds in which there are centaurs is just to say that *it could be that* there are centaurs. In general, to say that $X$ is the case in some non-real world is just to say "$X$ is the case" with some modifying prefix like "Greek myth-makers have said that", "Jones imagines that", or "It could be that". But to say that $X$ is the case in the real or the actual world, or that it is really or actually or in fact the case, is just to say that it is the case – flat, and without any prefix whatever. To say that there are centaurs in the real world, for example, is not to say that there are centaurs in some region of the universe in which we happen to have more interest than in others; it is simply to say that *there are centaurs*. Talk of the real world, in other words, is not a metaphorical fudging-up of talk in which our sentences have a special kind of prefix, but a fudging-up of talk in which the relevant sentences have no prefixes at all. "Really", "actually", "in fact", "in the real world" are strictly *redundant* expressions – that, and not any prejudice or provincialism, is their specialness.

So to say that although there are no centaurs in the real world there are some in the world of Greek mythology, is just to say that *although there are no centaurs Greek myth-makers have said that there are;* to say that although God does not exist in reality he exists in the mind, is just to say that *although God does not exist people may imagine that he does;* to say that although Sextus raped Lucretia in the real world there is a possible world in which he didn't, is just to say that *although Sextus raped Lucretia he need not have done so*. There is, if you like, no other place than the real world for God or centaurs to exist in or for Sextus to rape Lucretia in; for God or centaurs to exist in the real world, or for Sextus to rape Lucretia in the real world, is just for God or centaurs to exist, or for Sextus to rape Lucretia. Again, "Greek myth-makers have said that there are centaurs in the real world" is all one with "Greek myth-makers have said that there are centaurs", and so is "Greek myth-makers in the real world have said that there are centaurs".

And now the present. It is tempting to think of the present as a region of the universe in which certain things happen, such as the war in Vietnam, and the past and the future as other regions in which other things happen, such as the battle of Hastings and men going to Mars. But to this picture there is the same objection as to the picture of the "real world" as a box or region among other boxes or regions. It doesn't bring out what is so *special* about the present; and to be more specific, it doesn't bring out the way in which the present is *real* and the past and future are not. And I want to suggest that the reality of the present consists in what the reality of anything else consists in, namely the absence of a qualifying prefix. To say that Whitrow's lecture is past is to say that *it has been the case* that Whitrow is lecturing. To say that Scott's lecture is future is to say that *it will be the case* that Scott is lecturing. But to say that my lecture is present is just to say that *I am lecturing* – flat, no prefixes. The pastness of an event, that is to say its having taken place, is not the same thing as the event itself; nor is its futurity; but the presentness of an event *is* just the event. The presentness of my lecturing, for instance, is just my lecturing. Moreover, just as a real thought of a centaur, and a thought of a real centaur, are both of them just a thought of a centaur, so the present pastness of Whitrow's lecture, and its past presentness, are both just its pastness. And conversely, its pastness is its present pastness, so that although Whitrow's lecture isn't now present and so isn't real, isn't a fact, nevertheless its pastness, its *having* taken place, *is* a present fact, *is* a reality, and will be one as long as time shall last.

Notoriously, much of what is present isn't present permanently; the present is a shifting, changing thing. That is only to say that much of what is the case, of what is real and true, is constantly changing. Not everything, of course; some things that are the case also have always been the case and will always be the case. I imagine scientists have a special interest in such things. And among the things that not only are the case but always have been and always will be, are the laws of change themselves, I mean such laws as that if anything *has* occurred then for ever after it *will have* occurred (like Whitrow's lecture). These are the laws of what is now called *tense logic*, and the conception of the present that I have just been suggesting is deeply embedded in the syntax of that discipline. So that conception underlies, or anyhow seems to underlie, what is now a pretty flourishing systematic enterprise. Precisely for this reason, it seems to me important that we tense-logicians should realise that there are difficulties about this conception of the present, arising either from physical science or from the philosophy of physical science. So I want now to state as clearly and crudely as I can what this difficulty appears to be.

Suppose we have observed on some very distant body a regularly repeating process of some sort, say a pulsation. We have just observed one of these pulsations, and as the body is a very distant one, we know that the pulsation we are observing happened some time ago. We now consider the pulsation immediately after the one we are observing, and we ask whether this next pulsation, although we won't of course observe it for a while, is in fact going on right now, or is really still to come, or has occurred already. On the view of presentness which I have been suggesting, this is *always* a sensible question. At least if there are to be any further pulsation at all, then either the body is pulsating, or it is not the case but will be the case that it is pulsating, or it is not the case but has been the case that it is pulsating. The difference between pulsating – really and actually pulsating – and merely having pulsated or being about to pulsate, is as clear and comprehensible a difference as any that we can think of, being but one facet of the great gulf that separates the real from the unreal, what is from what is not. Just this, however, is what the special theory of relativity appears to deny. If the distant body is having its $n$th pulsation as we perceive it having its $n$-1th –

*is* pulsating, and not merely has been or will be pulsating – then the *n*th pulsation and the perception of the *n*-1th are simultaneous; not just simultaneous from such and such a point of view or in such and such a frame of reference, but simultaneous. And according to the special theory of relativity, such "absolute" simultaneity is in many cases just not to be had.

One possible reaction to this situation, which to my mind is perfectly respectable though it isn't very fashionable, is to insist that all that physics has shown to be true or likely is that in some cases we can never *know*, we can never *physically find out*, whether something is actually happening or merely has happened or will happen. I'm sure there *are* questions which are perfectly genuine and intelligible questions but which seem to be incapable of being answered. For instance, I know perfectly well what it would be for you to see what I would call purple wherever I see red, and for you to see what I would call blue wherever I see purple, and so on round the clock; but I cannot imagine any procedure which would conclusively show that our respective visual experiences are, or that they are not, related in this way. And there may well be a similar but more subtle systematic impossibility in finding the answer to questions like my one about the distant pulsating body.

Furthermore, when confronted with unanswerable questions, it is often good scientific practice to devise a language in which these questions cannot be even asked. And this usually involves a good deal more than just refraining from admitting certain words or longer expressions into one's scientific vocabulary; the very syntax of scientific language will be involved too. As far as our present subject is concerned, even before Einstein physical scientists not only eschewed the words "past", "present" and "future", but eschewed tenses too. Time enters physical science through intervals by which one event may be earlier or later than another. Whether the events are the case or merely have been or will be, is of no concern to the scientist, so he uses a language in which the difference between being and having been and being about to be is inexpressible. And this, as I've said, has been the case since long before the special theory of relativity. That theory, all the same, has made an important difference. Before it was devised, the relation between tensed language and the tenseless language of the scientist was pretty straightforward. It amounted to this: When a scientist said "The interval between an earlier event $A$ and a later event $B$ is n time-units", you could translate this as "It is or has been or will be the case that ($B$ is occuring and it was the case n time units ago that ($A$ is occurring))". But I don't think this is what a scientist now means by "earlier" and "later", and indeed a scientist is not now likely even to say that the interval between $A$ and $B$ is $n$ time-units, just like that; the only interval between a pair of events to which he will give a definite value is a space-time one.

# Instants and Intervals

C. L. HAMBLIN*

Many writers, in both philosophy and science, have been concerned at the remoteness of our time-language from the facts of actual observation. This was, in one way, Einstein's starting-point in developing relativity theory: he said he was inspired by reading Berkeley. In Philosophy there is a strong strain of interest in supposedly primitive observational or phenomenal languages, languages whose features can be related directly to features of our experience of the world around us. Logical positivists, owing something to Mach and to Carnap's early *Logische Aufbau*, have often assumed that the atoms of our experience have a form like "Red – here – now" or perhaps "Red $(x, y, z, t)$", and that the whole of what we meaningfully say can be represented as a logical function, albeit very complicated, of statements of this form. ([1], [2], [6]: the formulations given seem to be first found explicitly in [2].) But, whatever we think about the ultimate thesis, there is one incongruity at its very base, namely, the presence in the atoms of coordinates of position and time of an apparently purely Newtonian or Kantian nature, unsupported by any attempt to give a phenomenal account of them.

There are at least two things wrong with the "$(x, y, z, t)$" and, by association, the "here now". The first, dependence on external definition of coordinates, has already been the subject of comment by other speakers. But I want to draw attention to a different one, the suggestion that our elementary experiences are *extensionless* and *instantaneous*. It can hardly be said that we observe qualities like redness to exist at Euclidean space-time points. In fact, it is not clear what such a contention could mean.

The position is a little worse in the case of the temporal coordinate than it is in the case of the spatial ones. If my visual field has a fine spatial structure, so that I am ordinarily conscious of some kind of average effect, I may often *look closer*, and see more detail; and it can be imagined, as it once used to be, that this process could be continued without limit. In the case of time there is no way of taking a closer – as it were slower – look at a finely structured event.

Most of what I want to say about instantaneity applies equally to systems involving absolute Newtonian coordinates and to egocentric systems: it is,

---

* Professor C. L. Hamblin, School of Philosophy, University of New South Wales, Kensington, N. S. W. 2033, Australia.

or ought to be, as much a problem for the classical physicist as for the modern one and for the psychologist.

By an *instant* of time I mean roughly what Euclid meant by a point of space, a piece of time that has no parts. I shall here assume that time is, in some sense, continuous, since the alternative possibility that it consists of a succession of discrete instants seems to raise considerable problems both at the phenomenal level and at the level of physical theory. This said, there are at least two generic objections to the contention that phenomenal time is subdivisible into instants.

(1) Instants can have no content: it takes too many of them to make up a durable experience. The red book on my table can turn green for half a second or half a century but it cannot turn green durationlessly and instantaneously at the stroke of twelve, remaining red at all times earlier and later. To put the objection another way: the temporal continuum is richer than we need for the description of the world, in that it permits the description of phenomenally impossible states of affairs such as that my book should be red at all rational points of the time-scale and green at all irrational ones.

(2) If time divides into instants we can give no account of temporal relations. This is a complaint, among others, of Gestalt psychologists: how, asks Koffka ([5], p. 437), when we perceive the notes of a melody, do we come to perceive that they make up a tune? More generally, what do we mean by saying that time is a *succession* of instants rather than just a bundle or set of them? Our perception of this succession cannot be explained in terms of our perception of the instants themselves.

One traditional way of escaping these objections, at least so far as perception is concerned, is that attributed by William James to E. R. Clay. The present instant of observation, says James ([4], vol. 1, p. 609), is not a knife-edge but a broad saddle-back. It has a duration and we can sit on it comfortably. Or, to use a happier metaphor than James's, the present is like the window of a railway carriage in which we are sitting. If it were an infinitesimal slit we could not see out properly, and we could not see the countryside laid out with its features in their proper relations; but since it has a width light can enter and we can see each thing in relation to the next and so form for ourselves a picture of the whole – in Koffka's example, the melody, not just the single notes. The problem of content and the problem of temporal relations are both solved.

Yet the solution becomes more unsatisfactory the more we look at it. The problems break out afresh in the new model. Within the supposed present interval we need to be able to analyse the picture we see; but this means that we need to be able to discriminate earlier and later events, and relate them as such, even though they are allegedly perceived simultaneously. Moreover, we have not really got rid of instants, for, apart from the question of whether there are not instants within the present interval, the interval itself must be defined by the instants between which it stretches: the railway window has a left-hand and a right-hand edge. We need to be able in principle to specify the instant at which a given object, say a cow, first comes into view in the window, and the instant at which it leaves.

Associated with the problem of temporal relations is a problem that perhaps at first appeals only to logicians, but which is ultimately of central importance. It concerns the potential three-valuedness of our phenomenally primitive atomic statements or, rather, the quality-predicates that take the spatio-temporal coordinates as arguments. Properly speaking, if a thing is not red it is *non*-red; if it is not warm it is *non*-warm; if a note is not high-pitched it is *non*-high-pitched; and so on. This is not to deny that redness, warmth and pitch may be matters of degree. It is simply a question of how we are to speak and reason. If a given predicate fails to apply in a given situation the corresponding negative one applies. But if these qualities must be predicated of intervals of space and time rather than of points and instants, two values of them will not suffice. It does not follow from the fact that we do not have red-for-the-interval-*a* that we must have *non*-red-for-the-interval-*a*, since we may have red for part of the interval and non-red for the rest of it. Of course, if it is not *red-throughout*-the-interval-*a* it does follow that it is non-red for some part of the interval *a*, but not that it is *non-red-throughout*-the-interval-*a*. Let us write "φ" for "red" (or some other such phenomenal predicate) and, for simplicity, drop the spatial coordinates: now, in symbols, there are not just two possibilities

$$\varphi a, (-\varphi)a$$

but also a third, "neither φ*a* nor (—φ)*a*", or

$$-\varphi a \cdot -(-\varphi)a.$$

And it seems from this argument that if the world of time consists *basically* of intervals the predicates that represent the qualities occurring in it must be *basically* three-valued in this way.

Now I am going to show, I hope, that the theory that time consists of intervals rather than instants can escape all these criticisms. This theory can be built on a logically sound basis, and the objections usually urged against it all fail. Apart from this it may or may not have any major merits but, at least, it is a theory that is open to those that would like to adopt it. In what follows I am not going to discuss the general feasibility of a phenomenal language and I am not going to take up the question of whether such a language may use an external coordinate system; and I shall ignore the question of spatial coordinates as distinct from temporal ones. My thesis is, in any case, broad enough to be independent of many of these considerations.

First let me point out something that is usually not realised, namely, that the objection in respect of three-valuedness can be urged even more strongly against the instant-theory than it can against the interval-theory. Let us suppose that at 10 p.m. one evening there is an electricity failure and the room in which I am sitting is plunged into darkness. Before 10 p.m. it was light; after 10 p.m. it is non-light; but which is it *at* 10 p.m.? Of course, it is neither light nor non-light: it is in a third state, the state of changing from light to non-light. At the instant a whistle changes from high pitch to low it is neither high-pitched nor non-high-pitched; and if I destroy a letter by throwing it in the fire the moment at which I do so is a moment of its history at which it is neither in existence nor out of existence, but in limbo. One cannot, moreover, get rid of the problem by describing these processes as gradual ones instead of instantaneous, since the three-valuedness comes up at a higher level; and one cannot get rid of it by treating the moment of change as ill-defined, since one simply extends the instant of limbo into an interval. (These objections are discussed in more detail in my [3].) The best anyone seems able to do with the problem is to solve it by *fiat*, by specifying *a priori* that the instant of change shall be considered as the first instant of the subsequent state, or the last instant of the preceding one: the latter is the traditional solution attributed to Aristotle, based on a reading of *Physics* 236a7–15. The arbitrary and *ad hoc* nature of this decision makes it as unsatisfactory as the situation it is supposed to resolve.

It so happens that, if we take time to consist of intervals, a little logical sophistication enables us to avoid this dilemma. If time consists of instants I can see no way out of it.

In what follows I give a sketch of a logic of intervals. I shall show that it can be built within the lower predicate calculus without mentioning or otherwise depending on the concept of an instant, and that when properties are predicated of these intervals the relevant distinctions can be made without fundamental three-valuedness.

Let $a, b, \ldots$ be individual variables interpretable as *intervals* of time, and let $\varphi, \psi, \ldots$ be predicates taking these intervals as arguments, such that elementary statements will be of the form $\varphi a, \varphi b, \ldots, \psi a, \ldots$ There will be *one* special symbol "$<$", to be read "precedes", taking pairs of intervals as arguments: "$a < b$" will be interpreted as meaning that the interval $a$ wholly precedes the interval $b$. I shall assume the logical apparatus of the lower predicate calculus with identity. However, the predicates $\varphi, \psi, \ldots$ are not general predicates of this calculus, but will need to satisfy certain special axioms.

Other temporal relations between intervals can be defined using "$<$". In particular we have

$$aAb = \text{Df} \quad a < b \cdot -(\exists c)(a < c \cdot c < b) \quad (\text{``}a \text{ abuts } b \text{ on the left''})$$
$$aOb = \text{Df} \quad -(a < b) \cdot -(b < a) \quad (\text{``}a \text{ overlaps } b\text{''})$$
$$aCb = \text{Df} \quad (c)(cOa \supset cOb) \quad (\text{``}a \text{ is contained in } b\text{''})$$

There are thirteen logically distinct ways in which two intervals can be temporally related.

The following axioms are subjoined to those for the lower predicate calculus. (This set is a revision of the set given in my [3], which was found to be incomplete. Axioms 1–8 are an appropriate strengthening of this earlier set: axioms 9 and 10, which were considered in the earlier article, are here made explicit. On the necessity for 10 I am indebted to a number of discussions with A. N. Prior, mainly during the conference at Oberwolfach but also in an exchange of letters shortly before his untimely death in Norway. Whether 9 and 10 completely characterise the predicates that are admissible as "phenomenal" is a debatable matter, but not one to be settled entirely by calculation.)

1. $-(a < a)$
2. $(a < c \cdot b < d) \supset (a < d \mathbin{v} b < c)$
3. $a < b \supset aAb \mathbin{v} (\exists c)(aAc \cdot cAb)$
4. $(aAc \cdot aAd \cdot bAc) \supset bAd$
5. $(aAb \cdot bAd \cdot aAc \cdot cAd) \supset b = c$
6. $(\exists b)(a < b)$
7. $(\exists b)(b < a)$
8. $(\exists b)(bCa \cdot -(b = a))$
9. $bCa \supset (\varphi a \supset \varphi b)$
10. $(b)(bCa \supset (\exists c)(cCb \cdot \varphi c)) \supset \varphi a$

Axioms 1–8 give us a logic of linear order for intervals. Axioms 9 and 10 are to hold for each of the elementary predicates φ, ψ, ... representing phenomenal predicates.

The axiom that positively differentiates this logic from one of instants is axiom 3, which says that when one interval properly precedes another there is always an interval that fills the space between them. Nothing like this holds for instants. Axioms 6 and 7 specify that there are infinitely many intervals future to, and infinitely many past to, any given interval. Axiom 8 says that every interval has another contained in it, and gives us the equivalent of a density property: what might be called the "fundamental theorem of density" to the effect that any interval may be divided into a pair of abutting intervals, can be proved from it.

Under suitable circumstances it is possible to define the *join* and *intersection* of two intervals. Unless the expedient of a "null-interval" is resorted to these will not always exist, but the intersection will exist if the two intervals overlap, and the join if they abut or overlap. When they exist they have the usual Boolean properties. (For their definitions and properties see [3].)

Axiom 9 serves to specify of each of the elementary predicates that when it holds of an interval it holds of any subinterval: that is, "φa" is to be taken as asserting that φ is true *throughout* the interval a. To interpret axiom 10 it is convenient first to define the *negation of a predicate* in the form

$$(-\varphi)a = \text{Df} \quad (b)(bCa \supset -\varphi a).$$

# Instants and Intervals

We can read "$(-\varphi)a$" as "non-$\varphi$ throughout $a$", and it asserts that there is no subinterval of $a$ throughout which $\varphi$ holds. Now axiom 10 can be written in the alternative form

$$-\varphi a \supset (\exists b)(bCa \cdot (-\varphi)b)$$

and says that whenever $\varphi$ fails to hold throughout some interval $a$ there is a subinterval $b$ such that non-$\varphi$ holds throughout it. This axiom excludes the possibility of indefinitely finely intermingled periods of redness and non-redness, or of high and low pitch, or of whatever else it is that we take the elementary predicates to represent.

Now let us draw some conclusions from the existence of this system. In the first place it should be noted that we have no need, at this level, to postulate the existence of instants at all. We can suppose, say, that $a$ abuts $b$ and that some property $\varphi$ holds throughout $a$ and fails throughout $b$ – that is, we can suppose the truth of $\varphi a \cdot (-\varphi)b$ – without raising the inexpressible question of whether $\varphi$ or $-\varphi$ may be predicated of the point of abutment. Moreover, although instants, in the sense of points of abutment, can be introduced into this language by definition, it is necessary to move outside the bounds of the lower predicate calculus to do so, invoking for example set-theoretical concepts. There is an interesting duality here, for it is the more usual course to use set-theoretical concepts to define intervals in terms of the points defining or composing them.

Thus let $\alpha$, $\beta$ and $\gamma$ be the ordered pairs of intervals $(a_1, a_2)$, $(b_1, b_2)$ and $(c_1, c_2)$, such that $a_1 A a_2$, $b_1 A b_2$ and $c_1 A c_2$. Interpreting $\alpha$, $\beta$ and $\gamma$ as *instants* we define for them

$$\alpha = \beta = \text{Df} \quad a_1 A b_2$$
$$\alpha < \beta = \text{Df} \quad a_1 < b_2 \cdot -a_1 A b_2.$$

(The double use of "$=$" and "$<$" causes no ambiguity since we use Greek letters for instants and Roman for intervals.) Now let us write down "axioms" of dense linear order for $\alpha, \beta, \ldots$, namely

01. $-(\alpha < \alpha)$
02. $(\alpha < \beta \cdot \beta < \gamma) \supset \alpha < \gamma$
03. $\alpha < \beta \vee \beta < \alpha \vee \alpha = \beta$
04. $(\exists \beta)(\beta < \alpha)$
05. $(\exists \beta)(\alpha < \beta)$
06. $\alpha < \beta \supset (\exists \gamma)(\alpha < \gamma \cdot \gamma < \beta)$
07. $(\alpha = \beta \cdot \alpha < \gamma) \supset \beta < \gamma$
08. $(\alpha = \beta \cdot \gamma < \alpha) \supset \gamma < \beta$

We also require that "$=$" be reflexive, symmetrical and transitive under its new definition. These formulae can be translated into interval-language in accordance with the definitions, provided the appropriate abutment properties are prefixed to each formula in respect of its free variables and conjoined within the scope of each existential quantifier in respect of its variable: thus 01, for example, becomes

$$a_1 A a_2 \supset -(a_1 < a_2 \cdot -(a_1 A a_2))$$

and 04 becomes

$$a_1 A a_2 \supset (\exists b_1)(\exists b_2)(b_1 A b_2 \cdot b_1 < a_2 \cdot -(b_1 A a_2)).$$

Under these translations all of 01–08, together with reflexivity, symmetry and transitivity for the defined "=", can be proved from axioms 1–8. Proofs, nowhere very difficult, need not be given here.

It follows that if mathematical physics needs the language of instants to deal with, say, laws of motion it can get it eventually from an elementary logic of intervals. But we should not feel obliged, just because mathematical physics needs such a language – if in fact it does – to use it for every facet of our experience. The question of whether redness, warmth or high pitch can exist for an isolated instant is a case in point, since it arises entirely from the attempt to apply the mathematico-physical concept of an instant in a context in which it has no real relevance.

The solution of the problem of three-valuedness can now also be set out. It consists in distinguishing clearly between the internal and external negations of an elementary statement $\varphi a$, namely, between $(-\varphi)a$ and $-(\varphi a)$ where, in the second case, we can omit brackets (and have done so earlier). But we have seen that the first of these can be defined in terms of the second and need not be regarded as involving a new elementary idea. To be sure, for any predicate of any interval $a$ there remain the three possibilities $\varphi a$, $(-\varphi)a$ and $-\varphi a \cdot -(-\varphi)a$; but these can still be expressed in a fundamentally two-valued language, in effect by specifying what each of them involves regarding the possible subintervals of $a$. Actually we can introduce any "truth-function" of predicates, also by definition; for we have for conjunction

$$(\varphi \cdot \psi)a = \mathrm{Df} \quad \varphi a \cdot \psi a$$

and any truth-function can be built up from conjunction and negation. For example, for disjunction we have

$$(\varphi \vee \psi)a \equiv (-(-\varphi \cdot -\psi))a$$
$$\equiv (b)(bCa \supset (\exists c)(cCb \cdot (\varphi c \vee \psi c))),$$

that is, "for any subinterval $b$ of $a$ there is a subinterval $c$ of $b$ such that $\varphi$ throughout $c$ or $\psi$ throughout $c$". It is fairly easy to prove that $\chi a$ is a theorem for any complex predicate $\chi$ that has the form of a truth-functional theorem. It can also be proved that predicates that are "truth-functionally" compounded of elementary predicates are still elementary in the sense that they satisfy axioms 9 and 10.

The "law of the excluded middle" holds in both the forms

$$\varphi a \vee -\varphi a$$
$$(\varphi \vee -\varphi)a$$

and fails only in the mixed form in which the negation is internal and the disjunction external, namely

$$\varphi a \vee (-\varphi)a.$$

In these several respects the logic of intervals is more straightforward and closer to experience than that of instants.

It was drawn to my attention during the Oberwolfach conference by H. A. C. Dobbs that the definition of instants from intervals in set-theoretical terms has previously been discussed by A. G. Walker in [7].

*References*

1. Carnap, R.: *Der logische Aufbau der Welt*. Berlin-Schlachtensee: Weltkreis-Verlag 1928. English translation by Rolf A. George: *The Logical Structure of the World; Pseudoproblems in Philosophy*. London: Routledge and Kegan Paul 1967.
2. — *Die physikalische Sprache als Universalsprache der Wissenschaft*. Erkenntnis 2 (1931); English translation by M. Black: *The Unity of Science*. London: 1934.
3. Hamblin, C. L.: *Starting and Stopping*. Monist 54, 2 (July 1969).
4. James. W.: *Principles of Psychology*. London: Macmillan 1907; 2 volumes.
5. Koffka, K.: *Principles of Gestalt Psychology*. London: Kegan Paul 1935.
6. Mach, E.: *Beiträge zur Analyse der Empfindungen*. Jena: 1886. English translation, *Contributions to the Analysis of Sensations*. Chicago: 1897.
7. Walker, A. G.: *Durées et instants*. Revue des cours scientifiques, no. 3266 (1947).

# The Fiction of Instants

M. ČAPEK*

*Abstract.* The claim that the mathematical durationless instants do exist in a physical sense appeared always as both paradoxical and inevitable. Paradoxical, because it is difficult to find the concept of entity lasting for zero-time intelligible; inevitable, since both experience and logic suggested that it really exists. For all empirically available temporal intervals were, in the classical period at least, divisible into smaller and smaller sub-intervals unless they were durationless instants. Furthermore, the very denial of instants, as proposed by various theories of chronon, surreptitiously postulated their existence. But an attentive analysis of perceptual and, more generally, phenomenal continua shows that they consist neither of instants nor of contiguous atomic segments; their structure clearly transcends the disjunction "instants versus chronons". They exhibit a type of continuity which is different from mathematical continuity and which, as Poincaré's paradox shows, is extremely difficult to be conceptualized. There is a considerable circumstantial evidence that time on the microphysical level has a similar structure. This would make possible to deny the reality of instants without accepting the self-contradictory "atomization" of time.

In his *Essay on the Foundations of Geometry* Bertrand Russell analyzed the so called "antinomy of the point" and concluded that the concept of geometrical dimensionless point is "palpable contradiction, only rendered tolerable by its necessity and familiarity". Now, if we substitute in Russell's analysis the word "instant" for "point", "time" for "space" and "chronometry" for "geometry", we shall obtain "the antinomy of the instant" completely analogous to the antinomy of the point. The passage thus paraphrased will run as follows:

> "We saw, in dealing with measurement, how time must be regarded as infinitely divisible, and yet as mere relativity. But what is divisible and consists of parts, must lead at last, by continued analysis, to a simple and unanalyzable part, as the unit of differentiation. For whatever can be divided, and has parts, possesses some thinghood, and must, therefore, contain two ultimate units, the whole namely, and the smallest element possessing thinghood. But in time this is notoriously not the case. After hypostatizing time, as chronometry is compelled to do, the mind imperatively demands elements, and insists on having them, whether possible or not. Of this demand, all chronometrical applications of the infinitesimal calculus are evidence. But what sort of elements do we thus obtain?

---

* Professor Milič Čapek, Department of Philosophy, Boston University, 232 Bay State Road, Boston, Massachusetts 02215, U.S.A.

Analysis, being unable to find any earlier halting place, finds its elements in instants, that is, zero quanta of time. Such a conception is a palpable contradiction only rendered tolerable by its necessity and familiarity. An instant must be temporal, otherwise it would not fulfill the function of a temporal element; but again it must contain no time, for any finite interval is capable of further analysis. *Instants can never be given in intuition*, which has no concern with the infinitesimal: they are purely conceptual constructions, arising out of the need of terms between which temporal relations can hold. If time be more than relativity, temporal relations must involve temporal relata; but no relata appear, until we have analyzed our temporal relata down to nothing. The contradictory notion of the instant, as a thing in time without temporal magnitude, is the only outcome of our search for temporal relata."[1] (Italics added.)

Russell pointed out clearly the logical correlation between the concept of unlimited temporal divisibility and that of durationless instant: since every finite temporal interval is divisible, the indivisibility can belong only to the zero-intervals, i.e., mathematical instants. He conceeded, at least in this essay, that no empirical data correspond to instants; durationless instants cannot be experienced. Furthermore, they are logically impossible since, according to Russell, they should possess two mutually exclusive attributes – temporality since they are the constitutive elements of every temporal interval, and the negation of temporality since they are supposedly without temporal extension. At the same time, both experience and logic seemingly lead us to postulate their existence. In this way an antinomic situation arises.

Before we shall investigate the alleged empirical and logical reasons which were only in part sketched by Russell, let us say that Russell's attitude to this problem was far from consistent; it kept changing almost from one of his books to another. Only six years later, in his *Principles of Mathematics*, he vigorously upheld the concept of instant and found nothing contradictory in it. This was the main reason why he sided with Zeno against Bergson.[2] Yet, he came remarkably close to Bergson in his article in *The Monist* in 1915 in which he resolutely denied the existence of instants in psychology thus coming back to his original view quoted above (i.e., that nothing in our experience corresponds to instants.[3]) The coming of the quantum physics made him cautious toward the existence of durationless instants even in the physical world.[4] A detailed survey of Russell's successive and changing views about this problem would require a separate essay which is beyond the limits of this paper. In truth, it was not necessary to start our investigation with Russell's

---

1 Russel, B.: *An Essay on the Foundations of Geometry*. Dover: 1956, pp. 189–190.
2 *The Principles of Mathematics*. New York: W. W. Norton 1964, p. 347.
3 *On the Experience of Time*. The Monist XXV (1915) 217.
4 *The Analysis of Matter*. Dover: 1954, p. 341.

passage quoted above. Equally revealing texts, insisting on the logical correlation between the mathematical continuity of time and the concept of durationless instant can be quoted from Descartes, Galileo, Leibniz, Kant and others.

Two kinds of arguments have been used to justify the belief in the existence of extensionless instants. First the empirical one. Our macroscopic experience suggests that every temporal interval, no matter how short, is divisible into its sub-intervals. It is true that common sense has the tendency to distinguish between temporal intervals which are divisible (the duration of a day, divisible into hours, hours divisible into minutes, minutes into seconds) and *moments* which appear to our spontaneous perception as indivisible: a flash of light, a single sound, a contact of two bodies. This probably led some ancient and medieval thinkers to accept the limit to the divisibility of time; thus Aristoxenus postulated the primordial unit of time, ὁ πρῶτος χρόνος which Aristides Quintillianus called "tempus brevissimum" and Isidor of Seville "atomus temporis". Beda Venerabilis computed that each hour consists of 22560 atoms of time. Similar speculations about the atomicity of time can be found in Stoics.[5] In truth, as late as in the seventeenth century we see Gassendi holding the view that the continuity of time is merely a macroscopic illusion hiding to our imperfect perception the reality of the atoms of time, "insecabilia temporis".[6]

Needless to say, such a view proved to be unsatisfactory to the founders of classical physics and the view of Galileo that in every interval of time, no matter how small, there is an infinite number of durationless instants[7] fully prevailed and became the cornerstone of rational mechanics. Furthermore, the more refined experience showed that the distinction between "moments" and "intervals" is only that of degree; what we call moments are merely shorter intervals. Since the time when Wheatstone measured the duration of a lightning which was regarded as a momentary event *par excellence*, the technique of the measuring of small intervals of time increased enormously. It is more than thirty years ago since Magnan showed by the method of ultra-rapid kinematography that the so-called "momentary" event, like a drop of water hitting the ground or a single vibration of insect's wings in a slowed down motion picture appears as a very long history consisting of an enormous succession of shorter sub-events.[8] By a natural extrapolation it was believed that such divisibility belongs to any interval of time, no matter how short; in other words, that only durationless instants are indivisible.

---

5 Lasswitz, K.: *Geschichte der Atomistik vom Mittelalter bis Newton.* Braunschweig: 1890, I, pp. 31–4; Sambursky, S.: *Physics of Stoics.* New York: MacMillan 1959, p. 105.

6 *Syntagma philosophicum* in *Opera omnia.* Florence: 1727, p. 300.

7 *Le opere di Galileo Galilei, prima edizione completa;* XIII. Florence: 1855, p. 158: "in ogni tempo quanto, ancorche picolissimo, sono infiniti instanti."

8 Magnan, A.: *Cinematographie jusqu'à 12000 vues par seconde (Avec application à l'étude du vol des Insectes).* Paris: Herman 1932.

But far more powerful reasons for the affirmation of the existence of instants were, apparently at least, of logical kind. It was believed on a priori grounds, independently of experience, that the concept of instant is necessarily correlated with the very idea of temporal interval. For no interval of time is thinkable without its limits; and these limits must be without temporal extension, that is, point-like instants. For were they not, they would be just other temporal intervals; and the same reasoning as that above would require the existence of their own instant-like boundaries. The only way to avoid the affirmation of mathematical instants was to insist that they are mere conceptual limits, never attainable in experience, or, in Russell's words, "not given in intuition". This apparently was the view of Kant when in the section on *Anticipations of Perception* he – anticipating Bergson – insisted that points and instants are mere conceptual limits out of which no space and time can be constructed. In other words, what is given are temporal wholes while instants have a mere derivative existence of conceptual constructs. Yet, only a few pages before this passage, in his *Axioms of Intuition*, Kant holds an apparently opposite view according to which temporal intervals are always *aggregates* which their smaller constituent parts – that is, ultimately, instants – precede. This was also the view of Galilei and of Russell in *The Principles of Mathematics*.[9] The discrepancy between two of Kant's views was pointed out by Whitehead in his *Science and the Modern World*.[10] But bearing in mind that the concept of durationless instant and that of infinite divisibility of time are logically correlated, it is fairer to regard the difference between the two passages as that of emphasis rather than that of substance. From the same point of view Russell's vacillations also appear to be less serious; neither he nor Kant doubted the mathematical continuity of time and the existence of the mathematical extensionless present. From the time of Galilei this was one of the basic features of classical physics and one of the most fundamental ideas of rational mechanics.

The fact that the mathematical instant is never experienced did not shake the belief in its existence in any way. This explains why the psychological present which is the only present, the only "moment" which we can experience and which is always of non-zero duration, was called "specious present". The choice of this term is significant: it was coined to suggest that the only *true* present, the only true instant is without duration, a mere temporal point; in contrast to this true mathematical present our psychological present, which we experience as "minimum sensibile", is necessarily only "specious". Again experience seemingly substantiated this claim: are there not countless physical events whose duration was measured to be incomparably shorter than the duration of our psychological present? Does not this show that the unper-

---

9 Russell, B.: *The Principles of Mathematics*, p. 144. On Galileo, cf. Note 2.
10 Whitehead, A. N.: *Science and the Modern World*. New York: MacMillan 1926, pp. 183–84; Kant, I.: *Critique of Pure Reason*, transl. by Norman Kemp Smith. London: MacMillan 1953, pp. 198, 204.

ceivability of the true mathematical present cannot be used as a decisive argument against its existence?

The final and seemingly decisive argument for the existence of mathematical instants was the *apparent impossibility of their denial*. More specifically, their very denial could not be phrased without surreptitiously asserting their existence. This is what made "the chronon theory" so suspect in the past and it remains its main shortcoming even today. By "chronon theory" I mean the thesis according to which there are the minimum intervals of time which are not further divisible; they are the physical analogues of our "specious present" only on a vastly reduced temporal scale. This is the common feature of all atomistic theories of time from the Stoics to some contemporary physicists and it is of secondary importance whether the atoms of time are called ὁ πρῶτος χρόνος, *tempus brevissimum*, *tempusculum*, or *chronon*. Now the theory of the minimum temporal segments affirms that time consists *not* of the succession of durationless instants, but of the succession of finite temporal intervals. But since each interval has its own boundaries and since these boundaries are necessarily durationless, are we not surreptitiously re-introducing the very concept of instant which we purported to eliminate?[11] And since within each chronon no successive instants can be distinguished, do we not implicitly assert that within each chronon time is "standing still", in other words, that the minima of time are themselves "holes in time"? Thus the chronon theory seems to be burdened by the same shortcoming as the doctrine of instants: it tries to build time out of non-temporal elements and, furthermore, its very formulation implies the assumption of mathematical, durationless instants.

So much in favor of the doctrine of instants. Now let us survey the facts and reasons which make this doctrine extremely implausible, especially in the light of contemporary physics; and then let us explore whether it is possible to phrase the chronon theory in a language free of contradiction, i.e., that which would not involve a tacit assumption of what it denies.

In the first place, no matter how strongly the concept of instant imposes itself on our mind, its intrinsic difficulties remain. It is the same basic difficulty as that concerning the geometrical points and the extensionless atoms of Boscovich: how can we build temporal intervals out of durationless instants, spatial segments out of extensionless points, material volumes out of the atoms devoid of volume? It is true that from the standpoint of formal mathematics this difficulty is not insurmountable: elementary calculus taught us that the product of zero by infinity can be equal to any finite value. Thus in assuming an actually infinite number of instants any finite interval of time can be built. But we have to be on guard against the controversial concept of actual infinity

---

11 On the intrinsic difficulties of the chronon theory cf. Čapek, *The Philosophical Impact of Contemporary Physics*. Nav Nostrand, 1964, pp. 40–41; 231f. Cf. also G. J. Whitrow; *The Natural Philosophy of Time*. London, Edingburgh: Thomas Nelson 1961, Ch. "Temporal Atomicity", pp. 153–57.

by which various magical results can be obtained which are plainly absurd from the point of view of a physicist. This may be illustrated by the astonishing conclusions reached by Tarski and Banach from the discovery made by Hausdorff:

> "It is possible to divide a large sphere (say, of the size of the sun) into a finite number of mutually disjoint parts which together exhaust the volume of the large sphere, and to move each one of these parts (without changing its size or shape) into a small sphere (say, of the size of a pea) in such a way that the moved parts remain mutually disjoint and exhaust the volume of the small sphere. This statement means that, if a man could only break up the large sphere in the proper clever way, he could put the whole of it into his pocket".[12]

Karl Menger, who mentions this curiosity, is certainly right when he says that a physical application of this mathematical discovery is impossible and he points out explicitly the reason of this impossibility: the required construction involves an infinite number of operations and as such it is inapplicable to any physical object. Thus man will never be able to put the sun into his pocket, – not even the moon unless we mean it as a figure of speech. Since the concept of instant requires the actual infinity to be useful for physical science, its physical applicability is as suspect as that of the actual infinity itself.

The general trends of contemporary physics only strenghten the doubts stated above. In the whole area of physics the actual infinities, whether the infinitely large or infinitely small, are on retreat. Since the time of Riemann we know that the infinitely large space of Euclid and Newton is not the only logical possibility; beginning with Einstein a number of cosmologists believe that the cosmic space is *finite*, though without limits. We know today that Boscovich, despite his other remarkable anticipations was wrong in his conception of the extensionless atoms; all elementary particles apparently have a finite radius of the order of $10^{-13}$ cm. The assumption of the electron with the zero radius was always loaded with difficulties since it led to the infinite density of electric charge compressed into a rigorously mathematical point; it was this difficulty that led to the postulation of the minimum length or hodon. Similarly, infinitely large velocities which implied that a certain physical action can be literally *ubiquitous*, that is, present at one and the same instant in *all* the positions of its instantaneous trajectory, began their departure from physics in 1675 when Olaf Römer discovered the finite velocity of light. Gravitation seemed to be the only exception, but the recent discovery of the gravitational waves, if it is confirmed, would strengthen one of the central ideas of the relativity theory about the finite velocity of *all* physical actions.

---

12 Menger, K.: *Theory of Relativity and Geometry*. In: Albert Einstein: *Philosopher and Scientist* (ed. by Paul Schilp). Evanston, Illinois: 1949, pp. 469–70.

It is only logical that the physical applicability of the concept of instant is now equally questioned. The postulate of the minimum time-interval or chronon was frequently made jointly with the assumption of hodon; this was only natural since the discovery of quantum phenomena made doubtful the physical applicability of the concept of spatio-temporal continuity. More specifically, it was the same doubts about the possibility of the infinite density of energy which led Lévi in 1927 to postulate that the proper times of the electron consist of the succession of chronons whose duration is of the order of $10^{-24}$ sec.

Epistemological as well as psychological considerations make the concept of zero-duration even more suspect. For there is no problem how this concept psychologically originated. We already pointed out that the concept itself is correlated with that of spatio-temporal continuity, that is, infinite divisibility of both space and time. There can be no doubt that this type of continuity is, as Schrödinger observed and as even the mathematicians Hilbert and Bernays conceded, "an enormous, exorbitant extrapolation" of what is empirically given in our perception and experience.[13] Furthermore, there is no doubt that the concept of instant was created in analogy to the concept of point. As Bergson observed, "as soon as we make a line correspond to duration, 'segment of duration' must correspond to segments of line, and an 'extremity of duration' must correspond to an extremity of line; such will be the instant...".[14] One does not have to be a Bergsonian to see how much Bergson's observation about the spatialization of time is correct; let us only listen to the following words of Kant:

> "We represent the time-sequence by a line progressing to infinity, in which the manifold constitutes a series of one dimension only; and we reason from the properties of this line to *all* the properties of time, with this one exception, that while the parts of the line are simultaneous, the parts of time are successive. (Italics added.)
> Even time itself we cannot represent, save in so far as we attend, in the drawing of a straight line (which has to serve as the outer figurative representation of time), merely to the act of synthesis of the manifold whereby we successively determine inner sense...."[15].

From this to the affirmation of infinite divisibility of time and of the mathematical continuity of all changes in *The Anticipations of Perception* there is only a small step. The reality of the mathematical instant was accepted in both editions of *Critique of Pure Reason* not only on the physical, but even on the psychological level. That was only natural; if time is symbolized by a geometrical line and since every line contains points, time must consist of instants.

---

13 Hilbert, D., Bernays, P.: *Grundlagen der Mathematik*. Jena: 1931, pp. 15–17. Schrödinger, E.: *Science and Humanism*. Cambridge: University Press 1952, pp. 30–31.
14 Bergson, H.: *Durée et simultanéité*. 3me ed. Paris: 1926, pp. 68–69.
15 Kant, *op. cit.*, pp. 77, 167.

Bergson's diagnosis of the origin of the concept of instant in our tendency to geometrize is clearly illustrated by the above quotation from Kant. This is certainly one illustration among many. Even if we, for a moment, do not question the legitimacy of the concept of the geometrical point, the tacit assumption that within any temporal sequence there must be instants as the counterparts to the points is nothing but an effect of the misleading spatial associations. Bergson's lasting merit is to point out the distortions which spatializing habits cause in our representation of time.

The inadequacy of the concept of instant for physical considerations was pointed out after Bergson by Whitehead. In the very first pages of his *An Enquiry Concerning the Principles of Natural Science* he showed that "a state of change at a durationless instant is a very difficult conception".[16] Some essential physical quantities such as velocity, acceleration, momentum and kinetic energy are meaningless without some reference to the past and the future. (It is interesting to note that Russell, despite his vigorous defense of the durationless instant in his *Principles of Mathematics* in which he sided with Zeno against Bergson by saying that "we live in an unchanging world, and that the arrow, at every moment of its flight is strictly at rest", in another passage of the same book virtually agreed with Whitehead in rejecting the concept of *the instantaneous state of motion*.[17]) Whitehead clearly saw that the concept of instantaneous state applies even less to organism: "In biology the concept of organism cannot be expressed in terms of a material distribution at an instant. The essence of organism is that it is one thing which functions and is spread through space. Now functioning takes time." "There is no such thing as life 'at an instant'; life is too obstinately concrete to be located in an extensive element of instantaneous space." But in this respect the difference between the organic and inorganic nature is merely that of degree: "This is no special peculiarity of life. It is equally true of molecules of iron or of a musical phrase".[18] Here we have the root idea of Whitehead's organic conception of nature.

Whitehead's reference to musical phrase has a far deeper significance than it appears at the first glance. For the perception of melody, or of any succession of sounds, possesses a certain structure showing that there are certain continua which in spite of their conspicuous temporal character are radically different from what we call 'mathematical continuum'. The awareness of melody represents a phenomenal field whose temporal character is not only very pronounced, but essential; for the individuality of its constituting sounds exists

---

16 *An Enquiry Concerning the Principles of Natural Knowledge*. Cambridge: University Press, 1955, p. 2.
17 *The Principles of Mathematics*, p. 473. Contrast it with the passage on p. 347 referred to in Note 2.
18 *An Enquiry* ..., p. 3, 196.

only *because* of their succession. At the same time, melody is literally nothing at an instant as such diverse thinkers as Whitehead and Norbert Wiener observed; even a single tone has a certain duration.[19] *The auditory phenomenal field clearly is devoid of instants;* this is true, though less conspicuously, of other phenomenal fields. In this way, certain paradoxical features of modern physics can be made more intelligible. We know that the relativistic rejection of absolute simultaneity means that, to speak with Eddington, there are no "world wide instants"; in other words, that the physical world is nothing at an instant. This appears paradoxical only as long as we believe that *every* temporal continuum must be mathematically continuous. But this is clearly not true as the case of auditory continuum shows. In the light of what I would call the "melodic structure" of time the second form of Heisenberg's principle, according to which to pin down a microphysical event at a mathematical instant would make the corresponding energy completely undetermined, also becomes more intelligible. The only reason why the formula $\Delta E \cdot \Delta t \geq h$ appears to us arbitrary is that the three centuries of calculus and classical mechanics firmly established the belief that it is meaningful to speak of 'energy at a certain instant'. Now, as Whitehead observed, "nature is nothing at an instant" in a similar sense that "melody is nothing at an instant". "Instantaneous cross-sections" are as illegitimate on the microphysical as on the macrophysical level.

The structure of the auditory phenomenal field is revealing also in another sense; it shows clearly that the dichotomy "instants versus chronons" is *not* logically exhaustive; it is a false dichotomy. As pointed out above, one of the most decisive arguments for the theory of instants was that the only alternative view was the chronon theory which surreptitiously assumed the very concept of instant which it overtly denied. But if there is anything certain, it is the fact that the structure of any phenomenal field – not only that of auditory continuum – is *not* atomic. This is – or at least *should* be – known since the times of William James and of Gestalt psychology when it was convincingly shown that the basic fallacy of the associationist psychology was an artificial atomization of the introspective data. The phenomenal field, which James called "stream of thought" and Bergson "durée réelle", contains no instants and no edges, nor does it consist of the succession of well defined atomic entities; in other words, it is neither mathematically continuous nor atomic in the sense of the chronon theory. It is neither divisible *in infinitum*, nor divisible into finite segments; though its successive phases are qualitatively differentiated, their diversification must not be confused with the mutual externality of the contiguous segments as the geometrical symbolism of the chronon theory wrongly suggests. In the same way as, to use Wittgenstein's words, "the visual field

---

19 Whitehead, A. N.: *Science and the Modern World*, p. 54; Wiener, N.: *I am a Mathematician*. M. I. T. Press, 1964, pp. 105–107.

has no limits", the concretely intuited present has no boundaries.[20] Or, in the words of Bertrand Russell, unwittingly agreeing on this point with Bergson, "the present has no sharp boundaries, and no constituent of it can be picked out as the earliest".[21]

It was not only William James and the Gestalt psychologists, but also some outstanding mathematicians and physicists, who became aware of the distinction between mathematical continuum and qualitative (phenomenal) continuum. The doubts of Schrödinger and Hilbert about the universal applicability of the concept of spatio-temporal continuity had been already referred to. Herrman Weyl stressed the difference between what he called "intuitive" ("anschauliches") continuum and mathematical continuum; he even gave credit to Bergson for establishing this distinction. Poincaré used a less appropriate term – "le continu physique" – for what Weyl called "intuitive continuum", but he was equally aware of its difference from mathematical continuum. Both Weyl and Poincaré pointed out that the term "mathematical continuity" is misleading since in the continuum of real numbers the elements are as external with respect to each other as natural numbers. This again was in agreement with Bergson who correctly recognized that the so called "mathematical continuity" is nothing but "discontinuity infinitely repeated", i. e. infinite divisibility.[22] There is no doubt that much confusion could be avoided if the same term – "continuity" – were not used in two radically different senses.

But while both Weyl and Poincaré stressed the difference between mathematical and qualitative continuum, they were – unlike Gestalt psychologists – less explicit in stressing the non-atomic character of the intuited continua. In other words, they failed to stress that qualitative temporal continuum does not consist of the mutually external, contiguous segments. In truth, Poincaré in one of his last writings suggested the possibility of the atomic structure of time[23] which would be nothing but a contiguum of finite intervals in the sense defined above. Nor did he suggest the possibility that on the microscopic – or rather *microchronic* – scale the temporal continuum may have a similar structure as intuitive, qualitative continuum.

---

20 Wittgenstein, L.: *Tractatus Logico-Philosophicus*, 6.43311. Cf. also Ephron, R.: *The Duration of the Present*. Annals of the New York Academy of Science, vol. 138 (February 6, 1967), p. 714: "The onset of a perception cannot be perceived for it is not an object of perception ... Analogously, we do not *perceive* the 'edge' of our visual field or the 'borders' of our blind spot."

21 Russell, B.: *On the Experience of Time*, p. 223.

22 Weyl, H.: *Das Kontinuum und andere Monographien*. New York: Chelsea Publishing Co., n. d., pp. 65–71; Poincaré, H.: *Science and Hypothesis*. In: *The Foundations of Science*, transl. by G. B. Halsted. Lancaster: The Science Press 1913, p. 43; Bergson, H.: *Creative Evolution*, transl. by Arthur Mitchell. New York: Random House 1944, p. 170: "the intellectual representation of continuity is negative, being, at bottom, only the refusal of our mind, before any actually given system of of decomposition, to regard it as the only possible one."

23 Poincaré, H.: *Dernières pensées*. Paris, 1913, p. 188.

I suspect that there were two reasons why he did not suggest it. First, such suggestion would lead him to concede the existence of some rudimentary qualities even on the physical or microphysical level – and he was too deeply immersed in the classical habits of thought to do it. Second, Poincaré was the first who explicitly analyzed the paradoxical and in a sense "logically scandalous" structure of qualitative continua; the idea that something analogous to it could occur on the objective physical level simply did not occur to him. He showed that the most paradoxical feature of such continua is that *the transitivity of the relation of equality* apparently does not hold in them: while two contiguous terms are indistinguishable from each other, the non-contiguous are:

$$A = B, B = C, A \neq C.$$

He illustrated it by the perception of weight increase; while we do not perceive the difference between 10 grams and 11 grams, or between 11 and 12 grams, we do perceive the difference between 10 and 12 grams. This has been known since the time of Gustav Theodore Fechner. Similar examples can be found in different sensory or introspective continua, such as different shades of the same color or different intensities of the same pitch, etc. Bertrand Russell pointed out in 1915 that the perceptual temporal continuum exhibits the same structure: the relation of *psychological simultaneity* is not transitive:

> "Suppose that I see a given object $A$ continuously while I am hearing two successive sounds $B$ and $C$. Then $B$ is simultaneous with $A$ and $A$ with $C$, but $B$ is not simultaneous with $C$".[24]

There seems to be an obvious way out of this difficulty: to assume that qualitative continuum, whether its terms are simultaneous or successive, is merely "apparent", "phenomenal" or "illusory" and that the alleged logical difficulty stems from its intrinsic "haziness". In other words, the difficulty disappears when we consider the underlying physico-mathematical continuum as "the only real". Illustrated by a concrete example: when we gradually increase weight from ten to twelve grams, the only thing we have to consider is the continuous range of magnitudes through which the physical stimuli pass; within this continuum each term is sharply distinguished from each other and the logical absurdity of the non-transitivity of equality can never arise. The whole difficulty is removed if the above scheme $a = b$, $b = c$, $a \neq c$ is replaced by the following: "$a$ is *indistinguishable* from $b$, $b$ is *indistinguishable* from $c$, but $c$ is *distinguishable* from $a$". In other words, the paradox arises merely out of the limited capacity of consciousness to discern the minutely different stimuli. Contradiction is thus confined to our experience; it is not inherent in reality. The same is true of the intuited temporal continuum: the underlying mathe-

---

[24] Poincaré, H.: *The Foundations of Science*, p. 46; Rusell, B.: *loc. cit.* p. 228.

matical continuum of successive physical stimuli is *the only real;* in it the transitivity of simultaneity is fully preserved and the apparent non-transitivity of "temporal togetherness' is due entirely to the haziness of our experience, more specifically, to the fact that our psychological present is merely "specious", without sharp boundaries. The only true present is the mathematical, "knife-edge" instant of the physical world.

This explanation is in line with the centuries old philosophical tradition which from Parmenides to Bradley opposed the logically flawless "real world" to the "haziness", "confusions" and "contradictions" of our immediate experience. Yet, a closer scrutiny shows that this explanation faces two serious objections.

First, nobody claims that the non-transitivity of equality exists on the level of physical stimuli. Nobody denied that two stimuli whose difference is imperceptible are *physically* different, though indistinguishable psychologically. But it is clearly meaningless to call two sensations "different, though indistinguishable". The sensations resulting from two minutely different physical stimuli are qualitatively the same *qua* sensations; to postulate their difference despite their unperceivability, does not make sense. An unperceived difference, that is, that which is neither sensed nor felt, simply does not *psychologically* exist; if we continue to say that two sensations are "really different" in spite of their "apparent" identity, we are not speaking of sensations *qua* sensations, but of their external stimuli. In other words, we are unconsciously slipping from the language of perceptual data to the language of physical stimuli. The paradox of intuited continuum is not dismissed when we insist that it is absent on the physical level; it continues to exist on the psychological level whose paradoxical structure it reveals.

This leads us to the second objection. Are we really sure that the physical reality down to its deepest microphysical level is adequately described in the terms of mathematical continuum? We mentioned this before when we pointed out that the applicability of the concept of spatiotemporal continuity on the quantum level is more than questionable. In truth, there is a considerable circumstantial evidence that the physical world, at least in its deepest microphysical strata, does not possess such sharp edges and clear cut contours as the last century physics hopefully expected. This is perfectly compatible with the fact that nature on the macrophysical level of physiological stimuli *is for all practical purposes* continuous in a mathematical sense. The main reason why it is so difficult to give up the applicability of mathematical continuity to the microphysical level is that it would imply the admission that there is an irreducible qualitative element in nature which resists a complete mathematization or formalization. We are still unconsciously committed to the dogma of bifurcation of nature which relegates all qualities into the subjective realm and eliminates them from the allegedly homogeneous, "purely quantitative" realm of matter. If we give up this dogma, as Whitehead did in his organic philosophy of nature, our reluctance will disappear together with other difficulties which Cartesian dualism created.

Let us sum up. Mathematical continuity is *very approximately* applicable on the macroscopic – and macrochronic – level; it ceases to be applicable on the microphysical, i. e. quantum level. In other words, time on this level is not infinitely divisible; consequently, durationless instants have no physical existence. Nor do they possess psychological existence since the psychological present has a certain duration. This absence of instants on both the psychological and physical level suggests that perhaps the microphysical time may have the same paradoxical structure as qualitative temporal continua. From this point of view, the failure of the "chronon theory" would become as understandable as the failure of the associatinistic, atomistic psychology. Whether it will ever be possible to construct a consistent formal calculus which would adequately express the paradoxical features of such continua, is an open question. Karl Menger's "topology without points" as well as A. L. Zadeh's "fuzzy set theory" are serious attempts in this direction. But it is also possible that we are here reaching the ultimate limits of formal analysis.[25]

---

25 Zadeh, L. A.: *Fuzzy Sets*. In: Information and Control, 8 (1965) 338–353; Menger, K.: *Topology without Points*. Rice Institut Pamphlets, vol. 27 (1940), No. 1, p. 107; Bergmann, G.: *Duration and Specious Present*. Phil. of Science, vol. 27 (1960) 4f.

# On the Reality of Becoming

Eva Cassirer*

*Summary.* The paper discusses the concept of the present moment 'now'. Two views regarding the nature of the 'now', held by A. Grünbaum and G. J. Whitrow, resp., are contrasted in Part I. Grünbaum believes that the becoming of the present moment is totally subjective and observer-dependent, while Whitrow thinks that it has objective reality and that the moment 'now' is uniquely the same for the whole universe and marks a qualitative change between the actuality of the past and the potentiality of the future.

In Part II the question is raised – and rejected – whether the discrepancy is only a linguistic one. Both views are criticised, the first for its extreme subjectivism, the other for deriving ontological consequences from the accidental positioning of an observer. Part III attempts a dissolution of the disagreement by showing that the dichotomy derives from the notion of a static world, to which time is added as an extra dimension, while it does not arise from an Einsteinian-type description of the universe as dynamic.

I

One of the important metaphysical inquiries regarding the status of a theory of Time has been concerned with the concept of the 'now'. What we usually designate by the word "now" is that instant that divides the past from the future: the momentary present. Of course, the 'now', so defined, is a mathematical abstraction. We do not normally experience the present as an instant, "not as a knife-edge, but as a saddle-back"[1] of some duration, – long enough to enable us to have a sensation or perception. But since the length of the psychological present varies and is dependent upon the type of experience we are just undergoing – e. g. it can have the length of a musical note of a quarter beat, or of a full bar, or even that of a whole musical phrase – I shall refer, in this paper, to the 'now' or 'the present moment' in its mathematical sense – as a kind of Dedekindian *Schnitt* between the past and the future.

It has been thought that the 'now', as experienced by an observer or uttered by a speaker, is private to this person and relates intimately to his present perceptions. But it has also been asked whether it must not be possible to share this 'now' with other sentient beings to whom we can indicate the moment we are referring to (simply by uttering the word "now" in their presence), and whether, by extrapolating from those present to all other people, it would not make sense to say that the moment 'now' was shared by all persons alive at

---

* Eva Cassirer, Ph. D., Department of Logic and Metaphysics, University of St. Andrews, Great Britain.
1 William James: *Principles of Psychology*, Vol. I, p. 609.

that moment. In other words, it has been asked whether the world as such is not going through a continuous series of 'nows' so that each individual 'now' is exactly contemporaneous with the 'now' of all other living people; and, therefrom, whether it might not also make sense to say that there exists a common, objective 'now' in the world: – the universal present, the 'now' of becoming.

The most recent discussion of this point has been conducted by two well-known scholars of the problem of Time, G. J. Whitrow and A. Grünbaum. Both of them are, I believe, in an interesting way mistaken about some of the things they say, although one of them asserts, the other denies, the objective reality of the 'now'.

Let us take the denial first: Grünbaum, in his article "The Status of Temporal Becoming"[2] defines 'becoming' as a mind-dependent process of experiencing an event, while at the same time being aware of this fact (of experiencing). That is to say, he defines it as the overlapping of two simultaneous mental processes, one of which has experiential content, while the other has 'awareness that' (something is the case) as its content. This contemporaneous dual experience constitutes what he calls "the becoming of the now". He puts it this way: "M experiences an event at time t such that at t he is aware of having that certain experience simultaneously with the fact of being aware of having it at all."[3]

With this fairly awkward definition, Grünbaum is making the becoming of the 'now' not only mind-dependent (by using the word "awareness"), but also speaker-reflexive. In other words, he jumps from the fact that 'past' and 'future' are egocentric terms to the conclusion that there can be no independent (objective) physical occurrence called 'becoming', since *by definition* this involves consciousness. If he is right, it would be meaningless (and not just wrong) to postulate the existence of a universal 'now', since the moment so designated is then confined to a necessarily private experience of self-awareness of each individual. But since the concepts of the past and future are dependent upon the definition of the present, with which the 'now' has been identified, it becomes questionable whether we could still talk meaningfully about the world having a past and going into the future, since they, too, would become entirely subjective terms. There would then be only a private past and a private future for each observer. Time would therefore be characterised solely by the earlier-later relationship, with no qualitative distinction between its different parts. And since it can be shown that its quantitative properties (such as its infinite divisibility into equal units) are entirely arbitrary, since they depend upon definitions,

---

[2] Grünbaum, A.: *The Status of Temporal Becoming*. In: Interdisciplinary Perspectives of Time, Annals of the New York Academy of Sciences, Vol. 138, Art. 2, Feb. 1967. pp. 374–395. (Reprinted in: *Modern Science and Zeno's Paradoxes,* London: Allen & Unwin 1968. Chapter I.)

[3] *Ibid.,* p. 381.

– different parts of time, on Grünbaum's recent view, have no distinguishing properties at all, and the distinction between past and future cannot be derived or accounted for or explained in this way.

Before returning to an analysis of Grünbaum's view, let me take a look at Dr. Whitrow's discussion of this problem.

In his paper "On the Natural Philosophy of Time"[4], Whitrow says that we must distinguish between two things, viz.

(a) *Becoming*, i. e. the happening of events and *hence* the distinction between past, present, and future; and

(b) *the Anisotropy of Time* (Time's "arrow"), i. e., the before-after relation taken as irreversible.

Philosophers and physicists, he says, have been divided in their attitude towards these two alternatives. There are those who believe that:

(α) there is nothing in the external world corresponding to the distinction between past, present and future; these are therefore essentially subjective; or that

(β) there is nothing corresponding in the external world to our sense of time-direction (the "arrow"); this is due to memory and consciousness (and therefore also subjective); while in physics the concept of symmetrical time suffices.

Whitrow himself rejects both (α) and (β). That is, he believes that to both, becoming and time's arrow, there corresponds something 'real' in the external world. Grünbaum, he thinks, accepts (α) and rejects (β), while Mehlberg[5] accepts both.

Regarding the first point (a), Whitrow believes that the past-present-future relation "lies at the core of the concept of time". He thinks that *the 'now' is an objective moment of becoming,* namely that instant where something turns from indefiniteness to definiteness, i. e. to being determined. The future, he says, is the realm of possibilities; but actualities are only experienced in the present: the 'now' has objective reality.

The obvious objections (cf.[6]) that have been raised against this view, viz. that the 'now' is private to each observer, and that 'determined' means 'determined for me now', he meets by saying that the 'now' can be made intersubjective, can be shared. Better still, it can be objectivised in the following way: If I use a camera to take a picture of a certain state of affairs, i. e., a certain state (in one part) of the world, the picture will then have "frozen" a certain 'now', which agrees with the memory I have of that moment and with the perception I had at that time. But I find this argument not convincing: it only proves that the camera's click and my perception of the scene coincided temporally, which

---

[4] Whitrow, G. J.: *Reflections on the Natural Philosophy of Time*. In: Interdisciplinary Perspectives of Time, pp. 422–432.

[5] Henryk Mehlberg, Proc. Congress Philos. Sci., Jerusalem (1964), passim.

[6] Bergmann, H.: *Der Kampf um das Kausalgesetz in der jüngsten Physik* (1929), as cited by Whitrow and Grünbaum.

is not so surprising if one considers that the camera, if it takes the picture in my presence, will be *my* percept; it will be seen by me, and therefore be part of my 'now'. *Of course* it and I then have the same 'now'. And if it were not my percept, i. e. if it had taken the picture not in my presence, I would have grave doubts as to whether it was catching the same 'now' as my perception. Even if the scene were exactly the same on the photograph as I had seen it then and now remember it, this would not prove to me – and it could never be conclusively established later on – that it was taken at precisely the same moment that I had the perception.

So either our 'nows' are trivially the same, because one perception includes the other, or one can never really be sure that they are the same, without assuming what was to be proven.

II

Perhaps the apparent contradiction between the two views regarding the nature of 'becoming' and the 'now' is merely a linguistic one, and derives from different uses of the word "becoming". Whitrow uses the expression "the becoming of the 'now'" to designate that (one) member of the set of consecutive instants at which the future turns into the past. And since he believes that the future consists (at each moment) of unrealised possibilities of events and the past of determined events, the moment 'now' which separates the two is, for him, that moment at which the undetermined turns into being or becoming determined. For him, then, 'becoming' refers to events.

Grünbaum, on the other hand, says explicitly that "'becoming' means 'becoming for me now'"[7] which makes the word subjective and personal. Since the word "now" is strictly speaker-reflexive, the phrase "becoming for me now" also takes on ego-centric connotation, and the process or experience to which it refers therefore becomes a personal one. In other words, 'becoming', for Grünbaum, is an epistemological concept, while for Whitrow it carries ontological meaning. Their disagreement is therefore, I believe, somehow specious, since it is one of terminology.

It seems to me, however, that one can quarrel with both uses to which the word has been put.

Grünbaum's identification of the word 'becoming' with the expression 'becoming aware of' is clearly unusual, to say the least. Of course anyone can use any word in any way he wishes, providing he makes his intention clear. But he must then not be surprised if his unorthodox use of a key word gives rise to misunderstandings and results in disputes. By 'becoming' we do not usually mean a perceptual process, nor an experiential one, but an objective change in a matter of fact or a state of affairs – a process which takes time, or conversely,

---

7 Grünbaum, *loc. cit., passim.*

the time which a process takes. Grünbaum's use of the word is therefore certainly misleading.

On the other hand, Whitrow's description of the becoming of events as an objective process in the world which changes their qualitative character and marks a transition from a realm of possibilities to that of actualities burdens the concept with an ontological task which, I think, overtaxes its strength and under which it might well break down altogether. In his use of the word, everything that happens, has happened and will happen does or did so at the moment of becoming, and the word becomes a synonym for the word 'present'. When he says: "By 'becoming', we mean the happening of events, and *hence* the distinction between past, present and future...",[8] the word "hence" does not seem to me to be justified. There is no *logical* relation (as between antecedent and consequent) between the happening of an event and the distinction between future and past. The two concepts belong, I believe, to different categories, or at least to different modes of speech. Let me try to explain what I mean:

Every event, every change, every development that occurs in the universe can be considered as a process of becoming – whether it is positive, constructive, such as growth, increase in heat, the imprint of a foot in the sand; or negative, destructive, such as fire, radio-active decay, and the dispersion of the footprint by the wind; i. e. whether it de-creases or in-creases the entropy of the local system. (Actually, we use the word more commonly in the constructive cases, that is, those of entropy decrease.)

These events happen at all times: they have happened in the past and will continue to happen in the future. But the past-present-future series is logically independent of the concept of a series of events. The former refers to and requires the presence, "the special point of view"[9] of a sentient being, whose interaction with the world always takes place in what we call 'the present'. It is a contingent fact that my present happens to be the moment at which I am writing this, but it is by definition, by linguistic convention, and therefore in some sense a *necessary* fact that all action takes place in the present, because this is how we have defined the word 'present'. However, it is an *interpretation* that I am putting on the word 'present' when I identify it with the 'moment of becoming' (as Whitrow does). That at this moment my relative future and my relative past should suffer, qua temporal extensions, a qualitative change so grave as to alter their ontological status *sub specie aeternitatis* from one of mere possibility to that of actuality and determinedness is a claim whose truth has not yet been sufficiently substantiated by arguments.

Whitrow himself has drawn our attention[10] to the distinction McTaggart first made between the A-series (the past-present-future-series) and the B-series

---

8 Whitrow, *loc. cit.*, p. 422.
9 Cf. Ayer, A. J.: *The observer's special point of view* ... In: Problems of Knowledge, London, 1963. *Passim.*
10 Whitrow, *loc. cit.*, p. 425/6.

(the earlier-later relation), but he then blurs this distinction by claiming objective reality for both series. While it is defensible to maintain that of two consecutive events the earlier one will always remain earlier, *given the direction of time,* and so has at least a fixed relative objectivity in the series of events, their pastness and futurity is not a property which adheres to them, but requires the reference to a third "event", viz. the presence of a speaker, writer, or observer. The property of belonging to one group in the A-series is therefore not a distinguishing or uniquely defining characteristic within the system of events.

I presume it is this necessity of reference to an observer that gave rise to Grünberg's unhappy formulation (of "becoming" as "becoming aware of").

III

There is, however, a real dichotomy at the basis of this dispute. And that is the age-old question whether the temporal world is to be conceived as a stretch before (or below) God's eye (or *sub specie aeternitatis*) as something which is already "all there", and which we only "come across" (as it were) in time – something like Minkowski's 'block universe', a predetermined, static entity, which we, as humans, pass along or pass through in time, thereby getting to know it only successively. Or whether the world is "open" at the present moment, determined only up to now, and truly indetermined, not-yet-there, from the present into the future, genuinely "becoming" at every moment that passes. The latter would, I believe, be Whitrow's view, while the former view has been attributed to Grünbaum, who, however, denies holding it.

The trouble with both hypotheses is, I believe, that they are only descriptive models, attempts to describe our experience of the temporal characteristics of our and the world's existence. And I feel that we are being pushed by this apparent dichotomy into a choice between two alternatives which are not genuine ones. Let me try to analyse these two views in turn.

If we accept Wittgenstein's characterization of the world as "everything that is the case",[11] – a description which I find innocuous and vague enough to be unobjectionable – then, clearly, the future is not yet part of the world, because it is not yet the case. What-is-the-case is what happens, are the events, – and future events have not yet happened, are not yet events. Therefore, to say that the world is already all there and that we only come across it in time, is to attribute to both, time and the world, properties which neither of them possess.

The world of tomorrow is not yet "there" (or anywhere), because it has not yet happened, "its" events have not yet occurred and are therefore not events. To speak of them as such (as I have just done), is to misname them. When we make claims about the world of tomorrow, or of next year's, we do this on the basis of very justifiable assumptions, arrived at by inductive reason-

---

11 Wittgenstein, L.: *Tractatus Logico-Philosophicus,* (1922) (1. "The world is everything that is the case.")

ing, that most of the gross matter of today's world is going to persist in space until tomorrow (at least). That is to say: we speak of something that is not yet the case on the assumption that it will be the case. And in the majority of situations, our assumptions will be justified by tomorrow's occurrences.

We could, of course, identify the world, not with all the events happening in it, as we do in Einsteinian dynamics, but conceive it as static Newtonian conglomerate of matter. "The world" then means the same as "all the matter in the world" (i. e. all the stars, nebulae, galaxies, atoms, etc., but *not* their motion). And since the galaxies, nebulae, stars, etc. *are* composed of *moving* atoms, the world consists merely of atoms (or their parts). On this identification, "the future of the world" means merely, "the atoms as they will be arranged tomorrow, or next year", – and it becomes quite feasible to assume that they (or their parts) will still be there tomorrow – *sub specie aeternitatis* or not. But since this static world is void of motion, it *has* no events, and our "coming across it in time" would consist in an addition, a superposition, of a temporal element in all its forms – as change, growth, electron and molecular motion, etc. – unto an idealised isomorphic distribution of immobile particles which would then, I presume, have to be at zero temperature. Surely, this fantastic picture is not the one we have in mind when we speak of tomorrow's world. This shows that we cannot exclude the temporal element from the future world, even in thought; rather, if we come across a block universe in time, we would have to "come across" a dynamic, temporal world *in* time, or *with* time – and this, in turn, would involve a second or third dimension of time.

Also, to attribute to time some capacity to transport us from today to tomorrow (as we do when we say that the future world is already "there" and we only come across it in time) is to take the metaphor that "time flows like a river" too literally. Since spatial properties are perceptible, available to direct experience through our senses, while temportal properties are not, we have, from the early beginnings of language, become accustomed to using spatial models and metaphors to describe temporal properties. Moving bodies are perceptible objects which have temporal as well as spatial properties, since a moving object will pass different fixed objects in space at successive intervals of time. Therefore we often use moving objects (and sometimes motion itself) as convenient models to demonstrate temporal qualities: we use the flowing waters of rivers to describe the passage of time – while it is, in fact, time as a schema of ordering the succession of events that enables us to say that a river moves.

To say, therefore, that time is like a river, or rather, to think that it is like a boat on the river which we can enter and which will transport us to the next day or year – as it does to the next bridge along the canals of Venice – is to confuse the gondolieri's efforts with the motion of the boat that they produce. If it were true that the world's future is all there, *sub specie aeternitatis,* and we only come across it in time, then *we,* too, would have to be already there today in tomorrow's world – and how could we then come across ourselves tomorrow?

As we see, this descriptive model breaks down very quickly upon a little analysis.

Let me now take up the other horn of the dilemma. What do we mean when we say that time has two aspects, Being and Becoming, and that the present moment is the moment of becoming, – is that instant at which the world turns from not-being into being, from the realm of possibilities into determined actuality? It will seem to me that we are here subject to another illusion.

We can, again, identify the actual with "that which is the case", and the future with that which *might* be the case, that which is logically possible, but which is not (yet) the case. Now, to say that the future is that which *will* be the case is to be expressing a tautology: this is what we mean by the word "future". But if we say (with Whitrow) that the future is the realm of possibilities, we are giving the word a different meaning: we are considering it as designating, not that world of events that is going to be the case after now, but an empty temporal extension *without* events, as seen from our, the observers', present – to be filled with events whose nature we, at this moment, do not know. So considered, the present 'now', the "moment of becoming" can be seen as that instant at which our guesses or calculations about the future's innumerable possibilities of being turn into one definite actuality. But guessing and calculating are activities of the intellect, and their result is based on the information we have at the present time and on our powers of inductive and deductive reasoning. "Becoming determined for me now" would therefore mean no more than "becoming a definite bit of knowledge from an indefinite piece of extrapolated anticipation of the so-called 'realm of possibilities'" which is considered indeterminate only because we, in our ignorance, have not (yet) been able to determine it.

In the present state of our scientific and philosophical knowledge – that is to say: since Einstein – we believe that the world is best described by a dynamical model, as consisting of nothing but space-time events. It is, on this view, reasonable to assume that the world will consist of a similar group of space-time events tomorrow. Of this, we cannot be sure, but we can give fairly good arguments to defend this anticipation as a rational inductive inference (or, if you prefer, as a wellfounded conjecture). Temporal properties are an integral part of the world as conceived in this model: the world that is today, its existence or being, contains relative motion, acceleration, succession of events, rate of change, etc. In other words, its existence itself has temporal aspects. So will tomorrow's world, we assume, by tomorrow evening. What is it, then, that *becomes* between today and tomorrow – what could it possibly be? 'Becoming', I take it, is a dynamic word; it designates the change from one thing into another, or of some of its properties into others; even more: it is often taken to designate the coming-into-existence of a thing, its coming-into-being, from non-existence. But the world which is *everything* that is the case exists, has its being, today – complete with all its temporal properties (just named). Its being, today, is dynamic. Tomorrow's world, i.e. the world as today we believe it will be tomorrow, and as by tomorrow evening we shall *know* it to have been, is also dynamic in its being. What *dynamis* is there, I should like to know, between today and tomorrow that we could call 'the becoming of the world' or 'the becoming of events' which is not already accounted for in today's or in tomorrow's world? What moving force is there, what further change could be effected, that has not been accounted for in every moment of the past, as it will – if I am right – be accounted for at every moment of what we call (but what, tautologically, is not yet) the future?

If we can give any meaning at all to such metaphysical notions as 'Being' and 'Becoming', and if we try to consider them in the light of modern science at all, then we are forced to the conclusion that *the being of an event* and *the becoming of an event* are one and the same thing, since there exists no difference regarding the temporal (and spatial!) properties between an event that is and one that becomes. I do not see what insight regarding the nature of Time we can gain from making a distinction between them.

# Whitehead and the Philosophy of Time

W. MAYS*

*Summary.* In this paper I examine the approach of philosophers and others interested in describing our direct human experience of time, and the difficulties involved in reconciling such descriptions of time, with those given by scientists. The views of Merleau-Ponty and Husserl are compared with those of Whitehead. I consider Whitehead's attempt by means of his Method of Extensive Abstraction to bridge the gap between the time of human experience and that of science, and examine the cogency of Grünbaum's criticism of this method. I discuss Whitehead's account of congruence, which for him is connected with our recognition of sameness or uniformity. I next consider Whitehead's views on simultaneity and Northrop's and Grünbaum's criticisms of them, and point out that Whitehead was concerned with simultaneity in sense-experience rather than instantaneousness in physics, and that his account of simultaneity is an epistemological rather than a causal one. I conclude that Whitehead, unlike Grünbaum, does not believe that there is necessarily an isomorphism between the structure of the mathematical continuum and that of physical time.

I

Whitehead's account of time is still worth looking at. He was acutely aware of the problems which arise if one does not want to dismiss the time of our direct human experience as a merely illusory reaction on the part of our minds to the physical world. Although interested in time as a human phenomenon, Whitehead was also concerned to show its relation to the time of scientific thought, which he regarded as only dealing with certain formal relational aspects of our changing human experience.[1]

One finds Whitehead's position echoed among philosophers interested in the phenomenology of temporal awareness. Merleau-Ponty, for example, is highly critical of the abstract concept of time when it is taken as descriptive of the time of human experience. In ordinary life, he points out, "Everyone talks of Time, not as the zoologist talks about the dog or the horse, using these as collective nouns, but using it as a proper noun".[2] Time, he goes on, is sometimes even personified and regarded as "a single, concrete being, wholly present in each of its manifestations, as is a man in each of his spoken words".[3] Merleau-

---

* Wolfe Mays, Reader in Philosophy, Dept. of Philosophy, University of Manchester, Manchester 13, Great Britain.

1 The following books discuss aspects of Whitehead's philosophy of time: Hammerschmidt, William W.: *Whitehead's Philosophy of Time.* New York: King's Crown Press 1947. – Palter, Robert M.: *Whitehead's Philosophy of Science.* Chicago: University of Chicago Press 1960.

2 Merleau-Ponty, M.: *Phenomenology of Perception.* Translated by Colin Smith, p. 421. London: Routledge and Kegan Paul 1962.

3 *Ibid.,* p. 421.

Ponty believes that "There is more truth in mythical personifications of time than in the notion of time considered, in the scientific manner, as a variable of nature in itself, or, in the Kantian manner, as a form ideally separable from its matter".[4]

In this sort of approach Merleau-Ponty is putting forward a diametrically opposed view of time to that normally accepted in scientific discussions, one which agrees more with the descriptions given by literary men and artists than that given by physicists. Of course, much depends upon what is truth for whom. Obviously his remarks apply specifically to the human cultural situation rather than to the scientific context, in which the personification of time (as used in the physicist's sense) would be regarded as a gross anthropomorphism. I remember once in Manchester a physicist protesting that a true picture of time could only be arrived at by a study of such things as atomic clocks and other physical processes. The philosopher, historian, poet and artist could only be interlopers in such a field, especially as the individual and his cultural world were in the final analysis contained in the physical world, which could be studied by the precise mathematical and experimental methods of physics. As a final knock-down argument, he could assert that the physical world existed in time before human beings and human consciousness ever appeared on the earth.

The latter point has been taken up by Merleau-Ponty, who asks what could be meant by the statement "that the world existed before any human consciousness".[5] One reply to this might be, he says, "that the earth originally issued from a primitive nebula from which the combination of conditions necessary to life was absent".[6] But as against this he makes the point that "every one of these words, like every equation in physics, presupposes *our* pre-scientific experience of the world",[7] upon which the meaning of the above statement is based. "Nothing", he says, "will ever bring home to my comprehension what a nebula that no one sees could possibly be. Laplace's nebula is not behind us, at our remote beginnings, but in front of us in our cultural world."[8] One might counter such a remark by saying that one can at least conceive or imagine such a nebula, which is presumably what Laplace did, just as one can conceive multi-dimensional spaces without being able to have sense-awareness of them. But I suppose that what Merleau-Ponty really means here is that these concepts are not independent of man and the cultural environment in which he finds himself, and that they ultimately have to be cashed in terms of human meanings.

Whitehead, at least in his early work, attempts to give an analysis of scientific concepts in terms of actual experience. He would, however, rather base it on the data given in our sense-awareness than on the cultural world as Merleau-Ponty does, although their views have much in common. Whitehead

---

4 *Ibid.*, p. 422.   5 *Ibid.*, p. 432.   6 *Ibid.*   7 *Ibid.*   8 *Ibid.*

criticised traditional scientific theories because, as he puts it, they give no intelligible account of the meaning of such important physical concepts as "velocity", "momentum" and "stress".[9] Against Whitehead's position one can either put forward something like a Platonism – that physical concepts are intellectually intuited and are hence not reducible to (or derivative from) sense-experience; or accept a conventionalist approach, and say that scientific concepts are postulated and hence do not depend for their meaning on sense-experience, but on our freely created concepts or on our definitions of them. Whitehead's views on this question have a somewhat deceptive simplicity, as he uses some very sophisticated logico-mathematical machinery whose ontological status is not always clear, to relate scientific concepts to empirical data.

Merleau-Ponty, following Husserl, conceives time as a process of "self-production". This he distinguishes from what he calls constituted time, the series of possible relations in terms of before and after and which he claims is not time itself, but the ultimate recording of time.[10] This sort of time has a spatial character since its moments coexist in thought in the form of a linear series. On the other hand, in what he calls true time "with the arrival of every moment its predecessor undergoes a change". Hence, he claims, "time, in our primordial experience of it, is not for us a system of objective positions, through which we pass, but a mobile setting which moves away from us, like the landscape seen through a railway carriage window".[11]

Interestingly enough a similar position is taken up by the physicist Bridgman, for whom the time of experience consists of a blurred sequence of memories culminating in the budding and unfolding present having a unique apex with the possibility that everything may go awry. He points out that under the influence of the mathematical expression of time in scientific theorising, we nearly all think of time as a homogeneous unlimited one-dimensional sequence, all past time on the one side, all future time on the other, separated by the present which is in continuous motion from the past to the future. Because of this we tend to assume that the future has existence and is essentially predictable. Bridgman contrasts this approach with that of the Greek who thought of himself as facing the past with the future coming up over his shoulder, as the landscape unfolds to one riding back to the engine in a train. Although, he adds, even this picture did not get rid of the idea of the existence of the future, but it did emphasise that the future is unknown.[12]

---

9 Whitehead, A. N.: *An Enquiry Concerning the Principles of Natural Knowledge.* Cf. Chap. 1, Meaning. Cambridge: Cambridge Univ. Press 1919.

10 *Phenomenology of Perception,* cf. p. 415.

11 *Ibid.,* pp. 419—420.

12 Bridgman, Percy W.: *The Nature of Physical Theory.* Cf. pp. 29–32. Princeton: Princeton University Press 1936.

## II

One of the difficulties in understanding Whitehead's views on time is that of grasping his philosophical approach to experience. His refusal to bifurcate nature is really a rejection of Cartesianism, with its doctrine of psychic additions to nature. He refuses to divide the data of perception (what he calls the seamless coat of experience) into primary and secondary qualities; the former belonging to the perceived objects, the latter a product of mental excitement. For Whitehead everything perceived is in nature. "We may not pick and choose. For us the red glow of the sunset should be as much part of nature as are the molecules and electric waves by which men of science would explain the phenomenon".[13]

In his early work, at least, he is concerned with describing and analysing how these various elements of nature are connected. In this respect he conceives himself as adopting our immediate instinctive attitude towards perceptual knowledge which is only abandoned under the influence of theory. Whitehead's anti-bifurcationist approach has a certain resemblance to Husserl's phenomenological reduction. Husserl with his watch-word "Back to the things" argues that the analysis of meanings and opinions whether of common-sense or more sophisticated positions, is not the primary objective of philosophy. Philosophy, he argues, must begin with the phenomena themselves – all study of theory takes second place. Thus instead of trying to explain our perceptions by means of physical stimuli and changes in our nervous system, which involve a reference to sophisticated scientific theories, we should concentrate on describing the immediately observed.

In this connection it is of some interest to examine Husserl's phenomenological examination of time-consciousness. Husserl argues that we must first completely exclude any assumption, stipulation or conviction concerning objective (or public) time. Just as the objective real world is not a phenomenological datum, so also is not the time of natural science, in which he includes psychology. By objective time he then has in mind that in which all things and events – material things with their physical properties, minds with their mental states – have their definite temporal positions which can be measured by chronometers.

Although objective time might have its basis in immediate experience, Husserl is not primarily concerned with this problem. He makes it clear, for example, as Whitehead also does, that the sensed equality (or simultaneity) of phenomenological temporal intervals cannot be equated with the objective equality of intervals of physical time.[14] Husserl is primarily concerned with an epistemological analysis of temporal lived experience – with its meaning and descriptive content – and not with a causal study of mental states in terms of their development, formation and transformation according to natural laws.

---

13 Whitehead, A. N.: *The Concept of Nature*, p. 29. Cambridge: Cambridge Univ. Press 1920.
14 Husserl, E.: *The Phenomenology of Internal Time-Consciousness*, translated by James S. Churchill. Cf. p. 26. The Hague: Martinus Nijhoff 1964.

In his discussion of time-consciousness, Husserl endeavours to analyse some of the formal relationships which we may discern in our awareness of temporal passage. Examples are: "(1) that the fixed temporal order is that of an infinite, two dimensional series; (2) that two different times can never be conjoint; (3) that their relation is a non-simultaneous one; (4) that there is transitivity, that to every time belongs an earlier and a later; etc."[15]

His account of temporal experience bears a certain resemblance to that of Whitehead, except that the latter's is worded in terms of events rather than temporal intervals, and between which holds the relation of extension which describes the way a larger event extends over smaller events in our perceptual experience. In addition Whitehead's view that public time is of a much more abstract character than experienced time also has certain similarities with Husserl's position. Husserl, unlike Whitehead, is not primarily concerned to give an analysis of the structure of perceived temporal change. Further, since for Whitehead the basic units of experience are events which have both spatial and temporal components, he does not regard time itself as a primitive datum phenomenologically.

III

Whitehead rejects the classical Newtonian theory of absolute time, which is taken as self-subsistent and independent of its content, i. e. matter. On this theory time is regarded as an ordered succession of durationless instants, which are known to us as relata in the serial time-ordering relation. We are aware of this time-ordering relation in itself concurrently with our knowledge of the things occurring at these instants. In holding this view Newton was no doubt influenced by the order of numbers and their generality: points and instants are in this respect like numbers – they are indifferent to whatever is situated in them.

Whitehead also rejects the relational theory of time, which in the history of thought is primarily associated with the name of Leibniz. On this theory time is a set of relations having as relata either material things or sense-qualities. With Hume, for example, time is thought of as a certain order of our impressions conceived as passive in character. Whitehead would argue that in both cases such timeless endurances cannot give us the flux and the creative passage of the nature we experience.

Whitehead asserts that we never directly experience time as a succession of instants, and that one can only think of it metaphorically either as a succession of dots on a line or a set of values in certain differential equations. Although the postulation of points as ultimate or primitive elements is quite legitimate for mathematicians engaged in purely mathematical studies, it is another matter if the mathematical point continuum is given a role beyond its analytical

---

15 *Ibid.*, p. 29.

one, and regarded as a basis for our description of experienced events. The product of such an analysis, he asserts, is time "as a simple linear series of durationless instants with certain mathematical properties of serial continuity".[16]

The mathematical concept of time has, Whitehead notes, tacitly crept from books on mathematical physics into general scientific thought as expressive of the ultimate structure of space-time. Whitehead claims that on this view velocity, for example, cannot be defined by simple reference to one instant, since one needs a neighbourhood of instants to do this. Further, in the biological field every expression of life takes time, as nothing that is characteristic of life can manifest itself at an instant.

It is clear that Whitehead is right to maintain that we never perceive temporally unextended instants, as all our factual knowledge is confined to observations over a period of time. The belief that there is an instantaneous present directly experienced by us is, he says, a case of warping experience by theory. For Whitehead, what we are immediately aware of is a duration with temporal thickness, its earlier boundary being blurred by a fading into memory and its later boundary by an emergence from anticipation: the present is a wavering breadth of boundary between the two extremes.[17] He concludes therefore that the whole conception of nature at an instant is an abstract conception, involving as it does the belief that there is an ideal exactitude of observation, a belief which nevertheless is useful for the purposes of common sense and science.

One of the reasons why Whitehead would not have been sympathetic to the ordinary language approach to philosophical problems and especially to that of experienced time, is that he believes that ordinary language is designed to express clear-cut concepts and that not all sensed phenomena fall within the simplified classificatory criteria of ordinary language. He readily admits that the set of abstract concepts (including the linear concept of time) implicit in the structure of ordinary language has proved itself to be of great pragmatic value in enabling us to handle our commonsense world. Nevertheless, Whitehead believes that such language only gives us a useful abstract for the purposes of life.[18]

Whitehead's attempt to ground the concepts of science on our experience may be contrasted with the views of someone like Northrop who is strongly critical of Whitehead's anti-bifurcationist programme. Science, he argues, has to admit a difference between the postulated and the sensed, "the objective world and the events defined in terms of the scientific objects which compose it are not known by observation but by trial and error postulation, confirmed only

---

16 Whitehead, A. N.: *Time, Space and Material*, p. 44. Proceedings, Aristotelian Society. Supplementary Volume II, 1919.
17 *Concept of Nature*. Cf. p. 69.
18 Whitehead, A. N.: *Process and Reality*. Cf. p. 234. Cambridge: Cambridge Univ. Press 1929.

indirectly through ... the epistemic correlation of some of its events with the phenomenal events which are immediately sensed".[19] A somewhat similar position has been taken up by Einstein who has remarked that "the relation between scientific concepts and sensations is not like that of soup to beef, but rather like that of the cloakroom check number to the coat".[20]

Concepts for Einstein are then logically independent of sense-awareness, and he believes that "we can view only as miraculous that our sense-experience can be unified by our freely created concepts".[21] For Whitehead, however, the age of miracles is past and he would not be entirely sympathetic with the view that scientific concepts are freely created by pure thought alone divorced from our actual perceptions. Whitehead is concerned with the basic question, why do our abstract scientific concepts expressed in mathematical terms apply so well to the physical world? He would, of course, be the first to admit that this effort to harmonize thought and perception is largely a process of progressive approximation.

IV

In his philosophy of time Whitehead therefore believes that one should start one's analysis from the perceivable properties of nature given in sense-awareness, rather than with mental abstractions or with intuited or a-priori data, as Kant thought. For Whitehead, as we have seen, our observed field of experience (or duration) is made up of related events, unit factors which have the character of passage about them, and not of a succession of lifeless sensa as Hume, for example, thought. Each event is seen to develop into the event which is future relative to it; thus the event which is this morning passes into the event which is this afternoon.

Events for Whitehead are particulars which are lived through and cannot recur. Our mode of apprehending such events would seem to resemble the immediate unreflective kind of experience which Sartre has named the pre-reflective cogito – namely before it has been analysed out into objects and qualities. The objects and qualities we discern in events can, on the other hand, recur, i.e. they are recognized as self-identical through time – thus we recognize the same piece of chalk on different occasions. In the case of scientific objects, a much higher degree of intellectual elaboration of natural relations is involved. Whitehead's distinction between events and objects has much in common with the traditional distinction between particulars and universals or facts and essences.

---

19 Northrop, Filmer S. C.: *Whitehead's Philosophy of Science*. In: The Philosophy of Alfred North Whitehead. Edited by P. A. Schilpp, p. 202. Evanston and Chicago: Northwestern University, 1941.

20 Einstein, A.: *Physik und Realität*. Journal of the Franklin Institute. *CCXXI* (1936) 317. Quoted from, Palter, Robert M.: *Whitehead's Philosophy of Science*, p. 4.

21 *Ibid.*, p. 4n.

Further, time and space are conceived by Whitehead as variations on a single basic relation of extension, which he identifies with the passage of one event over another. Thus, he tells us, the event which is the passage of the car is part of the whole life of the street. Also the passage of the wheel is part of the event which is the passage of the car We are not, he says, accustomed to consider the endurance of the Great Pyramid throughout any definite day as an event. We are, he goes on, so trained by language and formal teaching to express our thoughts in terms of the materialist analysis of time, space and material, that we tend to ignore the fact that the primary concrete unit discriminated by us in nature is the event retaining its character of passage.

In his later work, especially in *Process and Reality*, Whitehead puts forward a theory of continuity of events, based on a more general relation than the whole and part relation of extension – i.e. that of extensive connection which the latter subsumes. Every event is regarded as a unit of becoming having an immediate successor. The immediate present gives rise to the novel occasion and thereby plays its part in the creation of the future. Since each unit of becoming is simple and cannot be divisible into parts, we thus seem to avoid the infinite regress implicit in the notion of a temporal interval.

V

An interesting feature of Whitehead's account of space and time is his discussion of congruence or the recognition of sameness in nature, as seen, for example, in our judgments relating to matching, etc., and which forms for him the basis of spatial and temporal measurement. The recognition of sameness is bound up with his view that experienced nature gives us some evidence that events exhibit some uniformity in their relationships to other events. Russell, Whitehead says, pointed out that apart from minor inexactitudes a determinate congruence relation is among the factors of nature which our sense-awareness posits for us. As against this, Poincaré, however, argued that it was doubtful whether there is a factor in nature which might lead any particular congruence relation to play a pre-eminent role. He therefore claimed that we were free to choose any criterion of congruence we wished.[22]

Whitehead sees no answer to either of these contentions if one accepts a materialistic theory of nature, as on this theory nature at an instant is an independent fact. If we have to look for our pre-eminent congruence relation amid nature in instantaneous space, Poincaré, he goes on, is right in saying that nature on this hypothesis gives us no help in finding it. Since absolute space and time are no longer acceptable, conventionalism seems to be the one remaining alternative. It needs to be noted, therefore, that Whitehead is fully aware of the fact that if we start with the concept of instantaneous space, then congruence becomes a purely conventional matter.

---

22 *Concept of Nature*. Cf. pp. 121–124.

On the other hand, Whitehead believes Russell is in an equally strong position when he asserts that as a fact of duration, we do find it and furthermore agree in finding the same congruence relation. It is a remarkable fact, he says, that all mankind without any assignable reason should agree on fixing attention on just one congruence relation. As to the nature of this relation, this is to be found in our ability to recognize the persistence of objects or qualities throughout a period of time. Examples of this are to be found in such material objects as a measuring rod or a pendulum, as well as in certain sets of physical conditions, such as the uniformity of the conditions for the uniform transmission of light, which is presupposed by Einstein's definition of simultaneity.

Such judgments of constancy Whitehead recognizes are, of course, not incorrigible. In practice we try to correct the readings of our measuring rods or clocks for effects of temperature, gravity, etc., rather than rely on direct perceptions of self-congruence from one instance of the use of the measuring rod to another. On the other hand, he points out, even such corrections depend in the last analysis on our own judgments of constancy.

The conventionalist, Whitehead continues, defines congruence by the requirement that Newton's physical laws of motion are true. As against this position Whitehead has argued that uniformity in change was directly perceived and the measurement of time was known to all civilised nations before Newton's laws were thought of: it is this time as thus measured that the laws are concerned with. As far as judgments of spatial congruence are concerned, it is, Whitehead goes on, a fact of nature that a distance of thirty miles is a long walk for anyone, and this does not seem to be a matter of convention. The process of measurement, he concludes, is merely a procedure to extend the recognition of congruence to cases where these immediate judgments are not available.

## VI

Whitehead attempts to bridge the gap between the time of our human experience and the precise concept of time as used in scientific theory, by the introduction of his Method of Extensive Abstraction.[23] The method can be applied both to the derivation of points of space as well as instants of time. We have already seen that events for Whitehead have both a spatial and temporal extension, and extend over each other to form a system. The most important relation from which the method starts is the asymmetrical relation of whole to part, or of larger event to sub-event. Using this relation as a basis for his analysis, he derives his concept of an abstractive set in terms of which points and instants are defined as sets of convergent volumes. In *Process and Reality,* however, Whitehead starts from the more primitive relation of extensive connection which includes that of extension as a special case. But unlike his earlier account of the method, the abstractive sets from which points are now derived are based

---

23 Cf. *Concept of Nature.* Chapter IV, *An Enquiry Concerning the Principles of Natural Knowledge,* Part III.

on the interconnection of abstract postulated regions and not on experienced events.

Whitehead's Method of Extensive Abstraction has been strongly criticised by Grünbaum.[24] Grünbaum points out that part of contemporary mathematics rests on the conception that an interval is composed of an infinite number of unextended point elements. Since every denumerable point has zero measure (i.e. it is unextended) an interval can be consistently regarded as an aggregate of points, only if this aggregate is super-denumerably infinite, namely, if it belongs to a higher class than the points which compose it: this super-denumerability guarantees continuity. Whitehead is therefore faced with the problem of discovering in sensed nature a corresponding super-denumerable infinity of abstractive sets which define these points. Since sense-awareness cannot exhibit the existence of such collections, Grünbaum concludes that Whitehead's attempt to deduce points and also instants from sensory data fails.

Grünbaum's criticism of the method stems to a considerable extent from the acceptance of the view that it is based simply on sensationalist foundations. On the other hand, Nicod, for example, suggested that Whitehead's contribution could be taken as a construction of a pure geometry rather than as an analysis of the real world, since Whitehead starts from an analysis of the terms and relations that nature presents, which already possess the properties of geometrical volumes. The function of the method is then to form a series of volumes which obey the laws of points, so that for every point in the geometry of points, we can substitute the limit of a unit aggregate of volumes and *vice versa*. For Nicod, then, Whitehead's whole account only applies to mathematical volumes for which, unlike points, a specific interpretation exists in sense-perception provided we start from a sufficiently large size of volume.[25] On this view Whitehead would be using a mathematical model to make clear certain relations appearing in sense-perception: although the regions of the model converge to definite limits, the sets of events only approximate towards such an ideal simplicity. We would thus have to postulate ideal events to arrive at such limits.

In this connection Russell made the comment that in psychological space the method does not yield continuity unless we assume that sense-data which have a minimum size below which nothing is experienced, always contain parts which are not sense-data. Russell concludes that the full employment of Whitehead's methods therefore belongs rather to physical space than to the space of experience.[26]

---

24 Grünbaum, A.: *Whitehead's Method of Extensive Abstraction.* Brit. Jl. Philos. Sci. IV (1953) 215–226.

25 Nicod, J.: *Foundations of Geometry and Induction,* pp. 40–43. London: Routledge and Kegan Paul 1950.

26 Russell, B.: *Our Knowledge of the External World.* Cf. p. 121. London: Allen and Unwin 1949.

It is clear, then, that one cannot simply interpret the Method of Extensive Abstraction as literally enabling us to deduce all geometrical entities from sense-experience; and this even though the volumes Whitehead postulates in *Process and Reality* are simple indivisible elements, and do not presuppose points as do ordinary geometrical volumes. One must also not overlook that Whitehead's whole machinery of abstractive sets and its employment of convergent series does itself presuppose Cantorian continuity.

## VII

Grünbaum has also criticized Whitehead's views on simultaneity and congruence. As he bases his criticisms on an earlier critique of Northrop, we need first to examine Northrop's objections. Northrop argues that Whitehead's view that we have a direct awareness of simultaneity, conflicts with the basic definition of the temporal relation of simultaneity for spatially separated events in terms of light waves upon which the theory of relativity rests. Both common sense and Einstein's physics, he goes on, when they admit a public time the same for all observers, bifurcate nature into the intuited relation of simultaneity varying from person to person, and the postulated simultaneity of physical theory, the same from person to person at rest relative to each other on the same frame of reference.[27] Northrop therefore concludes that science requires bifurcation and indeed cannot avoid it.

Northrop, however, fails to appreciate that Whitehead is concerned with sense-experience, before one applies the categories public/private to it. Our immediate experience for Whitehead has a subject-object structure in which we are aware of ourselves as related to other perceived events in nature, between which a relation of simultaneity holds. Although Whitehead believes there is an objective world of events which we share in common with others, and of which different percipients may have different perspectives, this public world of common sense already involves an element of interpretation for him. He would not deny that science requires bifurcation, and he would readily admit that science would not have advanced without it. What he objects to is the taking of the scientist's system of abstract concepts as valid for the rest of experience.

In any case Whitehead distinguishes just as clearly as does Northrop, between perceived simultaneity and what he calls instantaneousness (or nature at an instant) which forms the datum of science and corresponds to Einstein's concept of simultaneity. But Whitehead's concept of simultaneity, it is pointed out, is not concerned with light or any other type of physical signal and applies only to actual perceivable events. In addition Whitehead makes it clear that in a physical system employing the concept of instantaneousness, no such unique relation will exist. To some extent then, the conflict between Whitehead's account of simultaneity and that of the relativity physicist is a verbal one,

---

27 Northrop, Filmer S. C.: *Whitehead's Philosophy of Science.* Cf. pp. 200–201.

but not entirely, as he would regard the public time of the astronomer which Northrop falls back upon, as involving a large degree of conceptual interpretation.

Grünbaum has taken up some of the points made by Northrop. He regards the concept of congruence – our means of judging spatial or temporal equalities – as conventional. With regard to simultaneity, Grünbaum points out, Einstein assumes that within the class of physical events the readings of natural clocks do not define relations of absolute simultaneity under transport (i.e. they are non-synchronous). The failure of human signalling and measuring operations to disclose relations of absolute simultaneity is therefore only the epistemic consequence of the primary non-existence of these relations.[28]

Grünbaum does not think that Whitehead's historical observation that the human race possessed a metric prior to the statement of Newton's laws invalidates Poincaré's contention "that (1) time-congruence in physics is conventional, (2) the definition of temporal congruence used in refined physical theory is given by Newton's laws, and (3) we have no direct intuition of the temporal congruence of non-adjacent intervals".[29]

In the case of space, Grünbaum would claim that it too is deficient in such a metric. He does not therefore think that Whitehead is entitled to regard coincidence (or matching) as the only test of congruence; that measurement presupposes a criterion does not at all invalidate the conventionality of the self-congruence of our rods at different places. He tries to meet Whitehead's point that a distance of thirty miles is a long walk for anyone, by saying that this is due to our gait being tied to the yard (or metre) stick, "so that an interval which measures thirty miles in the metric of the yardstick will contain a great many of our steps".[30] But he does not believe that this shows that the self-congruence of the yardstick is non-conventional. What Grünbaum seems to overlook here is the feeling of fatigue that a normal person would have after he had walked a distance of some thirty miles. Usually fatigue feelings are a direct function of the number of steps taken, but as a phenomenological fact one usually experiences the fatigue rather than counting the steps, unless one is the sort of obsessional who is careful not to step on the cracks of pavements.

Grünbaum further argues that the agreement which obtains between the metric of psychological time and the physical time congruence defined by Newton's laws and a variety of physical processes, cannot be invoked against the conventionalist view. The ability of men and animals to make successful estimates of duration derives from the fact that the metric of psychological time is tied causally to the physical processes which may define time congruence in

---

28 Grünbaum, A.: *Whitehead's Philosophy of Science*. The Philosophical Review. *LXXI* (1962) 222–223.
29 Grünbaum, A.: *Philosophical Problems of Space and Time*, p. 53. New York: Knopf, 1963.
30 *Ibid.*, p. 62.

physics. He concludes that the fact that psychological time possesses such a metric gives us no information about physical time, since he has demonstrated that the latter has no such metric.[31]

Grünbaum, unlike Northrop, does not sufficiently realize that Whitehead is talking about sensed simultaneity and not instantaneousness (or simultaneity in physics). This is particularly the case when Grünbaum quotes experimental work from biology and psychology to show that psychological time as investigated empirically (in terms of chronometers, etc.) is merely a function of physical time and is hence no guide to it. Grünbaum would seem to be appealing to something like a causal theory of perception as evidence for the falsity of Whitehead's views. But such a theory for Whitehead gives what he considers to be abstract scientific notions a greater reality than the perceptually given. Further, psychological time as investigated by the psychologist in his laboratory would for Whitehead be just as much an interpretation of experienced time as are the accounts of time given by the physicist.

In any case Whitehead is not concerned with the causes of our knowledge. He is engaged in an epistemological rather than a causal enquiry. Whitehead would say that the congruence judgments of physicists, biologists and psychologists, ultimately depend on their recognition of self-identities in their experience of matching, etc.

Commenting on Whitehead's anti-bifurcationist philosophy of nature, Grünbaum says the verdict on it must be the same as that which Pauli gave on unified field theory, namely, "What God hath put asunder no man shall join!"[32] For Whitehead as far as our immediate experience is concerned, the reverse is really the case.

VIII

In a more recent work[33], Grünbaum tells us that James and Whitehead, with Zeno's paradoxes of the Dichotomy and the Achilles in mind, believe that time does not possess the structure of the linear mathematical continuum. According to Grünbaum, they argue "(1) the relations of temporal order among physical events *as they actually happen* are as known to us in our conscious awareness of their coming into being; (2) occurring *now* or happening is pulsational and not punctual" (i.e. our perceptions have a durational threshold) "and (3) the serial order of pulsational coming into being is *not* dense but *discrete*".[34] If we assumed, he goes on, that the temporal order of events is isomorphic with the discrete order of nows of awareness, we would have to deny that the serial order of time is dense.[35]

---

31 *Ibid.*, cf. p. 60.
32 Grünbaum, A.: *Whitehead's Philosophy of Science*, p. 229.
33 Grünbaum, A.: *Modern Science and Zeno's Paradoxes,* London: Allen and Unwin, 1968.
34 *Ibid.*, pp. 45–46.
35 *Ibid.*, cf. p. 52.

Grünbaum attempts to justify the denseness of physical time by denying that physical events as such come into being at all – that they merely occur tenselessly in a network of timeless separation. He therefore rejects the view that physical events must occur in the pulsational consecutive manner in which they are perceived to come into being. The coming into being of events is rather regarded by him as a mind-dependent quality. "Nowness" and "temporal becoming" are like colour and taste not properties of physical events, and have no existence apart from our minds.[36] Hence, he believes that perceived temporal change has no counterpart in the time of physics. The atomicity of becoming, he concludes, "thus turns out to be expressive of an organismic feature of our kind of nervous system instead of warranting the quantization of physical time".[37]

Although it is true that Whitehead uses such words as "now", "here", etc., to refer to the way we are aware of ourselves as related to the rest of our perceptual field, this is an entirely different relationship from the causal one involved in the theory of mind-dependence of colours and tastes, which is concerned with our mental reaction to physical stimuli. Presumably in the same sort of way 'temporal becoming' is to be regarded as a psychological reaction to the tenseless seriality of physical time. Grünbaum's account of temporal becoming is therefore a complete inversion of Whitehead's position, since Grünbaum accepts the view that the structure of the abstract mathematical continuum is isomorphic with that of public physical time, and consigns the quality of temporal becoming to the field of subjective illusion. Indeed the network of timeless separation which Grünbaum takes as the very essence of time would seem to be time as contemplated by God, *sub species aeternitas*.

As far as Grünbaum's argument relating to the inability of our nervous system to perceive continuity in the Cantorian dense sense is concerned, he overlooks the fairly obvious empirical finding that apart from physical quanta neural impulses are themselves pulsational. Whitehead would argue that as soon as we start talking of the biological events in our nervous system, we already presuppose a physiological theory involving the application of concepts and criteria, which involve procedures of intellectual interpretation. Whatever James's position might have been, Whitehead has always asserted that physical time only involves certain features of perceived temporal change, which we rationally reconstruct in terms of an irreversible temporal series.

Grünbaum would seem to believe that there is broad inductive evidence that space and time are continuous in the Cantorian sense. Whitehead has, however, pointed out that owing to the inexactness of measurement, we cannot tell whether a continuous physical quantity possesses the compactness of the

---

36 *Ibid.*, cf. p. 55.
37 *Ibid.*, p. 56.

series of rationals or the continuity of the series of real numbers; that although we usually adopt the latter hypothesis because of its simplicity there are no a-priori reasons in its favour. The continuity of space (and time) therefore rests according to him upon an assumption unsupported by a-priori or empirical reasons.

IX

What I have done in this paper is to begin by examining the approach of philosophers and others interested in describing our direct human experience of time, and the difficulties involved in reconciling such descriptions of time with those given by scientists. The views of Merleau-Ponty and Husserl were compared with those of Whitehead. I also indicated that Whitehead's theory of time is closely bound up with his refusal to bifurcate nature into primary and secondary qualities in the classical Cartesian manner.

Whitehead's criticism of the traditional theories of absolute and relative time is connected with his criticism of the way the mathematical concept of time has become absorbed into our everyday life and language. I compared Whitehead's views with those of Northrop and Einstein, who tend to regard the concepts of science as postulated or free creations of our mind. Whitehead is, however, vitally concerned with the question: Why do our mathematical concepts apply so well to the physical world?

Whitehead discriminates our immediate perceptual experience into two sets of entities: events and objects. Events are lived through, objects are recognized as self-identical and can recur. For Whitehead 'time' and 'space', at least in his earlier work, are considered to be a variation of a single whole and part relation – extension. In addition Whitehead accepts something like the notion of intersubjectivity; that we are aware of a common nature which each percipient grasps in his own way.

I then looked at Whitehead's views on congruence, which for him is connected with our recognition of sameness or uniformity in nature, and his criticism of the attempt to regard our congruence criterion as purely conventional in character. I next considered his attempt, by means of the Method of Extensive Abstraction, to bridge the gap between the time of human experience and that of science. Grünbaum has criticized the method on the ground that it is logically impossible to arrive at Cantorian continuity in this way. However, there seems some ground for believing that Whitehead regarded the method as a logical technique only having an approximate fit to sense-experience.

In discussing Northrop's and Grünbaum's criticisms of Whitehead's views on simultaneity and congruence, we saw that Whitehead was concerned rather with simultaneity in our immediate sense-experience than with instantaneousness in physics. Grünbaum in his conventionalist approach to the problem of congruence endeavours to show that since the metric of psychological time is ultimately dependent on physical time, it cannot serve as a guide to it. He thereby overlooks that Whitehead is concerned with an epistemological rather than a causal enquiry in his discussion of congruence and simultaneity. And as against the isomorphism which Grünbaum believes to exist between the structure of the Cantorian continuum and that of physical time, Whitehead has pointed out that we adopt this view because of its mathematical simplicity and not for any other intrinsic reason.

# The Deification of Time

S. G. F. BRANDON†

The thesis of this paper is that religion has stemmed from man's consciousness of Time, and that his reaction to Time has found a variety of expression, including the deification of Time.[1]

The basic premise of this thesis is that human self-consciousness connotes awareness of the three temporal categories of past, present and future. In unsophisticated minds this awareness is, inevitably, of an existential kind: the person concerned remembers past events, or concentrates on present action, or anticipates future needs and happenings. As consciousness becomes more rationally organised, the temporal categories are more sharply apprehended and manipulated, until in a highly sophisticated technological society such as ours the pattern of life is strictly controlled by clock and calendar. Indeed, this domination of Time is epitomised in our modern catch-word 'planning'.

The action of 'planning' involves operation in the three temporal categories: we draw upon past experience in the present, to deal with needs or events anticipated to occur in the future. This operation we carry out in both a personal and communal context. And, generally speaking, the most successful persons or societies are those that have been most capable in exploiting their time-sense in the planning of their affairs.

It is, indeed, this ability to exploit our keen awareness of Time that has enabled our species to dominate all others in the struggle for life. The Palaeolithic hunter, far back at the dawn of human culture, exhibited this ability when he sat at the entrance of his cave chipping out a stone hand-axe. His tool-making action involved the three temporal categories. As he sat making his axe, he envisaged future occasions when he would use it; and, in shaping the stone, he called upon his past experience of axes and how best to make them. By this planning he hoped to ensure that when the need arose, he would be equipped with a tool or weapon that would increase his own physical powers.[2]

---

[1] Many aspects of this study are dealt with at length and with full documentation by the author in his book *History, Time and Deity*, published by Manchester University Press, 1965, and abbreviated here *HTD*. See also his contribution entitled 'Time and the Destiny of Man', in *The Voices of Time*, ed. J. T. Fraser, New York; Braziller (1966), and chapter 4 of his *Religion in Ancient History*, New York: Scribners (1969).

[2] On other aspects of Palaeolithic Man's conscious of Time cf. *HTD*, pp. 13-18.

Our time-consciousness has, accordingly, delivered our species from that submersion in the here-now which seems to characterise all other forms of life. It has given us a wide-ranging spectrum of interest and concern. Our historical records inform us of the rich variety of the past experience of our race. And our acute appreciation of the inevitability of change warns us not to rest content with the security of our present situation, but to ensure its continuity and improvement in the future.

This time-sense, however, has a kind of debit side – mankind has to pay a price for this faculty that has made it so successful biologically in the struggle for existence. The price is the knowledge of mortality – each human being anticipates his own demise. Because of this ability to look forward in time, every man and woman is aware that they are subject to the process of decay and, ultimately, death.

Thus, human life is beset by a strange paradox – a paradox that profoundly affects its quality in comparison with other species. Although his time-sense enables man both to secure himself against contingencies that threaten his material well-being and to improve the material conditions of his life, it also makes him aware of his ultimate insecurity – that, despite all his planning, he is doomed to physical disintegration and personal extinction.[3]

Faced with this disturbing prospect, mankind has reacted in a variety or ways, most of which have taken the form of seeking for some kind of security or escape from such a fate. The history of religions records this reaction – in fact, its records can be reasonably interpreted as showing that the basic motive-factor in religion itself is the quest for security from death or the consequences of death.[4] This quest, as we shall briefly see, has assumed a variety of guises among the religions of mankind; but, on the final analysis, they can all be shown to be motivated by the desire for immunity from the effacing process of Time. The rest of this paper will now be devoted to illustrating the ways in which Time has been conceived and dealt with in the more notable religions.

The religion of the ancient Egyptians, which is the best documented and most comprehensive of the ancient religions, provides significant evidence of a positive and confident attitude towards the menace of Time. It concentrated effort on the two aspects of Time which seem particularly to have dominated Egyptian attention. For, envisaging death as the attack of a demonic being and shocked by the physical disintegration that resulted, the Egyptians strove to prevent or reverse this disintegration.[5] They consequently developed a technique of embalming the dead; but the process was not intended only to

---
3 Cf. *HTD*, pp. 5–10.
4 Cf. *HTD*, pp. 12 ff., 206–210.
5 On the ancient Egyptian conception of death cf. Brandon, *The Personification of Death*, in the Bulletin of the John Rylands Library. Vol. 43 (Manchester 1961), pp. 318–22, 333–5; Sander-Hansen, C. E.: *Der Begriff des Todes bei den Ägyptern* (Copenhagen 1942), pp. 8–9; 28–9; Zandee, J.: *Death as an Enemy according to Ancient Egyptian Conception* (Leiden 1960), pp. 23, 185 (B. 9 b).

arrest the chemical decomposition of the corpse, it was part of a magical ritual designed to restore to the deceased all their previous faculties, and it included revivification.[6] After embalmment, the body was placed in a tomb, significantly called the 'house of eternity', which was equipped with all needs of life. Everything was done to make the deceased person for ever secure in his 'house of eternity'. In the case of the pharaohs, mortuary temples were also built for their eternal service, being styled 'Houses of Millions of Years'.[7]

This mortuary ritual, as a practical technique, was aimed at resurrecting the dead and rendering them imperious to physical decay, so that they would live in their tombs for ever. The funerary texts that provided the libretto to this ritual reveal that the Egyptians, in this connection, were concerned also with Time itself. Thus in the following passage from the so-called *Coffin Texts,* which date about 1800 B.C., the deceased is represented as declaring: 'I am Tomorrow! Yesterday, it belongs to me!'[8] And in the celebrated *Book of the Dead,* dating some four hundred years later, the dead person who has been ritually assimilated to Osiris, the dying-rising god of Egyptian religion, exclaims: 'I am Yesterday, Today and Tomorrow'.[9] In other words, the Egyptians sought security from Time by becoming Time itself – by comprehending within their own being, through their ritual identity with Osiris, the three temporal categories of Past, Present and Future.

The Egyptians did not actually have a god of Time; but their iconography illustrates some of their conceptions of the nature of Time. Thus the figure of Ḥeḥ, as it appears on the back panel of the cedar chair of Tut-Ankh-Amun, shows how they symbolised unending Time, represented concretely as 'millions of years'.[10] A kneeling human figure is surmounted by the sun's disc, and holds two measuring staves, which were the hieroglyphic sign for 'millions of years'. In this particular depiction, Ḥeḥ is represented with the *ankh,* the symbol of life, suspended from his right arm, and about him are set the cartouches containing the titles of the dead king. In this context, Ḥeḥ is undoubtedly presented as ensuring to Tut-Ankh-Amun what was tantamount in Egyptian thought to eternal life.[11]

---

6 Cf. Brandon: *Man and his Destiny in the Great Religions* (Manchester: University Press 1962) pp. 35 ff., *The Ritual Technique of Salvation in the Ancient Near East,* in: The Saviour God, ed. S. G. F. Brandon (Manchester University Press 1963), where documentation is given.

7 Cf. Bonnet, H.: *Reallexikon der aegyptischen Religionsgeschichte* (Berlin 1952), pp. 257–60, 833–34.

8 Buck, A. de: *The Egyptian Coffin Texts* II (Chicago 1935), 93; cf. M. Guilmot in *Revue de l'histoire des religions,* CLXXV (1969), pp. 15–16.

9 Cap. 2 xiv. 2, *Papyrus of Nebensi.* In: Budge, E. A. W.: *The Book of the Dead.* Text I (London 1898), p. 177. Cf. Allen, T. G.: *The Egyptian Book of the Dead* (University of Chicago Press 1960), pp. 137, 139; Brandon, *HTD,* pp. 4, 24, 30 n. l.: Morenz, S.: *Aegyptische Religion* (Stuttgart 1960), pp. 74–84.

10 Cf. *HTD,* p. 56 and Fig. 3.

11 Cf. Piankoff, A.: *The Shrines of Tut-Ankh-Amon* (New York 1962), pp. 14–15.

Fig. 1. Ḥeḥ, Egyptian personification of millions of years or unending Time (from a chair found in tomb of Tut-Ankh-Amon). This figure should be reversed left to right

Fig. 2

Fig. 3

Fig. 2. Mithraic conception of Time as Zurvān-Ahriman
Fig. 3. Śiva (Naṭarāja) performing the cosmic dance (11th cent. bronze, now in Victoria and Albert Museum, London)

Fig. 4. The goddess Kālī. Indian representation of destructive nature of Time
Fig. 5. Time equated with Death. From the Dance of Death (after Holbein)

The assurance that seems to characterise the religious literature of the ancient Egyptians, namely, that they were possessed of the means of securing everlasting life, even to the extent of identifying themselves with Time, does not mean that they were incaple of scepticism. Indeed, there is evidence of a realistic evaluation of the effacing flux of Time, as the following passage from the so-called *Song of the Harper,* which was sometimes inscribed in tombs, shows:

> Generations pass away, and others remain
> Since the time of the ancestors.
> The gods who lived formerly rest in their pyramids.
> The beatified dead also, buried in their pyramids.
> And they who built houses – their places are not
> See what has been made of them!
> . . .
> Their walls are broken apart, and their places are not –
> As though they had never been![12]

---

[12] Translated by J. A. Wilson in *Ancient Near Eastern Texts,* ed. J. B. Pritchard (Princeton University Press 1955), p. 467 a.

Such pessimistic sentiments were induced by the spectacle of the decay wrought by Time, which was only too evident in the ruin of ancient tombs and mortuary temples in Egypt.[13] However, despite this strain of cynicism, it was the belief that security from the dissolution wrought by Time could be achieved which prevailed in Egypt until the end of its native civilization.

A very different evaluation of the human situation prevailed in the other great civilisation of the ancient Near East, namely, in Mesopotamia. There a fundamental realism about the constitution of human nature precluded any hope of a happy afterlife.[14] Destiny, it was believed, was decreed by the gods – for example, at Babylon it was thought that Marduk, the city's god, decided the destiny of the state annually at the *akitu* or New Year festival.[15] It was also believed that destiny was indicated by astral phenomena, the 'writing of the heavens'. Hence the development of the pseudo-science of astrology in Mesopotamia as a means of foretelling the future – evidence of the casting of an individual's horoscope dates from 410 B.C.[16] This astralism, which originated in Mesopotamia, gradually spread westward, and became a widely-held belief in Graeco-Roman society. Although Time was not actually deified in this system, the planets, that were believed to determine 'the times and seasons', were regarded as daimonic beings, sometimes called in Greek *archontes* (rulers), because they ruled the lives of the mankind dwelling in this sub-planetary world.[17] The Apostle Paul interpreted the salvation won by Christ as the deliverance of mankind from its enslavement to the planetary powers, which determined human fate and fortune.[18]

It is in the religions of the Indo-European peoples that the deification of Time found its most graphic forms of expression. There is evidence that already in the twelfth century B.C. a god with a name that can be identified with the Persian yord 'Zurvān', i.e. 'Time', was known.[19] Although the first reference to Zurvān as a personification of Time in native Iranian literature is comparatively late, the Greek scholar Eudemus of Rhodes, in the fourth century B.C., testifies to the fact Time was a basic concept of Iranian cosmology in this quotation his works: 'both the Magi and the whole Aryan race ... call by the name 'Space' (*topon*) or Time (*chronon*) that which forms an intelligible and integrated whole, and from which a good god (*theon agathon*) and an evil

---

13 Cf. Breasted, J. H.: *The Development of Religion and Thought in Ancient Egypt* (London 1912) pp. 179–81.
14 Cf. Brandon: *Man and His Destiny*. pp. 74 ff., with relevant documentation.
15 Cf. Brandon: *Creation Legends of the Ancient Near East* (London 1963), pp. 91 ff.
16 Cf. Saggs, H. W. F.: *The Greatness that was Babylon* (London 1962), pp. 490–1; Meissner, B.: *Babylonien und Assyrien*; II (Heidelberg 1925), pp. 256–7.
17 *HTD*, pp. 167 ff.
18 I *Cor.* ii: 6–8.
19 Cf. Zaehner, R. C.: *Zurvān: a Zoroastrianism Dilemma* (Oxford 1955), pp. 20, 88; Duchesne-Guillemin, J.: *Symbolik des Parsismis* (Stuttgart 1961), pp. 36–40.

daimon (*daimona kakon*) were separated out, or, as some say, light and darkness before these. Both parties, however, postulate, after the differentiation of undifferentiated nature, a duality of the superior elements, the one being governed by Oromasdes and the other by Areimanios.'[20]

The significance of this passage is very great. It means that Greek scholarship at this time was acquainted with the fact that Iranian cosmology was dualistic, the two opposing principles being deified as Oromasdes (or Ohrmazd) and Areimanios (or Ahriman).[21] And, further, that these two cosmic deities were derived from a primordial 'intelligible and integrated whole', that was equated with Space and Time.

It is evident that a tradition of cosmological speculation in ancient Iran must lie behind this interpretation reported by Eudemus of Rhodes. And there is later Iranian evidence to confirm it. For example, the primordial nature of Time is clearly defined in this passage from the *Rivāyat:* 'it is obvious that, with the exception of Time, all other things have been created. For Time no limit is apparent, and no height can be seen nor deep perceived, and it (Time) has always existed and will always exist. No one with intelligence says: "Time, whence comes it", or "This power, when was it not?" And there was none who could (originally) have named it creator, in the sense that is, that it (Time) had not yet brought forth the creation. Then it created fire and water, and, when these had intermixed, came forth Ohrmazd. Time is both Creator and the Lord of the creation which it created'.[22]

This philosophical presentation of the primal nature of Time seems to have found religious expression in a myth that deified Zurvān (Time) as progenitor of the Demiourgos. It was told how Zurvān desired to have a son who would create the universe. To this end he offered sacrifice for a thousand years. But, before the millennium was completed, Zurvān was assailed with doubt as to the efficacy of the sacrifices he had offered. This doubt, through momentary, had a fatal consequence; for it caused the conception of a second son who was to be Ahriman, the evil one. When the period of gestation was ended, two sons were born: Ohrmazd, radiant with light; Ahriman, foul and dark. Thus, mythologically, the origin of the Persian dualistic conception of the universe was explained; for Ohrmazd created that which was good and beautiful, while from Ahriman derived all that was evil in nature and action.[23]

It is possible that behind this myth lies the memory of an originally ambivalent conception of the Time-god Zurvān. As we shall note presently, the Indo-Iranian tradition of deity was basically ambivalent. In this myth, Zurvān is presented as the ultimate source of good and evil, and of life and

---

[20] Cited by Damascius (*Dubitationes et solutiones de Principis*, c. 125 bis; In: Bidez, J., Cumont, Fr.: *Les Mages hellénisés* (Paris 1938), II, pp. 69 (15–70). Cf. *HTD,* p. 38.
[21] Cf. Bidez-Cumont, I, p. 66; Zaehner: *Zurvān,* pp. 20, 49.
[22] Cf. Widengren, G.: *Hochgottglauben im alten Iran* (Lund 1938), p. 274.
[23] Cf. Zaehner: *Zurvān,* pp. 419–37.

death. Now, there is evidence in Iranian sources that two forms or aspects of Time were distinguished and personified. One was *Zurvān akarana*, which was Infinite Time, and which was conceived as ultimate reality, and identified with all that is good, true and beautiful. The other form of Time was designated *Zurvān daregho-chvadhātā*, i.e. 'Time of the long Dominion'. This was Finite Time which rules in this world, and which brings decay, old age and death to all living things.[24]

This dual conception of Time was incorporated into Mithraism, and was carried westward into the Roman Empire with the spread of that Iranian mystery-cult. It would appear that Zurvān *daregho-chvadhātā* was identified with Ahriman, the personification of evil, darkness and death, and was represented in Mithraic sanctuaries in the form of a lion-headed monster. The images that have been found usually show the monster's body entwined by a serpent and bearing the signs of the zodiac. Wings sprout from the shoulders, and the hands hold keys and a staff; the emblem of a thunderbolt is shown on the breast, and the figure stands on a sphere. All this symbolism indicates well-known aspects of Time; but the emphasis laid upon the horrific in these statues shows that it is Time as the Controller of man's destiny and as the Destroyer that is portrayed here.[25]

The significance of these images of Zurvān-Ahriman in Mithraic sanctuaries is not clear. But from a statement of Plutarch, that Mithras was the mediator between Ohrmazd and Ahriman, it is possible that the images symbolised the dominion of Time the Destroyer to which mankind were subject, and from which Mithras delivered his initiates and brought them to Ohrmazd, who was probably identified in this context with Zurvān *akarana*, 'Infinite Time' or Eternity.[26]

The Iranian conception of Zurvān *daregho-chvadhātā* can be traced in other cults of the Graeco-Roman world. The Orphic figure of Phanes, as a bas-relief in Modena shows, was influenced by the Iranian personification of Finite Time, as was that of the monstrous Jaldabaoth of Gnosticism.[27] The concept may also have found its way into pre-Islamic Arabia, and inspired the fatalism that Muhammad condemns in the *Qur'ān*: 'They say: "There is nothing but this present life of ours; we die and we live, and it is only Time which destroys us"'.[28]

The idea of Zurvān *akarana*, 'Infinite Time' inspired the Hellenistic conception of Aiōn, and, associated or identified with Sarapis, it became the tutelary

---

24 Cf. Zaehner: *Zurvān*, pp. 57, 87; Duchesne-Guillemin: *Symbolik*, p. 37; Biancho, U.: *Zaman i Ohrmazd* (Torino 1958), pp. 99–100.
25 Cf. *HTD*, pp. 43–5. See also Campbell, L. A.: Mithraic Iconography (1968), pp. 348–57.
26 *De Iside et Otsiride*, 46. Cf. Duchesne-Guillemin: *Symbolik*, p. 88; Brandon *HTD* pp. 45–6.
27 Cf. *HTD*, pp. 47–52.
28 Surah XIV. 23–4.

deity of the great city of Alexandria.²⁹ The emphasis in the concept was on eternity. From Alexandria the concept passed to Rome, where it contributed to the developing *mystik* of Rome's imperial destiny, presented in the twofold aspect of the *aeternitas populi Romani* and the *aeternitas imperii*.³⁰ How Aion or Aeternitas was visualised in Rome has been preserved for posterity in the sculptured scene of the apotheosis of the Emperor Antoninus Pius that adorned the column erected to his memory. The deity is represented in the form of an heroic male figure, nude, and equipped with the majestic wings of an eagle; in his left hand he holds a globe encircled by serpents. The figure is patently allegorical, and its abstract quality is unlikely to have rendered it a dynamic concept to the Romans.³¹

Indian religious thought has shown a great preoccupation with Time in relation to human destiny. It has found expression in theistic imagery in the concept of divine ambivalence, and in metaphysics it has inspired that interpretation of reality which has characterised the Indian philosophy of life. The most dramatic presentation of the deification of Time occurs in the famous *Bhagavadgītā*, or 'Song of the Lord', which dates from the third or fourth century B.C. The purpose of this composition was to inspire *bhakti*, i.e. devotion to Vishnu as the supreme deity. It takes the form of a revelation of the divine nature made to a prince named Arjuna. First, Vishnu displays to Arjuna the multiplicity and complexity of his being as the creator and sustainer of the universe. In this aspect, the deity is benign, but he hints that there is another side to his nature: 'I am the immortal and also death' ... 'I am death that seizes all, the origin of all that shall be'.³² Arjuna is not content to know only one aspect of God, and despite the warning, he asks for a complete vision. His request is granted, and then comes a most terrible revelation. Vishnu is seen as God the Destroyer – a monstrous being, into whose dreadful mouths all forms of life, in an unending stream, are shown as moving to their destruction. And Vishnu speaks in explanation to the terrified Arjuna: 'Know I am Time, that makes the worlds to perish, When ripe, and come to bring on them destruction.'³³

This revelation of the ambivalence of deity as Creator and Destroyer, and its identification with Time, is made in terms of the god Vishnu in the *Bhagavadgītā*. But a similar equation is made in respect of the other great god of Hindu theism, namely, Shiva. This deity also personifies the ambivalence of cosmic existence as the inhabitants of India have sensed it. On the one hand, Shiva embodies the dynamic persistence of life, in all its teeming abundance

---

29 Cf. Sasse, H.: In: *Reallexikon für Antike und Christentum*, I, 195; *HTD*, pp. 55 ff.
30 Cf Sasse in *op. cit.*, I, 197–200; *Kleine Pauly*, I (1962) 104.
31 Cf. *HTD*, p. 60.
32 *B.-G.*, IX. 19, X. 34.
33 *B.-G.*, XI. 32. Cf. *HTD.*, pp. 31–2.

and complexity of form. In iconography, this aspect is symbolised by the *lingam*, the mighty generative organ of the god, or portrayed in images of Shiva as Nataraja, the 'King of Dancers', who performs the cosmic dance, symbolising the energy of the universe, unceasingly creating and sustaining the forms in which it manifests itself.³⁴

In his other rôle, Shiva is Bhairava, 'the terrible destroyer'. He haunts cemeteries and places of cremation; he appears, serpent-entwined, adorned with a necklace of skulls. His titles are significant. He is Mahā Kāla – 'Great Time', or Kāla-Rudra – 'all-devouring Time'.³⁵ In the caves of Elephanta he is portrayed with a symbolism of ominous meaning – in one of his many hands he holds a human figure; in another, a sword or sacrificial axe; in a third, a basin of blood; in a fourth, a sacrificial bell; with two other hands he extinguishes the sun.³⁶

But Hindu thought was not content with this deification of Time, and it produced an even stranger imagery. It hypostatised the *shakti* or activating energy of Shiva as a goddess. In its destructive aspect it took the form of the goddess Kālī, the personification of Time; for kālī is the feminine form of the Sanskrit word *kāla* ('Time').³⁷ The dread of Time finds expression in the iconography of this goddess. She is black in colour; she wears a chaplet of severed heads, and her many hands hold symbols of her ambivalent nature – the exterminating sword, scissors that cut short the thread of life, and the lotus of eternal generation. Often she is repesented as trampling on the corpse-like body of Shiva, from whom she has emanated.³⁸

In this strange, and to Western taste repulsive imagery, Indian culture has sought to portray the cosmic rhythm of life, decay and death of which it has for ever been so acutely conscious. It finds the process synonymous with the process of Time, and so it proceeds to identify the supreme cosmic deity with Time, either in the form of Vishnu or Shiva or the *shakti* of Shiva.

In the mystical philosophy of Hinduism embodied in the Upanishads and the Vedanta, Time is virtually identified with the phenomenal or empirical world which the unenlightened soul (*ātman*) takes for reality. According to this metaphysics, the soul or self, beguiled into supposing that it is a distinct self-conscious entity, becomes enmeshed in the Time-process, which is cyclical. The self is thus subject to *samsāra*, i.e. an unceasing process of rebirth into the empirical world, which involves an unending succession of lives, with all their consequent sufferings and deaths.³⁹ Indian thought is very conscious of the

---

34 Cf. Zimmer, H.: *Mythus and Symbols in Indian Art und Civilisation* (New York 1962), pp. 151–7.
35 Cf. Zimmer, pp. 135, 155.
36 Cf. Zimmer, pp. 148–51; Gonda, J.: *Die Religionen Indiens* I (Stuttgart 1960), p. 261,
37 Cf. Zimmer, pp. 211–12; Eliade, M.: *Images and Symbols* (E. T., London 1961), pp. 64–5.
38 Cf. *HTD*, pp. 36–7 and Plate III.
39 Cf. *HTD*, pp. 97 ff.

misery of empirical existence as the following passage from the *Maitri Upanishad* shows: 'In this body which is afflicted with desire, anger, covetousness, delusion, fear, despondency, envy, separation from what is desired, union with the undesired, hunger, thirst, old age, death, disease, sorrow and the like, what is the good of the enjoyment of desires? And we see that all this is perishing, as these gnats, mosquitoes and the like, the grass and the trees that grow and decay'.[40]

Since it was also believed, according to the doctrine of *karma*,[41] that the self or soul carried over into each life the consequences of all its past actions, the prospect of unceasing existence in the empirical world was, to say the least, daunting – for since the Time-process was cyclical, no end could be hoped for but only a dreary repetition of the same pattern of things. However, the various Indian philosophical systems have offered the means of deliverance from this sorry state. The deliverance, thus offered, is essentially deliverance from Time, as the following passage from the *Chāndogya Upanishad* shows, in which the process of Time is marked by the succession of day and night: 'Verily, for him, who knows thus, this mystic doctrine of Brahmā, the sun neither rises nor sets. For him it is day for ever'.[42]

Buddhism, although it rejected the Hindu idea of an abiding self or *atman*, has similarly equated the misery of human life with subjection to Time, which signified Impermanence.[43] Consequently, salvation or 'Nirvāna' for Buddhists has connoted a state of being beyond Time. The Buddha is thus represented as describing Nirvāna: 'There, monks, I say there is neither coming nor going, nor staying nor passing away nor arising. Without support or going on or basis is it. This is the end of pain'.[44]

In Judaism and Christianity there could be no deification of Time, owing to the principle of monotheism that is the assumed basis of each of these faiths. However, both have been profoundly concerned with the significance of Time, or rather the Time-process, which is interpreted teleologically. In other words, Judaism and Christianity have elaborated what are virtually philosophies of history, in that they have identified the passage of Time with the revelation and achievement of the purpose of God – the doctrine is succinctly expressed in the lines of the well-known hymn: 'God is working His purpose out as year succeeds to year.'[45]

---

40 I, 3–4; trans. Radhakrishnan, S.: *The Principal Upanisads* (London 1953), pp. 796–7.
41 Cf. Dasgupta, S.: *A History of Indian Philosophy* I (Cambridge I), pp. 54–7; Gonda, I, pp. 206–8.
42 III, 11.3; trans, Radhakrishnan, *op. at.*, p. 386.
43 Cf. *HTD*, Plate XII; *Man and his Destiny*, pp. 336 ff.
44 *Udāna*, VIII. 1, trans. Thomas, E. J.: *Early Buddhist Scriptures* (London 1935), p. 110. Cf. Coomarasuamy, A. K.: *Time and Eternity* (Ascona 1947), pp. 37, 47.
45 Cf. *HTD*, chapters V and VI.

It is not within the terms of this paper to describe or analyse either the Jewish or the Christian philosophies of History. But it may be noted in passing that each religion looks forward to the ultimate cessation of Time or the end of the existing world-order, which represents the same thing. The ultimate establishment of the Kingdom of God or the Beatific Vision, to use two forms of imagery current in these faiths, implies an eternal state of being, beyond the change and decay of Time.[46]

Although Time could not be deified in Christianity, we may end our survey by briefly noting how Time came to be personified in Christian thought and imagery.

Two traditions, at first quite distinct, gradually converged in Western Christendom to produce in Renaissance art the figure of Father Time, with which we are all familiar. One of these traditions derived from pagan mythology. The ancient Greek god Kronos had been fused with the Latin god Saturn to form a grim deity connected with Time. His sickle or scythe, traditionally signifying his original association with agriculture or the act of castration recorded by Hesiod, acquired a fatal significance.[47] The importance of Saturn in astrology assured his survival in mediaeval Christian culture, and also invested him with a baleful character.[48] In iconography he was depicted as a morose and aged man.[49] The other tradition was that of the representation of Death. The concept had already been dramatically personified in the Johannine Apocalypse.[50]. In mediaeval art it was represented by an animated skeleton, variously equipped with scythe, axe, or bow and arrows, with which it conceivably dealt the blow or wound of death.[51] By at least 1514, as the famous etching of Albrecht Dürer shows, Death was depicted with the hour-glass, thus denoting its association with Time.[52] The two traditions finally coalesced in the figure of Father Time, conceived as an aged man, winged, and bearing scythe and hour-glass. However, although the figure has become the accepted symbol for Time in the Western World, it has never acquired the deep-emotional significance that the depiction of Time seems to have had in other religions – and the figure of Father Time, let us note, is a personification, not a deification of Time.[53]

This survey has been confined to the deification of Time in the major Near Eastern religions and Hinduism and Buddhism. It is essentially an outline account, designed to indicate some significant conceptions of Time in the reli-

---

46 Cf. *HTD*, pp. 207–8.
47 On the various conceptions of Kronos cf. *Kleine Pauly*, III (1968), pp. 355–64.
48 Cf. Seznec, J.: *La survivance des dieux antiques* (London 1940), pp. 50 ff., Plates xxii (45), xxix (61), xxx (64), xxxii (67).
49 Cf. Panofsky, E.: *Studies in Iconology* (New York, 1962), pp. 73–4, 77, 82; Plates xxi-x 2.
50 vi: 7–8.
51 Cf. *HTD*, pp. 61–2.
52 Cf. *HTD*, Plate VIII.
53 Cf. Panofsky, chapter III.

gions concerned. The various conceptions need to be studied further both in depth and with comparative reference. And the field of investigation should be extended, particularly to the cultures of China and pre-Columbian America. Indeed, the exploration of the cultural and religious significance of Time is a task of immense importance that awaits those scholars equipped to undertake it.

# The Darwinian Revolution in the Concept of Time*

Francis C. Haber**

*Introduction*

The mathematical view of time as a uniform line of successive instants extending from a beginning to some indefinite point in the future is embedded in the metaphysical assumptions of historians today. All historical chronology is based upon it. Although the historian might insist that he is only concerned with existences in the past, rather than with time, and that if the record were complete, he could reconstruct the order of events by putting them edge to edge, the fact is that he accepts the public time given to us by the astronomers as a coordinate for the ordering of events. Whether he thinks of this time as a reified existence or as a heuristic tool, the result is the same for the practice of history at the empirical level of establishing dates and chronologies for events. It is against this mathematical time that he measures rates of change in history and applies terms such as revolution or evolution for series of events. Furthermore, no matter how relational the historian in his philosophical moments may think of chronological time, to some degree it is reified in his thinking when he is doing historical research and writing. In a period of revolution, for instance, it would never occur to him that time had accelerated or was foreshortened, even though he might use some such expressions.

That the modern professional historian accepts time as a straight line is borne out by his resistance to all attempts to put the events of history into cycles, rhythms, or periodicities. He quickly brands those who attempt it, like Oswald Spengler and Arnold J. Toynbee, as metahistorians.[1] He may divide history into periods, but they are only segments in the continuum of time. The uniformitarian character of this time is also part of his metaphysical faith, and

---

\* Research used in this paper was supported by a grant from the Penrose Fund of the American Philosophical Society.
\*\* Dr. Francis C. Haber, University of Maryland, Department of History, College of Arts and Science, College Park, Md. 20742, U.S.A.
1 Oswald Spengler: *Der Untergang des Abendlandes, Gestalt und Wirklichkeit* (Munich: 1919) and Arnold J. Toynbee: *A Study of History*, 10 vols. Oxford: 1934–54. For the extent of reaction by historians to Toynbee's work, see John C. Rule and Barbara Stevens Crosby: *Bibliography of Works on Arnold J. Toynbee, 1946–1960*. History and Theory IV (1965) 212–233, and Toynbee's *Study of History*. Vol. XII, "Reconsiderations". Oxford: 1961 and pp. 227–229 for metahistory.

where gaps in the record appear on the continuum, he assumes that the gap is in the record and not in the events. Interpretational debates of historians often center on how the gaps are filled, that is, whether the historian extrapolates a continuity or a change between documented events. Historical catastrophists debate the intensity of change with historical evolutionists, but unlike the geological catastrophists in the early nineteenth century, they do not attempt to measure the scale of time by means of the events indicated in the record, for the historian's scale of time is taken on faith from the public calendar.

Metaphysical assumptions about time are also present in the historian's explanation of causation. The modern professional historian believes that all historical events are unique. Like the instants in mathematical time, they can never recur again. Even though events may have similarities, their happenings are irreversible. In establishing connections between events it is first necessary to ascertain their before-and-after relationship, for it is not an acceptable procedure to attribute to events a cause and effect relationship that violates the before-and-after time order. Putting events in a chronological sequence is a legitimate method of descriptive historical narrative, but all too often, the historian has an unstated theory of causation in history which leads him to select only those events as significant which fulfill the theory and then by stringing them on a chronological thread lets their sequential time order pass as a causal order.

Empirical historians decry all metaphysics in history[2], but assumptions about the nature of time cannot be regarded in any other light, regardless of how dogmatic the historian's faith may be about their truth. This becomes even more apparent when the history of assumptions about the nature of time in history is examined, for the present point of view is relatively recent. It is the thesis of this paper that although this mathematical view of time in modern history may be traced back to the ancient Greeks or earlier, it did not hold the field as the principal mode of understanding temporal reality in history until after Darwin's *Origin of Species* in 1859.

Prior to Darwin, teleological time flourished along with mathematical time in historical thought. The mathematical time was used for establishing the chronology of actual historical events, but teleological time set the pattern in which the events had their temporal existence. As long as it was believed that God had created the world for a purpose and according to a plan that would be unfolded and reach completion in a preordained course of time, the time of history was transcendental and teleological. This belief had been maintained in Christian cosmology from the Patristic period, especially through the influence of St. Augustine. It was strongly reinforced in Natural Theology during the seventeenth and eighteenth centuries through the Argument from Design, the

---

[2] Chester G. Starr: *Historical and Philosophical Time*. In: History and the Concept of Time, Beiheft 6 of History and Theory (1966).

argument for the existence of God based on the hypothesis of an ultimate design and purpose in the universe. As science brought ever new proofs from nature of the complexity in the Design of the Creation, the conclusion seemed inescapable that only the intelligence of God, and not the play of chance, could have wrought such intricacies as were manifest in the world. If Nature revealed such careful planning, it was easy to assume that history too was the unfolding of a design. Such had been the received view of Christianity, at least since St. Augustine, but even in the secularization of the Enlightenment when salvationist history was under attack, it was easy to retain the concept of a transcendental time pattern in history through a divinized Nature.

By presenting the first plausible alternative to the Argument from Design to explain the history of organic life on the earth, Darwin's theory of evolution made it possible to think of historical time in non-teleological terms. In that sense, I maintain that Darwin introduced a revolution in the concept of time. In addition, as teleology was removed from history it left the nature of historical facts on one plane of temporal significance. Henceforth, causation in history had to be sought in historical events, and transcendental purposes could not be invoked to explain why things happened when they did. A whole vocabulary of expressions remained as mere figures of speech: "the times called men forth," "the time was ripe," "in the fullness of time," etc.

Naturally, these shifts in the understanding of historical time and the nature of historical events did not take place all at once nor through Darwin alone. I propose, then, to give a brief historical sketch to place the nature of Darwin's displacement of teleological time in perspective.

*Sacramental History*

Part of what I have described as the metaphysics of the modern historian has a long history. The idea of the uniqueness of events and their sequential linearity of time in history are often traced back to the biblical view of time and history, and while this is not incorrect, it is by itself inappropriate in the context of early Christianity. It must be balanced within that context which had as its overriding integrating model the drama of salvation. Any view on time or history had to be harmonized within that model or remain outside the Christian cosmology.

Whatever questions St. Augustine may have had about the nature of time, there is no doubt that his concern was to resolve them in a way that would illuminate the workings of God in relation to man and the world. A mathematically uniform or a geometrically inspired metaphor of a line for his view of history is most inadequate. His cosmological time was periodized around the days of creation, each day prefiguring an epoch of history.[3] It was an archetypal view of time, and a closer analogy than a line to represent it would be

---

3 St. Augustine: *The City of God,* XXII, 30.

a book with identical covers. Opening the book completely until the covers were back to back would splay out the span of historical existence with seven chapters listed in the table of contents, the pages standing out in radial form and covered with print. Man moving from word to word on the pages could only read them in succession, although after a few chapters and with the aid of the table of contents, he could dimly discern that there was a story unfolding with many clues along the way prefiguring the future.

Prefiguration was part of a larger "figuralist" or typological view of reality which linked the profane with the sacramental.[4] It had many planes of reality. It was possible to think of Adam or Seth smuggling a branch of the Tree of Knowledge out of the Garden of Eden, and wood grown from the branch serving as the rod of Moses, a beam in the Temple of Solomon, and the cross upon which Christ was crucified.[5] Certain people and events in the Old Testament were seen as figures or paradigms to be reincarnated repeatedly in history; they were shadows of things to come, as well as having had concrete historical existence. History and time in the Bible were to be read and understood literally at one level, figuratively and prophetically at another. It was in congruence with this mode of thinking that the whole universe of correspondences was worked out in medieval and early modern times.

The typological mode of perceiving reality bears a strong resemblance to the archaic or primitive mode of viewing time, with the many planes of perception of reality all interacting, the sacred with the profane, the temporal with the atemporal. It was a union between the historical and perpetual incarnation[6], and this sacramental view of time and history remained as an important mode of thought well into the seventeenth century[7], in many aspects well into the nineteenth century, and for the pious it has never disappeared. There is nothing evolutionary about history in this cosmological frame of reference; it was instead an unfolding in the succession of time. History was typological, and it had a didactic function of using the profane as a means of vivifying the awareness of the atemporal through the temporal senses. A strong survival of this mode of using history to intensify the imagination can be clearly seen in the meditational literature of the seventeenth century, as when St. Ignatius Loyola

---

4 I use the definition of E. O. James: *Sacrifice and Sacrament*. London: Thames and Hudson 1962, p. 16. "To think of the universe as the expression of the divine acting dynamically and immanently in all the processes of nature is a sacramental conception inasmuch as it assumes divine action and control through material forms to effect some cosmic purpose and spiritual reality."

5 Alan W. Watts: *Myth and Ritual in Christianity*. London: Thames and Hudson 1953, p. 55.

6 Jackson I. Cope: *The Metaphoric Structure of Paradise Lost*. Baltimore: The Johns Hopkins Press 1962.

7 *Ibid.* Also see J. A. Mazzeo: *Cromwell as Davidic King*. In: J. A. Mazzeo (ed.): *Reason and the Imagination: Studies in the History of Ideas, 1600–1800*. New York: Columbia University Press 1962, pp. 29–56.

writes that we must see "the places where the thinges we meditate on were wrought, by imagining our selves to be really present at those places; which we must endeavour to represent so lively, as though we saw them indeed, with our corporall eyes. . . ."[8] The meditational force of this use of history was still apparent in the nineteenth century as in William Hazlitt's description of his father lovingly poring over the history of the Old Testament.[9]

I would suggest, then, that the most important impact of St. Augustine on cosmological time was not in connection with its linearity, but its atemporal archetypal pattern, an impact that resulted not from his own originality, but from his dominance in the Christian tradition. This impact comes from his support of a conception of creation, the idea that God had the whole plan of creation in his mind before His Word brought forth the realization and unfolding of the plan. It was an archetypal conception, and it was consistent with the typological or figuralist sacramental view of history. It was also symbiotic with Platonic ideas of all things developing out of seeds or primordial archetypes. Predestination was only one manifestation of the whole model of Augustinian creation, and Calvinism was only a more extreme position amongst sixteenth century Augustinians.

Although the archetypal views about the Creation persisted, particularly in biology and geology, down to Darwin, there was a shift in the typology away from a sacramental view of things in nature in the seventeenth century as nature was secularized and matter was reduced to dead corpuscles under the rule of secondary causes. Even this was an uneven progress and fossils were seen as medals of creation or relics of the Deluge well into the eighteenth century, while the sentimentalizing of nature as a spectacle glorifying the power of the Creator passed into Romanticism. The main shift, however, was to divinize the Design itself, and the clock was a favorite analogy for explaining on new mechanical principles how the universe worked.

*The Clock As Cosmological Metaphor*

In 1825, the Right Honourable Earl of Bridgewater willed £ 8,000 to be administered by the Royal Society of London for the writing and publishing of works illustrating "the Power, Wisdom, and Goodness of God, as manifested in the Creation." During the next few decades a distinguished series of works was published in this Bridgewater series, such as John Kidd's *On the Adaptation of External Nature to the Physical Condition of Man*, Charles Bell's *The Hand, Its Mechanism and Vital Endowments as Evincing Design*, William Buckland's

---

8 Richard Gibbon (tr.): *The Spiritual Exercises of Saint Ignatius*, quoted in Louis L. Martz: *The Poetry of Meditation: a Study in English Religious Literature*. New Haven: Yale University Press, rev. ed., 1962, p. 27.
9 Francis C. Haber: *Age of the World: Moses to Darwin*. Baltimore: The Johns Hopkins Press 1959, pp. 204–205.

*On Geology and Mineralogy* and William Whewell's *Astronomy and General Physics Considered With Reference to Natural Theology*.

The phrase, "the Power, Wisdom, and Goodness of God as manifested in the Creation," was a formula passed down through the centuries and was the remnant of a complex integration of Christian, Platonic, and Aristotelian ideas worked out in a typological manner by theologians of the twelfth century.[10] Plato's Demiurge, or Will, became the "Power" of the Father in the Holy Trinity, Plato's Idea, or Intelligence, became the "Wisdom" of the Son, Plato's Good became the Spirit of the Holy Ghost, or "Goodness." In the Creation, Aristotle's efficient cause was identified with the Power of the Father, his formal cause with the Design, or Wisdom of the Son, and his final cause with the Goodness of the Holy Ghost which sustained the Creation. The Creation, then, "manifested" the Trinity and its attributes. In a sacramental mode, not a sparrow falls, that does not manifest the Power, Wisdom, and Goodness of the Creation. All nature thus becomes a second Holy Writ from the hand of God, to be studied, subjected to exegesis, treated with reverence, and searched for typologies.

Somewhere between St. Thomas Aquinas and the Earl of Bridgewater, the search for manifestations became a search for designs evincing intelligence in the Creation. The more successful man was in making designs and realizing them through art, architecture, and technology, however, the more compelling the Argument from Design became. Furthermore, there were numerous scriptural statements in the same vein and many classical allusions to the Creator as an architect or maker. Clockwork with automatons became an especially convincing analogy for showing design by an intelligence, as well as reinforcing the archetypal concept of creation espoused by St. Augustine.

The clockwork and automaton tradition was closely related to attempts to mechanically simulate natural phenomena. Derek J. de Solla Price has emphasized the "deep-rooted urge of man to simulate the world about him through the graphic and plastic arts."[11] This has been manifested in paintings, figurines, dolls, statues and attempts to simulate the movements of man, idols, and animals. Professor Price points to the importance of cosmological simulacra in the development of automata amongst the Greeks, and in connection with clockwork, he observes: "It would be a mistake to suppose that waterclocks, or the sundials to which they are closely related, had the primary utilitarian purpose of telling time. Doubtless they were on occasion made to serve this practical end, but on the whole their design and intention seems to have been the aesthetic or religious satisfaction derived from making a device to simulate the heavens."[12]

---

10 J. M. Parent: *La doctrine de la création dans l'école de Chartres.* Paris: Publication de l'Institut d'études médievales d'Ottawa 1939.

11 Derek J. de Solla Price: *Automata and the Origins of Mechanism and Mechanistic Philosophy.* Technology and Culture V (1964) 10.

12 *Ibid.,* p. 13.

To some extent any of the technologies can be a source of explanatory analogies for how things work in nature – glassmaking, weaving, spinning, woodworking, shipbuilding, architecture, agriculture, or hydraulics – but the practical arts are only incidentally involved in simulation and therefore lack the same degree of explanatory power as the simulacra, whose principle business is the imitation of how things work. In the Middle Ages, the making of biological simulacra, cosmological simulacra, and sacramental simulacra were developed separately and in combination as automatons in the monumental astronomical clocks.[13] By the middle of the fourteenth century, when escapements for weight-driven clocks had come into use, the art of clockmaking was the most sophisticated of the medieval technologies.[14]

Simple weight-driven clocks for the telling of time, could be made with sufficient skill that a clock such as that in the Salisbury Cathedral, made around 1386, ran for some 500 years. But the craft's sophistication lay in other areas. The astronomical clock, or astrarium, of Giovanni Dondi, completed in 1364, was so elegant and complicated in its reconstruction of the heavenly movements that it became legendary.[15] The 1354 clock in the Strasbourg Cathedral not only had a calendar and astrolabe, but also had brilliantly executed automatons built into the mechanism. At the stroke of the hours, the three Magi came before a statue of the Holy Virgin, who was carrying the child Jesus in her arms, bowed, and then passed on while a carillon built into the clock played melodies from sacred songs. Its most celebrated automaton, however, was the still extant cock, about a meter in height, which at the stroke of twelve, flapped its wings, lowered its tail, and crowed.[16]

When biological and cosmological automatons had reached a stage of development to attract public admiration, they began to be used by scholars for illustrations and analogies. The clockmaker began his journey towards making God in his own image.

St. Thomas Aquinas, when he explained why animals do not have rational souls, invoked the analogy of the clock as an example of something made by reason that was itself lacking in reason, although it moved as though it were endowed with reason. He made the Argument from Design explicit in this analogy by saying that as artificial things are in comparison to human art, so

---

13 Silvio Bedini: *The Role of Automata in the History of Technology.* Technology and Culture V (1964) 24–42.

14 The complexity of geared planetaria in the early fourteenth century, or earlier, is fully illustrated by J. David North in his reconstruction of an astronomical clock from three manuscripts in *Opus Quorundam Rotarum Mirabilium.* Physis VIII (1966) 337–372. Another example is the astronomical clock of Richard of Wallingford (c. 1330), for which both Derek J. de Solla Price and J. David North have discovered texts. See Silvio A. Bedini and Francis R. Maddison: *Mechanical Universe, The Astrarium of Giovanni de' Dondi, Transactions of the American Philosophical Society,* n. s., Vol. 56, Part 5 (1966), pp. 5–9. The motive power of these clocks is not known.

15 Bedini and Maddison, *op. cit.*

16 Alfred Ungerer: *L'horloge astronomique de la Cathédrale de Strasbourg.* Paris 1922.

are all natural things in comparison to the Divine art.[17] Nicole Oresme (d. 1382) in his work on astronomy, discussed how angels turned the heavenly spheres, but also threw out the suggestion that perhaps God established the proper proportions of the celestial movements so that they were moved without further interference from God, "just as when man makes a clock which moves by itself when properly made and prepared."[18]

The archetypal view of Augustinian creation was elaborated with especial force and skill by Nicole Cusanus and he did not fail to employ the analogy of the clock. For Cusanus, God was unknowable, but we could get some awareness of his attributes through the resolution of opposites, such as eternity and temporality. In the *Vision of God,* written in 1453, he portrayed the temporal world as the material copy of a divine archetype. In the material world man can only see things in the succession of time, but all that happens in succession is present in the mind of God and was present in the idea of creation itself. Yet God who knows all that will happen still follows the events taking place in succession, reading over our shoulder as we read. A resolution of the apparent opposition between this eternity of God, and the succession in time of the Creation, was suggested by the analogy of the *concept* of the clock in the maker's mind and the *running* of the clock after it was made. Cusanus writes: "Let then the concept of the clock represent eternity's self; then motion in the clock representeth succession. Eternity, therefore, both enfoldeth and unfoldeth succession, since the concept of the clock, which is eternity, doth alike enfold and unfold all things."[19]

Cusanus also illustrated the archetypal view in another connection with the use of a tree analogy. Looking at a tree, he saw it as big, spreading, colored, and laden with branches, leaves, and seeds. But, in the eye of his mind, he perceived that the tree existed in its seed, not as he now beheld it, but potentially. And not only did that tree exist potentially in the seed, but all trees of the same species, as well as all other species of trees, existed potentially in their seeds. The generative power that existed in these seeds could never be fully explained in any time measured by the motion of the heavens, but he could imagine a Power behind all the generative powers of particular seeds of species, a Power embracing the pattern of all species.

The convergence between the clock analogy and the tree analogy emphasizes the unfolding aspect of the plan of creation. The hand of the clock is a symbol of the unseen or invisible hand. We have to proceed in a temporal succession, awaiting the future, but the turning of the hands calls forth the succession of

---

17 St. Thomas Aquinas: *Summa Theologia,* First Part of the Second Part, Article 13, Reply to Objection 3.

18 Nicole Oresme: *De proportionibus proportionum* and *Ad pauca respicientes* (ed. Edward Grant). Madison: University of Wisconsin Press 1966, p. 53 n.

19 Nicholas of Cusa: *The Vision of God* (tr. Emma Gurney Salter). London: J. M. Dent and Sons 1928, p. 52.

events for history, it sets the times and periods, all is enfolded in God's plan, even as it is unfolded. In nature, too, the species themselves unfold from their archetypal seeds. The time that Cusanus had in mind, like the time of St. Augustine, was not the time measured by the heavenly bodies, but the time of the Design in which moving bodies were themselves given their periods and the unfolding of archetypal seeds were given their periodicities. That the unfolding has nothing to do with development, but is simply a form of predestination, is made clear when he says: "Naught of that which appeareth in succession departeth in any way from the concept, but tis the unfolding of the concept, seeing that the concept giveth being to each; naught in consequence hath existed earlier than it appeared, because it was not earlier conceived that it should exist."[20]

## The Strasbourg Clock

The development of clockwork within the framework of simulating nature and expressing symbolic aspects of the salvationist drama continued through the sixteenth and early seventeenth century. The pinnacle in the development of these monumental astronomical clocks was reached in the second clock built in the Strasbourg Cathedral in 1574 to replace the earlier one which had ceased to function. It was the technological wonder of its age and was called "the Phoenix of all the clocks of Christendome."[21] It was celebrated in song and verse, its architect, Conrad Dasypodius, published descriptions with a woodcut of the clock in German and Latin, small replicas and partial imitations of it were constructed elsewhere in Europe, and tourists flocked to see it.[22]

The astronomical clock was indeed monumental, standing on a base about twenty-five feet wide, with a central tower about sixty feet high. The weight tower was surmounted by the cock from the earlier clock and had panels painted by Josias Stimmer, including a copied portrait of Copernicus and a picture of Urania as the symbol of astronomy. Panels on the clock portrayed scenes of the salvation drama; the Creation, Christ judging the world, the Resurrection, the Last Judgment, and Vice and Innocence. There were also paintings showing the four monarchies of the ancient world – Assyria, Persia, Greece, and Rome – and another set of scenes with representations of the four seasons. The automatons consisted of remarkably life-like figures. Symbolic figures seated in chariots representing the days of the week rotated in such a way that at noon

---

20 *Ibid.*
21 Thomas Coryat: *Coryat's Crudities* ... Glasgow: 1905, II, p. 191.
22 *Cunradi Dasypodii Heron mechanicus: seu De mechanicis artibus atque disciplinis: Ejusdem Horologii astronomici* ... (Strasbourg 1580); *Cunradi Dasypodii Warhafftige Auslegung und Beschreybung des Astronomischen Uhrwercks zu Strassburg* ... (Strassburg 1580). Alfred Ungerer: *Les horloges astronomiques et monumentales* ... Strasbourg: 1931, p. 163, states that about 400 works and descriptions have been published on the clock.

each day the proper figure faced outward. Figures representing the four ages of life – a child, an adolescent, a warrior, an old man – successively tapped the quarter hours, Christ moved out to meet each as he went by to signify redemption, but the hours were struck by Death himself carrying his own bell. The grand show, however, took place at midday, when, in addition, the ancient cock from the clock of 1354, beat its wings, shook its tail, raised its head, opened its beak and crowed, to the accompaniment of mechanically produced psalm songs, which was repeated three times as the apostles passed by. The cock, of course, signified the betrayal of Christ by the Apostle Peter.

Besides all this, there were moving dials or disks showing the holy days, the real and apparent movement of the sun around the earth, the apparent movement of the moon about the earth, the solar and lunar eclipses, sunrise and sunset, and a clock giving the meridional time of Strasbourg and the civil time of the city.

All this vast concatenation of movements was powered by a descending weight out of sight, as were all the other driving mechanisms, behind the façade. What one saw was a Theater of the World with a morality play taking place through mechanisms and automatons. Or, one saw a mechanized cosmic poem, such as the one written in 1578 by Du Bartas, *La Semaine, ou Création du Monde*. Du Bartas, incidentally, had a reference to the Strasbourg clock when he sang the praises of man, who with his mortal hands could make a model of the Great Architect's Creation.[23] The position of the clock inside the Cathedral was fitting indeed, for it was a huge visual aid, or mechanized teaching machine, rehearsing the meaning of life and the epitome of the microcosm-macrocosm relationship.

## The Clockwork Universe

Michel Foucault has argued that there was a major shift in the structural interrelationships of theories of language, life, and economy at the beginning of the seventeenth century with the emergence of the classical age.[24] He points out that the natural history of Aldrovandus at the end of the sixteenth century included myth, history, and symbolic emblems in the classification of animals, but all that disappeared in the next century. Alexandre Koyré maintains that one aspect of the Scientific Revolution was replacing the hierarchical and value-determined conception of the world with one that was indefinite in size and bound together by the identity of its fundamental components and laws, and in which the components were all placed on the same level of being.[25] Certainly

---

23 *Bartas: His Devine Weekes and Workes* (tr. Joshua Sylvester, 1605), ed. Francis C. Haber. Gainesville, Fla: Scholar's Facsimiles and Reprints 1965, p. 222.

24 Michel Foucault: *Les mots et les choses: une archéologie des sciences humaines.* Paris: Editions Gallimard 1966.

25 Alexandre Koyré: *From the Closed World to the Infinite Universe.* Baltimore: The Johns Hopkins Press 1957, p. 2.

there was a shift to the explanation of natural phenomena by means of secondary causes. If the whole material world was nothing more than moving inert matter occupying space, as Descartes insisted, it was difficult to maintain the sacramental view of nature. The same emptying out of the metaphorical and symbolic modes took place with historical events, and increasingly only the literal meaning of historical facts was left. This can be seen in Sir Walter Raleigh's *History of the World* as he tries to pinpoint exactly at what geographical place the Ark came to rest.

In the complex and many-faceted process of literalizing and secularizing history and nature, the sacramental aspect was transferred to the Design itself, while the "artifacts" of the natural world were sentimentalized as witnesses of the skill of the Great Architect. Two corollaries were also played up in the rhetoric of Natural Theology: the great variety of contrivances with which the Architect had furnished the world and the great economy with which He had executed the Design.[26]

During the transition, the metaphor of the clock became the master metaphor of the new mechanical philosophy and of Natural Theology. Correspondence of Descartes in 1629 reveals that he was familiar with the automatons of the Strasbourg clock.[27] He may have had them in mind as he developed the idea that animals were merely soulless automatons. The use of the clock analogy was prominent in the writings of Cartesians, both in physiology and physics[28], and in the writings of the new mechanical philosophers, and most effectively in the works of Robert Boyle.[29]

The pious Robert Boyle had no doubt about the rule of Providence in the moral affairs of man, but he emphasized that God relied on a "general concourse" to regulate the machinery of the world of nature. The general concourse was more than natural laws. It was the whole Design being sustained in the Creation by God but left to run by itself like a perfectly executed clock. He asserted that God, all of whose works were known to Him from the beginning, having resolved before the creation to make a world such as this one of ours, divided up the matter that he provided into corpuscles and put them in motion so that by the assistance of his "ordinary preserving concourse," the phenomena

---

26 For example, William Derham: *Physico-Theology, or a Demonstration of the Being and Attributes of God, from His Works of Creation,* 15th ed. Dublin: 1754, pp. 38, 82, 265.

27 Descartes to Mersenne, Oct. 8, 1629, in Charles Adam and Paul Tannery (eds.): *Œuvres de Descartes.* Paris: 1897, I, 25.

28 For Descartes, see Part V of his *Discourse on Method* (1637) and additional remarks on the passages of the *Discourse* in *Ibid.,* II, 39–40. For the ramifications of his thinking of animals as automatons, see Leonora Davidson (Cohen) Rosenfield: *From Beast-Machine to Man-Machine: Animal Soul in French Letters from Descartes to La Mettrie.* New York: Oxford University Press 1941. In Physiology, Sir Kenelm Digby: *Two Treatises . . .* (1658), and Thomas Willis: *Pharmaceutice Rationalis . . .* Second Part (1684), Preface and pp. 97, 124.

29 E.g., *The Works of the Honourable Robert Boyle,* ed. Thomas Birch, 6 vols. London: 1772, II, 7, 22, 30, 39–40, 45–46, 48, V, 163, VI, 331.

must as orderly follow this concourse and its motions as if each of them had a design of self-preservation and were furnished with knowledge and industry to prosecute it. The harmony between all the bodies in nature made it look as though an intelligent being was diffused throughout nature to administer it and preserve it. But the "curious engine" of the world was like the Strasbourg clock. "And the various motions of the wheels and other parts concur to exhibit the phaenomena designed by the artificer in the engine, as exactly as if they were animated by a common principle, which makes them knowingly conspire to do so, and might, to a rude Indian, seem to be more intelligent than *Conradus Dasypodius* himself, that published a description of it; wherein he tells the world, that he contrived it, who could not tell the hours, and measure the time so accurately as his clock."[30]

By the end of the seventeenth century the universe of Design had been thoroughly adapted to the clock analogy. As a cosmological model it was capable of absorbing the new knowledge from science and history easily, for every new illustration of variety, diversity, or complexity that was discovered in nature or history simply extended the evidence of intelligence in the Design and made its Architect more praiseworthy. The more science revealed the precision of the world machine, the more firmly the Argument from Design was demonstrated. In his popular *Sacred Theory of the Earth* (1684), Thomas Burnet, writing about God and His Design of the Creation, argued: "We think him a better Artist that makes a Clock that strikes regularly at every Hour from the Springs and Wheels which he puts in the Work, than he that hath so made his Clock that he must put his Finger to it every hour to make it strike."[31]

The Argument from Design was eulogized in the rhetoric of eighteenth-century Natural Theology, but it was equally prevalent in the rhetoric of Deism. Voltaire, for instance, put the Argument from Design forward dogmatically. "When we see a fine machine," he wrote, "we say there is a good machinist, and that he has an excellent understanding. The world is assuredly an admirable machine; therefore there is in the world, somewhere or other, an admirable intelligence. This argument is old, but is not therefore the worse."[32]

*Development and Progress*

The mechanized universe rationalized around the analogy of the clock was one extension of the archetypal idea of the Augustinian Design. It tended to make the universe into a stationary engine, like a clock that was in movement, but not changing in its structure. The unfolding aspect of the paradigm of Design faded rapidly in celestial mechanics after Newton, but was retained

---

30 *Ibid.*, II, 39.
31 Thomas Burnet: *The Theory of the Earth.* 3rd ed. London 1697, Bk. I, p. 72.
32 Voltaire: *A Philosophical Dictionary.* London: 1824, I, 9.

in the eighteenth century in the earth sciences, biology, and history. Repeated attempts were made to correlate the epochs of biblical history and geology, theories of an unfolding from the original seeds or archetypal patterns were repeatedly put forth to explain the diversity of species and propagation, and in history, if the teleology shifted from a divine eschaton to Reason, or Freedom, or Happiness, there were still invisible hands, the cunning of reason, and the wisdom of nature to realize them according to some plan. There was a family resemblance to the model given by Nicole Cusanus in all of the later elaborations about unfolding.

When we approach theories of evolution in the seventeenth and eighteenth centuries, it is important to note that they were either advanced as fiction or received as fiction. The usual literary form in which classical theories of evolution had been presented was the poem, and while poetry may have been the highest form of knowledge in antiquity, it had been defined as fancy by the seventeenth century. In a serious scientific innovation such as the corpuscular theory, every attempt was made to disassociate it from the Epicurean cosmology of evolution by chance and to ground it in Design.[33] When an author wanted to present a new evolutionary speculation, he often chose to advance it as fiction. Descartes did this in his *Le Monde*.[34] De Maillet's theory of evolution was put in fictional form by his editor.[35] Buffon advanced his evolutionary *Theory of the Earth* (1749) and *Epochs of Nature* (1778)[36] as fictional. Erasmus Darwin put his evolutionary views forward in poetry.[37] The Baron d'Holbach's evolutionary system was received as fictional.[38]

There were many practical reasons, such as theological and scientific opposition, for the resort to fiction, but from the reading public's perspective these speculative cosmologies were seen as fictional. The appeal of speculative evolution was powerful, for it was a flight of the imagination into new and wonderful worlds of fancy, but in the sober world of scientific reality, it was necessary to come back to the detailed work of filling out the mysteries of the Great Design.

Mircea Eliade has pointed out that in the Middle Ages germs of a linear progress of history can be recognized, but that the tendency which gained increasing adherence was the immanentization of the cyclical theory, and even in the astrological treatises and scientific astronomy of the sixteenth and seventeenth century, the cyclical ideology survives side by side with the new con-

---

33 E.g., Walter Charleton: *The Darkness of Atheism* (1652), pp. 53–55, and *Physiologia: Epicuro-Charletoniana* (1654); Robert Boyle, Works, III, 42–49.

34 Descartes: *Le Monde, ou Traité de la Lumière*. Published posthumously in 1664.

35 Benoît de Maillet: *Telliamed* (1748). The editor's role has been discovered by Professor Albert V. Carozzi, University of Illinois.

36 Buffon was forced by the Faculty of Theology at the Sorbonne to publish a retraction of statements in his *Theory of the Earth*, even though it was presented as fiction. *Œuvres complètes de Buffon* (ed. Pierre Flourens). Paris: 1853–54, XII, 350–353.

37 Erasmus Darwin: *Zoonomia; or, the Laws of Organic Life* (1794–96).

38 Baron d'Holbach: *Système de la nature* (1770).

ceptions of linear progress professed by a Francis Bacon or a Pascal. "From the seventeenth century on," he writes, "linearism and the progressivistic conception of history assert themselves more and more, inaugurating faith in an infinite progress, a faith already proclaimed by Leibniz, predominant in the century of enlightenment, and popularized in the nineteenth century by the triumph of the ideas of the evolutionists."[39] I think that Eliade, whose research has not been primarily in the modern period of history, has failed to realize the force of his own insights about myth and archetypes in the later period, for the teleological time in the Design, though not cyclical, was in a sense prefigured and closed.

The mathematical view of time that spread out to history and cosmology during the seventeenth and eighteenth centuries gives a deceptive picture of linearity, for the idea of a progression of time and history according to the preordained Design remains very much in force during the period.

We need to take a more careful look at ideas of progress that were developed in the eighteenth century. For the purposes of this paper, let me turn to the brilliant article of Ronald S. Crane, *Anglican Apologetics and the Idea of Progress, 1699–1745*.[40] In this article, he shows text by text how Anglican churchmen moved from defending Christianity against seventeenth century freethinkers, such as Anthony Collins and Charles Blount, to becoming apostles of progress at the end of the eighteenth century.

The old question was again raised by the freethinkers at the end of the seventeenth century as to why God had waited so long to reveal the Gospel. Crane presents some of the defenses given by the Church Fathers for the late coming of Christ into the world. Tertullian linked the history of religion with the analogy of education, so that the gradual process of revealing divine truth was found to be dependent upon the stage of growth in the pupil. As man reached new stages in his capacity to understand divine truth more was revealed to him.[41] The rate of revelation, then, depended upon man's progress in learning. The idea of the progressive improvement of man's comprehension with the course of time had found reaffirmation in St. Thomas Aquinas, who wrote that "a thing is not brought to perfection at once from the outset, but through an orderly succession of time; thus one is at first a boy, and then a man."[42] Crane then proceeds to show how this defense was developed into a theory of progress and human perfectibility.

However, a distinction made by St. Thomas needs to be noticed. Two causes are needed for actual generation, an agent, and matter. "In the order

---

39 Mircea Eliade: *The Myth of the Eternal Return* (tr. Willard R. Trask). New York: Pantheon Books 1954, p. 145.
40 Ronald S. Crane: *Anglican Apologetics and the Idea of Progress, 1699–1745*. Modern Philology XXXI (1934) 273–306; 349–382.
41 *Ibid.*, p. 274–275.
42 *Ibid.*, p. 279.

of the active cause, the more perfect is naturally first; and in this way nature makes a beginning with perfect things, since the imperfect is not brought to perfection, except by something perfect already in existence. On the other hand, in the order of the material cause, the imperfect comes first, and in this way nature proceeds from the imperfect to the perfect."[43] The knowledge of faith had to proceed from imperfection to perfection, in this point of view, but it was still a simple unfolding until the plan was fulfilled. It was a closed period, not an evolution, and the stages were to be completed at their proper times.

In the wealth of illustrations given by Crane about the development of human knowledge into a doctrine of human perfectibility, it can also be seen that the idea is retained of a completion or final term in the future when the Design will have been fulfilled.

Using some of Crane's examples, Joseph Glanvill in 1661 says: "But *Philosophy* and Arts commenced *Embryo's*, and are compleated by Times gradual accomplishments."[44] John Edwards writing in 1699 says: "There is a gradual Subordination of these several Oeconomies, and upon the Cessation and Extinction of one that is inferiour, a more Sublime and Perfect one arises in its Room: and it is God's Will and Pleasure that these divers Administrations shall take place in their Order, and that one shall not anticipate the other."[45] William Worthington in 1743 talks of the Scheme planned for the recovery of man, which "consists of a Series of Dispensations, each of which tallied exactly with the Circumstances of the World, at the Time it was made; to every Period of which it was wisely accomodated, and was the best fitted to promote its End, . . .". The scheme was "opened and unfolded by degrees."[46]

Edmund Law, in Crane's view, "one of the turning points in eighteenth-century thought,"[47] whose work went through seven editions between 1745 and 1784, integrated Locke's psychology into his own developmental theory. Everything in our minds is built up by a process largely mechanical from particular associations, and these grow almost mechanically from the environment we are in, so that progressive development goes on somewhat mechanically through acquisition. Nevertheless, the growth of each degree of knowledge in Law's view everywhere displayed the Divine Wisdom and Goodness of the original constitution of things.[48]

Edward Young in 1769 urged his contemporaries to abandon imitation and strike out on new paths in literature, for he saw no reason why gradual progress should not take place in the intellectual sphere, "*since,* as the moral world

---

43 *Ibid.,* p. 278.
44 *Ibid.,* p. 280.
45 *Ibid.,* p. 284.
46 *Ibid.,* p. 302.
47 *Ibid.,* p. 349.
48 *Ibid.,* pp. 362–365.

expects its glorious Milennium, the world intellectual may hope, by the rules of analogy, for some superior degrees of excellence to crown her latter scenes..."⁴⁹ John Gordon, used by Crane as an illustration of indefinite progress in the future, says progress will go on from one degree of advancement to another, "till it has reached the highest point, for which it was designed; when it will yield up its inhabitants to other worlds, and greater bliss, than it could give them."⁵⁰

Joseph Priestley urged that the excellence of human nature, which lies in the variety of which it is capable, should be given greater freedom for development.⁵¹ He appeared to think that human nature was capable of infinite improvement, but since that improvement was under the uniform intention of divine providence to lead mankind to happiness in a progressive method, it seems implicitly understood that man will only reach the perfection described by St. Thomas in order that the curtain can at last be rung down on the drama. Time will have come to the end of its term.

It is obvious that the sources which Crane used to illustrate the development of the idea of progress can be read in another way. While they can be seen as forerunners of a developmental view of history, there is little indication that the authors in their own minds were doing more than explaining how man would arrive at the eschaton of perfection to fulfill original Design. This view of history was fully congruent with the clockwork model, with the various periods appearing on the scene like the automatons of the clock when their time had come. Some of the wheels were geared to turn very slowly, and Henry Power, a seventeenth-century enthusiast of experimental philosophy had even suggested that the "great Automaton of the World," in all likelihood would not be destroyed by its maker until the slowest motion therein has made one revolution.⁵² The expectation of the eschaton shifted steadily to a more distant future, but the eschaton was still visible by the nineteenth century in time and history.

*Darwin and the Argument from Design*

Much had developed prior to Darwin to support a theory of evolution, so much in fact that the older model was hard pressed to absorb it. A Time Revolution had already begun early in the nineteenth century and was in full tide in the middle of the century.⁵³ The closed temporal world of biblical chronology was stretched into vast epochs of geological time. The enormity of time in the history of life on earth that was revealed from the reconstruction of fossils was as staggering to the imagination in the early nineteenth century

---

49 *Ibid.*, p. 374.
50 *Ibid.*, p. 377.
51 *Ibid.*, pp. 379–382.
52 Henry Power: *Experimental Philosophy* (1663).
53 Haber: *Age of the World*, pp. 1–10; 290–292.

as the spatial magnitude of the universe had been in the seventeenth century. The temporal space between the historical record of man and the fossil record of species was filled in by the discovery of pre-historic man. The culture of man was historicized as language, literature, religion, the arts, legal institutions, and even history were brought under the idea of a development. The rate of change in the state of knowledge during the decades just prior to Darwin was almost explosive.

Yet, for the closest colleagues of Darwin, as well as the public at large, the experience of reading the *Origin of Species* was a shock, for those who went with Darwin, a shock of recognition, for those who went against him, a shock of alarm. In spite of all the romanticization of time and the process of development in time, in spite of all the detailed reconstructions of history, pre-history, and natural history, the genetic idea had not made a direct challenge to the Design Argument and its clock analogue, at least in terms of the hard-headed scientific view of the age.

One of the by-products of the Design cosmology was a conception of what constituted an explanation. It was clearly stated by Robert Boyle, for instance, in terms of the clock analogy:

"For to explicate a phaenomenon, it is not enough to ascribe it to one general efficient, but we must intelligibly shew the particular manner, how that general cause produces the proposed effect. He must be a very dull inquirer, who, demanding an account of the phaenomena of a watch, shall rest satisified with being told, that it is an engine made by a watchmaker; though nothing thereby declared of the structure and co-aption of the spring, wheels, balance, and other parts of the engine, and the manner, how they act on one another, so as to cooperate to make the needle point out the true hour of the day."[54]

Boyle made the point that it was possible to explain a watch without assuming from its outside appearance that it must be a living body or endowed with a soul. Those theories of evolution which had to fall back on a vitalism for their driving force were, from this theory of explanation, simply theories unexplained. Those theories, such as the Epicurean one of D'Holbach, which fell back on an aimless chance encounter of atoms could not be convincing without some additional mechanism to explain the already known complexities of construction in the world, the most telling one being the human eye. The Bridgewater treatises alone had shown these complexities in sufficient detail that the chance argument was absurd. Logically, of course, Hume had bankrupted the Argument from Design in his *Dialogues on Natural Religion,* but the logic lay elsewhere. There had to be a more satisfying alternative before the Argument from Design could be displaced. Darwin was the first to supply

---

54 Boyle: *Works,* V, 245.

one which conformed to the conception of an explanation sufficiently to be taken seriously.

Almost every component of Darwin's theory had been familiar to his peers when Darwin wrote his *Origin of Species*. The uniformitarian geology of Lyell had established the idea that small changes of no greater intensity than those now in existence could transform the surface of the earth if given time enough. There was no directionality of time descernible in these processes, but the extent of time required for uniformitarianism would force the abandonment of any biblical scale of time. Still it was possible to fit the rest of the mechanics within the Design, as Hutton clearly illustrates.[55] The paleontologist put the arrow of time in geology, but most of the leading paleontologists worked within the Design framework, even trying to retain something of the Days of Creation as Epochs. Richard Owen, who became one of Darwin's formidable opponents, not only was a conventional believer in the Design, but his theories of paleontology were permeated with archetypal ideas. Death as a population control was an old idea in Natural Theology for keeping species in balance in the oeconomy of nature. Natural selection was used by Edward Blyth in defense of the fixity of species.[56] Malthus was a fairly orthodox supporter of Design.

As for the transmutation of species, Darwin himself wrote in a letter to Baden Powell in 1860: "The only novelty in my work is the attempt to explain *how* species became modified, and to a certain extent how the theory of descent explains large classes of facts; and in these respects I received no assistance from my predecessors."[57]

Darwin's own assessment of his work seems to be a correct one. He was able to explain the *mechanism* of the modification by descent by putting all the parts that had been lying around like extra wheels into a new construction with natural selection as the driving mainspring. Much of the scientific opposition came to bear on where he had failed to do this.[58]

Darwin was himself an opponent of the Design Argument. He was not in sympathy with Platonism, archetypes, or vitalistic theories of evolution associated with an idealistic view of reality. His new model avoided archetypal seeds, invisible hands calling things forth, natural cycles such as youth, maturity, and decay, and the teleology of an eschaton. There was no final term for the completion of man's perfection. Perfection itself was redefined to mean the ability to adapt most advantageously within the ecology. The process of time

---

55 Haber: *op. cit.*, pp. 164–173.

56 Loren C. Eiseley: *Charles Darwin, Edward Blyth, and the Theory of Natural Selection*. Proceedings of the American Philosophical Society Vol. 103 (1959) 94–158.

57 Sir Gavin de Beer (ed.): *Some Unpublished Letters of Charles Darwin*. Notes and Records of the Royal Society of London XIV (1959) 32.

58 The extent to which Darwin substituted rhetoric for demontration is pointed out by Walter F. Cannon, *Darwin's Vision in "On the Origin of Species"*. In: George Levine and William Madden (eds.): *The Art of Victorian Prose*. New York: Oxford University Press 1968, 154–176.

was left open-ended. As historians adopted the Darwinian view of time and process, teleological time in history gave way to the simple linear mathematical view. Eschatons of one sort or another, such as the perfection of man have lingered on, but in critical history they were eschewed as speculative, philosophical or metaphysical, and were eliminated from the proper business of the historian.

# Temporal Attitudes in Four Negro Subcultures

HELEN BAGENSTOSE GREEN[*]

*Summary.* Similar attitudes toward time are found in the Negro subcultures of low socio-economic status in West Africa, Brazil, the Caribbean, and the Unites States. Contributing and interrelated factors in the formation of these temporal attitudes are traditional West African cosmology and religious practices, the priority given to communal activities, and the absence of delayed reinforcement in child-rearing. In addition, colonial subjugation or slavery has influenced the selection of defense mechanisms which reduce the pressure of time through denial, rationalization, and compensation. In these four Negro subcultures, temporal attitudes appear in the temporal arts, the nature of festivity, the ephemeral materialism, the access to spontaneity, and the sensitivity to social resonance.

Temporal attitudes are useful in analyzing the similarities and differences between cultures. By temporal attitudes is meant the concepts and behavior related to time. Considerations of time can be applied as a common denominator across all societies, however unlike or widely separated. Customs within any one culture, such as religion, politics or family structure can be separately selected for examination on the basis of temporal attitudes. Furthermore, the various socio-economic levels, as for example the lower-classes, can be compared within and between cultures for aspects of time.

This essay will describe the temporal attitudes of four lower-class Negro subcultures: those of northeastern Brazil, the Caribbean islands, the Black belt of the American South, and present-day West Africa. In these societies comparable orientations toward time indicate the survival of large complexes of similar cultural characteristics despite environmental differences. The original West African culture from which these four contemporary subcultures are derived is the basis for their present relatedness.

In emphasizing the differences between Negro cultures, scholarship has failed to recognize the extent of their similarities. The first professional ethnographers in West Africa were impressed by the exotic aspects of individual tribes and had too narrow experience to assess the underlying communality. Only recently has it become apparent that, like Europeans, West Africans differed in nationality and language while sharing a general common culture.

There are a number of reasons why recognition of West African culture as a closely related entity has been slow to evolve and, in particular, to reach the

---

[*] Dr. Helen Bagenstose Green, Ph. D., Wesleyan University, Middletown, Connecticut 06457, USA.

English-speaking world. French and German publications have preceded those in English on such subjects as West African cosmology, doctrine, philosophy, and religion. However, the basic character of West African thought now appears in English in the works of Bascom [2] on the Yoruba, Field [13] on the Ga, Griaule [20] on the Dogun, and Jahn [27] from sources for the Bambarra, Bantu and Congo tribes. This material on West African ideological systems demonstrates a continuum with behavioral manifestations in the social, political, economic and religious aspects of life in West Africa. Further evidence of underlying consistency appears in the art forms. Music, dance, sculpture and textile design have certain common characteristics found across vast regions of varied economy and colonial influence. The validity thus afforded by the concurrence of related belief systems, behavior, and artifacts indicates a strength and unity in West African culture which has heretofore been unrecognized.

The same lag has occurred in understanding the degree of carry-over and continuance of West African culture by Negroes in the New World. The largest body of research on this subject is Brazilian which is published in the Portuguese language and lacks translation. In northeastern Brazil, a number of institutes are investigating the nature of the relationship between the African tradition and its present expression in Brazilian life. Experts on Afro-Brazilian customs – such as Verger [48], Bastide [37], Ribeiro [39], and Rodriguez [41] – are convinced that the study of the Brazilian Negro shows what African life was like before the Arab infiltration and colonial conquest of West Africa. This approach is justified on the ground that Moslem and Christian proseletizing and European exploitation may have reduced the culture heritage of present-day West Africans to a greater extent than the autonomous living conditions and superficial conversion to Western culture of slave groups in the Western hemisphere.

The lower-class Negroes in these four cultures are considered to have the heaviest determinant from the original culture of West Africa because of their density and remoteness from white contact. Brazil, which is roughly equal to the area of the United States, received double the number of slaves than did the United States. They were settled in greater concentration in the agricultural and mining regions. In the cities of Brazil, the extreme disparity between the poor and the very rich left a vacuum at the middle level into which talented, Arab-educated, or manumitted Negroes were able to move [17]. This created a Negro leadership group which pridefully maintained West African customs. These efforts were, and still are, continued in the hundreds of socio-religious units of northeastern Brazil which have a self-conscious allegiance to West African tradition.

The most important work on the background of Negroes other than Brazilians has been done by Herskovits. He began the documentation of 'continuing africanisms' among Negroes in the Guianas [26], Surinam [26], Trinidad [24], Haiti [23] and some regions of the American South [25]. His assessment is open to criticism on two grounds: first, he was not sufficiently familiar with

West African practices to always find or interpret them correctly among New World Negroes. Second, he understated the similarity between West Africans and other Negroes in their larger orientations in his professional attention to exact similarities in minor habits. Bascom's study of friendship patterns among the sea island Negroes of South Carolina [1] and R. F. Thompson's study of African derivatives in American slave art [46] show that work on this subject can still be fruitful.

The psychological investigation of cross-cultural relatedness in this field has been hampered by the American emphasis on conditioned learning. This focuses upon the formation and modification of behavior within the lifetime of the individual. The lifetime of a culture is obviously more difficult for research although it follows a similar learning process wherein cultural transmission is repeatedly imposed in general form by each adult population upon each younger generation. Evidence for this kind of continuity now rests for the most part on the integration of clusters of anthropological observations to which psychological considerations are gradually being added. The cross-cultural comparisons of this paper are offered as a thrust in this direction.

The Negro subcultures in the Western hemisphere have shown a more facile adaptation to modern technology than the Amerindian societies of North and South America, or the thousands of East Indian indentured laborers imported to South America and the Caribbean areas from India. Because of this greater partial adjustment, the Negro subcultures have been assumed to have lost all of the behaviors with which they arrived from West Africa. Much research attests the remarkable tenacity with which Amerindians [31] and East Indians [29, 33] persist in adhering to their traditional concepts and attitudes. Negro subcultures, often contiguous with the Amerindians and East Indians, have, by contrast, been considered totally acculturated and stripped of their West African carry-overs by the effects of slavery and poverty. In many respects, however, both the Negroes, the native red Indians, and the 'coloured' people from India have all suffered greatly from low socio-economic status, color prejudice, and lack of educational opportunity. Yet despite the nearly similar deprivations, the behavior of indigenous Indians, and of East Indians does not today resemble the living patterns of Negroes, even in the innumerable communities where they live and work side by side. This suggests that the Negroes are maintaining some large behavioral patterns which came with them originally from West Africa.

In just those ways in which West African life fitted best with New World requirements, the Negroes have been most advanced in their adjustment. West African social and political practices have fitted well with the democratic expansion of new communities. West African religious beliefs have been easily incorporated into Christian teaching and membership. West African emphasis on the benefit of large group cooperation in work has aided the economic development of new regions. The balanced separation of West African responsibilities in the roles of men and women has furthered their independent action under new conditions. But the great disparity between West African and Western time orientation has slowed the fuller integration of Negro subcultures into the modern life of the Americas and of colonial Africa. Advanced in many other adjustments, the four Negro subcultures, including contemporary West Africans, have been slowest to change their temporal attitudes.

*West African Ideology*

The work of Griaule [20] and the teams carrying on his seventeen years of investigation into the ideology of tribes now living in the great bend of the Niger reveals a massive, ancient, closely-knit and complex structure of West African thought. This body of material is consistent with findings in other areas of West Africa, such as studies by Dieterlen [8] for the Bambara, Tempels [45] for the Congo, Field [11] for Ghana, Kagame [28] for the Bantu, Van der Plas [47] for the Mandingoes, and Jahn [27] for West African philosophy as a whole. Furthermore – and very importantly – it is corroborated in general as having existed in much the same form and effect by the amateur observers of past centuries. The narratives of explorers, missionaries, and resident factors at the time when slaves first began to be sold for shipment from West Africa describe beliefs and practices at that time which are congruent with the material now being recorded. For example, three early accounts from different regions – the journals of Mungo Park [35] for his travels through upper West Africa, the accounts by Wilson [52] of his twenty-five years as supervisor of missions in West Africa, and the reports of Sir Richard Burton [5] from Dahomey – indicate essentially the same doctrine and customs as the contemporary professional publications.

There are a number of reasons why this body of doctrine has been difficult to ascertain in its entirety. Primarily, it was the responsibility and property of the priesthood, a secret professional order maintaining great strength and mysterious power as part of its prerogative. European colonizers did not deal with this group but rather with the secular leaders, such as the chiefs and their advisors who handled the political, economic, and outside contacts of their people. Even today, village affairs are divided between the headman who are democratically sensitive to community opinion and the priests who are autonomous from local influence. Furthermore, the missions refused recognition of the 'pagan' priests which forced the latter into underground practice. And finally, the priesthood was composed of older men unsuitable for slave labor and protected as holy persons by their people from slave purchase. A conquering tribe did not usually sell the priests among its captives because the assimilation of these men was thought to increase the good will of the spirit world. In the slave settlements of the New World there were, therefore, very few religious leaders among the slaves, and untrained men had to assume some sort of religious leadership. Consequently, practices and beliefs were perpetuated which lacked the explicit structure of West African doctrine.

There are three principles which can be formulated to describe the concept of time as it is understood and practiced by West Africans and informally exhibited in some of the behavior of New World Negroes. The first is the principle of the presence and action of life-force. The essence of being, which life-force is, is considered to move unevenly in time by a serpentine progress toward infinity. There is no end. The movement of life-force varies constantly from straight to zig-zag, from large to little, and from fast to slow. Held or released, life-force is dynamic in its eternal urgency to move and to cause everything to move with it. The world-order sprang from the investing of life-force by a non-anthropomorphic deity into the affairs of men who in turn reaffirm and release this supernatural force. Time is not separate from life-force. It stays when life-force is arrested, and it flows fast when life force is most active. It is like psychological time (or perceived time) rather than actual time in the Western sense. Abstract or equal units of time are not possible in this concept. Measuring one unit of time by a change in space is not equal to measuring another unit of time because the space has changed also during the measurement. Space and time are thus a unity depending on the amount and direction of the life-force which is involved.

Life-force is not necessarily in material and immaterial things, there is only the potentiality for its investment under certain conditions. It may or may not be active in a piece of wood, it may or may not be active in the spirit of a child, or a grain of rice, or a dead man. The determination of its presence is highly pragmatic, depending on the results.

The second principle is the concept of dual unity. In this concept, opposites which are complementary to each other form a dual identity in their function. Multitudes of complementary polarities underlie Negro thought. For example, there is the complementary and necessary relationship between masculine and feminine, object and subject, day and night, concrete and abstract, work and rest, organic and mechanical, comedy and tragedy, old and young. This concept of dual identity extends to unify the actual with the illusory, the dead with the living, the personal with the communal, and the individual psyche with the individual soma. These polarities are not warring dichotomies nor does one lose its individuation in its reciprocal function with the other. The equality is maintained with no dominance.

The third principle in that of dynamism between the dualities which gives a beat or pulse to life-force. While one opposite surges the other lapses, so that there is always an equipoise which is at the same time unstable. Alternation of this instability creates the rhythm. Because the opposites are essentially equal, one accent always equals the next, producing a heart-beat type of rhythm.

These three principles in life-force, i.e., movement, duality, and rhythm, are necessary to the understanding of the Negro subcultures regard for time. The three concepts are themselves temporal because interflow produces constant change. Mathematically, life has been described by Griaule [20] as a metaphorical helicoid which moves and also has a certain type of movement within it while maintaining a unitary shape containing opposites. The movement of this helicoid toward eternity undulates and hesitates like a screw-thread because it must subsume a vast number of pulsating dynamics within it.

Behavior in accordance with these concepts can be seen in the attitudes of the Negro subcultures toward the past, present and future. The past, whose rhythmic beat has been, is regarded as indissoluble from the present. There is no division between past and present because the past does not fade in the present but exists fully in the present. Thus the past is viewed as releasing rather than constraining. The Negro peoples want most to enter fully into the unity, rhythm and balance of the moment. Involvement in an enormous present therefore reduces Negro concern with the time-extended ramifications of past and future. Unlike some other peoples, Negros are not anchored in their past nor haunted by their future.

There are only two possible attitudes toward the future, both of which reduce long-term planning: one is a fearful concern based on the unpredictable mysteries inherent in the complex of life-forces always being exerted from the dead and by the living. The other is the cynical supposition that, since the present will soon be the past of the future, the next phase of living will not be very different. When a person dies, that individual moves with his aura of life-force, his same personality and his typical concerns, to a spiritual track somewhat parallel to that of those still living in the known world.

*Formative and Inter-Related Factors in Temporal Attitudes*

*I. Religion.* The potentiality of life-force to enter or leave anything is a central concept to West African religions. Persons (living or deceased), animals, places and objects may have, lose, or regain the activity of life-force. Life-force may become inactive or disorderly in effect. Therefore the basic objective of religion is to get life-force to move in a desirable way, one which will impart power where and when it is needed. This is a religion of augmentation of life-force at times and under circumstances when it is felt to be needed. God or the lesser dieties do not punish for sin but only for neglect. The minor gods, who cut across the temporal and the eternal to enter the here-and-now, effect good in return for the attention given them.

Life-force can be activated under specified religious conditions. It was imparted first by a unitary God in begetting lesser gods through intercourse with earth. It was and is imparted by a hierarchy of lesser gods to man and in some instances to the animal and vegetable kingdoms. It is imparted by man in creating offspring. It is imparted by the living to the dead, thereby keeping the dead vitalized. The 'undead' dead can impart their augmented and augmenting life-force to the living.

Sacrifice has the effect of giving and receiving life-force within an invoked closed circle – i.e., it is given from the sacrificer to the power called upon by the act of calling and offering gifts. This enables the power called upon to release its greater life-force to the sacrificer.

The constant and reciprocal interaction between the natural and spirit worlds may modify any facet of life. The real and the spiritual together support a precarious relation in which apparently 'acausal' events find a *post hoc* explanation. Since West African religion is pragmatic, illogical causality is inferred to be logical by the nature of developments. The resulting anxiety for the Negro subcultures is allayed by an effort to offset or 'top' undesirable life-forces by a greater life-force derived religiously in the following ways:

1. through ritual as the redistribution of force
2. throught sexuality as the creation of force
3. through dance as the rhythm of the movement of the world
4. through speech as the medium for communication between the living and the dead.

In each of these behaviors, life-force is projected by the pulse of alternation of life-force flowing between the giver and the receiver. Negro religion is therefore not one of worship but of invoked give and take. It is not one of debasement and humility but of placation and appeal. It is also one of fear rather than confidence, for there is uncertainty about the outcome of interaction between the supernatural and natural worlds.

Negroes in the New World, forced to become Christians, effected in many ways a syncretism with the Trinity, saints, rites and tenets of Christianity. The candombles of north-eastern Brazil [48, 3], the macumbas of Congo-Angolan pattern in central Brazil, the religious practices of Negro societies in the Guyanas [26], the Shango and Voodoo cults of the Caribbean [27, 24], and the occult in many American religious Negro groups indicate much relationship to West African religion. The uniformity between these religious practices in widely separated regions lies in the following characteristics:

1. the persistence of ritual sacrifice as gifts to a reciproating agent
2. the refusal to denounce or ignore sexuality as a legitimate concern of religion
3. rhythmic behavior as an integral part of religion in such forms as dancing, clapping, swaying or antiphonal response
4. address to the spirits and exhortation presumably received from them.

Although these features have, of course, varied from country to country with the Christian pressure for their disuse, most of them are still found in some form.

*II. Subjugation.* The reaction of West African, and particularly, the New World Negroes to colonialism, slavery or, disadvantaged freedom has been to increase their focus upon the present. Conquest and stress induce in any people the formation of psychological defenses to facilitate coping and survival. At least six or seven million West Africans were purchased at great expense and transported at great hazard to be slaves in Brazil, the Caribbean and the U. S. They were then settled and trained in a country when the manpower of Indians might have been exploited and was not sought. Early experience with the redmen showed that they did not adapt, or died, or failed to reproduce under conquest to such an extent that it was necessary to obtain another labor force. From the beginning, the Negroes had great survival strength. Until recently it has been assumed that a 'docile' or (from a Western point of view) child-like mentality enabled Negroes to bend without breaking. Examination of Negro behavior patterns under adversity suggests that they used those sources of psychological defense which particularly build endurance or the ability to continue in time. Much of their characteristic behavior as unfortunate minorities in the past and the present falls into three categories which use time as a servant rather than a controller of life.

The first defense against time is the *denial* of time's pressure by measures which ignore or defy time. Procrastination and unplanned living deny the exigency of the future. Refusal to hurry, maintaining 'cool' and other techniques of disassociation from urgency are forms of aggression against time. These behaviors have been the bane of colonial rulers and the complaint of employers of Negroes over centuries and across continents. The right to be slow, to appear indifferent to harassment, to act by 'African time' or 'coloured people's time' is essentially to behave in defiance of time and the consequences of flaunting it.

The second defense against time is the *rationalization* of failure by which personal responsibility for unfortunate developments is given creditable but false explanation. West African failure is typically attributed to external circumstances and to other individuals and groups, without self-reproach. The causes of failure are externalized. This behavior makes the present more bearable because the person avoids blame by attributing his failures to other than actual causes. Data comparing white and Negro samples in the Carribbean show a significant difference in the number of Negroes citing external factors as the reason for failure (in contradistinction to factors of self-reproach) [18]. A sample of West Africans given the same questionnaire by the author also replied with a large proportion offering non-intropunitive explanations for failure. This same tendency has been noted in West Africans by anthropologists [22] and by psychiatric observers [12]. When circumstances do not yield reasonable excuses for inadequacy, Negroes in the four subcultures frequently offer bad luck or supernatural events as explanations for failure [19]. Other rationalizations, characteristic of Negro slaves or servitors, are the adoption of 'liking-to-serve' attitudes, the elevating of their masters to a superior plane accompanied by reduction of self-regard to a level commensurate with the lower status of Negroes [6], and the self-conviction that heaven will bring a reward for suffering [15]. All of these psychic manœuvres aid the individual in obscuring reality and abandoning self-destructive behavior in favor of adaptive behavior. These mechanisms of psychological defense preserve the balance and on-going movement of inner life-force through the disregard or disguising of actuality. Rationalization is thus often in the service of time because it frees rather than inhibits, and therefore does not stop the flow of life-force into future time.

The third defense against time is *compensation* whereby small extraneous behaviors are added to give satisfactions which counteract the effects of work, anger and deprivation. These extraneous behaviors have four relationships to time. First, they delay the time between beginning and finishing a goal, thereby cheating time's pressure. Second, they cause the past and the future to be temporarily forgotten because they expand consciousness of the present. Third, these activities build the staying-power of people by supplying temporary involvements and pleasures to offset fatigue, frustration and despair. They keep frustration from mounting to an explosive pitch or from sinking to a suppressive stoicism. Finally, these activities make time pass faster, for monotonous time is longest [14].

These compensatory behaviors can be grouped under a number of categories. There is an interpersonal category characterized by much time given to talk, humor, playfulness, and dramatic involvement in friendship and romance. The next category is paliative because it serves to blanket unpleasant experience by the seeking of new scenes, drugs or alcohol. A final category is somatically expressive because it makes much use of the kinaesthetic and auditory sensoria in rhythm, music, dance or sexual activity.

These behaviors may best be described as rapid psychic refreshers. They are rapid because they are easily introduced under most circumstances and can be quickly terminated. They are psychic in the sense of having the pull of psychological involvement. And they are refreshing in that they leave the performer revitalized by a little pleasure before returning to his main activity.

All of these compensatory activities are available in some form in every culture. It is not suggested that they are unique to Negro subcultures. What is unusual in their use by Negro people is the frequency and ease with which the average Negro interpolates these activities into his everyday affairs. When the opportunity arises, time is temporarily given to rapid psychic refreshment and then as easily ended. These diversions are introduced as necessities of the moment, and are not thought of as luxuries or rewards. Comparison might be made to the eating habits of various cultures throughout the world: some societies take a main meal once during the day when work is over, whereas other peoples eat a little at frequent intervals throughout the day and never have a very large meal.

*III. Child-Training.* If one agrees that the child is father to the man, it is reasonable to examine child-rearing practices for any formative influence on adult orientations to time. In the Negro subcultures, children do not receive precepts for the future because most teaching is kept specific to the occasion [51]. Injunctions are of the nature of 'do not do that now' rather than 'do not ever do that.' Training is not over-all shaping as much as it is little-by-little control, and principles for action are not as likely to be formed by parents or by children.

Experimentally, frustration appears by delay to produce a lengthened experience of duration [14]. Therefore the absence of frustration may contribute to an unworried attitude toward time as well as a short and undifferentiated time sense. This is corroborated by a number of psychoanalytical, anthropological and correlational studies which indicate a relation between infant satisfactions and adult serenity and optimism about extended time [51, 53]. Some of these findings bear on temporal attitudes in the Negro subcultures. Babyhood is usually a period of complete indulgence, which is succeeded, as the child grows a little older, by general freedom and absence of restraint. Delay in reinforcement is not practiced by the majority of Negro parents, many of whom assert the validity of acting on an initial impulse to reward or punish as best for both adult and child [18]. There is avoidance of deliberately prolonging the time between the reward and punishment of children. This, of course, does not mean that the amount of frustration is any less than in other cultures although its faster resolution may have some bearing, as children grow up, on their habitual attitudes toward time.

Several experimental studies of Negro adolescents show a preference for immediacy, even when the odds for delay are raised [32]. For example, a significant number of Negro school children in Trinidad chose to receive as the reward for taking some tests, a tiny candy bar at the time rather than wait a week for a candy bar three times as large. The paired sample of East Indian adolescents had a significantly larger proportion of youngsters who chose to wait for the later but bigger reward. Adult groups, both in West Africa and Jamaica, have stated that they would rather accept a small sum of money at the time than wait one year for ten times the amount [9].

Additional symptoms of preference for immediacy are observable in the living patterns of lower-class Negroes throughout West Africa [30], Brazil [37, 38], the Caribbean [7, 4, 16, 43] and the U. S. [36, 42]. Work is done when need pinches, pay is wanted in short intervals – by the day or week – and it is spent relatively soon. Sex must have a quick return or the person moves on in the erotic role. A maximizing of the present appears in the West African proverb 'Suck the honey as it drips' and in the Caribbean saying 'Now for now, man'.

Another interesting aspect of child-training in relation to time is the presence or absence of anal-retentive characteristics. This inglorious term is used by psychoanalysts for all children's first physical and psychological tendencies to control time by holding back what belongs to them. Freud showed that as the child becomes an adult, these tendencies often continue in the more appropriate forms of holding back such as hoarding or planning revenge. Presumably the cycle is set in motion in infancy when rigid demands and heavy discipline cause the child to try to find these safe and indirect forms of hostility toward others. In the Negro subcultures both hoarding from others and holding malice over a period of time are culturally proscribed or denounced [10]. Current sayings in West Africa are 'Beat the snake now rather than the track after it passes' and 'strike the iron while it is hot'. The result of this non-delay in expressing aggression towards others leads to much conflict in all these Negro subcultures. However, studies of Negro crime patterns made in British West Africa [47], Bahia in Brazil [34], Trinidad in the Caribbean [18], and many regions of the United States [4] show that violence which is delayed or planned (such as premeditated murder or suicide) is much rarer in Negro societies than among contiguous groups.

*IV. Communal Priority.* In the regulation of society and the relation of the individual to it, these subcultures illustrate best the three Negro time principles of life-force, dual unity and dynamism. Their social institutions have unusual elasticity, that is, adjustment according to the needs of individuals. Laws are not formalized into standing law because each case is different. Regulation and justice are seen as necessarily moving with the times, and must therefore be specifically formed for each new instance. Heroes and leaders do not acquire fixed status, and bureaucracy does not form because the forward surge of time is thought to bring ever new problems. The efficiency with which societies have been run without legal codification or permanent bureaucracy is described in the early accounts of Park [35] on Senegambia, and of Wilson [52] on the Guinea coast, in the modern accounts by Field [42] on Ghana, and by John Stevens [44] for British Honduras in the 1840's. Candombles and macumbas are similarly organized present-day groups in Brazil which manage in small units the business of social control. The prevailing temporal attitude is that of the Latin adage 'Tempora mutantur, et nos mutamur in illis' (times change and we are changed in them).

Qualities most admired among Negroes are the ability to formulate opinions which persuade others, the good fortune to be liked, the ability to understand and cooperate with others, and the gift of contagious geniality. In short, well-being on the social dimension is the most telling evidence of the possession of life-force. These qualities are in contrast to the ideal person described in Max Weber's Protestant Ethic [50] or David Riesman's inner-directed man [40]. These works describe a typical man who labors to increase his talents to the greater glory of God through industry, enterprise, frugality and planning. Only after working well does he allow himself leisure or luxuries. Such a man in Negro cultures would become an isolate, disdained for self-centeredness unless this could be offset by demonstrations of compatibility and generosity. The worst thing that can happen is to be excluded on an individual basis –

in short, permitted to have privacy and secrecy which is considered to be at variance with the unity of society.

The dynamic interflow between the polarities of the individual and his group is seen in the high incidence of communal activities. In West Africa these follow the traditional patterns of tribe, clan, village, age-set, compound and extended familiy. Although removal of these people to the New World disrupted such formal groups, affiliative behavior continued. Preference for group work, close and shared living conditions, and the many membership groups of associations, lodges, cults, cooperatives, and recreational organizations show the habitual inclination of Negroes for group activity. Although the Negro typically finds support and protection in such group membership, he also finds release for his individuality in the multifarious roles and subdivisions of responsibility with which these activities are organized. Most Negro functions have an extraordinary number of minor parts and duties which give many persons a share in the undertaking.

When such groups formulate a goal, the arrangements for accomplishing it are adjusted in such a way that everyone shares in the pace and in the end result. The screw-thread of progress revolves so that everyone's pressures and abilities are accomodated at once. The individual therefore has little introception of causal satisfaction or guilt for his part in such an undertaking. The outcome of group action is not referred by the participant back to himself but, rather, he diffuses responsibility for it into the group. The relevance to time here lies in the difference between individual accomplishment and group accomplishment. Research on motivation for achievement suggests that the longer ranges of achievement grow mainly out of self-reward and individual task orientation [21]. In these Negro subcultures, group projects reduce both self-reward and personal responsibility for the task. The future therefore is not specifically created by the individual because he merges his attitudes about goals with his group belongingness.

*Results of Temporal Attitudes*

*1. Temporal Arts.* The static arts of architecture, painting, sculpture, pottery and the related minor arts have not engaged the Negro people of West Africa or the New World to the same extent as the performing arts. In West Africa the paucity of iron ore and the damages by the white ant were contributing factors. The bronze art of Benin is an exception possibly due to Portuguese or Arab influence. In the New World, European forms of art were both dominant and alien to the Negro subcultures. Whatever the reason may be, the evidence shows a much greater investment of creativity by Negroes in the impermanent arts of social expression. It may be said that the static arts do not employ or manipulate time, they only exist in time. The performing arts, however, depend on the timing of successive and simultaneous action. These arts have fascinated Negroes for centuries. The most obvious are their rhythmic music, group dances, drama, oratory, team sports, and the timely coordination of physical and psychic which are necessary to love-making. These arts demand a functioning of the whole performer in a unity of body and mind. They also depend for creative impetus on duality with a respondent, either as band, team or audience. These arts have a dynamic in that they necessitate a skillful control of changes during rendition. They do not exist from time to time because when a play is over or a dance finished, there is nothing to measure and nothing left to preserve or to decay. These arts are also temporal in the sense that, while there may have been painstaking rehearsal, the essential style is lost unless the final production has an economy of time.

*II. The nature of Festivity.* The frequency with which festivities are organized in West African subcultures has been noted by observers from ancient times to the present. Negro festivities generally are characterized by rhythmic music, dancing, and playfulness between the sexes. In Brazil and the Caribbean Negroes have made *Carnival* a great joyous mass phenomenon of very rhythmic percussion and dance. *Carnival* is a moratorium from adversity and guilt which releases the common people into a frenzy of unity. In the United States, jazz, together with the dance forms that accompany it, originated with the Negroes and are constantly progressive in creative changes that the Negroes initiate. Plato said that when you give a people a music, you give them a culture. The Negroes have in this way unified a youth culture, for the Big Beat of their music affects the behavoir of young people around the world.

*III. Ephemeral Materialism.* The possession of life-force and the ability to receive or give it to others is more important in lower-class Negro subcultures than the accumulation of material objects. Housing and furnishings tend to be casual or minimal whether the location be Miami, Harlem, Rio, Beliz or Timbuctu. Without cues from road conditions, cars, and telephone poles, one often could not tell whether one is in a lower-class Negro settlement in West Africa, Brazil, the Caribbean or the U. S. Housing is much the same: small, and crowded together, unsymmetrical, and unimproved. Contiguous lower-class subcultures of other origin provide a contrast. On the other hand, articles of apparel receive greater expenditure and care as expressive of individuality. They are valued by Negros as necessary to enable the person to impart his life-force to others and attract others to impart theirs to him.

*IV. Access to Spontaneity.* Impulsivity and improvisation are the by-product of early training for independence and non-delay. They illustrate the critical dynamic between two polarities – moving too far or not at all – because spontaneity is effective in a situation only within the practical context. The opportunistic use of temporal vacancy appears in Negro humor, slang, emergency readiness, and the elaboration of time within a musical metre or a dance measure.

*V. Social Resonance.* Drawn by their temporal attitudes away from dependence on the material world, the Negro subcultures are very sophisticated in social apperception. By this is meant their grasp of the underlying mood or climate of interpersonal exchange. Social atmosphere has considerable importance for them. The ethnocentrism and feeling of superiority on the part of whites and Indians is the basis for the growing resentment in Negro subcultures. In the United States, and increasingly in Brazil [49], the realization that there is continued prejudice despite freedom is leading to reactive alienation of Negroes toward other ethnic groups. It also brings about closer atunement within the Negro subcultures. In the United States this has produced a 'Soul culture' characterized by the cohesion and human-heartedness which American Negroes feel for one another. These values run counter to competition and single-minded planning because the movement of the whole is something more than the sum of the parts. It has a Negro dynamic of unity between everyone and his group.

*Conclusion*

Taken together, the contributing and the resulting factors in the temporal attitudes of these Negro subcultures reflect their concept of time. The contributing factors are religion, subjugation causing psychological defense mechanisms, child-rearing practices and communal priorities. The resulting factors are temporal arts, festivity, ephemeral materialism, spontaneity and social resonance. All of these are related to the endurance of their social and metaphysical helicoid moving unevenly and persistently through time, balancing and revolving dichotomies in its progress.

*References*

1. Bascom, W. R.: *Friendship patterns of Sea Island Negroes.* J. American Folklore 78 (303) (1965) 3–20.
2. — *Ifa divination; communication between gods and men in West Africa.* Bloomington, Indiana: Indiana University Press 1969.
3. Bastide, R.: *Le condomble de Bahia (rite Nago). Le Monde d'Outre-Mer Passe et Present.* Premiere Serie: Etudes V. Paris: Mouton & Co. 1958.
4. Blake, J.: *Family structure in Jamaica.* New York: Crowell-Collier 1961.
5. Burton, R. F.: *A mission to Gelele, king of Dahomey.* London: Tylston & Edwards 1864.
6. Clark, K. B.: *Prejudice and your child.* Boston: Beacon Press 1963.
7. Clarke, E.: *My mother who fathered me.* London: Allen & Unwin 1957.
8. Dieterlen, G.: *Essai sur la religion bambara.* Paris: Presses Universitaires de France 1951.
9. Doob, L. W.: *Becoming more civilized.* New Haven: Yale University Press 1960.
10. Field, M. J.: *Social organization of the Ga People.* London: Crown Agent for the Colonies, 4 Milbank 1940.
11. — *Akim-Kotoku, an Oman of the Gold Coast.* London: Crown Agents for the Colonies, 4 Milbank 1948.

12. — *Search for security*. Evanston, Ill.: Northwestern University Press 1960.
13. — *Religion and medicine of the Ga people*. London: Oxford University Press 1961.
14. Fraisse, P.: *The psychology of time*. New York: Harper and Row. 1963.
15. Frazier, E. F.: *The Negro church in America*. New York: Schocken Press 1963.
16. Freilich, M.: *Serial polygyny*. Amer. Anthrop., 63 (1961) 955–972.
17. Freyre, G.: *New World in the Tropics*. New York: Knopf 1959.
18. Green, H. B.: *Socialization values in the Negro and East Indian subcultures of Trinidad*. J. Social Psychology 64 (1964) 1–20.
19. — *Values of Negro and East Indian school children in Trinidad*. Social and Economic Studies 14, 2 (1965) 204–215.
20. Griaule, M.: *Conversations with Ogotemmeli*. London: Oxford University Press 1965.
21. Heckhausen, H.: *The anatomy of achievement motivation*. New York: Academic Press 1967.
22. Herskovits, M. J.: Freudian mechanisms in Negro psychology. In Evans-Pritchard, Ed.: *Essays presented to C. G. Seligman*. London: Kegan Paul 1934.
23. — *Life in a Haitian Village*. New York: Knopf 1937.
24. — *Trinidad Village*. New York: Knopf 1947.
25. — *The myth of the Negro past*. Boston: Beacon Press 1958.
26. —, Herskovits, F.: *Rebel Destiny*. New York: McGraw-Hill 1934.
27. Jahn, J.: *Muntu*. New York: Grove Press 1961.
28. Kagame, A.: *La philosophie bantu-rwandaise de l'Etre*. Bruxelles 1951.
29. Klass, M.: *East Indians in Trinidad: a study of cultural persistence*. New York: Columbia University Press 1961.
30. Lystad, R. A.: *Tentative thoughts on basic African values*. In: *Africa and the United States*, U. S. Commission for Unesco, Boston 1961.
31. McAndrew, J.: *The open-air churches of sixteenth century Mexico*. Cambridge, Mass.: Harvard University Press 1965.
32. Mischel, W.: *Preference for delayed reinforcement*. J. Abnormal and Social Psychology 56 (1958) 57–61.
33. Niehoff, A., Niehoff, J.: *East Indians in the West Indies*. Milwaukee Public Museum Publications in Anthropology, 6. Milwaukee, Wisc. 1960.
34. Oliveira, W.: (Commissioner of Police for San Salvador, Bahia) in a personal communication to the author.
35. Park, Mungo: *Travels in the interior districts of Africa*. Philadelphia: James Humphreys 1800.
36. Pettigrew, T. F.: *A profile of the Negro American*. Princeton, N. J.: Van Nostrand 1964.
37. Pierson, D.: *Negroes in Brazil*. Chicago, Ill.: University of Chicago Press 1942.
38. Ramos, A.: *The Negro in Brazil*. Washington, D. C.: Associated Publishers 1951.
39. Ribeiro, R.: *Projective mechanisms and the structuralization of perception in Afro-Brazilian divination*. Revue Internationale d'Ethno-psychologie Normale et Pathologique 1 (2) (1965) 3—23.
40. Riesman, D.: *The lonely crowd*. Garden City, N. Y.: Doubleday 1956.
41. Rodriguez, J. H.: *Brasil and Africa*. Berkeley: University of California Press 1965.
42 Silberman, C. E.: *Crisis in black and white*. New York: Random House 1964.
43. Simey, T. S.: *Welfare and planning in the West Indies*. Oxford: Clarendon Press 1946.
44. Stevens, J. L.: *Travels in Central America, Chiakas, and Yucatan*. New Brunswick, N. J.: Rutgers University Press 1949.
45. Tempels, P.: *Bantoe-Filosofie*. Antwerp 1946.
46. Thompson, R. F.: *African influence on the art of the United States*. (In: Black Studies, A Symposium). New Haven: Yale University Press 1969.
47. Van der Plas, C.: *Socio-economic survey of the Gambia*. New York: United Nations, Feb. 9, 1956.
48. Verger, P.: *Notes sur le culte des Orisa et Vodun a Bahia, la Baie de tous les Saints au Bresil et a l'ancienne cote des esclaves en Afrique*. Memoirs de l'Institut Francais d'Afrique Noire, 1957, LI.
49. Wagley, Ch.: *Race and class in rural Brazil*. New York: Columbia University Press 1963.

50. Weber, M.: *The Protestant ethic and the spirit of capitalism.* Translated by Talcott Parsons. New York: Scribner 1930.
51. Whiting, B. (Ed.): *Six cultures: studies in child rearing.* New York: Wiley and Sons 1963.
52. Wilson, J. L.: *Western Africa.* New York: Harper 1856.
53. Zern, D.: *The influence of certain developmental factors in fostering the ability to differentiate the passage of time.* J. Soc. Psych. 72 (1) (1967) 9—17.

# On Hegel - A Study in Sorcery

Eric Voegelin*

When the gods are expelled from the cosmos, the world they have left becomes boring. In the seventeenth century, the *ennui* explored by Pascal was still the mood of a man who had lost his faith and must protect himself from the blackness of anxiety by *divertissements*; after the French Revolution, the *ennui* was recognized by Hegel as the syndrome of an age in history. It had taken a century-and-a-half for the lostness in a world without God to develop from a personal *malaise* of existence to a social disease.[1]

## I

*Die Langeweile der Welt*, the boredom of the world, is Hegel's symbol for the spiritual state of a society to whom its gods have died. The phrase appears in the so-called *Fortsetzung des "Systems der Sittlichkeit"*, written about 1804–1806 while Hegel was working on the *Phänomenologie*.[2] According to the MS the state of *Langeweile* has occurred twice in Western history. Once in antiquity, in the wake of the Roman imperial conquest; and a second time in modernity, in the wake of the Reformation. Hegel describes the state of boredom in the two cases as follows:

The expansion of the Roman Empire had destroyed the free states of the ancient world and with them the vitality of their gods in whom the spirit had become objective; with the living individuality of their gods and cults the peoples of the Empire had lost their morality; and over their singularity had spread the empty generality of imperial rule. In this diremption of the world into a singularity that is not linked to the spirit and a generality that is wanting in divine life, the "primordial identity" had to rise with its "eternal force" to overcome the "infinite pain" and to reconcile in a new wholeness what had been torn asunder – or mankind would have perished within itself (D 318). Christ

---

\* Professor Dr. Eric Voegelin, Hoover Institution on War, Revolution and Peace, Stanford, California 94305, U.S.A.

1 For the modern history of melancholy and boredom *cf.* the recent study by Wolf Lepenies: *Melancholie und Gesellschaft*. Frankfurt a. Main 1969.

2 The MS, now lost, was partly excerpted partly reported by Rosenkranz and Haym. A critical edition, based on these reports, was published by Johannes Hoffmeister: *Dokumente zu Hegels Entwicklung*. Stuttgart 1936, 314–325. In the following the *Dokumente* are quoted as D. *Die Langeweile der Welt* on D 318.

became the founder of a religion, because he could articulate the "suffering of a whole age" from the innermost depth, through the divine power of the spirit, through the absolute certainty of reconciliation he carried in himself, and because by his own confidence he could generate confidence in others (D 319).

The reconciliation achieved by the "primordial identity" through the Incarnation of God in a man was preserved by the Church. The initial reconciliation of the spirit with reality through the resacralization of man was even expanded to embrace society and nature; sacredness was extended to the ruling power of the monarch; and in every country the messengers of God left their traces, so that each had its own sacred history of reconciliation. The whole world had become a "temple of reawakened life" (D 322).

The great diremption of the new reconciliation was caused by the Reformation. Protestantism has abolished "the poetry of sacrality" by tearing the new fatherland of man asunder into the inwardness (*Innerlichkeit*) of spiritual life and "an undisturbed engagement (*Versenken*) in the commonness (*Gemeinheit*) of empirical existence and everyday necessity". "The Sabbath of the world has disappeared, and life has become a common, unholy workday" (D 323).

The "beauty and sacrality" of the pre-Reformation world is lost for good; history cannot be turned back; we have to advance toward a new religion which understands the former reconciliation as an "alien sacralization" and replaces it by a sacralization through the spirit that has become "inward": "The Spirit has to sacralize itself as Spirit in its own form". The diremption will be overcome when "a free people" has the audacity, not to receive a religious form, but to take one for itself "on its own soil and by its own majesty" (D 324). In Protestantism, this relation between Spirit and reality has achieved its breakthrough to consciousness through Philosophy. The new Philosophy restores "its aliveness to Reason and its spirit to Nature". The Philosophy that emerges from the Protestant diremption is destined to follow Catholicism and Protestantism as the new, third Religion (D 323).

Ages of diremption (*Zerrissenheit*) and boredom (*Langeweile*) do not just happen, nor will new religions simply emerge. The eternal force of the primordial identity operates concretely through such human beings as Christ and Luther. If Philosophy is to be the third religion, succeeding to Catholicism and Protestantism, who will succeed Christ and Luther as the founder of the new religion? Perhaps Hegel?

The question had Hegel so badly worried that its pressure formed his existence as a philosopher. In order to gauge its import it will be apposite first to distinguish the various strata of the question:

(1) As a philosopher in the classic sense, Hegel knew that he could not diagnose the diremption of the age without exempting himself somehow from its boredom. Some degree of reconciliation had to be realized in his own existence or he could not have recognized the diremption for what it was;

as a philosopher he had to be spiritually healthy enough to diagnose the spiritual state of society as diseased; even more, the analysis of the social disease had to become to Hegel, as to every philosopher, the meditative action by which the physician, who is born as the child of his age, heals first of all himself. Only when through the diagnosis of the evil surrounding him he has arrived, God's Grace permitting, at insight into the truth of his own existence as a man can he become effective as a reconciler and restorer of existential order to his fellowmen.

(2) The second stratum is represented by the pneumatism of the inner man and the inner light. The sectarian spirituals of the Middle Ages and the Renaissance, the godded men and *homines novi,* were followed in the eighteenth century by the occultists, visionaries, and *Schwärmer,* by the illuminists and theosophs, by the Swedenborg, Martinez, Saint-Martin, and Cagliostro, by Lavater, Jung-Stilling, and so forth. Beginning with the French Revolution, then, a cloud of new Christs descended on the Western world – Saint-Simon, Fourier, Comte, Fichte, and Hegel himself. The life-time of Hegel (1770–1831) runs parallel with the period studied by Auguste Viatte in his *Les Sources Occultes du Romantisme, 1770–1820* (1927; 1965). Hegel's own inwardness is firmly related to Jacob Boehme and the German Pietists.

(3) The third stratum is the imaginative construction of ages that will permit the imaginator to anticipate the future course of history. By means of this construction, the imaginator can shift the meaning of existence from life in the presence under God, with its personal and social duties of the day, to the rôle of a functionary of history; the reality of existence will be eclipsed and replaced by the Second Reality of the imaginative project. In order to fulfill this purpose, the project must first of all eclipse the unknown future by the image of a known future; it must further endow the construction of the ages with the certainty of a science – of a *"Wissenschaftslehre"*, a "system of science", a *"philosophie positive"*, a *"Wissenschaftlicher Sozialismus";* and it must, finally, conceive the future age in such a manner that the present imaginator becomes its inaugurator and master. The purpose of securing a meaning of existence, with certainty, in a masterly rôle betrays the motives of the construction in the imaginator's existential insecurity, anxiety, and *libido dominandi.* This is megalomania on the grand scale. Still, the Messiases of the early nineteenth century have left so deep an imprint on the socalled Modern Age that we have become accustomed to their madness; our sensitivity for the element of the grotesque in their enterprise has become dulled. In order to sharpen it somewhat, let us imagine a Jesus running around and announcing to everybody the good news that he is the man from whom the era of Christ will be reckoned – as Comte announced *urbi et orbi* that with the completion of his work in 1854 the era of Comte had begun.

The interaction of the three strata in Hegel's existence makes him a characteristically modern thinker. There is a sensitive philosopher and spiritualist, a noetically and pneumatically competent critic of the age, an intellectual force of the first rank, and yet, he cannot quite gain the stature of his true self as a man under God. From the darkness of this existential deficiency, then, rises the *libido dominandi* and forces him into the imaginative construction of a false self as the Messias of the new age. The interaction of the strata, thus, cannot be brought on a simple formula. In the construction of the system, it is true, the Second Reality of the third stratum prevails and badly deforms the existence of the philosopher and spiritualist. But Hegel does not always construct his system. He can write brilliant common sense studies on politics, as well as literary essays which reveal him as a master of the German language and a great man of letters. Moreover, the systematic works themselves are filled with excellent philosophical and historical analyses which can stand for themselves, unaffected in their integrity by the system into which they are built. Hence, the modernity of Hegel can be characterized as the co-existence of two selfs, as an existence divided into a true and a false self holding one another in such balance that neither the one nor the other ever become completely dominant. Neither does the true self become strong enough to break the system, nor does the false self become strong enough to transform Hegel into a murderous revolutionary or a psychiatric case.[3]

II

The existence of a modern man is complicated. In Pascal's language, Hegel's System of Science is a *divertissement*. The philosopher who wants to heal the disease of society is not able to achieve the truth of his own existence but develops a further diremption between the philosophical intention of his true self and the actual pursuit of the false self and its rôle in the imaginative project of history. A second diremption in the philosopher's existence, thus, is stocked on the first one which he has correctly diagnosed as the spiritual disease of society. The result is the intricate pattern of relations between the two stories of diremption that in our time falls apart in violent social and personal catastrophies without the redeeming catharsis of tragedy. As far as society is concerned, the spiritually sensitive revolt against its unsatisfactory state is conducted by existentially deficient men who add themselves as a new source of disorder to troubles which are bad enough without them. As far as the rebels are concerned, the rôle in which they cast themselves is not easy to act; and rarely do they shoulder its burden with such conscientiousness as Hegel's.

Because of his conscientiousness as a thinker, Hegel's case acquires the quality of a paradigm for the vicissitudes of the multitensional existence we call modern. As a philosopher Hegel is bound by the tradition of philosophizing from

---

[3] On the question of the two Selfs *cf.* R. D. Laing: *The Divided Self. An Existential Study in Sanity and Madness* (1960; Pelican Edition 1965).

antiquity to the present which he knows superbly well. By a philosopher's existence, however, Hegel would feel frustrated; for philosophers, even of the highest rank, are not the historical figures who put their signature on the millennia; we live in the era of Christ, after all, not in the era of Plato. Hence, in order to accomodate a *libido dominandi* that cannot be fulfilled by a philosopher's existence, philosophy must be dressed up as "religion". In Hegel's earlier conception, philosophy is a state of consciousness that emerges reflectively as a "third religion" from Protestantism; in his later conception, it absorbs "religion" into itself. Philosophy becomes the ultimate revelation of the new "primordial identity", and the old God of revelation is declared to be dead for good. In order to legitimate these strategic changes of meaning, Hegel must then develop an imaginative project of immanent history, with a construction of ages that will include an ultimate age to be inaugurated by himself. This immanentist apocalypse engendered by the thinker's *libido dominandi* has the purpose of eclipsing the mystery of meaning in history expressed by the Christian symbolism of eschatological events. That the construction has been thrown up by an outburst of libidinous imagination, however, must not be admitted; the philosopher's true self is much too strong in Hegel to let imagination be enthroned as a source of truth superior to reason, as is done by André Breton and the surrealist young revolutionaries in our time. On the contrary, the libidinous aroma that attaches to the construction and impairs its legitimating function, must be doubly covered up: The new philosophy is declared to be, not a mere love of wisdom like the old one, but a final possession of knowledge; and this knowledge is further enhanced by the new symbol "science" which began, in the wake of Newton, to acquire its peculiar modern magic. And finally, if the imaginative history is not to clash with historical reality, events must be found in contemporary history that look promising as the wave of the future of which the philosopher wants to become the Messias. If he does not want to be laughed out of court as a *Schwaermer* or a crackpot, the philosopher must tie his Messianic ambitions to a reasonably successful looking political force of his time.

I shall now document the tortuous path Hegel winds through this pattern of existence by some of his accounts of the experience.

The great event that impressed the young man of twenty as the opening of a new age was the French Revolution. Forty years later, in the *Philosophy of History,* the old Hegel recollects the impact and its nature:[4]

> "As long as the sun stands in heaven and the planets revolve around it, has it not happened that man stood on his head, that is on his thought, and built reality in conformity to it. Anaxagoras had been the first to say that Nous governs the world; but only now has man gained the insight that thought

---

4 Hegel: *Philosophie der Geschichte.* (ed. F. Brunstaedt), Stuttgart: Reclam 1961, 593.

should govern spiritual reality. This was a splendid sunrise; all thinking beings shared in celebrating the epoch. The age was ruled by a sublime emotion, the world trembled as the enthusiasm of the spirit (*Geist*) pervaded it, as if only now the divine had been truly reconciled to the world."

At this late date, the utterance of Hegel has become liturgical – the language symbols are used with the meanings they have acquired in his work of a lifetime. In the thought (*Gedanke*) and spirit (*Geist*) that interpenetrate in the Revolution we recognize the "philosophy" and "religion" which Hegel has absorbed into his "science" of *Das absolute Wissen;* and the relation of thought and spirit to the Nous of Anaxagoras is not intelligible without his imaginative construction of history. In the 1790's Hegel certainly would not have articulated the experience in the language of the passage that was written (or spoken) after the July Revolution of 1830. Nevertheless, there is no reason to doubt the validity of the account. The impact of the Revolution was indeed the experience that fundamentally formed Hegel's existence. The fact that toward the end of his life he still could accept the experience as valid, and did not have to reject it as a youthful aberration, that he even could express it by the symbols he had developed in the existential process that had started from it, is the best proof for the authenticity of the account. The conventional reflections on the status of Hegel as a philosopher of Enlightenment, or the last Christian philosopher, or the reactionary glorifier of the Prussian state, become irrelevant in the light of his self-declaration as the philosopher of the French Revolution.

Nearer to an original articulation of the experience are the pages of the *Fortsetzung des "Systems der Sittlichkeit"* from which I have previously quoted. There Hegel speaks of the "reason (*Vernunft*) that has rediscovered its reality as moral spirit"; of the spirit that now again "can sacralize itself as spirit in its own form"; of the Protestantism that has taken off (*ausgezogen*) "the alien sacralization" – leaving it unclear whether the "alien" refers to an ultramontane Papacy or a supramundane Divinity; and of the "free people" that will give its religious form (*religiöse Gestalt*) to itself "by its own majesty" (D 324). The accents, thus fall on an enterprise of self-salvation, with overtones of a "Nordic subjectivity" which alone is capable of the feat (D 323). The passages have already the flavor of Nietzsche's advice to modern man to redeem himself by extending Grace to himself instead of waiting for a divine Redeemer by the Grace of God. The new freedom and activism of self-salvation is experienced by Hegel as the core of meaning in the great events that shook the world.

A disturbingly unsatisfactory situation – for Hegel had not started the French Revolution, and the battles of the Napoleonic wars raged around him while his own existence as a *Dozent* in Jena was distinctly non-combatant. He was worried in these days by the question how a philosopher could participate in the meaning of the bloody events which to him were the only meaningful reality in the world. Rosenkranz reports his answer on the basis of the original

MS: Philosophy is necessary to a people as the ideal supplement of war. More specifically (D 314):

> "Death alone is absolute work (*absolute Arbeit*) as it abolishes (*aufhebt*) the determinate singularity. Courage brings its absolute sacrifice to the State. The humiliation not to have died, however, is the lot of those who do not die in battle and still have the enjoyment of their singularity. Hence, there is nothing left to them but speculation, the absolute knowledge of truth, as the form in which the pure (*einfache*) consciousness of the infinite is possible without the determinateness of an individual independent life."

One must not cheapen this passage by psychologizing on the bad conscience of the non-combatant. Hegel is serious about the equivalence of death in battle and philosophy – provided the battles are conducted to establish a "free people" and the speculative process results in "absolute knowledge". In order to gain its form, the "free people" needs the supreme sacrifice as well as the absolute spirit. Hegel's philosophy is not the Socratic practice of dying – it is the equivalent of death on the battlefield of the Revolution.

To philosophize in such a manner that the philosopher's work integrates itself meaningfully into the process of history is a demanding task. Fortunately we have Hegel's own text for his reflection on this issue (D 324.):

> "Every single man is but a blind link in the chain of absolute necessity by which the world builds itself forth (*sich fortbildet*). The single man can elevate himself to dominance (*Herrschaft*) over an appreciable length of this chain only if he knows the direction in which the great necessity wants to move and if he learns from this knowledge to pronounce the magic words (*die Zauberworte*) that will evoke its shape (*Gestalt*)."

This passage reveals the intense resentment of Hegel's as well as its cause. It is a key-passage for the understanding of modern existence. Man has become a nothing; he has no reality of his own; he is a blind particle in a process of the world which has the monopoly of real reality and real meaning. In order to raise himself from nothing to something, the blind particle must become a seeing particle. But even if the particle has gained sight, it sees nothing but the direction in which the process is moving whether seen by the particle or not. And yet, to Hegel something important has been gained: The nothing that has raised itself to a something has become, if not a man, at least a sorcerer who can evoke, if not the reality of history, at least its shape. I almost hesitate to continue – the spectacle of a nihilist stripping himself to the nude is embarrassing. For Hegel betrays in so many words that being a man is not enough for him; and as he cannot be the divine Lord of history himself, he is going to achieve *Herrschaft* as the sorcerer who will conjure up an image of history – a shape, a ghost – that is meant to eclipse the history of God's making. The imaginative project of history falls in its place in the pattern of modern existence as the conjurer's instrument of power.

Hegel concludes his reflection with the statement (D 325):
"This knowledge – which means including the whole suffering and the conflict which for several thousand years has ruled the world and all forms of its manifestation (*Ausbildung*) in oneself, and at the same time elevating oneself above it (i.e. the conflict) – this knowledge only *Philosophy* can give."

"This knowledge", we remember, is the knowledge from which its possessor can learn the magic words that will evoke the shape of things to come. Regarding its contents, "this knowledge" must be the all-inclusive book of the suffering and conflict in the world's process, for only if it is all-inclusive can the possessor of "this knowledge" elevate himself above the world's suffering and conflict. The theme of diremption and reconciliation is resumed. The all-inclusive knowledge must be achieved in order to make an end of the world-process, of this nightmare of suffering and conflict, and to inaugurate the age of reconciliation. A shape is evoked indeed by Hegel's program: The shape of the Christ who takes the conflict and suffering of this world on his shoulders and thereby becomes its redeemer. This redemptive knowledge is the knowledge that only Philosophy can give. "Philosophy" becomes the *grimoire* of the magician who will evoke for everybody the shape of the reconciliation that for himself he cannot achieve in the reality of his existence.

III

Hegel has carried out his project. In 1807 he published his *grimoire* under the title of *System der Wissenschaft. Erster Theil, die Phänomenologie des Geistes*.[5]

Form and language of the work reflect the complexity of modern existence whose presentation is its purpose. As a genus of philosophical literature, the *Phaenomenology* is a treatise on Aletheia, on truth and reality, and a very important one indeed; no philosopher can afford to ignore it. Nevertheless, the diremption of Hegel's existence into the true self of the philosopher and the false self of the Messianic sorcerer imposes itself on the work, so that its philosophical excellencies become subordinate to the anti-philosophical *Ziel*, to the goal of enabling philosophy at last "to give up its name of a love of knowledge and to become real knowledge (*wirkliches Wissen*)" (Ph 12).

No modern propaganda-minister could have devised a more harmless sounding, persuasively progressivist phrase as a screen for the enormity transacted behind it. For philosophy, though its insights can advance, cannot advance beyond its structure as "love of wisdom". In Plato's exegesis of the "name", philosophy denotes the erotic tension of man toward the divine ground of his

---

5 Hegel: *Phaenomenologie*. (ed. Hoffmeister). Hamburg 1952. Quoted in the following text as Ph.

existence. God alone has *sophia*, the "real knowledge"; man finds the truth about God and the world, as well as of his own existence, by becoming *philosophos*, the lover of God and his wisdom. The philosopher's eroticism implies the humanity of man and the divinity of God as the poles of his existential tension. The practice of philosophy in the Socratic-Platonic sense is the equivalent of the Christian sanctification of man, it is the growth of the image of God in man. Hegel's harmless sounding phrase, thus, covers the program of abolishing the humanity of man; the *sophia* of God can be brought into the orbit of man only by transforming man into God. The *Ziel* of the *Phaenomenology* is the creation of the man-god.

The technical difficulties Hegel has to surmount, in order to reach his goal while camouflaging what he is doing, are enormous. More about them presently. The principle of construction in the *Phaenomenology*, however, is so simple that it will not be unfair to call it a sleight-of-hand. As it would prove impossible even for the constructive genius of a Hegel to grind the real God and real man through the machinery of dialectics and come out with a man-god, he roundly does not concede the status of reality to either God or man. The *Phaenomenology* admits no reality but consciousness. Its phenomena range from the consciousness of sensation (I–III) and self-consciousness (IV), through reason (V), spirit (*Geist*) (VI), and religion (VII), to absolute knowledge (VIII). Since consciousness must be somebody's consciousness of something, and neither God nor man are admitted as somebody or something, the consciousness must be consciousness of itself. Its absolute reality is, therefore, properly defined as "the identity of identity and non-identity". The substance becomes the subject, and the subject the substance, in the process of a consciousness that is immanent to itself. Of course, Hegel does not state his principle of construction as baldly as I have done it now, or the enterprise of his *grimoire* would be self-defeating. The reader would justly ask what a consciousness that is nobody's consciousness could possibly be? And if he received no answer at all, or were more or less politely put off with the suggestion that it was his fault, if he did not understand what is crystal-clear, he might become suspicious. No, the *Phaenomenology* has 564 pages; and it ranges with an incredible wealth of observations over such phenomena as the relation of master and servant; Stoicism, Scepticism, the unhappy consciousness; existentialist attitudes such as the hedonist and the moralist; apolitical and political man, revolutionary and loyal citizen; classical tragedy and Christian religion; alienation, education, faith, intellectualism; enlightenment, superstition, freedom, and terror; the French Revolution and the Napoleonic Empire. In Hegel's construction, all of these phenomena are meant to be stages in the dialectical process of immanent consciousness toward its goal of "absolute knowledge", but the reader, living in his common sense habits, will understand the frequently brilliant observations as a philosopher's reflections on phenomena in the real world of personal existence in society and history. The *Phaenomenology* is a *divertissement* in the pregnant sense of an imaginative game, masterly devised so close to reality that the excited spectator may forget that what he is watching is no more than a game.

The ambiguity of the game must be isolated and recognized as a structure in the *Phaenomenology*, in order to avoid futile debate. One can concentrate on the consciousness suspended in a void as the principle of construction, and dismiss the *Phaenomenology* as a piece of nonsense. One can concentrate on the studies of enlightened intellectuals, or of reductionist psychology, or of terrorist masses, and admire Hegel as a profound analyst of existential aberrations. Both sides can be well defended; and yet, the argument would miss the game of replacing the first reality of experience by the second reality of imaginative construction, and of endowing the imaginary reality with the appearance of truth by letting it absorb pieces of first reality. Moreover, the game is played by a master whose imperatorical intellect can indeed organize such vast amounts of historical materials into his construction that even the not so innocent reader may be well enough diverted to overlook the gaps and inconsistencies, and to believe the ostensible *Ziel* of transforming the love of wisdom into a system of science to be reached. Hegel's cunning in coining a phrase which disguises an existential enormity is matched by his ability to put the deception over. The ambiguity of the game, its cunning and conning, must be recognized as a phenomenon in its own right – a phenomenon which does not appear among the phenomena of the *Phaenomenology* but has been well established by it as the prototype of the great confidence game played by modern man in his diremption existence under such titles as advertisement, propaganda, communication, and, comprehensively, as ideological politics.

The structure of the game must be isolated and recognized, but it must not be torn out of the context of the *grimoire*. Hegel does not want to play games for their sake, he wants to find the *Zauberworte* that will give him power over reality. And in its context, the game is not the diverting escape from reality as which it appears to the critical reader, but the necessary means for the end of establishing the "real knowledge" that will enable Hegel to evoke the shape of the future. Since this cannot be achieved in reality but only in an act of metastatic imagination, and the imagery of the act has to be consistent within itself, history must be transformed into the dialectical process of a consciousness that will come to its self-reflective fulfillment in the metastatic "consciousness" suspended in the void of Hegel's imagination. In order to break the chain to which he imagines himself to be linked, Hegel must link history to the imaginary chain of the dialectical process. The metastasis of the lover of wisdom into the possessor of knowledge requires the metastasis of history into the dialectics of the *Phaenomenology*.

The construction of a *grimoire* is a violent destruction of reality. In historical reality, a philosopher's truth is the exegesis of his experience: A real man participates in the reality of God and the world, of society and himself, and articulates his experiences by more or less adequate language symbols. But however compact, incomplete, and in need of further revision his experience and symbolization of reality may be, it has its dignity as a real man's image of the divine reality of the cosmos surrounding and embracing him. Moreover, the philosopher knows that his own, noetically controlled experience of participation, though it achieves more differentiated insights into the truth of reality than are possible in the more compact medium of the myth, is the same experience of participation in the same reality that has engendered the noetically less controlled symbolisms. However important his advance of insight may be, as a man he is as far from, or near to, the divine *sophia* as his mythopoetic predecessor; advances of insight can sharpen a man's understanding of his humanity, but they do not abolish his human condition. However widely they may differ regarding the historical state of their insight, the *philomythos* and the *philosophos,* the believer in salvation through Christ, the ancient gnostic, the medieval alchemist, and the modern sorcerer are all equal regarding the equidistance of their humanity from God. The equivalence of symbolisms as the expression of man's search of truth about himself and the ground of his existence is the principle, established by Aristotle, that guides the philosopher's inquiry concerning the historical manifold of truth experienced and symbolized.[6]

To imagine the search for truth not to be the essence of humanity but an historical imperfection of knowledge to be overcome, in history, by perfect knowledge that will put an end to the search, is an attack on man's consciousness of his existence under God. It is an attack on the dignity of man. That is the attack Hegel commits when he replaces the concrete consciousness of concrete man by the imaginary "consciousness" that runs its dialectical course in time to the absolute consciousness of Self in his System. He sustains the construction by imposing a closed network of relations on the symbols *Geist*, history, time, space, and world. There is no history before the *Geist* starts moving in the Asiatic empires of China, India, and Persia; there will be no history after the *Geist* has come to its consciousness of Self in the Napoleonic empire and Hegel's System of Science. "This last shape of the spirit (*diese letzte Gestalt des Geistes*)" gives to its "complete and true contents at the same time the form of the Self". "The spirit, *appearing* in this element of consciousness (or, what is the same, produced in it by it) *is Science*". "This is absolute knowledge" (Ph 556). Before however the *Geist* has achieved its conceptual form (*Begriffsgestalt*) it has already existence (*Dasein*) as "the ground and concept in its unmoved simplicity, i.e. as the inwardness or the Self of the *Geist* that has not yet existence (*noch nicht da ist*)". There is an experience and knowledge of the *Geist* as substance, i.e. as "a felt Truth, as an inwardly revealed Eternal, as a believed Holy, or whatever expression else one may use". But this experience (*Erfahrung*) of the *Geist* as substance through "religion" has the character of hiddeness (*Verborgenheit*) rather than revelation (*Offenbarkeit*), because the substance is not yet fully

---

6 On Aristotle's principle of equivalence *cf.* my *Anamnesis* (Munich 1966) 297–9.

revealed as a moment in the dialectical process as which it can be revealed only in retrospect from the position of absolute knowledge achieved. This movement of the *Geist,* from its hidden conceptual shape as substance, to its revealed conceptual shape in the self-reflective consciousness of absolute knowledge, is the contents of the dialectical process (Ph 557–8). The imaginary process of the imaginary "consciousness" must, then, be shielded against the reality of history by transforming time into an inner dimension of dialectics. "*Time* is the *concept* itself in its existence (*der da ist*), as it presents itself to consciousness as an empty intuition (*Anschauung*); that is the reason the Geist appears of necessity in time, and will appear in time as long as it has not conceived (*erfaßt*) the pure concept of itself, i.e. as long as it has not abolished time". Time is "the pure Self, seen from the outside but not yet conceived by the Self". The concept, by conceiving itself, "abolishes its time-form (*hebt seine Zeitform auf*)." Hence, "time appears as the fate and necessity of the *Geist* that has not yet found its fulfillment in itself". And it cannot reach its fulfillment as "self-conscious *Geist*" before it has not run its course as *Weltgeist*. "The movement by which it brings forth from itself the form of knowing itself, is the labor it performs as *real history*" (Ph 558–9). The significance of the construction will become clear, if one realizes that Hegel applies to the time of the *Geist* in history the argument Plato and St. Augustine applied to the time of the world: Time is a dimension internal to the reality of the world; there is no time in which God created the world; there is no time before time. Hegel's "real history" of the *Geist* is the history of a world with an inner time dimension. Its beginning and its end lie before the God who made it; there was no time before the time set by Hegel for its beginning; and there will be no time after the time-form has been abolished by Hegel through his System. Hegel is the Alpha and Omega of "real history".

Only a master of philosophical technique could have devised the construction of "consciousness" just analyzed; but then again, no philosopher would ever indulge in such a construction. The author of the *Phaenomenology* suffers so badly from the existential conflict between his two Selfs that it almost makes no sense to ask what Hegel *really* meant. The interpreter must be alert to the games of the divided Self. He must put Hegel into quotation marks, because no statement concerning "Hegel's" intentions can be valid, unless it takes into account the intricate movements of his Selfs. In the preceding paragraph, for instance, I have characterized Hegel's construction flatly as an attack on the dignity of man. But is it really? If we place ourselves inside the construction, no attack on man and his dignity occurs, because "Hegel" excludes real man's consciousness from his imaginative construction of "consciousness". The movement of dialectical knowledge "is the circle that runs back into itself, presupposing the beginning it reaches in the end" (Ph 559). Once you have entered the magic circle the sorcerer has drawn around himself you are lost.[7] And yet,

---

[7] Students of Hegel have noted this problem; in Joerg Splett: *Die Trinitaetslehre*

the attack on the dignity of man really occurs, because "Hegel" intends his construction, not to be a private amusement, but an eminently public proclamation of the "scientific" truth about the reality of man in society and history. One cannot simply shrug off the "Hegel" of the construction as a cranky imaginator, because there is the other "Hegel" who means his construction to be a treatise on Aletheia. And then there is the third "Hegel" who comprehends the other two, the potent sorcerer who imposes his *opus* on an "age" that is all too willing to find the way out of its diremption through sorcery.

The game of the Selfs must be watched with particular care, if one wants to understand "Hegel's" declaration that God is dead. The issue is still alive, as attested by the recent wave of Death-of-God revivalism; and it hardly could be alive, if its addicts had ever undergone the admittedly unpleasant discipline of reading Hegel closely. For in the context of the *Phaenomenology*, the death of God is inseparable from the life of God that comes to its fullness in "Hegel's" System. The issue must be formulated rather as the alternative whether Hegel has become God, or whether God was Hegel from the beginning and only took the time of "real history" to reveal himself fully in the System. Let us go through the various moves of the game:

(1) "Consciousness" is absolute reality; its process is a theogony; and when it is completed the God is fully real and present. As a matter of fact, at the end of Chapter VI Hegel introduces the *Ich* that is "assured of the certainty of the *Geist* within itself" as the *erscheinende Gott,* as the God who fully reveals himself "in the middle of those who know themselves as pure knowledge" (Ph 472). Chapter VII, then, removes the God of Christian Revelation by putting him in his place as a *Gestalt* of consciousness now obsolete and dead; and in Chapter VIII, finally, the consciousness as "absolute knowledge" is alone with itself. Since these Chapters were written by Hegel, and presumably he was not unconscious when he wrote them, we must conclude that in 1807 Hegel has become God.

(2) This conclusion, however, is no more than the first word in the matter. There must be taken into account the problem of the "circle": What is reached by the circle of the construction in the "end" is the "beginning" that has been presupposed (Ph 559). If God reveals Himself fully in "Hegel's" System in the "end", we must conclude that God was "Hegel" even in the "beginning" – only a simpler, more substantive, less self-reflected "Hegel".

(3) The matter is further complicated by a little uncertainty about "Hegel's" position in the Trinity. I have found no indication in Hegel's works that "Hegel"

---

G. W. F. *Hegels.* Freiburg/München 1965, 150–4, the author gives a survey of various, sometimes evasive or embarrassed, responses, to it. The best statement is Gadamer's: "The Archimedian point for unhinging Hegel's philosophy will never be found in (Hegel's) reflection itself. That is precisely the formal quality of reflection philosophy, that there is no position which could not be integrated into the reflection movement of the consciousness that comes to itself" (Hans Georg Gadamer: *Wahrheit und Methode.* Tübingen: 1960, 326).

was ever God the Father; this rôle is reserved to the "primordial entity". But he seems to have been God the Son. In the *Logik* (1812), there is no doubt that Hegel is the Logos, the Son of God, only bigger and better – but about that problem presently. In the *Phaenomenology* (1807), one can find a definite intimation that "Hegel" did not consider himself God the Father but only the Son: At the end of Chapter VI, the *erscheinende Gott* is present, incarnate, among us who know ourselves as pure knowledge (Ph 472). Without a doubt, however, throughout the *Phaenomenology* "Hegel" is the Holy Ghost.

(4) To the end of his life, Hegel insisted on his Protestant orthodoxy; and still in 1830 he delivered the speech celebrating the Tercentenary of the *Confessio Augustana*. At first sight, the orthodox "Hegel" seems to be incompatible with the "Hegel" who declares God to be dead. The positions will be in conflict, however, only to fundamentalist interpreters who understand the death of God as an atheistic counter-dogma to the theism of the Creed. To "Hegel", God was very much alive, as I have suggested, revealing Himself in the System more perfectly than He had ever done before in *Gestalten* which, in the Hegelian construction, are a "hiddenness" rather than a "revelation". To the "Hegel" of the imaginary consciousness, orthodoxy was a valid phase in the dialectical process of the *Geist*, though now superseded, through the labor of the *Weltgeist*, by its last *Gestalt* in the System. God is dead only in relation to Hegel's System. One cannot have Hegel's death of God without entering the System, just as one cannot have Nietzsche's death, and even murder, of God without turning into the *Übermensch*. Wallowing in the death of God, or drawing from it atheistic conclusions, or repairing it by social action would have been adjudged, by both Hegel and Nietzsche, pastimes beneath contempt.

(5) The "death of God", finally, is unintelligible without the "death of Hegel". In the *Fortsetzung des "Systems der Sittlichkeit"*, Hegel had postulated speculation as the alternative to death in battle. "Absolute knowledge" was to be the form "in which the pure consciousness of the infinite is possible without the determinateness of an individual, independent life" (D 314). In the *Introduction* to the *Phaenomenology*, written after the main body of the work had been finished, Hegel resumes the problem of speculation as the death of individual life: "The *Ziel* is set to knowledge as necessarily as the progress toward it; it (*vid.* the goal) is realized (*es ist da*), where it (*vid.* knowledge) has no longer to go beyond itself, where it finds itself, where the concept corresponds to the object and the object to the concept. What is restricted to natural life cannot, by itself, go beyond its immediate existence, but is driven beyond it by something other than itself, and this being-pulled-beyond-it (*Hinausgerissen werden*) is its death. Consciousness, however, is its *concept* for itself; it is, in its immediacy, the being-beyond-the-limited and, since the limited is part of itself, beyond itself; with its singularity, there is given to it the beyond of singularity, even if it were only a beyond *by the side* of the limited, as in spatial intuition.

Hence, consciousness suffers the violence of its limited contentment being destroyed (*verderben*) from itself" (Ph 69). Moreover, Hegel reflects on the anxiety aroused by the death of limited consciousness through the effort of speculation. Man will shy back from "truth" and try to preserve what threatens to be lost; but it will not be easy for him to find his peace of mind in "thoughtless inertia" or in the "sentimentality of finding everything good in its way", for the restlessness of thought disturbs thoughtless inertia as well as sentimentality. Hegel concludes the list with the fear of truth that hides behind a zeal for truth so fervent that no truth can be found but the truth of the *trockene Ich*, always cleverer than any thought, be it his own or somebody else's (Ph 69–70). From this catalog of escapes, there emerge the unrest of thought, as well as the trust in a reality that will prove amenable to self-reflective conceptualization, as the existential qualities of the thinker who strives to go beyond the natural limit of existence into the death of absolute knowledge.

With the last Chapter of the *Phaenomenology*, Hegel's "determinateness as an individual, independent life" has died. God is dead; and now Hegel is dead, too. Something, like the last scene of an Elizabethan tragedy.

The death of Hegel must not be separated from the death of God. They both together are, in the medium of speculative sorcery, the equivalent of a *theologia mystica* which recognizes the symbolism of positive theology as valid, while knowing about the experience of meditative participation in the divine ground, the *unio mystica,* beyond it. Hegel was a mystic *manqué*.[8]

By way of a postscript: The death of God is a dangerous plaything for epigonic intellectuals and confused theologians.

IV

Nobody can heal the spiritual disorder of an "age". A philosopher can do no more than work himself free from the rubble of idols which, under the name of an "age", threatens to cripple and bury him; and he can hope that the example of his effort will be of help to others who find themselves in the same situation and experience the same desire to gain their humanity under God. Hegel, however, wanted to become, not a man, but a Great Man: The Great Man whose name marks an epoch in history was his obsession. Moreover, he did not want to become just any Great Man in history, preceded and followed by others, but the greatest of them all; and this position he could secure only by becoming the Great Man who abolishes history, ages, and epochs through his evocation of the Last Age that will forever after bear his imprint. The Great-Great Man in history is the Great Man beyond history. To gain power over

---

8 On the problem of mysticism and its deformation through Hegel *cf.* the excellent Note in Alexandre Kojève: *Introduction à la lecture de Hegel. Leçons sur la Phénoménologie de l'Esprit.* Paris: 1947, 296.

history by putting an end to history with its diremption and boredom was the driving force of Hegel's sorcery.

What induced a potential philosopher to go on the rampage of becoming the Great-Great Man, is impenetrable. As in the case of Hegel's great successors in sorcery, of Marx and Nietzsche who wanted to evoke the *Übermensch,* the spiritual disease of refusing to apperceive reality, and of closing one's existence through the construction of an imaginary Second Reality, is a secret between man and God. One can do no more than describe the phenomenon. In Hegel's case, the five or six years preceding the publication of the *Phaenomenology* in 1807 were the critical period in which the magic project crystallized. Though one would like the documentation of the process to be more complete, what has been published from the manuscripts so far is sufficient to permit a reconstruction.

What crystallized in the critical years was, first of all, the symbolism of *Geist* (spirit), *Gedanke* (thought), *Vorstellung* (conception), and *Idee* (idea) – the instrument for eclipsing the reality of Myth, Philosophy, and Revelation. Its nature and function become apparent in Hegel's criticism of Plato's myths: The myths have charm and are pedagogically useful, they make the dialogues attractive reading, but they betray Plato's inability to penetrate certain areas of the *Geist* by *Gedanke*. "Myth is always a presentation which introduces sensual images, appealing to conception, not to thought; it is an impotence of the thought which cannot yet get hold of itself. In mythical presentation, thought is not yet free; sensual *Gestalt* is a pollution of thought as it cannot express what thought wants to express ... Frequently Plato says it is difficult to set forth a subject-matter by thought and he will, therefore, render it by a myth; easier it certainly is."[9]

The passage sounds as if Hegel had never become even fleetingly aware that Plato's introduction of the myth manifests, not his failure as a thinker, but his critical understanding of philosophical analysis and its limits. The philosopher can clarify the structure and process of consciousness; he can draw more clearly the line between the reality of consciousness and the reality of which it is conscious; but he can neither expand man's consciousness into the reality in which it is an event, nor contract reality into the event of consciousness. Plato knows quite well that his myth – of Eros, of the Psyche as the site of man's search for the divine ground of his existence, of the immortality of the soul, of its pre- and post-existence, its guilt and purification, of the Last Judgment, of the demiurgic origin of the cosmos – symbolizes experiences of the *Geist,* but he also knows that man's *Geist* is not identical with the reality in which it participates consciously through experience. The experience of participation in a divinely

---

9 Hegel: *Vorlesungen über die Geschichte der Philosophie.* Vol. II, (Jubiläumsausgabe, ed. Glockner) 188–9. The lecture course on the *Geschichte der Philosophie* was delivered for the first time in the winter-term of 1805/06 in Jena.

ordered cosmos extending beyond man can be expressed only by means of the myth; it cannot be transformed into processes of thought within consciousness. Moreover, Plato was so acutely aware of man's consubstantiality but nonidentity with divine reality that he developed a special symbol for man's experience of his intermediate status between the human and the divine: He called the consciousness of this status the *metaxy*, the In-Between of existence.

The In-Between of existence is not an empty space between two static entities, but the meeting-ground of the human and the divine in a consciousness of their distinction and interpenetration. This consciousness of the *metaxy* is in historical flux. The differentiation of noetic consciousness through the philosophers is an event in history; and when it has occurred, man's insight into consciousness and its beyond has advanced. The old myth had been adequate to a more compact experience of the cosmos; when consciousness becomes noetically luminous, a new myth is required. Plato opposed his own myth sharply to the Homeric myth. This real advance of noetic insight, with its concomitant adjustment of the myth, Hegel has extrapolated into the postulate of a metastatic demythisation that will absorb the beyond of consciousness into consciousness itself. To what degree Hegel was conscious of his transforming the *metaxy* of existence into the dialectics of an imaginary consciousness must remain uncertain; in the nature of the case he could not give a full exposition of his fallacious procedure, or he would have had to abandon it. There is no lack of certainty, however, about the purpose of postulating the metastasis: Only if "reality is made to coincide with the concept, the Idea will come into existence." "To rule means to determine the real state, to act in it according to the nature of things. And that requires consciousness of the concept of things ... In history, the Idea is to be accomplished; God rules the world, the Idea is the absolute power that brings itself forth".[10] The philosopher can achieve identity with the divine power of the Idea that rules the world, if he achieves the metastasis of philosophy into "a movement in pure thought", if indeed he can absorb reality into the concept.[11]

Plato's philosophizing had, in Hegel's interpretation, the same object as his own: When Plato wants the philosophers to be rulers, he means that "the whole of a society should be determined by general principles".[12] Under the backward conditions of a Greek polis that was impossible; under the progressive conditions of the modern state the philosopher's object must be modified, because the immediate goal of Plato has to a large extent been realized. For ever since the Migration period, when the Christian had become the general religion, it has become the accepted purpose of government to build the *übersinnliche Reich* into the reality of society and history. In a modern state, as it is governed by general principles, the Platonic program of the philosopher-king has been

---

10 *Op. cit.* 193.
11 *Op. cit.* 196.
12 *Op. cit.* 195.

realized, as for instance in the rule of Frederick II in Prussia. With this advance of the modern state beyond the ancient polis, the role of the philosopher has changed. The philosopher-king has become so much "a custom, a habit" of the political scene that "the princes are no longer even called philosophers".[13] Hence, the philosopher need no longer worry about building general principles into the order of society; the chore of realizing the Idea in the process of world-history can be safely left to such figures as Frederick II. The philosopher has to advance beyond mere kingship; he has to merge with the absolute power of the Idea in history.

Hegel's obsession was power. If he wanted to be the sorcerer who could evoke the shape of history, he had to penetrate the political events of the time with thought until the events and thought would coincide. The events of the years preceding the publication of the *Phaenomenology* were impressive indeed. On May 18, 1804, Napoleon was proclaimed Emperor of the French; on August 11, Francis II reacted by assuming the title of Francis I, Emperor of Austria; and then the new Emperors recognized one another. In 1805 followed the War of the Third Coalition, with Trafalgar and Austerlitz, concluded on December 26 by the Treaty of Pressburg. 1806 brought the Napoleonic reorganization of Europe through the Federative System; on July 12, the Confederation of the Rhine was organized under the protectorate of Napoleon; on August 6, Francis II resigned the dignity of a Roman Emperor and declared the Roman-German Empire extinct. These were the events to which Hegel responded on September 18, 1806 by concluding his *Collegium* on his so-called speculative philosophy with the following address to the students:[14]

> "Gentlemen, this is speculative philosophy as far as I have come in its elaboration. Consider it a beginning of philosophizing to be continued by you. We find ourselves in an important epoch of time, in a ferment; the *Geist*, with a sudden jerk, has moved to advance beyond its previous *Gestalt* and to assume a new one. The whole mass of hitherto accepted conceptions (*der bisherigen Vorstellungen*) and concepts, the bonds of the world, have been dissolved and collapse like a dream image. A new epiphany (*Hervorgang*) of the *Geist* is preparing itself. It is becoming to Philosophy to greet its appearance and to recognize it; while others, in impotent resistance, adhere to what belongs to the past; and the majority is no more than the unconscious mass of its appearance. But Philosophy, recognizing it (the *Geist*) as the eternal, must give it the honor due to it. Recommending myself to your kind remembrance, I wish you a pleasant vacation."

---

13 *Op. cit.* 194–5.
14 *Aus Jenenser Vorlesungen* (*Dokumente*, ed. Hoffmeister, 335–352), 352.

Four weeks later, on October 14, came Jena and Auerstaedt. On the day before the battle, Hegel hat the pleasure of seeing Napoleon in the flesh. He recorded his response to the event in the famous letter to his friend Niethammer, dated "Jena. Monday, October 13, 1806, the day when Jena was occupied by the French, and the Emperor Napoleon arrived within its walls". Hegel wrote:[15]

> "I have seen the Emperor – this World-Soul – riding through town, and out of it, for a reconnaissance; – it is a wondrous feeling indeed to see such an individual who, concentrated in one point, sitting on a horse, reaches over the world and dominates it."

The passages give a fairly good picture of Hegel's state of mind in the critical years. There are the shrewd observations on the masses who never know what hits them; on the die-hard traditionalists who cannot believe that a house they have left to decay for centuries at last comes crashing down; on the duty incumbent on man to understand what is going on around him, and to find his bearings in a situation of revolutionary change. The actual response to the challenge, however, betrays the pneumopathological confusion of a man whose philosopher's self is disintegrating while the sorcerer's self begins to crystallize. The Aristotelian *thaumazein* has become the "wondrous feeling" aroused by the sight of an Emperor; God has disappeared behind a Neoplatonic world-soul, which in its turn has taken to sitting on a horse; the rider on horseback who might have stirred memories of the Apocalypse, then, has become a new *Gestalt* of the *Geist*, reaching over the world to dominate it; and the honor given by Philosophy to the new *Gestalt* is not exactly what is meant by rendering unto Caesar the things which are Caesar's, and unto God the things that are God's.

When a philosopher hastens to pay his respects to an imperial conquest in progress, he invites the suspicion of being an unwise opportunist; when he declares it the duty of philosophy to give the new *Gestalt* in history due honor, he sounds as if he were degrading philosophy to an *ancilla potestatis*; when he admonishes his students to follow his example, he appears engaged in the very corruption of youth which Plato considered a crime second in foulness only to physical murder; and when, after Waterloo, he transfers the honors from the fallen conqueror's empire to the Prussian state, he seems to put the finishing touch to the portrait of a detestable character. Though no portrait could more insidiously distort Hegel's personality than this one, it must be drawn, because it faithfully renders the public appearances of the modern deformation of existence. Spiritual disease is not a man's private affair, but has public consequences; the man who deforms himself does not live in a vacuum but in a society; and the Second Reality he has created for himself impinges on the First Reality in which he lives. The character traits assembled in the portrait result from the friction

---

15 *Briefe von und an Hegel* (ed. Hoffmeister), Vol. I, 1952, 120.

between Second and First Reality – they maliciously distort the truth as seen from the position of the spiritually diseased man who engages in such actions, but they are the regrettably true consequences of existence in untruth.

Hegel's response to the *translatio imperii* he witnessed will appear confused if judged by the standards of Myth, Philosophy, and Revelation, because these symbolisms express reality as experienced by a man whose soul is open toward the divine ground of the cosmos and his own existence. It will not appear confused at all if judged by the standards of the sorcerer's existence which Hegel develops and casts into language symbols during these years in Jena. If a man lives in openness toward God, Bergson's *l'âme ouverte,* his consciousness of his existential tension will be the cognitive core in his experience of reality. If a man deforms his existence by closing it toward the divine ground, the cognitive core in his experience of reality will change, because he must replace the divine pole of the tension by one or the other world-immanent phenomenon. The deformed cognitive core, then, entails a deformed style of cognition by which the First Reality experienced in open existence is transformed into a Second Reality imagined in closed existence. Hegel's choice of an imaginary absolute pole was "Empire", understood as the ecumenic organization of mankind under the Idea in history; and the deformation of the cognitive core imposed the deformed style of cognition which produced the imaginary history of the Idea. This style, however, has a rationale of its own. Though the genesis of Hegel's Second Reality will appear confused, and even nonsensical, if confronted with cognitive procedures in First Reality, it is intelligible on its own premises. The following reflections on this deformed procedure of cognition are based on the section "*Aus Jenenser Vorlesungen*", from which I have quoted already the concluding address to the students, in Hoffmeister's *Dokumente*.

Hegel was interested, not in political power, but in the power of the Idea. "The Idea is the absolute power that brings itself forth". "The pure Idea is the *power of the divine mystery*, from whose untroubled self-containedness (*ungetrübte Dichtheit*) nature and conscious *Geist* are set free to exist for themselves" (D 348). "The immanent dialectics of the absolute is the life-history (*Lebenslauf*) of God" (D 348–9). "The creation of the universe is the *speaking* of the absolute *word*, the return of the universe to itself is the *hearing* (of the *word*), so that nature and history become the *Medium* between speaking and hearing that will disappear as Other-Being (*Anderssein*)" (D 349). The *Medium* between speaking and hearing of the Word is, in Hegel's language, the equivalent to Plato's *metaxy*, to the consciousness of existence in the In-Between of divine and human. To Hegel, however, philosophy is not the consciousness of the *metaxy*, its exploration and man's ordering of his existence by the insights gained, as it is to Plato, but the enterprise of abolishing the *Medium* through the magic act of speculation. "Truth as conceived by revealed religion" must be purified through cognition (*Erkennen*). Consciousness, "as it elevates itself to the last possible standpoint", has done "with the metamorphoses of its *Gestalten*". As it elaborates the "System of Science", consciousness achieves the equality of its certainty with the truth of revealed religion. By its realization of absolute essence, the contents of Science has become to self-consciousness (a) "the general self-consciousness", (b) "all reality or essence (*Wesenheit*) in itself", and (c) "this individual self-consciousness to itself" (D 329–30). Under the title of Science, thus, philosophy penetrates the divine mystery and converts it into the self-consciousness of the individual man who has achieved the penetration, i.e. of Hegel. The *Medium* of the world has come to its end through the apocalyptic event of the Word hearing itself spoken in Hegel's System of Science.

Though the universe returns to itself as a whole, not every participant in the abolition of the *Medium* acts with the same degree of consciousness. On the level of politics, the apocalyptic return is more sensed than reflected, it remains semiconscious; in order to raise the apocalyptic meaning of the events to the level of full consciousness, philosophy is needed. Napoleon is the Great Man, because he is the world-historic servant of the Idea as it comes to its fulfillment; Hegel is the Great-Great Man, because his System of Science puts the seal of self-reflective consciousness on the metastasis of reality. Napoleon's imperial expansion and Hegel's elaboration of speculative philosophy belong together as two levels of consciousness in the apocalyptic return.

As a consequence of this imaginative construction Hegel could not be, from his own position, a political opportunist. The deformation of his cognitive core had blinded him to the First Reality of pragmatic history and the vicissitudes of power politics. He had already transformed the events of First Reality into symbolic events in the apocalyptic drama of his imagination; or, to be more exact, in the years in Jena the transformation was in progress. This metastatic growth can be diagnosed in the previously quoted language of "the *ganze Masse der bisherigen Vorstellungen* which now collapse like a dream image" (D 352). The phrase cannot be adequately Englished, because the English language has not absorbed modern apocalyptic symbols to the same degree as the German. The adjective *"bisherige"* lumps all conceptions of history and social order up to Hegel's writing together as a dream, now to be superseded by the truth of reality. This adjective of Hegel's has entered, as the operative apocalyptic symbol, the first sentence of the *Communist Manifesto: "Die Geschichte aller bisherigen Gesellschaft ist die Geschichte von Klassenkämpfen";* and in German it has remained operative in such phrases as *"das Ende der bisherigen Geschichte"* (Alfred Weber) after the upheaval of the Second World War. English renderings like "all previous conceptions", or "the history of all hitherto existing society", do not carry the apocalyptic weight of *"bisherige"*. By transforming pragmatic history into apocalyptic drama, the imaginator transforms himself from an ordinary man in open existence, from the "link in the chain", into the intellectual guide of mankind on its way to metastatic liberation.

I have singled out the symbol *"bisherige"*, because it concentrates the magic power of speculation on the realm of society and history. Its survival in the combinations of *alle bisherige Gesellschaft* and *Geschichte* is proof of its pungency. Nevertheless, the later simplifiers and vulgarizers tear the symbol out of the speculative context from which it derives its magic power. A first-rate sorcerer who knows his business would not for a moment consider a metastasis of history without the metastasis of the cosmos of which history is a part. Hence, Hegel provides the background for his *bisherige Vorstellungen* by pronouncing the *Zauberworte* which transform the divine cosmos: The return of the universe to itself; the hearing of the creative word that hitherto has been only spoken; the relegation of history to the past of the *Medium* now to be abolished; the reconciliation of dirempted reality through the power of the Idea now achieving fulfillment in self-reflective consciousness; and the liberation from the bonds of the old world, now collapsing like a dream image, through man's entrance into the divine mystery. Only in a Second Reality, imagined as a cosmos in metastatic change, obtain the relations between Empire and Philosophy on which Hegel depends for his magic operation of evoking the *Gestalt* of the new world. Only if by an act of metastatic speculation the renewal of the cosmos is imagined as real, can "renewal" become the common factor by which a conqueror's new order of power is linked to a philosopher's meditative renewal of insight.

Hegel has established "renewal" as the common factor in Empire and Philosophy by his fascinating reflections on the relation between Alexander and Aristotle (D 345–6): The Great Man in history appears in the epochs of transition when "the old moral form of the nations (*die alte sittliche Form der Völker*)" is to be radically overcome by a new one. When the time is ripe, the "perceptive natures" who accomplish the transition have "only to speak the word and the nations will follow them". But in order to be capable of the feat, these "great spirits" must have cleansed themselves of "all singularities of the preceding *Gestalt*. In order to accomplish the work in *its* totality, they must have comprehended it by *their* totality." If a man can advance only part of the work, nature will topple him and bring other men to the fore until the whole work is done. If it is to be the work of *One* man, however, this man "must have understood the whole and through such understanding have purified himself of all limitation (*Beschränktheit*)." "The terrors of the objective world, all bonds of moral reality, and together with them all *outside support (fremde Stützen)* to stand in this world, as well as all confidence in a firm bond within this world, must have fallen from him, *i.e.* he must have been formed in the school of philosophy. By virtue of his formation in this school, he can raise the *Gestalt* of a new moral world from its slumber, and can enter the lists against the old forms of the world-spirit as Jacob wrestled with God, with the assurance that the forms he *can* destroy are an obsolete *Gestalt* and that the new one is a new divine revelation". In the pursuit of this purpose, he is entitled "to consider all mankind in his path a substance (*Stoff*) to be appropriated by him and to be built into the body for his great individuality, a living substance that will form, more inert or more active, the organs of the great city". "Thus Alexander of Macedon went forth from the school of Aristotle to conquer the world".

Hegel was in his mid-thirties when he mixed the metaphors he drew from Sleeping Beauty and the Prince, Jacob wrestling with God, and the Mystical Body of Christ, in order to explain to his students what happens if one goes to the school of Philosophy. Napoleon was *ante portas*. And the *grimoire* was about to be finished.

The *Phaenomenology* "was finished in the night before the Battle of Jena." In a letter to Niethammer of April 29, 1814, Hegel reminded his friend of the historic night when the World-Soul prepared the climax of its revelation in both Empire and Philosophy. But Napoleon had not gone to the school of Philosophy, he was no Alexander, on April 11, 1814, he had abdicated. "Great things have happened around us," writes Hegel. "It is an immense spectacle to watch an enormous genius destroying himself. – This is the *tragikotaton* there is. The whole mass of mediocrity, with its absolute, leaden gravity, relentlessly and implacably presses on, until it has what is higher brought down to its own level and underneath itself. The turning point of the whole, the reason the mass has power and remains as the chorus on top, lies in the great individual itself as it has given the right to this turn and destroyed itself." Hegel, then, "wants to pride himself" of having predicted this "whole upheaval" in the *Phaenomenology*. The imperial enterprise had been vitiated from the beginning by the absolute freedom of Enlightenment, *i.e.* by an "abstract freedom" which destroys itself. Napoleon was one of the great individuals, characterized in the page on Alexander and Aristotle, who could fulfill the task set by the epoch only in part. Others will have to carry on the imperial side of organizing the ecumene. For the time being, Hegel is left without a partner; the burden of revealing the World-Soul now rests on his shoulders alone.[16]

The perspective of Second Reality in which Hegel places his work in 1814 faithfully reflects the perspective he actually developed in the years in Jena and in the finished *grimoire* itself. In the *Preface* to the *Phaenomenology* he dwells on purpose and technique of his metastatic sorcery. His purpose is the abolition of *Zerrissenheit*, of diremption. While in previously quoted contexts "diremption" had been predicated of "ages", it is now more carefully conceived as a fundamental characteristic of the human condition. If the diremption, present at all times, is experienced more acutely by people at large, it can become the characteristic of an "age"; and such an "age", then, is ripe to be overcome by a new *Gestalt* of the *Geist* in history. But Hegel, though he wants to overcome the diremption of his own age, does not want to see the Alexander-Aristotle *Gestalt*, or Church and Empire, now followed by a Napoleon-Hegel *Gestalt* which in its

---

16 Hegel: *Briefe* (ed. Hoffmeister), Vol. ii, 28.

turn would have to decline; he rather wants to abolish the fundamental diremption of man, so that the age inaugurated by the *Phaenomenology* will be the last age of history. Since Hegel however, cannot admit in the language of open existence that he wants to change the nature of man by writing a book, at this point the statement of purpose has to slide over into its execution through sorcery. In effecting the transition, Hegel uses the principles of construction which I have set forth earlier in this essay: God and man are eliminated from the universe of discourse; their place is taken by the imaginary *Bewußtsein* or *Geist;* and the symbols of philosophy developed in open existence are transferred into the power field of the new Second Reality. *Zerrissenheit,* thus, need no longer be predicated of man or his soul, but has become the property of the *Geist;* its abolition is a process immanent to the *Geist;* and the embarrassing spectacle of Hegel tampering with the nature of man is avoided (Ph 29–30).

In order to be effective as a magic *opus,* the System of Science had to satisfy two conditions:

(1) The operation in Second Reality had to look as if it were an operation in First Reality.

(2) The operation in Second Reality had to escape control and judgment by the criteria of First Reality.

Only if he satisfied these two conditions, could the author of the System hope to make the imaginary results of his operation acceptable as real resolutions to real problems in First Reality. Hegel fulfilled the first condition through the use of philosophical symbols as the conceptual units of his construction. The *bona fide* reader may find the book indigestible, but he will not doubt that he is reading a philosophical work when he is overwhelmed by the vocabulary of intellect, reason, and spirit, being and not-being, analytical and dialectical logic, consciousness, science, history, life and death, and so forth. Hegel fulfilled the second condition by never presenting the experiences of reality which had engendered the symbols as their means of expression, and by hardly ever mentioning the philosophers who had created them. By this technique Hegel can break the bond between the symbols and the First Reality in which they have their place and meaning. No questions must be asked regarding the origin and meaning of the symbols used; they are somehow there; they constitute a self-contained realm, waiting for the *Geist* to organize them into a System. "The *Geist,* as by unfolding it comes to know itself, is *Science*"; and inversely, "Science is the reality of the *Geist,* and the realm it builds for itself in its own element". "Pure self-cognition", "this Ether *as such*", is the soil in which Science grows. When philosophy has become Science, it does not begin from anywhere but from itself; its beginning is an In-the-beginning of divine absoluteness. "The beginning of philosophy presupposes or demands that consciousness has placed itself *(sich befinden)* in this *Element*" (Ph 24). "Because this Element, this immediacy of the *Geist,* is its very substance, the immediacy is transfigured essence *(verklärte We-*

*senheit*); it is pure reflection, the immediacy as such for itself; it is Being which is reflection in itself". Science demands of "self-consciousness that it has elevated itself into this Ether" (Ph 25). All criticism by appeal to reality experienced, finally, is precluded by the rule that Hegel's "insight" has to justify itself through nothing but "the presentation of the System itself" (Ph 19 ff). Toward the end of the *Phaenomenology*, Hegel summarizes this selfcontained circle of reflection in the sentence: "The *Geist*, appearing in this Element to consciousness or, what is the same thing, being brought forth in this Element by consciousness, *is the Science*" (Ph 556).

The protection from close scrutiny is especially important for the mythical In-the beginning. The purpose of the *Phaenomenology* is the abolition of *Zerrissenheit*; and diremption is predicated of the *Geist*. Hegel introduces the *Geist* as "the sublimest concept; it belongs to the modern age (*neuere Zeit*) and its religion" (Ph 24). The *Zerrissenheit* is not introduced at all; it just happens along as the property of the *Geist* (Ph 30). Both the *Geist* and the *Zerrissenheit* are "absolute" (Ph 24; 30). From this scanty information nobody would gather that *Zerrissenheit* is part of a Neoplatonic body of symbols centering around the problem of *tolma, i.e.* the audacious restlessness of the soul which causes it to forget its divine origin. As without knowledge of this source it is impossible to understand either the transmogrification of the Neoplatonic symbol into the Hegelian concept, or Hegel's resolution of the problem, I shall quote the key-passage from Plotinus' *Enneads* on *tolma* (V, i, 1):

> "What really has brought it about that the souls have forgotten God-Father, though they are parts coming from Him and wholly belonging to Him, and no longer know either themselves or Him? Well, the origin of the evil for them was restlessness (*tolma*), becoming (*genesis*), primordial otherness (*heterotes*), and the will to belong to themselves. Once they have gained appearance, they enjoy their self-rule; make ample use of their self-movement to run in the opposite direction (from God); and having reached a far distance, they no longer know from where they have come, like children who, taken away from their father and brought up a long time far from him, no longer know themselves or their father."

Further symbolizations of an original state of stillness (*hesychia*) in the One and a disturbance of stillness through curiosity for action (*polypragmosyne*) and a desire for self-rule (*archein autes*) occur in III, vii, 11, where Plotinus tries to clarify the relation of time and eternity. He is fully aware of developing a myth of the Platonic type, when he tells the story of a fall within the divinity, closely related to the fall of *sophia* in Gnostic texts, in order to make intelligible the experience of restlessness in self-assertive activity, the sense of stillness being the proper state of existence, and the desire of returning to a home that has been lost. The disease of existence, then, can be cured by the initiation of the countermovement: The consciousness of the disease as a state of lostness must be awakened, so that the soul can turn around (*epistrophe*, the Platonic *periagoge*) toward the divine ground from which it has moved away; the recollection (*anamnesis*) of the state of stillness lost must be aroused; until the return movement (*anagoge*) through the meditative ascent to the One comes under way. This rhythm of outgoing self-assertiveness and meditative return, as well as the dynamics of boldness, curiosity, discovery, and polypragmasy, of joyous independence and self-rule, of restlessness, lostness, and alienation (*allotriosis*), of search (*zetesis*), turning around, and so forth, are processes and moods of the soul, hardenings and softenings of the tension in man's existence. This tension of existence is the human condition. There is no way of abolishing it but death.

Hegel was familiar with the experience of existential tension and its Neoplatonic symbolization. His preoccupation with the state of self-assertive lostness is proven by the long catalog of symbols he uses for distinguishing its various aspects: *Anderssein, für sich zu sein, eigenes Dasein, Bewegung des Sichselbstsetzens, selbstbewußte Freiheit, abgesonderte Freiheit, Entzweiung, Geschiedenheit, Verschiedenheit, Zerrissenheit, das Konkrete, Härte, Negativität, fremd, Entfremdung, Unwirklichkeit, Tod*. The enumeration does not claim completeness. Moreover, Hegel proves his comprehension of the problem by rejecting any philosophy which looks away from the "negative" and presents nothing but "positive truth". A philosopher must "look the negative in the face"; he must get hold of the horror of non-reality (*Unwirklichkeit*), of the horror of "death, if that is what we want to call that non-reality" (Ph 29). The philosopher is not allowed to settle down on the positive pole of the existential tension; only the tension in its polarity of real and non-real is the full truth of reality. Hegel's true self was that of a great mystic-philosopher indeed.[17]

The suffering from existence in non-reality, the knowledge of his death, is the tomb from which Hegel rises as the sorcerer and ascends to the Element of Ether. The purpose is clear: It is not the healing of lostness and alienation through return to the One, but the metastasis of existential tension as a whole. No longer will there be movements and counter-movements within the In-Between of

---

17 A critical study of the *Phaenomenology* is seriously hampered by the poor state of the *Sachregister* attached to Hoffmeister's edition. Some of the omissions can be explained by negligence, as for instance the omission of one of the important passages on the death of God, or of *Entfremdung*. But when of the seventeen alienation symbols enumerated above only *Negativität* and *Tod* appear in the Register; or when the symbols *das Beschränkte, Jenseits, Ruhe, Unruhe, Trägheit, Gedankenlosigkeit, Angst vor der Wahrheit, Furcht vor der Wahrheit*, which Hegel uses to characterize the existential state of the enlightened intellectual, are altogether omitted; one begins to wonder whether the makers of the Register had a very clear idea of Hegel's problems. And when, then, one does not find the key symbols of *Äther, Kreis, Zauberkraft, Zerrissenheit, Ziel* at all; or *Element* only in the instances where it refers to air, water, fire, but not where it refers to the Element of Ether; one wonders whether this suppression of all references by which the reader could become aware of Hegel's state of alienation is sufficiently explained even by a lack of comprehension. Whatever the case may be, in future reprints of the *Phaenomenology* the making of the Register should be handed over to younger scholars who are up-to-date in methods of science, especially of comparative religion.

existence; the existential tension itself, together with its poles of God and man, must be dissolved in the dialectical process. Hegel is an energetic thinker, and as the data of the problem are thoroughly familiar to his true self, the technical task of performing the metastasis is not too difficult: (1) Since man experiences the tension of existence from within, as the reality by which he has to orient his humanity, first the tension must be transformed into an object on which the sorcerer can operate. To this purpose he creates the hypostases of *Bewußtsein* and *Geist*. (2) He must create, second, a basis from wich the operation can be performed. Since he has no other means for building the basis but his own humanity, his own state of lostness and alienation must be transformed into the absolute position from which he can operate. The Neoplatonic symbols of restlessness, becoming, self-movement, self-assertion, and so forth, which express a man's distance or remoteness from reality, are now used to symbolize reality in the eminent sense. In particular, Hegel transforms *Selbst, Ich,* and *Subjekt,* which in the context of First Reality symbolize the carrier-force of the alienating movement, into hypostases which are meant to replace the reality of God. Even more, the "energy of thinking" which is a property of "the pure Ego" (*des reinen Ichs*) is recognized as the "immense power of the negative" and, in this quality, elevated to the rank of "absolute power". (3) And third, the operational hypostases of *Selbst, Ich, Subjekt* must be related to the substantive hypostasis of *Geist*. That is done by attributing to the substance *Geist* the property of *Werden*, of Becoming, and making the operational *Subjekt* the moving force in the *Werden* of the *Geist*. The *Geist*, thus, is transformed into a *Substanz* in process of coming to self-reflective consciousness as the *Subjekt;* and the *Subjekt* arrives at its selfknowledge as the operative force in the *Werden* of the *Substanz*.

The technique of Hegel's sorcery is simple enough to be reduced to the three rules enumerated. But Hegel does not formulate these rules by which he transforms the state of alienation into true reality; he is engaged in the metastatic act itself. A few passages will make the actual situation clear: "Only because the concrete separates itself and makes itself the non-real, is it the self-moving. The activity of separating is the force and labor of *intellect* (*Verstand*), of the most wondrous and greatest, or rather of absolute power.... Death, if that is what we want to call the non-real, is the most awsome (*das Furchtbarste*), and to hold fast what is dead requires the greatest force.... But not life that is afraid of death and wants to keep itself pure of desolation, but life that can bear death and hold its own in it, is the life of the *Geist*. The *Geist* can gain its truth only, if in this absolute *Zerrissenheit* it finds itself.... This power it is only, if it looks the negative in the face and dwells with it. This dwelling (*Verweilen*) is the magic force (*die Zauberkraft*) which converts the negative into Being. This *Zauberkraft* is what we have formerly called the *Subjekt*" (Ph 29–30). The reader who does not know that the symbols appearing in this passage are, in First Reality, the symbols of alienated existence, will hardly understand what is going on.

The sorcerer's magic force is, at last, identified as the *Subjekt*. But how does the Subject come by its *Zauberkraft*? The following passage gives the information (Ph 53–4):

"Besides the sensually perceived or conceived Self, it is primarily the name as name which denotes the pure Subject, the empty non-conceptual One. For that reason it can, for instance, be helpful to avoid the name *God*. For this word is not itself concept, but properly name, the firm tranquillity of the *Subjekt* intended. Terms such as Being, or the One, or Oneness, the *Subjekt*, and so forth, will immediately suggest concepts."

The *Zauberkraft* accrues to the *Subjekt*, because the *Subjekt* is the metastasis of God. And what has become of Christ? That he must be thrown out has become clear from the page on Alexander and Aristotle where the Great Man is defined as the man who has renounced all *outside support*. The phrase "*alle fremden Stützen*", underlined by Hegel, refers to all support that is not world-immanent (D 346). Now this point is clarified too; for the *Subjekt* is "the true Substance, Being or Immediacy itself, which has no mediation outside itself, but is itself this mediation" (Ph 30). The *Subjekt* has taken over the rôles both of the One and of the Mediator. The sorcerer has drawn into himself the power of both God and Christ.

From the texts there emerge the outlines of a spiritual biography. By his true self of a mystic, Hegel experiences his state of alienation as an acute loss of reality, and even as death. But he cannot, or will not, initiate the movement of return; the *epistrophe*, the *perioagoge*, is impossible. The despair of lostness, then, turns into the mood of revolt. Hegel closes his existence in on himself; he develops a false self; and lets his false self engage in an act of self-salvation that is meant to substitute for the *periagoge* of which his true self proves incapable. The alienation which, as long as it remains a state of lostness in open existence, can be healed through the return, now hardens into the *Acheronta movebo* of the sorcerer who, through magic operations, forces salvation from the non-reality of his lostness. Since, however, non-reality has no power of salvation, and Hegel's true self knows this quite well, the false self must take the next step and, by "the energy of thinking", transform the reality of God into the dialectics of consciousness: The divine power accrues to the *Subjekt* that is engaged in self-salvation through reaching the state of reflective self-consciousness. If the soul cannot return to God, God must be alienated from himself and drawn into the human state of alienation. And finally, since none of these operations in Second Reality would change anything in the surrounding First Reality, but result only in the isolation of the sorcerer from the rest of society, the whole world must be drawn into the imaginary Secondary Reality. The sorcerer becomes the savior of the "age" by imposing his System of Science as the new revelation on mankind at large. All mankind must join the sorcerer in the hell of his damnation.

Hegel can evoke the shape of ecumenic mankind beyond diremption, because this last age lies beyond history. The *Geist* has unfolded in the time of world-history to its fulfillment in the complete penetration of the *Substanz* by the *Subjekt*. Before Hegel this process could not be understood because it had not yet been completed; but now "the movement of the concept will embrace the complete worldliness of consciousness in its necessity" (Ph 31). The "world-*Geist*" has undertaken the "enormous work" of bringing forth its *Gestalten* in "world-history"; now that the task is finished, the "individual", though it has to penetrate the same substance with thought, can do it with "much less labor", because "the immediacy of reality has already been conquered and its *Gestaltung* been reduced to the abbreviation of a simple term of thought (*Gedankenbestimmung*)". The contents of world-history is thought that has been thought (*ein Gedachtes*) and thereby become "the property (Eigentum) of the *Substanz*". Hence, it is no longer necessary to convert existence (*Dasein*) into the form of being-in-itself (*Ansichsein*); all that is left to be done is the conversion of the "remembered itself" into the "form of being-for-itself (*Fürsichsein*)" (Ph 27–8). Translated into more intelligible language, these passages mean: The process of world-history has run its course. Every *Gestalt* in history is a thought in the unfolding of the *Geist*; and every thought of the *Geist* has become *Gestalt*. The "individual", *i.e.* Hegel, who wants to understand history today, does not have to play *Weltgeist* and perform history all over again. The task is finished. Nevertheless, he must re-perform history in the mode of anamnesis by making the events of history intelligible as steps in the thinking of the *Geist*. The events must be converted into the *Gestalten* of unfolding consciousness that has come to its self-reflective luminosity in the consciousness of the "individual" Hegel. Men and events in history lose their presence under God; they are transmogrified into phases in the process of consciousness, that is the *Subjekt*, that is God, that is Hegel.

Thus far Hegel goes in explaining his technique of magic transformation, but no farther. Moreover, his explanatory formulae require translation to make their bearing intelligible. The *Phaenomenology* is indeed an "unintelligible" book, because Hegel cannot go too far in exhibiting his *modus operandi*. In the present instance, he cannot simply say: I am going to falsify history in open existence until it fits into my history in closed existence. Just as in an earlier instance, he could not say: I take symbols of alienation from various Neoplatonics, Gnostics, and Mystics, and shall use them as the starting point for my magic enterprise of self-salvation. The effectiveness of the *grimoire* depends on the transformation of First into Second Reality as a *fait accompli*. The book is written in magic code which the reader, if he does not want to be taken in, must decipher. This process of decoding the *Phaenomenology*, however, is always difficult, and sometimes next to impossible, especially when political events have been put into code. As an example, I shall give a passage for which fortunately we have Hegel's own decodification (Ph. 422):

"As the realm of the real world goes over into the realm of faith and insight, thus absolute freedom goes over from its self-destructive reality into another land (*Land*) of the self-conscious Geist where in its non-reality it (viz. the freedom) is accepted as the true, of which the thought, so far as *it is thought* and remains thought, edifies the *Geist* which knows this being, enclosed in self-consciousness, as the perfect and complete essence. The new *Gestalt* of the moral spirit (*des moralischen Geistes*) has come forth (*ist entstanden*)."

In the previously quoted letter to Niethammer, of April 29, 1814, Hegel writes on this passage:

"For the rest, I want to pride myself that I have predicted this whole upheaval. In my work (finished in the night before the Battle of Jena), I say on p. 547: "The absolute freedom (previously described; it is the abstract, formal freedom of the French Republic as it emerged, as I have shown, from Enlightenment) goes over from its self-destructive reality into *another land* (when I wrote this, I had a *country* in mind) of the self-conscious *Geist* where in its non-reality it is accepted as the true, of which the thought, so far as *it is thought and remains thought,* edifies the *Geist* which knows this being, enclosed in self-consciousness, as the perfect and complete essence. The new *Gestalt* of *the moral* spirit is present (*ist vorhanden*)."

Translating the interpretation: The essential freedom of the selfconscious *Geist* was vitiated in the version of the French Republic by an abstractness which derived from Enlightenment. Because of this abstractness, the freedom degenerated into violence and terrorism. Still, there is a truth even in abstract freedom which the *Geist* will appreciate as long as it remains a thought, a moment in the dialectical process. However, this thought must not be translated into action. Because of its vitiation through abstract freedom in France, the *Geist* will move on to Germany. The transition becomes effective through Hegel's inclusion of the French freedom as a "thought" in the process of his own self-consciousness completed. Through the *Phaenomenology,* the *Geist* in its *moral* form (underlined by Hegel) has come forth (*ist entstanden*) in the other "country". With a slight variation of the original text, the letter can therefore say that this moral form is now present (*ist vorhanden*) in Germany. – I see no reason to distrust Hegel's decipherment of 1814, but in reading the passage I never would have conjectured this interpretation.

The difficulties just exemplified make it impossible to understand the purpose of the *grimoire* without a code at hand that will permit decipherment page after page. Such a code, paralleling the "thoughts" of Second Reality with the persons and events in First Reality that have been converted into "thoughts", was elaborated by Alexandre Kojève in the course of his lectures on the *Phaenomenology* and published as an Appendix on *"Structure de la Phénoménologie"* in his *Introduction à la Lecture de Hegel* (1947). I am giving Kojève's decipherment for the critical page, where Hegel formulates his relation to Napoleon.[18]

Hegel's text opens and closes with the following passages (Ph 472):

"The two spirits (*Geister*), certain of themselves, have no other purpose but precisely this pure Self. But they are still different; and the difference is absolute, because it is posited in this element of pure concept. . . . The reconciling Yes whereby the two Egos (*Ich*) surrender their opposed existence (*Dasein*) is the existence of the Ego that has expanded into a dyad (*Zweiheit*); this Ego remains, as dyadic, the same with itself: – it is the manifest God *(der erscheinende Gott)* in the midst of those who know themselves as pure knowing."

In Kojève's interpretation, this page has the following meaning:[19]

"Finally one arrives at a duality: the Realizer – the Revealer, Napoleon – Hegel, Action (universal) and Knowledge (absolute). On the one side is *Bewußtsein*, on the other side *Selbstbewußtsein*.

Napoleon is turned toward the external world (social and natural): he understands it, and therefore acts in it with success. But he does not understand himself (he does not know that he *is* God). Hegel is turned toward Napoleon: but Napoleon is a man, he is the 'perfect' Man by virtue of his

---

18 Alexandre Kojève's Courses on the *Phaenomenology* were given, in the years 1933–1939, at the École pratique des Hautes Études under the title La Philosophie Religieuse de Hegel. The Notes taken of these Courses by M. Raymond Queneau, revised by Kojève, were published in 1947, under the title Introduction à la lecture de Hegel. Leçons sur la Phénoménologie de l'Esprit. To Queneau's Notes several papers by Kojève were added, as well as the Appendix III: "*Structure de la Phénoménologie*", pp. 574–595. To this analytical table of contents, paralleling the sections in the *Phaenomenology* with persons and events in First Reality, I refer as the "Code". The Code is indispensable to every serious reader of the *Phaenomenology*; it should be appended to every future edition of the work – just as the new Index of Subjects I have urged in Note 17. – By stressing the importance of Kojèves work as a break-through in the understanding of Hegel, I do not want to detract from the value of the comparable attempt by Jean Hyppolite: *Genèse et Structure de la Phénoménologie de l'Esprit de Hegel*. Paris 1946. Hyppolite's work is the most thorough of the "conventional" interpretations of the *Phaenomenology*. By "conventional" I mean an interpretation which places Hegel's work in the context of his predecessors and contemporaries who dealt with the same problems; in particular, I mean the presentation of Hegel's problems as a development in German idealistic philosophy. The presentation of the *Phaenomenology* as a meaningful development in the "conventional" context is indispensable, too. But in the light of Kojève's work, it inevitably raises the problem: How far back in Western history must the growth of Sorcery be traced that comes to its climax in the *Phaenomenology*? As far as I know, nobody has yet dared to tackle the question. – In the *Introduction*, Kojève confined himself to the decipherment of the *Phaenomenology* and the construction of the Code. As he was a Marxist, *i.e.* the disciple of another great sorcerer, he did not study the problem of sorcery in Hegel. On the contrary, in 1968 he published a piece of sorcery of his own, the *Essai d'une histoire raisonnée de la philosophie païenne, Tome I, Les Présocratiques*. The Volume is an Hegelian transmogrification of pre-Socratic philosophy. I recommend it warmly to every student of contemporary sorcery.

19 Kojève: *Introduction* 153–4.

total integration of History; to understand him means to understand Man, to understand oneself. By understanding (*i.e.* justifying) Napoleon, Hegel achieves his consciousness of *self*. That is how he becomes a Sage, an 'accomplished' philosopher. If Napoleon is the revealed God (*der erscheinende Gott*), it is Hegel who reveals him. The absolute Spirit . . . realized by Napoleon and revealed by Hegel.

Nevertheless: Hegel and Napoleon are two different men; *Bewußtsein* and *Selbstbewußtsein* are still separate. And Hegel does not like dualisms. Must this final dyad not be suppressed?

That could happen, if Napoleon "recognized" Hegel as Hegel has "recognized" Napoleon. Did Hegel perhaps expect (1806) to be called by Napoleon to Paris, in order to become the Philosopher (the Sage) of the universal and homogeneous State who would have to explain (justify) – and perhaps direct – Napoleon's action?

Ever since Plato, such an association has tempted, the great philosophers. But on this point, the text of the *Phaenomenology* is (deliberately?) obscure. Whatever the case may be – History has come to its end."

As in the case of Hegel's own decipherment of Ph 422 in the letter to Niethammer, I must say that this interpretation would probably never have occurred to me. As a matter of fact, the pencilled notes on the margin of my copy show that, before becoming acquainted with Kojève's work, I had explored the possibility of an Hegelian equivalent to trinitarian speculation.

Though it must be accepted on principle, the interpretation is not convincing in every detail. There remains the incongruity of a dyad in the work of a thinker who certainly "does not like dualisms". Kojève, it is true, senses the puzzle, but the solution he suggests cannot be supported by evidence. Even more: that Hegel should have considered an *imitatio Platonis* and toyed with the idea of becoming Napoleon's court-philosopher contradicts, not only the conception of his rôle as the Great-Great Man of history, but also all that we know about his shrewd understanding of political processes in First Reality. The suggestion of Plato's example is particulary unfortunate, because Hegel has been quite outspoken about Plato's attempt to use "the individual as a means" to realize his "Ideal of a State", with a side-swipe at the choice of Dionysius as a pupil.[20] But even if one replaces Plato-Dionysius by Aristotle-Alexander as the model, one does not fare better, because Hegel makes it a special point that Aristotle confined himself to the formation of Alexander's personality and did not influence his policy – the expansion into Asia was a plan of the Macedonian court and its staff even under Philip.[21] And then, one must respect Hegel's assurance that he had predicted Napoleon's *débâcle* in the *Phaenomenology* with

---

20 Hegel: *Geschichte der Philosophie*, Vol. II (Jubiläumsausgabe Vol. 18) 302.
21 *Op. cit.* 303 ff.

good reasons – all the more to be believed as the year of Austerlitz had also been the year of Trafalgar and Hegel was not alone in his estimate of the situation. No, Hegel had no intention of acting the leading rôle in a Shavian comedy *The Emperor's Philosopher*.

The appearance of an incongruity can be dissolved through a closer reading of the text. In the first place, the dyad is not a static entity but a relation between moments in Hegel's Consciousness, on the point of transition into the monad of the *Ich*. The two Egos are about to surrender their opposed existences and to become, through "the reconciling Yes", the one existence of the one *Ich* that has expanded into *Zweiheit*. That would be an astounding feat in First Reality; but as it is performed in the Second Reality of the Consciousness that is nobody's consciousness anyway, a competent sorcerer can do it quite easily by pronouncing the words: *Das Dasein des zur Zweiheit ausgedehnten Ichs*. Even so, the result is extraordinary inasmuch as the monad, reconciled to its expansion into a dyad, is the revealed God in our midsts who know ourselves as pure knowing (Ph 472).

The technical question of the dyad being resolved by elementary magic, there remains the question why the bothersome dyad should have appeared in the first place? In order to understand the appearance of the dyad, one must remember the empirical condition, set by Hegel himself, for the construction of his System of Science: World-history must have come to its end; the *Geist*, working its way through the *Gestalten* of history, must have completely penetrated its *Substanz* by its *Subjekt*; and the way of the *Gestalten* is the way of the self-assertive *Anderssein*, of the *Böse*, of the *Härte* in the *Geist*. The end of history means the ultimate reconciliation of *Sein* and *Anderssein* in the *Geist* through their mutual recognition and forgiveness (*Verzeihung*). This end of history, however, must have really happened, or be in the process of happening, recognizable as an event in First Reality, or the System would be a fancy without relation to the First Reality in which its magic is meant to operate. This event is for Hegel the French Revolution and the Napoleonic Empire. All other events belong, regarding their *Anderssein*, to the past; only their memory needs conversion into the "thought" of Consciousness. In the present in which history comes to its end, however, in Revolution and Empire, self-reflective Consciousness is confronted with the *Anderssein* in the reality of its *Härte*. This hardness of the *Ich* in its *Anderssein* requires recognition by the *Ich* that is engaged in the converting action. In the present in which Napoleon and Hegel happen, the difference between the two *Geister* becomes "absolute, because they are both posited in the element of pure concept". They both together, in their difference and opposition, are the inwardness of the *Geist* in its perfection (*das vollkommen Innre*) "as it confronts itself and comes forth into existence (*Dasein*)" (Ph 472). "The word of reconciliation is the *existing Geist* . . . It enters existence (*Dasein*) only at the height (*auf der Spitze*) at which the pure knowledge of itself becomes the opposition to, and mutual exchange with, itself" (Ph 471). The tension of *Sein* and *Anderssein* arrives at its consciousness as the dyadic structure of the *Ich* in the metastatic present when the *Geist* confronts itself in the dyad of Empire and System.

The dyadic monad is the *Geist* in the act of metastasis. "The *Geist* has gained its concept; in this Ether of its life it unfolds its existence and movement; it becomes Science" (Ph 562). Its depth has been revealed; and this revelation is the abolition of its depth (Ph 564). By raising the *Anderssein* to the Element of Ether, Hegel has healed the diremption, not only of the "age" but of mankind, and evoked the "New World" of existence. The magic act is completed and Hegel is now free to move as the philosopher in the Ether of the Concept.

What it means to be a philosopher who has determined the shape of history can be gathered from the following passage:[22]

"The philosophers are closer to the Lord than those who live by the crumbs of the Spirit; they read, or write, the cabinet-orders of God in the original; it is their duty to write them down. The philosophers are the *mystai* who have been present at the decision in the innermost sanctuary."

The contemptuous allusion to the Parable of the Rich Man and Lazarus in Luke 16 reveals, better than lenghty explanations could do, the direction in which Hegel's *libido dominandi* is moving: In order to be the revealer of divine cabinet-orders he must replace Christ at the right hand of God. And Christ he removes indeed. In the *Wissenschaft der Logik* (1812), we can admire the new Christ installed in his position in due form, issuing the new Gospel:[23]

"The *Logik* is to be understood as the System of pure reason, as the Realm of pure thought. *This Realm is the Truth as it is without veil and for itself.* It is permissible therefore to say that its contents is the *presentation (Darstellung) of God as He is in His eternal being, before the creation of nature and any finite spirit.*"

In the beginning was the Logos; and the Logos was with God; and God was the Logos. The In-the-beginning of Hegel's construction in the Element of Ether has, at last, become the *en arche* of the Logos before creation in the Gospel of St. John. The *erscheinende Gott* turns out to be the sorcerer who has transmogrified himself into Christ.

---

22 Hegel: *Geschichte der Philosophie*, Vol. III (Jubiläumsausgabe Vol. 19) 96.
23 Hegel: *Wissenschaft der Logik* (ed. Lasson), Vol. I, 31.

# Time and the Modern Self: Descartes, Rousseau, Beckett

G. SEBBA*

*Summary.* When the self tries to become "pure" by casting out all non-self, its nature and identity becomes obscure. Descartes found the right symbolism for this residual self which knows nothing except its own existence. The last sixteen years of Rousseau's life were one great effort to attain this state of pure selfhood where "time stands still" and the *fuga temporum* becomes mythical *durée*. Samuel Beckett pierces Rousseau's last illusions: what is left when a self reaches its "pure" state is not the "eternal moment" of bliss but hopeless suffering just this side of non-existence, and loss of all certainty except the certainty that this consciousness must go on moving without end. Time has become the "eternity" of unwanted existence, an invisible prison without walls and without exit.

The breakdown of the medieval world view initiated a radical change in the Western concept and experience of self. The question: "What am I?" can no longer be answered by defining man's status and destiny within "the primordial community of being."[1] Now a sovereign self stands in isolation over against all that is outside it, and the question must read: "But I, detached from all other human beings and from everything – what am I myself?" This is how Jean-Jacques Rousseau had formulated it when he was almost at his life's end. It is the question of the modern self.

This is of course not a "new" self. Man's nature has not changed. What has changed in Western consciousness is the felt relationship between self and world. In rising towards "mastery and possession of nature," as Descartes had called it, the self as subject confronts all non-self as an object for investigation and action; it experiences itself as the *Wholly Other*, to use Rudolf Otto's term in a different context. This affects the understanding of self in three ways. (1) As *ego, Ich, moi* the self feels itself to be unique, singular, central to itself. (2) Self as *consciousness* has but one immediately and incontrovertibly certain knowledge, the knowledge of its own existence. (3) Self as *mind, Geist, pensée* "hath no other immediate object but its own ideas, which it alone does and can contemplate" (Locke). What, then, is the nature or essence of this wholly other self, that without which it would not be what it is?

---

\* Professor Gregor Sebba, The Graduate Institute of Liberal Arts, Emory University, Atlanta, Georgia 30322, U.S.A.

1 "God and man, world and society form a primordial community of being. The community . . . is knowable only from the perspective of participating in it." Eric Voegelin: *Order and History*, Vol. 1. Louisiana University Press: 1956, p. 1.

Since self can no longer define itself through its relations to what now has become "outside," it must seek the answer within itself. This it can do in three ways. The first way is rational speculation – mind as subject looking at itself as object.² But since this object is "my" own self, there is another road towards the same goal: the descent of the self into itself in search of the *expérience vécue* of pure selfhood. The memory of this experience can then be lifted into the light of reflection, to be described and analyzed in the language of objective discourse. But – and this is a third way – the experience itself can be brought into language directly, in the disciplined, imaginative mode of noetic poetry. All three ways are ways of noetic search, and all of them must express the knowledge they found in symbolic form because the self cannot express and make·communicable its consciousness of itself in any other form.³ These are indeed the ways which the quest for the nature of the modern self has consecutively taken; the landmark names are René Descartes, Jean-Jacques Rousseau, and Samuel Beckett.

Descartes must stand at the beginning although he himself had no part in the search. Radically responding to the new situation, he discovered the symbolic form of the problem of the modern self and anticipated the method which the search for its nature was to take. According to him, the self as "res cogitans" *can* clearly and distinctly know its own substantial nature – can indeed know it better and more easily than it can know the substance of body and physical world.⁴ This assurance was to be put to the test. One generation later already, Malebranche re-asserted the older view: "Certainement l'âme n'a pas d'idée claire de sa substance . . ." "[L'âme] n'est à elle-même que ténèbres, sa lumière lui vient d'ailleurs . . ." The self knows itself only by a confused "sentiment intérieur"; the presence of this inner feeling is its assurance that it exists, "parce que . . . le néant ne peut être senti."⁵ With Rousseau the search for the nature of the detached self becomes a central concern. He begins by taking the road of speculation. Then, in a singular experiment on his own self, he struggles for the actual experience of wholly detached selfhood, going as far as a man of his temper and of his century could. It took a twentieth-century poet, Samuel

---

2 Marcel Raymond lists five "very evident reasons" why Rousseau's quest for the self had to be "a vain, chimeric, hopeless enterprise" which was nonetheless inevitable, once the Cartesian revolution had passed from the realm of ideas into the "sentiment de l'existence." *J.-J. Rousseau: La quête de soi et la rêverie*. Paris: Corti 1966, pp. 191 f., 161, 169.

3 "Consciousness cannot be defined: we may be ourselves fully aware what consciousness is, but we cannot without confusion convey to others a definition of what we ourselves clearly apprehend. The reason is plain: consciousness lies at the root of all knowledge." Sir William Hamilton: *Lectures on Metaphysics and Logic*. Vol. 1, Lecture XI. Boston: 1859, p. 132. But if consciousness lies "at the root of all knowledge," how can we ourselves without confusion be "fully aware" what consciousness "is"? – The theory of symbolization used in this paper is Eric Voegelin's: *op. cit.,* p. 1 ff. ("The Symbolization of Order," which I briefly summarized in The Southern Review 3 (1967) 289 f.) and his *Anamnesis*. München: Beck 1966.

4 *Discours de la méthode*, Quatrième partie. AT VI, p. 33, 3–11.

5 Nicolas Malebranche: *Recherche de la vérité* IV, xi, § 3. O. C. (Paris: Vrin), Vol. II, pp. 98, 99, 103.

Beckett, to give voice, in the mode of noetic art, to a consciousness that truly knows nothing but its own inexplicable, unbearable, inescapable condition.

Time, seemingly peripheral in Descartes, becomes central in Rousseau's enterprise. It is again left to Beckett (who had spoken early of "the poisonous ingenuity of Time in the science of affliction"[6]) to reveal the hidden role of Time in the catastrophe of the wholly detached self.

*Descartes*

Nihil nisi punctum petebat Archimedes[7]

If self and non-self are wholly different from one another, where does the boundary between them run? Of what can it be said with absolute certainty that it is pure self without any trace of non-self? The exodus from the community of being does not destroy the web of connections of which the self is a node. How then can the pure self be grasped in its essence? Descartes asked a structurally identical question with regard to knowledge; his method for finding primal certainty is also the germane symbolism for the quest of the self, expressed in three of the four cardinal features of his system:

(1) *The Cartesian split: substance dualism.* For Descartes, substance is, by definition, the wholly other. The two substances, one "inside," the other "outside" the self, have nothing in common (except in the eyes of God). On these terms, self-knowledge becomes what Hegel was to call "das reine Selbsterkennen im absoluten Anderssein," pure cognition of self in absolute otherness.[8]

(2) *Cartesian method: dépouillement.* The method for finding the nature of the detached self consists in stripping away, step by step, what is not clearly self, until nothing but the pure self remains. It is the Cartesian method of hyperbolic doubt, the method of residuals.[9]

(3) *The Cartesian cogito: minimal consciousness.* What remains as the irreducible minimum after hyperbolic doubt has run its course, is the certainty of *cogito, sum.* This pure feeling of existence is the irreducible residue left in a consciousness that is conscious of nothing but itself.

Where is Time in this symbolization? Time does not explicitly appear in it, but it is implicit in the *mouvement de la pensée* which yields the cogito. At the very moment when universal doubt overwhelms the last surviving beliefs of the

---

6 Samuel Beckett: *Proust* [1931] and *Three Dialoges with Georges Duthuit* [1949]. London: Calder, 1965, p. 15.

7 "Nihil nisi punctum petebat Archimedes, quod esset firmum et immobile, ut integram terram loco moveret; magna quoque speranda sunt, si vel minimum quid invenero quod certum sit and inconcussum." Descartes: *Meditatio secunda,* sub init.

8 Hegel: Vorrede zur *Phänomenologie des Geistes* (ed. Lasson³), p. 24. This pure cognition of self, "dieser Äther als solcher, ist der Grund und Boden der Wissenschaft oder das Wissen im Allgemeinen."

9 Georges Clémenceau exemplified Cartesian method in 1919. Pointing to a map of the proposed successor states to the Habsburg dual monarchy, he said: "Le reste, c'est l'Autriche."

doubter, a new, unassailable certainty stands revealed to him: the certainty that there is a doubt – somebody's doubt – at that moment. This certainty exists only while consciousness is filled with doubt; it rises out of the doubt and dies with it. When mind resumes the business of speculative thinking to utilize this new-gained certainty, the drama of the cogito is over, the unshakeable conviction is lost, and only the memory is left of having had an unshakeable conviction. But this memory is not unshakeable; it is subject to the assault of that very doubt which had produced the cogito in the first place. We conclude that the Time of the cogito and the time of the Cartesian "long chains of reasoning" are incompatible. But the Time of the cogito is the Time of self knowing nothing but its own existence.

## Jean-Jaques Rousseau

Ici commence l'œuvre de ténèbres[10]

"We must admit that the destiny of this man is strikingly singular: his life is cut into two parts which seem to belong to two different individuals, such that the period which separates them, namely the time when he published books, marks the death of the one and the birth of the other."[11] This is "Rousseau", speaking of "Jean-Jacques," in the third person. The statement is factual. The time "when he published books," 1750–1762, is the period during which Jean-Jacques the Innocent died and Jean-Jacques the Ostracized was born. It was the time of Rousseau the Thinker who sought a theoretical solution for the problem of the modern self in modern society. The last sixteen years of his life, 1762–1778, are one sustained and painful effort to attain the living experience of pure selfhood, recorded in the autobiographical writings which he left for posthumous publication.

The speculative solution, a philosophical anthropology of the self, came almost at the beginning of his "public" period: in the *Discourse on Inequality* of 1754. The method is speculative *dépouillement*,[12] its result presented in reverse order. Rousseau begins with the pure, wholly detached self, the self of the hypothetical man-beast. Then, step by step, he adds the elements of non-self which enter this self, denature, debase, deprave it until civilized man ends in hopeless contradiction with himself, having passed the point of no return.

The man-beast "as he must have come from the hands of nature" lives wholly within himself, in total isolation, unparticipating, without thought, language, memory, or foresight. He is purely himself, all else is the wholly other, including even the other members of his species whom he does not recognize as being at all

---

10 Rousseau: *Confessions* XII. O. C. (Pléiade) I, p. 589.
11 Rousseau: *Rousseau juge de Jean-Jacques*. Premier dialogue. O. C. I, p. 676.
12 Rousseau: *Discours sur l'origine, et les fondemens de l'inégalité parmi les hommes*, O. C. III, p. 134 f.: "En dépouillant cet Etre ... de tous les dons surnaturels qu'il a pu recevoir, et de toutes les facultés artificielles, qu'il n'a pu acquérir que par des longs progrès ..."

like him. What is outside him passes across the mirror of this self without leaving traces. His soul, which nothing can agitate, is conscious only of existing in the timeless instant, is filled with "le seul sentiment de son existence actuelle." His Time, then, is the Time of the cogito, a succession of unconnected states, each of them an eternity while it lasts. Such a being has only one object of love, itself. Its *amour de soi*, the innate drive for preservation of self, is perfect love that knows no obstacle since the lover and the beloved are one. What distinguishes the man-beast from other animals is the fatal gift of self-perfectibility, "le don funeste de la perfectibilité" – fatal because he is already perfect. Any improvement will take him out of himself, out of the Time of pure consciousness into irreversible historical time.

We need not follow this process from its probalistic beginnings to the stage where, with the creation of a new social structure, it is carried forward by its own momentum and feeds on itself.[13] What concerns us is the first small step towards the corruption of the natural self. Recognition of species – the beginning of socialization – leads to social competition: "Each one began to look at the others and to want to be looked at himself, *and public esteem was prized*."[14] Something has imperceptibly split away from the pure self: the self's image in its own eyes, derived from the image which this self projects – ist image in the eyes of others. Natural *amour de soi* will become *amour propre* or vanity in esteeming oneself higher than all others: the center of the self will be transferred from "within" to the social image "outside." The time of historical change has come. But there is a strange moment when this time stops: at the stage of "primitive", "savage" civilization which maintains a precarious balance between social advantages already acquired and naturalness still preserved. At this stage, social life repeats itself unendingly: God has thrown away his watch,[15] man remains what he now is. This moment might still endure, had God's watch not begun to tick again among some peoples who upset the equilibrium and were relentlessly driven on to that higher civilization whose merchants, missionaries, and mercenaries were to invade the world of the "savage" Golden Age and initiate its destruction.

---

13 "Very trivial causes acting without interruption" initiate the fall of the self and slowly push man to the stage of primitive society. Enterprises requiring collective labor will convert "savage" into "civilized" society which then autonomously generates forces making changes of self not only cumulative and irreversible as before, but unstoppable.

14 *Sur l'origine de l'inégalité*, O.C. III, p. 169 and Jean Starobinski's note, ibid. p. 1344.

15 Rousseau describing his "reform": "I gave up gold lace and white stockings . . . and sold my watch, saying to myself with incredible joy: Heaven be thanked, I shall never need to know what time it is." *Confessions* VIII. O. C. I, p. 363. For a psychoanalytical interpretation of this text see John Cohen's report in "Disorders of the Inner Clock," in J. T. Fraser (ed.): *The Voices of Time*. New York: Braziller 1966, p. 271; the diagnosis is "a peculiar type of temporal derangement." But this is, I think, to misunderstand that inveterate symbolizer Rousseau; it also fails to distinguish between the Inner Clock and Inner Time. Cf. the Beckett section below.

Since there is no return to nature, how can civilized man recover his true self *within* civilization? This is the theme of Rousseau's three great works of the period, the *Nouvelle Héloise,* the *Contrat Social* and the *Émile,* three attempts to test specific solutions by the method of thought experiment. None of the solutions proves viable.[16] This ends the rational quest: Rousseau has exhausted the speculative possibilities open to him. The time of his "publishing books" is over, too. In 1762 he is driven into exile: "le pauvre Jean-Jacques," as he calls himself, embarks on the search for the living experience of the wholly detached self. *Dépouillement* turns from a speculative exercise into an operation which he must perform upon himself; and if the surgeon is to obtain the knowledge he seeks, the patient must suffer vivisection without the benefit of anaesthesia. This is how we read those last sixteen years of his life, years of persecution as he sees it, of madness as others saw and see it.

Outwardly Jean-Jacques behaves like a typical neurotic who bemoans an adversity of his own subconscious making: "Disgrace and misfortunes fall upon me as of by themselves and unseen. When my torn heart groans, I give the appearance of a man who complains without cause . . ."[17] Bitter about being treated as an outcast, he rudely repels those who treat him otherwise, that the word may be fulfilled: "The most sociable and the most loving of human beings has been outlawed by unanimous consent."[18] The "outside" must be cast into incomprehensible darkness if the light "within" is to shine in purity.

The grimness of this unconscious plunge into utter isolation, the sufferings and indeed the blindness of a consciousness that has given up its bearings, are documented in that strangest, supposedly maddest, still little understood work of the years 1772–1774, the *Dialogues* that carry the schizophrenic title *Rousseau juge de Jean-Jacques.* Outwardly the work continues Rousseau's defense against character assasination which began in the letters to Malesherbes of 1762 and

---

16 In the Preface to the *Discours on Inequality* Rousseau asks: "What experiments would be necessary to achieve knowledge of natural man? And what are the means for making these experiments in the midst of society?" O.C. III, p. 123 f. His answer, at this stage, was: the thought experiment. In *Julie* he tries out the model of the "société intime", splendidly analyzed in Jean Starobinski's *J.-J. Rousseau: La transparence et l'obstacle*. Paris: 1967, Ch. 5. Julie must die if the survivors are to return to themselves in resigned recollection of the past. In *Émile*, Rousseau uses the developmental model of the Discours on Inequality to study what J. H. Broome has called the creation of "an artificially produced man [in] an artificially produced natural society" (*Rousseau: A Study of His Thought*. New York: 1963, p. 104.) In their first encounter with actual civilized society, Emile and Sophie suffer moral shipwreck; the planned sequel was to bring a resolution parallel to that in *Julie:* death of one partner, resignation of the survivor in the tranquil acceptance of the irrevocable. About the third model, the *Contrat Social,* Rousseau himself has said the necessary: "Forcé de combattre la nature ou les institutions sociales, il faut opter entre faire un homme ou un citoyen; car on ne peut faire à la fois l'un et l'autre." The best institutions are those best designed to denature man "et transporter le *moi* dans l'unité commune." *Emile,* O.C. IV, p. 248 f. The self must die if the common self is to come into existence.

17 *Confessions* XII. O.C. I, p. 589.

18 *Rêveries du promeneur solitaire,* Première Promenade. O.C. I, p. 995.

continued in the *Confessions*. Once more the defense takes the strange form of self-accusation, of proving his goodness by revealing what custom and self-interest would urge him to keep charitably concealed. What is he fighting for? His reputation? But what is *public esteem* to a thinker who had proclaimed this very same thing to be the hidden source of all that is fatal to happiness and innocence? This inconsistency holds the clue to the strange work. Jean-Jacques is battling for the preservation, not of his true and pure self, but of his *unique* self, and the battleground is Time: How can the unique self survive death?

This self begins to die the day it is born. Only as an image in the eyes of others can it survive the death of the body. There is no other immortality for the unique historical self, for the man as he really is. This immortality, Rousseau believes, is what his unseen enemies try to destroy. They can no longer tamper with the ideas which Jean-Jacques implanted in the memory of mankind during "the time when he published books." But there is a subtler, deadlier way of killing this immortality even in the life-time of Jean-Jacques: by falsifying his true image. The author of Rousseau's noble works is to go down in history as a criminal, and nobody will ever know the true Jean-Jacques: existential murder by defamation. "Defamation" here means more than slander, it means distortion. If this unique self is to live on, it must live on in the memory of the race with *all* its traits, all its faults, all its vices, and in all its natural goodness. Hence the need to disclose what other men would hide: the man's most intimate character, the full history of his self, the deeds that shaped it, be they bad or good. We are in the world of George Orwell's *1984:* the Jean-Jacques of the *Dialogues* is fighting against being made an unperson.

What weapon can an ostracized, lonely man wield against the men in high places whose secret agents, he believes, observe his every step, who invisibly control his destiny on earth? Only one: the complete, relentlessly true written portrait of himself, protected against falsification and suppression. And so the manuscript of the *Dialogues* becomes for this great symbolizer the physical object around which the spiritual battle rages. There is nothing insane in this idea, except this desperate misanthrope's thrust in the social order and its guardians; it is his last link with reality. The concluding "Histoire de cet Écrit" records the snapping of that link. To prevent the manuscript from falling into the enemies' hands after his death he decides to place it secretly on the High Altar of Notre Dame de Paris, addressed to Providence. This will come to the attention of the King who in his goodness will take the manuscript under his high protection. On the appointed day, Rousseau finds himself barred from the altar by an iron grille he had not noticed before. He flees in panic. Providence itself has rejected him. There is no recourse against the verdict.

Now and now only does this self feel completely alienated, isolated, divorced from its species: "Here then I am, alone on earth, having no brother, neighbor, friend, society any more except myself." Thus begins his last work, the *Rêveries du promeneur solitaire.* Now only does he ask the real question: "But I, detached from them and from all, what am I myself?" He has ceased struggling. Like his St. Preux, like his Émile, he can find true selfhood only in the tranquillity of resignation after shipwreck: "Here I am, calm at the bottom of the abyss, a poor, unfortunate mortal, yet unmoved *(impassible)* like God himself."[19] He is ready for the last, the inner, *dépouillement*. The *Rêveries* are the rambling record of this stripping away what has insidiously infiltrated the self, posing as part of it, perturbing its equilibrium, drawing it away from itself: blocks of past experience, passion, will rebelling against destiny. Hardest of all is the ultimate step: recognizing that all his struggle had been self-deception, that his supposed goodness, his *amour de soi,* have only been amour-propre, vanity, fighting for his image in the eyes of others.

In the midst of this labor of freeing the pure self rises the memory of the far-away days on the Island of St. Peter when, listening to the rhythm of the waves, or out on the lake in a boat gently rocked by them, he had been truly and only himself. In this state of bliss one feels "nothing exterior to oneself, nothing except oneself and one's own existence: so long as this state lasts, one is sufficient unto oneself like God." This, then, is the experience of pure selfhood: *"le sentiment de l'existence dépouillé de toute autre affection,"*[20] the feeling of existence, stripped of any other attachment.

Once more, Time stands still in the moment "where the heart can truly say to us: *I wish this moment would last forever.*" It is the moment for which Goethe's Faust will strive. For Rousseau it is the moment of the *arché*, the beginning time of man: the state where the soul can "gather in the whole of its being without any need for recalling the past or encroaching upon the future: *where time is nothing to the soul, where the present lasts always,* without yet marking its duration and with no trace of succession, without any other feeling of privation or enjoyment, pleasure or pain, desire or fear, except [the feeling] of our existence, and that this feeling alone can fill the whole soul."[21] Jean-Jacques has come home into the mythical time of the Cosmic World Egg rocking on the waves of the primordial ocean before Time began. So he tells us, *post festum*, findings in the memory of an experience long past what he had been seeking in the future: the euphoric bliss of pure selfhood, the eternal moment recollected in the tranquillity of a dying heart.[22]

---

19 *Ibid.*, p. 999.
20 Cinquième Promenade, *ibid.*, p. 1047.
21 *Ibid.*, p. 1046.
22 "I feel already my imagination ice over, all my faculties get feebler. I expect to see my reveries become colder from day to day until boredom will rob me of the courage to continue; and so my book will naturally end, if I continue it, when I approach the end of my life." Note jotted down on a playing card, O. C. I, p. 1165. Again in the second Promenade: ". . . un tiède allanguissement énerve toutes mes facultés, l'esprit de vie s'éteint en moi par degrés . . .", *ibid.*, p. 1002.

## "Jean-Jacques" and "Murphy"

All these Murphys, Molloys and Malones do not fool me[23]

Rousseau's account of the eternal moment closes on a curious note. The descent to the pure self, he says, has its mechanics; it requires fine adjustment of soul and bodily environment. The heart must be appeased, and the body needs to be attuned to the inner state. This requires setting it in a specific kind of rhythmic *motion*. What is required is "neither absolute repose nor too much agitation, but uniform and moderate without jerks and without intervals of rest," since life without motion is but lethargy: "Absolute silence leads to *tristesse*. It offers an image of death." Happy are those whom Heaven has endowed with a sunny imagination (*une imagination riante*) which alone can protect them.[24] Jean-Jacques has sensed a hidden threat to the wholly detached self. He even names it. But the courage of this radical explorer has its limits, despite his thunderous thruth claims: "Lebenbleiben wie das Sterben / Für das Vaterland ist süss," as Heinrich Heine was to say. This euphoric happiness is too useful and too precious to be exposed to the shock of confrontation with that lies around the last corner of the self's road. After all, *amour de soi* as he has defined it is the self's innate urge towards self-preservation. And so, *tristesse* and the image of death coming into sight, his "atrabilious passion," as Malesherbes called it, miraculously turns into a sunny imagination which, aided by a prudent dietetics of the soul, will shield Heaven's favorite child from the sight of ultimate truth.

It is by no means unusual for an explorer of the soul to anticipate the next advance beyond his own range, and to warn of it. This was still the eighteenth century, though Rousseau was already close to twentieth-century sensibility. But the consequences of the object-subject split had not yet become massive fact. If "ces doux extases" is a characteristic Rousseau word, the Beckett word is "consternation," which the dictionary defines as "Amazement and terror such as to prostrate the faculties."[25] But such is the logic of the quest that one has to begin where his predecessor had stopped. The last illusion had to be pierced before Beckett's staring eyes could see his writer's task: "One can only speak of what is in front of him, and that is now simply a mess," as he reportedly said.[26] And so it is no accident that on the very first page of Beckett's hilarious,

---

23 Samuel Beckett: *The Unnamenable*. We cite the New York edition: *Molloy. Malone Dies. The Unnameable*. Three Novels. Grove Press: 1957, p. 420. Quotations from Beckett's *Murphy* are from the first two pages and from the brief Section VI.

24 Rousseau: *Rêveries*. Cinquième Promenade. O. C. I, p. 1047.

25 In a "non-interview" with Israel Shenker (New York Times, June 5, 1956) Beckett is reported to have made the non-statement that Kafka's form "seems to be threatened the whole time – but the consternation is in the form. In my work there's consternation behind the form, not in the form. In the last book – L'Innommable – there's complete disintegration. No 'I', no 'have', no 'being'." Beckett on the painter Masson (in 1949): "Though little familiar [with his past problems] . . ., I feel their presence not far behind these canvasses veiled in consternation." *Proust*, p. 109.

26 So cited by Tom F. Driver, "Beckett by the Madeleine." Columbia Forum, Summer 1961, p. 23.

still conventional first novel "Murphy," Jean-Jaques in his boat, transmuted and transmogrified, appears as Murphy in his rocking chair. Whether or not Beckett even fleetingly thought of Rousseau when he put Murphy there does not matter. Murphy, feeling himself "split into two parts, a body and a mind," acts out the scenario of his more illustrious predecessor as he sits stark naked in his rocking chair to which he has tightly strapped himself, rocking away in uniform and moderate motion without jerks and intervals of rest, performing the *dépouillement* in search of "pleasure, such pleasure that pleasure was not the word," his descent into himself sardonically observed by Samuel Beckett reporting from inside Murphy's mind.

This mind pictured itself as a "large hollow sphere, hermetically closed to the universe without" – the split. Inside were three "zones": light, half-light, and dark, being three steps of dépouillement, of alienation from a hurtful outer world, each a source of rare gratification. In each zone, specific constituents of reality have been stripped away. Up in the light are the "forms with parallel," the mental forms of things and events "outside", a replica of the "universe without" stripped of *causality* and the *irreversibility of events:* a universe "broken into the pieces of a toy" and so made docile. Here Murphy can "correct" the outside system which is out of joint, here he can reverse the direction of the kick he received: the special pleasure here is "reprisal."

In the half-light, *memory* is gone, and *will*. Here are the "forms without parallel" (nothing to be recalled), offering the "Belaqua bliss" of will-less, unremembering contemplation of pure change.

In the dark, the object-subject split is gone: "nothing but commotion and the pure forms of commotion . . . without love or hate or any intelligible principle of change." The self, "caught up in a tumult of non-Newtonian motion," has contracted into a nondimensional point, "a mote in the darkness of absolute freedom." This is the ultimate pleasure: consciousness conscious only of its being *in unintelligible motion,* will-less, deathless, until further notice.

The epigraph to this Section Six knocks the supports from under this euphoric state: *Amor intellectualis quo Murphy se ipsum amat.*[27] This takes us back, not only to Spinoza, but to Rousseau's *amour de soi.* The climax in Rousseau's drama of the soul came when he gained the bitter final insight that his righteous *amour de soi* had only been vanity cunningly disguised, that he had never been truly himself except in those rare moments of timeless detachment, moments known only to him. Beckett's epigraph raises the fatal question about these moments: where are the limits to the poisonous ingenuity of *amour-propre* in disguising itself? Could it be that even these moments of true *amour de soi* had been moments of sublime self-deception, of love for his image in his own

---

27 "Deus se ipsum Amore intellectuali infinito amat." Spinoza: *Ethica* V, propos. 35. "Mentis Amor intellectualis erga Deum est ipse Dei Amor, quo Deus se ipsum amat . . .", propos. 36.

deluded eyes? Was this euphoric happiness an opiate administered by the false self to preserve Jean-Jacques's ignorance of the true state of pure selfhood?

Yet – what purer state can the self attain than that of reduction to a pure feeling of its own existence in the timeless moment? None. Jean-Jacques had his hand on it: ". . . sans aucun autre sentiment . . . que celui seul de notre existence." Only the accent was wrong. For him it fell on *"sentiment,"* not on *"existence."* This was the work of *amour de soi:* a self truly detached from *everything* knows its condition and does not love what it knows. Neither does it love what it does not know: itself. This is the ultimate insight. It informs Samuel Beckett's greatest work, the novel *The Unnameable,* where the last word is said about all these Murphies: "What rubbish all this stuff about light and dark. And how I have luxuriated in it."[28]

*Samuel Beckett*

I of whom I know nothing[29]

*The Unnameable* is the threnody of the modern self, the murmured, unending monologue of a voice de profundis. We do not know who, or what, the Unnameable is: the patient, his condition, its cause, or all of them. The modern self is at its end, in the paradoxical world of the certainty of uncertainty. The first words we hear restate Rousseau's question:

> "Where now? Who now? What now? Unquestioning. I say I. Unbelieving. Questions, hypotheses, call them that. Keep going, going on, call that going, call that on." [401]

"Who now?" – "I say I." This is the answer which Descartes had given, "unquestioning," since he had no proof. His experiment of hyperbolic doubt (histrionic doubt, Santayana would say) left behind a residue of awareness that there is doubt, in existence, now. *This* certitude is unshakeable. But then he immediately identifies this non-vanishing residual as his own self (le reste, c'est *moi*), adding a piece of knowledge of different origin and with different epistemological status. As a philosopher he can do that, though not with impunity. He knows that there *is* a residual. He also knows *where* it is, by prior knowledge. He philosophizes in the first person, his voice being the voice of the philosophizing self. There is no other locus of knowledge. Hence the awareness can only be "his". Take away this assurance, and the voice becomes the voice of a consciousness that has taken the doubt one step further than Descartes, despite his warning, and is irretrievably lost, as he had predicted.[30]

---

28 *The Unnameable,* p. 419. Subsequent page references in square bracketts refer to this work in the New York edition.

29 *Ibid.,* p. 420.

30 The resolve "to strip oneself of all opinions and beliefs he had received is not an example for everyone to follow." The world is almost made up of people who should not: those, we might say, who have no independent judgment and those who have too much of it.

The residual consciousness is something that has no name and can therefore have any name, just as a clothes-hanger can have any kind of clothes hanging on it, without necessarily being a clothes-hanger. The "unhappy consciousness"[31] has lost even the certainty of self-identity, can no longer say: "I = I". The formula must now read: "? = ?".

This voice may be the voice of some self speaking, or perhaps somebody's voice "saying" that self, or the echo of a voice that was and is no more: "perhaps it's done already, perhaps they have said me already . . . I don't know, I'll never know, in the silence you don't know . . ."[577] Yet the voice goes on, interminably, speaking about itself (that is: about nothing one can speak of), asking questions, inventing and rejecting stories about itself, remembering what is cannot remember, in command of an esoteric vocabulary where called for, knowledgeable beyond the knowledge of the best Beckett critics: where does all this come from if this is a consciousness emptied of all content?

The answer, like the whole situation, is paradoxical. When the surgeon's knife has carved away even the identity of the self, when the very existence of the self is in doubt, what difference does it make *where* the line between self and non-self is, or was, or might be? *De minimis non curat praetor.* And so everything that Descartes and Rousseau had painfully stripped away can stream back again, unhindered, once more at the free disposal of consciousness, not as certainties arrived at by Descartes' "long chains of perfectly simple and easy reasonings," but as equally impossible possibilities or equally improbable probabilities, this being a radically democratic situation where anything goes because nothing matters. It is the *ne plus ultra* of Cartesian certainty, and that's the humor of it: certainty of nothing. Or, as Beckett said, somewhere, sometime, "only nothing is real."

But when one moves on from the last page of *Malone Dies* to the first page of its sequel, *The Unnameable,* this "Where now? Who now? What now?" sounds more like the voice of a consciousness suddenly disembodied by death, trying to orient itself in its new state, having a lifetime's knowledge, also disembodied, "broken into the pieces of a toy," to play with. But this hypothesis is futile, too. There is no criterion left to distinguish life from after-life or pre-life, and where there is no answer, there is no question.

If this be Pyrrhonism, why does it not end in silence, as all true Pyrrhonism must? "Worüber man nicht sprechen kann, darüber muß man schweigen." Yet

---

The latter ones must not turn away from the common road, even for the sake of a thought experiment, because they could never stay on the narrow trail that will lead them back to inner certainty, and so they "would stay lost all their life." This warning accompanies the very first enunciation of Cartesian doubt. *Discours de la méthode*, Seconde Partie. AT VI, p. 15.

31 "[Das] schmerzliche Gefühl des unglücklichen Bewußtseins, daß Gott selbst gestorben ist . . . [ist] der Ausdruck des innersten sich einfach Wissens, die Rückkehr des Bewußtseins in die Tiefe der Nacht des Ich=Ich, die nichts mehr außer ihr mehr unterscheidet und weiß." At this point the opposition between substance and consciousness, the split, is gone. Hegel, *Phänomenologie des Geistes*, p. 545 ("Die offenbare Religion.")

falling silent is the one thing this voice cannot do, this human voice, compassionately recorded by an intelligence which fully comprehends the paradoxical logic of this situation: "... strange pain, strange sin ..." Words, mere words, perhaps; nothing is felt but the pain, nothing is certain but the compulsion to break the silence: "... where I am, I don't know, I'll never know, in the silence you don't know, you must go on, I can't go on, I'll go on." [577] This is how the book ends. The voice will go on breaking the silence as best it can, invention following invention, question following question, empty talk to mark time between inventions: "The fact would seem to be, if in my situation one can speak of facts, not only that I have to speak of things of which I cannot speak, but also, which is even more interesting, but also that I, which is if possible even more interesting, that I shall have to, I forget, no matter. At the same time I am obliged to speak. I shall never be silent. Never." [402]

Never? Never, indeed. For to be silent is to be dead, and consciousness cannot be conscious of its being dead. Consciousness dead is conscious of nothing. But since consciousness alive is always consciousness of something, how does a consciousness emptied of all content know that it must go on, how does it know that it *exists*?

The answer is: by being conscious of itself in Time; since Time is experienced only in change, consciousness must be in motion so long as it exists, *pensée* following *pensée,* the term taken in its widest Cartesian sense of all that can enter consciousness: thoughts, feelings, sensations, volitions. And if consciousness has nothing to provide itself with, when there is nothing to say, then it must invent content to be able to know itself as existing. It *must*, it cannot stop, for motion stopped does not just evoke an image of death, as Rousseau thought, it *is* death. Death, for which this consciousness longs and which it can never attain.[32] Death and birth are beyond its pale. So is the outer world, so is the Time of that world. After the ultimate *dépouillement* there is no reality left for this consciousness to orient itself by, except the reality of Inner Time. It is the only reality that cannot be stripped away so long as this consciousness exists, that is: moves. And so, ironically, Time becomes the prison without walls from which this self cannot escape: "Enormous prison, like a hundred thousand cathedrals, ... and in it, somewhere, perhaps, riveted, tiny, the prisoner, how can he be found, how false this space is ..." [569 f.] This is what "le seul sentiment de son propre *existence*" really is, once *amour de soi* is gone, together with identity. This is what pure consciousness of existence *feels* like: "I, of whom I know nothing, I know my eyes are open, because of the tears that pour

---

[32] At the end of *Malone Dies,* the interior monologue is interrupted by temporary lapses of consciousness which occur with increasing frequency until death stops it altogether. After each lapse, consciousness resumes its monologue exactly where it had stopped, repeating the last word or thought before the break or going on as if nothing had happened. Inner Time did not stop, the interval was non-existent. Beckett's art, last perhaps but not least, is the art of knowing.

from them unceasingly..." [420] This is one of the self-images standing in this Dantesque Inferno, erected by this self: The Unnameable as the Colossus of Memnon whose myths are as contradictory and eloquent as those of the Unnameable's own invention, versions of the same lament.

This self, prostrate, crippled by incomprehension, has a symbolic body equally defective. For if "he who possesses a body fit for many things possesses a mind of which the greater part is eternal," as Spinoza says,[33] then a merely everlasting mind, lacking intellectual love toward God (the love with which God loves Himself), will have a body fit for few, very few, things. Consciousness is in everlasting motion, but the body has a bad leg, must crawl, on hands and knees, on its belly, is truncated, stuck, in a jar, in an ash can, in the sand, in the mud. Defiled, truncated, rocked by the endless flux and reflux of small hopes and expected despairs, proceeding "by affirmations and negations invalided as uttered, or sooner, or later" [401], this thinking thing crawls through the corridors of its Time, intensely human, comical in its affliction, noble in its lack of self-pity as it shores up nothingness against its ruin.

Thus ends what began with the Cartesian split: with the *tristesse* of the soul longing for absolute silence and unthinkable death, with the paradoxes of the ultimate state where knowledge is ignorance, certainty is uncertainty, where everything is there because nothing matters. Spinoza has written the epigraph, or epitaph, to this story. It is on the last page of the *Ethics:* The ignorant man "never enjoys true peace of the soul, but lives also ignorant, as it were, both of God and things, *and as soon as he ceases to suffer he ceases to be.*"

## Conclusions

The quest for the pure self has been a protracted, subdued affair, as befits its character. One might dismiss it as a matter of some highly sensitive thinkers and writers adding to the already abounding collection of specimens, factual and imaginary, of exquisitely rare states of the soul, were it not for its significance in the sharp turn which Western civilization and society took at the dawn of the modern era. Let us briefly consider this aspect, keeping in mind that we are dealing with symbolizations, not with causes.

The new self-understanding manifested itself most patently in the creation of a type of science capable of cracking the codes of nature, thus increasingly "render us masters and possessors of nature" and, in the event, transform the technological, social and human character of civilization. The modern self slipped easily into the stance of subject-object differentiation between the observer and the observed, the knower-doer and the material on which he operates. In striving for "man's" mastery over "nature," however, the fact is easily lost that "nature"

---

[33] "Qui Corpus ad plurima aptum habet, is Mentem habet, cujus maxima pars est aeterna." *Ethica* V, propos. 39.

also includes "man" as the human object of research and thus, on principle, of increasing mastery and possession by the knower and transformer. This fact, commonly hidden under a cloud of illusions, comes into the open whenever the potential master acquires new knowledge rendering him capable of remodeling human society and human beings. Where can the controlling self find guidance in determining what it should do and what it must not do? Nature is silent under such questioning. The answer can only come from "within" the self. "Man" on the other side of the split is equally bereft of guidance. If he feels himself to be the victim of an alien, hostile "environment," this universe without offers him no clues when he seeks to understand his suffering. His modes of response – submission, withdrawal or rebellion – are modes of non-participation.

The "master" can of course adopt self-denying ordinances that put certain purposes and methods out of bounds; but no such ordinance has the force of an imperative: its acceptance is an autonomous act, revocable at will. The "victim", again, may dream of dispossessing the possessors and making the environment the property of all, but only the first purpose can be accomplished. After that there will again be the cleft between those operating the environment and those on whose behalf they operate it – between actual "masters" and potential "victims."[34]

All this can be equally well stated in evolutionary terms. Once the processes of biological evolution have produced a *res cogitans* capable of autonomous intervention, blind evolution becomes goaldirected within this thinking thing's environment. From the viewpoint of "nature" this is no change. Evolution remains "blind". The community of being (a term that makes very good sense here) continues to change in the mode of interaction, indifferent to the emergence of "new" agents and processes of change which have the same status as the "old" ones. It is otherwise when we shift to the position of the new agent. Here again the term "exodus from the community of being" makes good sense; it means precisely what biologists mean when they speak of change from biological or organic to cultural evolution. Man has not left nature, he could not do that; he has walked out of a community of interacting beings of equal status, he injects purposive change into an environment hitherto governed by adaptive change, he conceives the distinction between what *will* be and what *should* or should not be, and since he can to some extent foresee the effects of his purposive actions, he could not relinquish the task of setting goals if he wanted to. Once you can foresee the consequences of not acting as well as those of acting,

---

[34] The point is not that a complex social enterprise requires qualified decision-making leaders who have the necessary power but that this enterprise with its inbuilt dynamism needs minds capable of analysis in the subject-object mode. The "victim's" dream could come true only if kinesis would again become stasis, if all problems would become routine problems, if innovation would slow down to almost a halt – in other words, if God would throw away his watch once more. This is the point where Lévi-Strauss returns, not to nature, but to Rousseau.

the decision not to act, too, is goal-directed. But a mind so constituted can find the ground for its decisions only within itself.

The symbolic inquiry, then, has quietly exposed the critical problem in the evolutionary enterprise undertaken by the new agents, grasping the intimate connection between self and civilization at a time when the fastest travel machine in this civilization was the horse. What the symbolic exploration has yielded can be summarized briefly.

A self that steps away from the community of being and feels itself to be the Wholly Other gains access to operationally powerful modes of knowing, and loses access to vital modes of participation. In setting goals for its transformatory operations it must fall back upon its own resources which are not there. The certainties it can attain are all "outside"; if it seeks certainty about itself, the radical pursuit of this quest must plunge it into irremediable uncertainty about itself. Ignorance of its true state gives this self that complete detachment from which its enormous cognitional and transformatory powers spring; knowledge would debilitate and cripple it.

This is the most important result of the inquiry. It should not be misread as implying that the operational and the participatory modes cannot coexist. They can and they do. But one governs this civilization, the other does not. We must leave the problem at this point of merging into the larger problem of order and history. Let us consider, briefly, the two remaining questions: What accounts for the obvious contrast between the result of this analysis and the empirical image offered by technological civilization? and: Is the Western quest for the self unique?

The states of self described in the symbolic analysis are by no means fictitious; they occur, but they are exceptional.[35] Looking generally at the thinkers, researchers and doers who have shaped and shape this civilization, one is struck by their self-assurance; corrosive inner doubt, though not unknown, is no professional disease with them. The contrast with the pervasive uneasiness among the best creative minds outside the progressive enterprise is striking. It struck already Rousseau who, not finding another one like himself, saw it as a contrast between civilized man on the one side, himself on the other. His greatest insight was indeed, not that civilization denatures and debases the self, but that it so cements the false self into the social power structure that only catastrophe can wrench a human being away and make the return to true selfhood possible, if only in resignation and exile. And only an act of trumpery on the part of a Legislator (meaning a successful impostor) can temporarily convert a civilized society as a whole into a "good" society whose citizens are literally self-less. The shapers of a civilization are bonded to it: "Im ersten sind wir frei, im zweiten sind wir Knechte."

---

[35] The symbolic mode is not to be mistaken for psychologizing. But it is not accidental that psychologists tend to use symbolic language when naming and describing such states.

The same confidence that everything is normal reigns with regard to the standing of the detached self in its relation with the other selves. In its operational mode, this self sees and treats them as the wholly other; yet it feels warmly, at times almost aggressively, at one with the rest of mankind. The most obvious explanation of this contradiction need not detain us, but we do want to look at one device, at least, which permits the detached self to save appearances. Rousseau who had a flair (though not Nietzsche's genius) in such matters knew what he was doing when he built into his man-beast two primordial instincts where only one was needed: *amour de soi* had to be put in because it protects and preserves the self; *pity* was added because it enables a completely closed self to help the suffering creature and feel good about it, while happily remaining what it is: a completely closed self.

As to the drop-outs from the burden of civilization, there is the half-way house where doubt readily comes to rest, where "Jean-Jacques" and "Murphy" can fill up on happiness when life in the "universe without" becomes too much for them. Self-love will protect them there in the illusion of true selfhood and in the deceptive bliss of the eternal moment. But this half-way house is uncomfortably close to the house of death, as both these heroes know. And if doubt should corrode and break the dam built by self-love, the venture will end in the abyss of certain incertitude.

This, however, is not the final insight yet. That crippled thing without hope, crawling in the mud of existence, more human than those who walk in the light, knows its abysmal condition but does not know the real absurdity of it. The radical attempt to strip the self from all non-self is destructive, but it can not succeed, for the best of reasons: because *there is no such thing as the pure self*. Even that "I of whom I know nothing" is still *en règle*, is still within the order of being which no longer speaks to it.

Our last question concerns the status of this modern Western enterprise within the course of human history. The answer can only be that such self-understanding, unprecedented in its transformatory power and global sweep, is but one of many speculative and practical attempts to understand and, if possible, attain a state where "the soul, purified of all that is not itself, comes into the possession of its own timelessness." The quotation describes the state of samādhi or turīya according to Patanjali's system. It could as well be a Western formula. There is the split, the *dépouillement*, the timeless state. Time and Timelessness are indeed the universal symbol for the tension between self and non-self which arises wherever "compact" or cosmological understanding of the great order of being becomes differentiated in consciousness.[36] For then existence is experienced as a process in the "inbetween", in the uncertain relationship with

---

36 For this and the following see E. Voegelin: *Order and History* (Introductions to Vol. I and II), *Anamnesis* (the analyses of consciousness and "Eternal Being in Time"), and his forthcoming book *The Drama of Mankind*.

Time and Timelessness. This inherently dynamic experience can seek resolution in quite different directions, depending on how human existence in time is understood. The search for self and non-self, then, is universal within the reign of differentiated consciousness. It is the tension between "the soul" and "all that is not itself" which initiates the search. The differences arise in the interpretation of "not itself". If consciousness conceives of "non-self" as the wholly other, when it radically adopts the non-participatory, operational mode of understanding, then and only then do the Western consequences follow. If the soul seeks to come into the possession of its own timelessness with a different understanding of its nature and status, then it will do different ways, not only in its search for the pure state, but also in the search for social order. Such differences in the understanding of existence in time then explain how human beings in their given society act in Time, and what they seek in Timelessness.

# Time and the Modern Self: A Change in Dramatic Form

T. Ungvári*

In a well-known paragraph of his *Wilhelm Meister*, Goethe uses a metaphor characterizing the achievement of Shakespeare. A *crystal world-clock*, demonstrating the passage of time – that is the simile; a comparison, the origin of which can be traced back and forth in German as well as in some other European literature. A crystal world-clock: This reminds us what great impact classical science has had on distinguished minds of literature in earlier ages, and further, how such an ingenious remark can get worn down to a commonplace and truism through constant use. But even in its frozen and almost empty form it has a deep message. For it is quite conspicuous in the history of letters that time comes into every relevant definition of dramatic form. Aristotle and, even more, his followers formulated their definitions of tragedy in a framework of time and space. The question is not whether any unity of space and time was postulated, as in French classicism. The distinguishing quality of all definitions of tragedy is the reference to time. We may therefore rightly state that it is possible to enumerate some valid definitions of lyric and epic poetry that simply leave out the time aspect, whereas there is no definition of dramatic art which refrains from making a reference to time. The study of the famous Goethe–Schiller correspondence clearly shows us that the time aspect of epic poetry emerges only in an opposite position to that of tragedy. Epic poetry relates past events, in contrast to tragedy, which introduces us into the world of action, to the world of the *now*, the becoming, the present.

In working out the aspects of literary types and genres (the Germans have a better word for it: "Gattungen"), drama has a time emphasis. For we may rightly assume that when in a series of definitions preference is given to a quality, it only reflects the deep core of the thing itself – it mirrors the heart of the matter. There is a time-preference in every definition of drama, because drama has a preoccupation with time. To draw just a short parallel with physics: time seems to be as much an attribute of drama in the literary field as it is of motion in the natural sciences.

What is the reason for this preoccupation with time? Drama is an imitation of action and action is a form of motion. It is not by chance that Aristotle, who so clearly defined time as an attribute of motion, gave a preeminent place

---

* Dr. Tamás Ungvári, University of Budapest, Budapest, Hungary, II. Apostol u. 9/a.

to the equivalent of motion in human behaviour: that is to action. There is even a striking concordance between Book IV of his *Physics* and his *Poetics*. This relationship was curiously neglected by scholars of antiquity, although in his *Physics* Aristotle outlines what we should understand by the idea of the *now*, i. e. the present. He identifies it as the *link of time*. Here is a possibility of connecting this conception to that of his definition of drama. For that coherent unity of action which is the chief postulate of tragedy bears a close relationship to the definition of the *now*, the present. The unity of action is nothing else than the *now* mentioned in the *Physics* as the point which links past and future.

I will return to the idea of this present, to this link, to this *now*, as one of the most important time-aspects of drama. But before that I have to draw attention to the effort of Aristotle to change the respective terms when speaking about time in his *Physics* and his *Poetics*. He is – it seems to me – venturing on that risky road of translating a natural phenomenon into patterns of human behaviour. Motion and action are synonyms for him, used in their due places. With this he consciously reveals an existing correspondence between the world of nature and that of art – but [in the meantime] he is quite aware of the need to reformulate the terms for essentially the same quality when used in different contexts. What he is doing is nothing more than *humanizing nature's laws*, seeking corresponding or identical values in different fields, showing equivalents like that of *motion* in the physical world and *action* in human life.

His experiment should be appreciated with humble respect at a conference where scientists are meeting scholars of the arts. Aristotle was the first who implicitly showed that science and art are joining the common human endeavour to measure the outward world of experience. *Measurement* – briefly defined in practical terms – is in itself a humanization of our sense data; it is a special syntax of humanity based on analogies uniquely representing our capacity for abstraction.

Now – if art is a kind of abstraction – it shares a common attribute with measurement. It is also a human syntax for understanding the world, a geometry of feeling, action, viewpoint – basically a supposition parallel to that of the first Alexandrian scholar who, instead of orbiting the earth, had to walk only eight hundred kilometers to give a fair appreciation of a parallax to find our distance from the sun.

Art and scholarship – both are enterprises of understanding and therefore measurement. The grammar, the syntax, is different. Sometimes they share no common word. But since they are both efforts of communication, they are all languages – translating sense data into ideas, measurement into analogies, experience into intelligible, commonly agreed upon signs and symbols.

Time: the word meant originally *division* in ancient Greek. Division is measurement and it is not by coincidence that the Latin word derived from the

Greek, *tempus*, seems cognate with *templum*, the *church* which is a model of the universe. *Tempus* measures *motion* as *templum*, as *templum* divides off a place of worship and *emotion,* the place dividing the regions of power, earth from sky and time from eternity.

These preliminary remarks indicate a simple truth. Since Lessing we have been reminded of the fact that poetry has a special capacity to mirror *motion in time*. Confronted with other grammars of knowledge – not only with such artistic grammars of space as painting or sculpting – poetry is not time's fool but rather its master. It is an epistemology which penetrates into time. One is tempted to repeat the remark of Goethe: poetry is a huge crystal world-clock, similar to that of Newton. But here there is also a difference. Mechanical clocks measure a uniform flux of time whereas the clock of art rearranges time into articulate patterns, into aspects of tenses, into visions of past, present and future. If a mechanical clockwork bears a face of God-like objectivity, the passionless impartiality of a pure mathematic idea, in poetry "time must have a stop," to quote that great clock-master Shakespeare, *with every work of art.*

Within this broader context there is a chance to understand drama's unique position among other types and genres of poetry, to wit: drama cuts out only a small section of life. The Aristotelian definition of tragedy stresses this point with a certain strength: drama should focus on a closed, circumscribed action. *Mythos* – the greek word for story – does not mean any freely chosen topic at hand but a certain plot (in German a *Fabel*) that has a beginning, a middle and an end, and transform a set of events into a coherent pattern, linking past and future together.

*The smaller the section, the larger the part that is given to the calculus of events: time.* Since action is a strictly woven pattern of events, the thread must be spun from an everlasting idea, i.e. from something universal which is relevant, common without the help of immediate explanation. This assures the effect of drama which performs the magic act: it elevates a curiously limited story to a level of universal validity. This universal idea, which gets its immediacy through artistic means in drama, is once more and again: time.

Time, but in a particular aspect, that of the present, is used as a link between past and future. It transforms forgotten myths and fables into current action and paramount dialogue, both of which are means of invoking *a presence.* This dramatic present – confronted with the epic – implies more than the common usage of the word indicates. It is not the inconceivable present of the razor's edge. It should be interpreted in the deeper, Aristotelian sense: being a *link,* this present it is also a division of time. A focus point – this present – will join only those parts of past and future extensions that may have a particular function in a *presentation.*

When Teiresias tells us the story of the birth and childhood of Oedipus Rex, this might in another context be a rather boring story of false prediction. But here it changes a whole series of events; it is transformed from a past fable into immediate action. To describe this transformation, Georg Lukács has a German phrase: drama "muß alles vergegenwärtigen" – it has to transform everything into a present, otherwise it has no bearing on the *mythos,* the story, the plot.

The magic of all this transformation has a deep connection with its time aspect. The act of transformation offers all the chances for great artistic delicacy. It is not only by tradition that killing, murder, duel, or battle are related but never presented in Greek tragedy. They are related, because they are parts of a strictly woven plot, because they have an immediate bearing on action, they have a function in it – with which they become the part of that *present* which needs no further *presentation.*

We may agree therefore that drama's unique quality is the vivid evocation of a present. But this perception does not answer the important question – why does all this give satisfaction and enjoyment?

There is one interesting solution: that of Frank Kermode who in his brilliant *The Sense of an Ending* analysed the Aristotelian ideas of *catharsis* and *peripateia.* Kermode's premise is that every plot tries to establish a certain coherence between its beginning and end. It seeks concordances, links that mold a manifold course of events into one unit. To achieve this the plot rearranges reality according to the need of coherence – but in doing this, it is not estranged from reality. It works with a humanizing tendency: *rearrangement* means a *readjustment* to reality, it means a penetration into its hidden layers of existence. In short: imitation implies rearrangement, namely a readjustment to the course of events.

This process of rearrangement may be expressed through a particular time-aspect. More than that: this time-aspect is a uniquely relevant means of distinction between poetical types and genres, i.e. drama, epic or lyric poetry. Poetic forms are presentations – they are focusing on that Aristotelian edge of the *now,* the linking or separating point of view out of which they are rearranging reality. This opens a deeper interpretation of Goethe's remark about epic poetry evoking the images of the past. It has a certain Apollonian sense of distance, conceiving of a story-teller sitting on the dividing link of the chain looking back on past events in a reflective mood of nostalgia. Thus he is rearranging a *plusquam perfectum:* his own *now* being the dividing, rather than the linking point. Drama on the other hand uses the linking quality of this *now,* sewing the threads of past events into the present action.

It is a temptation, not easy to resist, to rewrite the whole history of poetics from this angle – showing how some transitional forms, like the ballad, owe their distinguishing qualities to a certain combination of various time-aspects.

For further clues and evidences one should turn once more to Aristotle. Although there is no clearly established connection between his allusions to the unity of action, space and time, on the one hand, and introducing the idea of *catharsis*, on the other, we may rightly assume that this is due only to the fact that his *Poetics* is a fragment. Frank Kermode states that a well-wrought plot, let it be in epic or drama, gives a peculiar feeling of coherence and order and all our satisfaction arises out of the understanding of the issues. A rearranged reality imposes a far more subtle order on things than the most common wish-fulfillment. And although it releases the forces of *éleos* and *phobos* (*Furcht und Zittern*) it has also a relieving element, a cathartic one, which is implicit in the vision of every coherent order.

This order is introduced to every literary type by its time-aspect. To condense all courses of events into a single action is an ingenious and arbitrary vision which can be evoked only when there is a common denominator for it and the ordinary world of normal chronology. Time is transformed in art but only to a point where there is a remaining link to experience.

It follows that poetry has a time-imposing force. But with its cathartic element it has also a *time-redeeming* one. You are freed from time for a couple of hours, released from its grasp just by entering into another fictitious time-scale. Art in this meaning is a double-faced God: it is an *in-time – out-of-time* experience.

Up until now I have carefully tried to avoid any proper definition of art in general and tragedy in particular. I am dealing here only with one or two qualities of the form – namely its coherence due to its time-aspect of the *now* and its capacity to focus attention on a rearranged pattern of a given action and then redeem us by its vision. This is also a contextual aspect of tragedy. To be condemned and to be redeemed – this is a common feature in all tragedies. How this condemnation and relief are molded into one clear vision, this paradoxical interdependence may be understood from a beautiful example often cited by historians and with a special emphasis by Marx. This example recalls the day of Saint-Just's execution in the Conciergerie. The condemned suddenly noted, pinned on the wall, a copy of the commandments of human rights. And then, in a clear moment of challenge he exclaimed: "C'est pourtant moi qui ai fait cela..."

The decisive word is *pourtant*. In spite of all, *trotz allem, trotzdem*. In this word you see the Aristotelian link of past and future: in this time-bound and time-freed moment of a cry: *"C'est pourtant moi..."*

I had to deal with the form of tragedy in general just to show that, whatever changes tragedy has undergone in centuries of transition and change, the time-aspect of the form remained basically the same. It was a form that explored the field of action, conceived of action as a *"medium inter aeternitatem et tempus"*, connecting past and future in a sense that gave sense to suffering by ultimately linking a set of rather horrifying deeds and actions into an intelligible comprehensible time-scale of succession. A beautiful line of Dante's elucidates this: tragedy focuses on the point – *il punto a cui tutti li tempi son presenti* –

the point which we call the linking present where every sacrifice is justified because its time-scale has a higher meaning than the mere repetition of the *before* and *after*. In other words: tragedy as a form finds a transcendental image of time, peculiarly relating all the aspects of it to one or more coherent action. to one or more character with identity and personal integrity.

It would be dubious to assume that it is particularly our age which changed all this and began a new phase of development or disintegration. In history it is always dangerous to proclaim a new-age doctrine. Our assumption is nevertheless verified by a simple fact – that somehow tragedy is a vanishing form of human self-expression. The more tragic our age seems to be, the fewer proper tragedies are written. They have become almost an extinct species in a new biology of letters.

Going now in pursuit of the vanished form of tragedy, I am convinced that the whole phenomenon has something to do with a change of the time aspect in the modern world itself. Hans Meyerhoff in a paper contributed to the anthology *The Broken Center* enumerates three major factors that altered our modern time-sense: (a) the lost consciousness of eternity, (b) the quantitative measurement of time in modern science, and (c) the tendency to view truth itself as a function of time, the historical process.

These three factors are relevant to our investigations. It could be easily illustrated that, just as the *"silence éternel"* of infinite space scared Pascal, the newly revealed time aspects scare the poet, the writer, the dramatist. The time of this modern age of new physics and uncontrollable history emerges as a *monster* as early as the poetry of Gautier at the end of the last century in his "La montre"; it is echoed in Baudelaire's *L'Horloge*. There you find the following remarkable lines about the clock: *Dieu sinistre, effrayant, impossible*. The word *effrayant* is clearly borrowed from Pascal, from his famous sentence on space: "le silence éternel de ces espaces infinis m'effraye."

So we have here a one-faced monster-God: time. And not only the Clock is the monster: – it is the time of mathematics, and the time of history too. Before offering positive examples, I shall mention the decisive negative one. To my knowledge there is no poem, novel, drama or any piece of art written from the end of nineteenth century onwards, which glorifies, praises, celebrates time. On the other hand a shiver, a shudder, an "effroi", a horror, a dread, a terror is present, when Monster-Time is mentioned.

In previous ages – starting with Hesiod – you may find symbols, where time's wheel is rotated by the Goddess *Diké*, by Truth and Justice. Tragedy and philosophy, whenever they saw *time out of joint*, rushed to repair it. What has really changed in our century is that the work or attempt of reparation is thought hopeless. Time was for Camus "the cruel mathematics that condemned our condition"; time's embodiment, History was for James Joyce a "nightmare from which I am trying to awake" and even in the song of the Grasshopper in "Finnegan's Wake" you find the line "But Holy Saltmarine, why can't you beat time?"

This vision of time destroys even the possibility of conquering a time-image in a coherent pattern of an ordered set of values. A rearrangement, if not in the pattern at least in the values, has affected tragedy. The same set of events, the same course of action, evokes a totally different kind of reaction. What until now had a tragic impact provokes laughter. In terms of literary history, there is a change in outlook, consequently a change of roles between genres. In contemporary drama, black or dark comedy, the grotesque takes over the task of tragedy to express mankind's suffering.

This transformation of a genre is aimed at consciously by some modern authors. Friedrich Dürrenmatt made a sparkling remark on the hidden parallels between his play *The Physicists* and that of the *Oedipus Rex*. He assures us that the plots of the two plays are the same. Möbius as well as Oedipus tries to avoid all the prophesied traps of life but both of them are falling into precisely the traps they were so eager to avoid. Here there is a striking similarity: the skeletons of the two plays are almost identical. But there is a difference – apart from the artistic merit: namely, *the ancient play is a tragedy whereas the modern one is a comedy*. Oedipus is a hero, Möbius is not. The first has grandeur and his sacrifice is justified – the place of his death, as Sophocles carefully mentions, will be worshipped forever. No such dignified end is reserved for Möbius. Although he has our compassion, he is rather a character who falls flat on his face. There is no future perspective in his fall, he cannot reach even the relieving moment of death. His present is desolation and not a link between past and future. His former life was a series of false predictions, his present is in any asylum – and there is no extension here to the future. What he really lacks is the final justification of his effort to avoid the unavoidable because there is a severed time-aspect in the outlook of the author: that of the future.

Tragedy on the other hand implies the aspect of future. You can endure its *now* of blood, death and peril because there is a glimpse of a final equilibrium. There is no Hamlet without Fortinbras, no Richard III without Richmond. Beside the dark Dionysian imagery of destruction there is an Apollonian radiance. This radiance shines from the cathartic perception of a higher order of things which has a hidden justification in its unfolding necessity.

But, by virtue of its existence, comedy denies *necessity*. The deeper the conflict here, the easier the solution, the greater the *Spielraum* of chance. In tragedy chance reveals the hidden force of necessity. In comedy the only necessity is that of another chance.

Applying this to the modern transformation of dramatic form: modern drama uses the blind mechanism of chance to express visions of tragic situations, physical and mental disintegration. That is the experiment of Dürrenmatt, transforming the Oedipus story into a comedy, depriving man's destiny of its justification. Möbius is not falling into the trap because he wanted to avoid it, as in the case of Oedipus. He would fall *anyway*.

This is a conscious effort of Dürrenmatt's, among others. In his *Theaterprobleme* he states explicitly that our age is not that of fate, destiny, grandeur. Its tragedies resemble *traffic accidents,* where the victim is usually chosen by chance. And where we have traffic accident instead of provoked fate of challenged destiny, we have *comedy instead of tragedy.*

I could perhaps have shown in other authors as well as in Dürrenmatt that chance without necessity is somehow a common denominator of an agreed vision. Tennessee Williams, commercial as ever, in the *Sweet Bird of Youth* even calls one of his characters *Chance*. This character confesses to hearing the ticking of a clock which is being attached to a dynamite construction that blows up the world.

We are back again to the point. The blown-up world where there is no human future seems to be the dominant image of drama nowadays. If in the former structures of tragedy the hero perished and the world survived, today the hero perishes only along with the whole world. Under this constant shadow of desolation, a dark comedy displays the life of the hopeless victims of time and chance.

I am not going to depreciate the artistic merit of all those modern dramatists who tried to face the problem of a futureless mankind. Samuel Beckett, for instance, draws most sharply the consequences of this *Weltanschauung* and offers a strong compassion, even love, in his desolated universe. His *dramatis personae* are sitting on some randomly chosen world-points in a Minkowski diagram, forgetting to communicate clearly. They are never entering into reasonable action, but being forlorn in the doomed timelessness of their present severed not only from its future, but consequently from the benefits of the memory of the past.

This bleak world cannot link any past to any future but it does something strikingly new: it links tragedy and comedy in a new manner. And since every new artistic experience has its time aspect, it works out a curious link between *present and eternity.*

In one of Beckett's plays there is a dialogue: "Have you not tormented me with your accursed time!" The reply to which is: "– When! When? One day, is that not enough for you?" The second sentence is that of a consolation. *One day* represents all eternity where eternal doom is just one day in a nightmare of continuous suffering.

This new genre of literature crushes all the means of the former tragedy as well as comedy. The plot is no longer a span between a *terminus a quo* to a *terminus ad quem;* the dialogue is nothing more than a sophisticated arrangement of two parallel-running monologues; action does not play a role any more than the equally monotonous unchanging scenery of environment.

Is this gloomy picture never cleared? Or: by what means is it cleared? By the very suggestion that our everyday life is equal to all eternity. It is cleared with its cry – it never offers a false consolation of a promising future. It is a discovery of our real here and now, which is ultimate, in a "condition humaine", *a destiny without destination.*

If ancient tragedy was a form for depicting heroes, these tragi-comedies are dedicated with a heart full of compassion to the victims. Inclined to laugh at the clown, your laughter is frozen – you suddenly discover that in an archetypal situation of comedy, there is the defeated hero, man, who in his quest for life could not conquer his deadliest enemy: time.

What dying Hotspur said becomes true in the black comedies of Ionesco, Beckett, Pinter or Dürrenmatt. With them, all those Shakespearian lines are explored to the bitter end. "That life's time's fool, and time that takes survey of all the world, must have a stop."

Lacking the force of Shakespearian poetry, a very prosaic conclusion must do here. Contemporary dramatic art – in a reaction to history on the one hand and scientific optimism on the other – is trying to find its own time-redeeming element. In earlier centuries this effort was realised in imposing a coherent pattern on events, which never denied the price of conquering time, but on the contrary, reenacted fight, death and desolation in a ritual which finally elevated their grasp on the human soul by means of catharsis, with a restoration of the eternal order of things. Modern drama's most vigorous trend tries to do the opposite. It shapes a final, *immanent* present where you live and die. It transforms every single day into a doomsday and every minute reflects the hopelessness of eternity. It takes us out of time, evades the Monster and conducts us into the motionless *now* where we are all time's fools and *time must have a stop.*

To pass a judgment on this trend or movement would be easy if we could find some counter-examples at hand of beautifully flourishing tragedies. But even such a dogmatically optimistic author as Brecht dismisses tragedy as inadequate to demonstrate our belief in the future. Personal sacrifice, effort and fight no longer promise a reasonable hope, not even to achieve the integrity of personality. As Dürrenmatt puts it, we do not even have personal responsibilities, since our greatest crimes are not personal either.

I believe that this gloomy picture never sought its justification in the advance of science. It rather casts a view on the events of history, on contemporary events. And this is the explanation why its creators lost faith and expectation: they simply gave up the hope to conquer time. T. S. Eliot has an enigmatic line to challenge this aspect, and the only consolation I can offer is to echo his conviction. However gloomy are the chances of overcoming or defeating this monster-time, "only through time, time is conquered."

# The Study of Time

J. T. FRASER*

*Introduction*

This essay is a retrospective look at the proceedings in the context of the large-scale questions which prompted the whole venture.

That a field of knowledge concerned with time should be developed has been suggested by several contemporary thinkers. Prior work in exploring the possible content, limits, and methodology of a study of time has revealed the necessity of a multi-disciplinary approach.[1] But one may hope for coherence in such an approach only if there exists some explicit theory to hold together the variety of views. In turn, such a theory can be arrived at only upon agreement on certain precepts. Two working assumptions which I regarded as necessary and sufficient for interdisciplinary arguments I have called "the principle of unity of time"[2] are these: 1. when specialists speak of time they speak of various aspects of the same entity; 2. this entity is amenable to study by the methods of the sciences; it can be made a meaningful subject of contemplation by the reflective mind, and it can be used as proper material for intuitive interpretation by the creative artist.

The principle of the unity of time received its first critical evaluation when Haber wrote in a review of *The Voices of Time* that

> "When used as an entity or force in the universe, time may very well prove to be a reified concept. Moving from discipline to discipline, the conviction grows that the actual use of the idea of time is to express a function or measurement of the processes under study, even though the writer may believe in the existence of an entity of time. However, even if the idea of time is an operational concept rather than a real structure in the universe, it is still a meaningful focus for research. In fact, a part of that research should be ascertaining whether it makes any difference to believe in the unity of time."[3]

---

* Dr. J. T. Fraser, P.O. Box 164 Pleasantville, N.Y. 10570, USA.

[1] For a summary of the most notable suggestions in this respect see *The Voices of Time* (ed. by J. T. Fraser). New York: Braziller 1966, p. 597, Note 5. This work will be abbreviated in subsequent references as *The Voices*.

[2] *The Voices*, p. xxi. This "unity of time" is to be distinguished from that of Aristotle, meaning the identity of dramatic and real time and that of Heidegger, meaning the coexistence of permanence and change.

[3] *On the Unity of Time* essay review in Science 152 (1966) 632.

The formal writing of an interdisciplinary work is not easy. Whitrow's pioneering book *The Natural Philosophy of Time*[4] offered unity and coherence that is possible only in a work which is the result of the labors of a single, well informed scholar. *The Voices of Time* while it could afford to deal with matters which can be best handled by specialists, had to derive its coherence from editorial balancing and from connecting expositions. In contrast to both, the present volume is a record of a conference. As such, it is the latest in a series of proceedings of earlier symposia on the same subject, held by various groups at different places: *Interdisciplinary Perspectives of Time* (Roland Fischer, ed. 1967); *Das Zeitproblem im 20. Jahrhundert* (R. W. Meyer, ed. 1964); *Man and Time* (J. Campbell, ed. 1957); *The Problem of Time* (U. Calif., Philosophical Union, 1935) and possibly others. However, since this is the proceedings of a Society specifically formed to encourage multidisciplinary exchange concerning time, *The Study of Time v. 1.* is something of a test of the viability of the principle of the unity of time. Namely, based on papers in which the authors followed their own guiding lights only and abided by the rules and methods of argumentation characteristic of their fields of learning, can we discern a useful pattern for a possible, future integration of our knowledge of time?

The answer to this question does not reside in the material as such; it will come from the reasoned reactions of the educated reader who may or may not perceive new challenges to his own scholarly domain.

In Part I of my paper I have selected seven of the many possible topics which are in need of multidisciplinary exploration. In these I have attempted to specify the place of the pertinent papers in the whole complex of questions, with the following caveat: what in my dealings with these topics emerge as relevant are *not* necessarily the chief points that the various authors had intended to make. Each paper is a contribution in its own right and must stand on its own ground independently of any other essay in this volume.[5] In Part II, I attempted to take stock of the study of time viewed as an intellectual enterprise, a field of knowing in its own right.

In spite of the universal role that time plays in thought and in daily life, a quidditic definition of it is difficult. Yet, the concept does display a unity sufficient for interdisciplinary studies because in all contexts of behavior it is associated with certain experiential qualities. Above all, time is a subtle but pervasive ordering principle employed in the establishment of personal and communal identities. Time conjoins the ideas of life and purpose; it links the individual to the astronomical universe; it forms the substratum of the great

---

4 G. J. Whitrow: *The Natural Philosophy of Time*. London: Nelson 1961. This work will be abbreviated in subsequent references as *Nat. Phil. Time*.

5 Proper names without identification by footnoting, and often in parentheses, are those of our Authors and stand as references to articles herein included. Other footnotes and references are handled in a conventional manner.

continuities of civilizations: magic, religion, the sciences and the arts; and it subsumes certain paradigms of tension, such as free will and determinism, contingency and necessity, being and becoming, permanence and change. These existential qualities will appear in many forms as we briefly consider a few topics in the study of time.

I. *Selected Topics in the Interdisciplinary Study of Time*

1. The Space-Time Syndrome of Physics

The concept of velocity understood as the ratio distance/time, is an integration of space with time: the metrizeable variables of distance and time are united in velocity. That this union is not at all self-evident was shown by Piaget[6] when he demonstrated that ideas of space and of time arise out of the child's perception of motion, not that the idea of motion is synthesized from prior perception of distance and time.

In the world of Newtonian kinematics, which is the study of motion in terms of velocities, the answers to the inquiries "When is then?" and "Where is there?" are easily found because of the assumption of a framework of absolute rest and a scaffolding of absolute time to which all events may be referred. In Relativity Theory answers to the same questions are, however, complex and embody two major discoveries: 1. the existence of distant space-time domains of unverifiable simultaneity to the here-now and 2. the dependence of clock readings on the relative motion (North) and gravitational environment of the observer.

These restraints, or limits of time do not arise from the space-time representation of motion even though such a representation is essential to the language of that theory. The limitations arise, instead, when the Principle of Relativity and that of the finitude and constancy of light velocity are inscribed into four-space, using Einstein's instructions on the calibration of distant clocks.

The visual model most often used for the demonstration of relativistic kinematics is not even a four but a three-dimensional construct, the Minkowski diagram. This is a geometrical representation of "plane-time" and not space-time, which does nevertheless yield generalizations valid for four-space. An indirect warning against taking the Minkowski diagram too seriously comes from McVittie when he notes that coordinate systems are to be regarded as labeling of events and that they can only be selected a posteriori by appeal to prior observation. That is, physical models of space-time are to be devised to make our experience of watching motion amenable to mathematical formulation, without any demand for philosophical insight.

Geometrical representations can accommodate but cannot account for the creative and unpredictable element of experience, hence they cannot represent

---

6 *The Voices,* p. 202 ff.

the totality of time. If one wishes to identify the time axis of four-space with time experienced, difficulties arise: time must be imagined as crawling along that axis in again another time dimension (Park) or one may be forced to seek in the mathematically imaginary nature of that axis some understanding of our perceptual apparatus (Dobbs).

Park believes that the concept of becoming cannot be geometrized. Similar convictions echo from Meredith's "local thawing out of the frozen framework of the Minkowski continuum" and Watanabe's "temporally telescoped Minkowski" in that both of these are attempts to join the creative or contingent element of experience to the deterministic, stationary aspects of time. Watanabe stresses the necessity when interpreting time in physics, of distinguishing between a world to be acted upon, one to which the observer may apply his sense of time and a world to be contemplated, one which presumably includes the observer. The Minkowski representation, in his judgment, insofar as it is assumed to include everything, belongs to the world to be contemplated.

Philosophical views of time which derive entirely from Relativity Theory may be called the "space-time syndrome" of physics. They are a group of expressions of scientific idealism[7] impressive in their abstractness but rather remote from experience. They exemplify mankind's dream of escaping change (Brandon) by seeking the permanence of lawfulness. Emphasis on the geometrical or static aspects of time is thus seen as another instant in the long search for constancy already represented by the earliest artifacts for magic means of controlling the future, by the immutability of the Platonic Forms, by the Scholastic concern for unchanging universals, by the idea of Providence in history and by the very idea of Laws of Nature.

The strength of mathematized science resides, however, not simply in the discovery of unchanging features in the world but by the pragmatic combination of being-like statements with contingencies in form of variables or boundary conditions. Thus, a complete representation of time can derive from the Minkowski diagram only if the experimenter who places the beginnings and ends of causal chains into the diagram is also included as complementary to the diagram.

A primitive temporality which did not exist in Newtonian physics may be found in the relativistic view of the world: the finite propagation velocity of information guarantees an asymmetry between future and past (Watanabe). Certain events in the future of here-now, even in a deterministic world, remain unpredictable because their causal origins cannot be reached from the "I-here-now". The asymmetry resides in that there is no analogous indeterminacy for events in the past of the here-now. In this asymmetry we encounter indeter-

---

[7] On this see K. GÖDEL: *A Remark about the Relationship Between Relativity Theory and Idealistic Philosophy*. In: *Albert Einstein: Philosopher–Scientist*, ed. by P. A. Schilpp. New York: Tudor 1949, p. 557.

minism changing into determinism in the instant of the present and observe with Eva Cassirer that in the "now" determinism and contingency, being and becoming cannot be told apart.[8]

The abstract beauty and technical usefulness of associating succession in time with points along a line and the power of mathematized physics employing this static image led some philosophers of science to the view that the idea of time is "mind-dependent" and not a feature of the world that includes man.[9] In one sense this view is trivial because all utterances are mind-dependent; in another sense, it is thought-provoking but irresponsible unless one can show how we are fooled into believing in the inexorability of the passing years. Hund suspects that temporality originates from sources more fundamental than can be accommodated in static imagery. If that is the case, then the contribution of space-time physics to the study of time cannot be evaluated from knowledge of physics alone but must be interpreted as part of the general flow of thought about time in intellectual history.

Yet the question remains: why do our static constructs (mathematical and geometrical laws) correspond with such stunning accuracy to the way we perceive nature? Eugene Wigner brought this into new focus in his Nobel lecture when he observed that "the surprising discovery of Newton's age is just the clear separation of laws of nature, on the one hand, and initial conditions, on the other. The former are precise beyond anything reasonable; we know virtually nothing about the latter."[10]

Perhaps today, through an interdisciplinary approach, we can think of the partial correspondence between the time of space-time and some of our experiences of time as having its origins in the evolutionary interaction between the organism and the environment, and seek the opinion of psychologists, anthropologists and sociologists with training in philosophy and in natural science. Meredith has been addressing himself to questions such as the one just implied; David Bohm made a brave attempt when he wrote about physics and perception in the context of the invariance requirements of Special Relativity Theory.[11]

In his monumental work, J. D. Bernal[12] has attempted to document how scientific views are shaped by social necessities and economic forces; neo-Freudians have traced the origins of the sciences to the externalization of certain instincts; social psychologists would seek the cultural determinants of

---

8 Cf. a discussion on the asymmetry in the physical determinism of the here-then. In: J. T. Fraser: *Time as a Hierarchy of Creative Conflicts.* Studium Generale 23 (1970) 620 ff. This work will be abbreviated in subsequent references as *Time as Conflict.*

9 See e. g. A. Grünbaum: *The Status of Temporal Becoming.* In: *The Philosophy of Time* (ed. by R. M. Gale). New York: Doubleday 1967, p. 322.

10 E. Wigner: *Symmetries and Reflections.* Bloomington: Ind. Indiana U. Press 1967, p. 40.

11 D. Bohm: *The Special Theory of Relativity.* New York: Benjamin 1965, p. 185 ff.

12 J. D. Bernal: *Science in History.* 4 volumes. Cambridge: M. I. T. Press 1970.

scientific theories in general while idealist thinkers might hold that physics can be derived from cerebration, without recourse to experiment.

It is clear that scholars of various training and persuasion would tend to interpret the correspondence between our experience of time and the time of space-time on the basis of varied truths, outside the findings of physics.

Mircea Eliade had remarked that in order to be delivered from the threat of passing, the believer in Indian mythology must accept the ontological unreality of time.[13] In an analogous fashion, regarding the ontological status of becoming as unreal makes it possible for the philosopher of science to enter the security of predictability through mathematical physics. Also, according to Eliade, myths are outside time, in eternity and are, for that reason, sacred. In my view, mathematical time may also be described as outside time for it is reversible, hence sacred, in contrast with the profane and irreversible time of existential knowledge. To the primitive mind myths are true because they are sacred; in much of the philosophy of physics the reversibility of time is sacred because it seems to derive from laws that are scientifically true.

I suspect that the question is this: Is the space-time syndrome a social myth? More specifically, is the belief, that the success of relativistic kinematics supports an idealistic view of time a myth, that thrives on the general fascination with quantity and technology? At this writing I know of no generally accepted answer to this question.

## 2. Physical Correlates of Time

Although the view that the passage of time is virtual has been maintained by many philosophers of physics,[14] others considered several possible physical correlates: 1. the Second Law of Thermodynamics namely, that available energy of a closed system would, on the average, always decrease (Costa de Beauregard, Denbigh, Landsberg, Park, Taylor and others passim); 2. interaction of the quantum systems with the observer (Denbigh, Taylor, Watanabe) and 3. the monotonic expansion of the universe.[14a] Various combinations of these three sources are held responsible for certain properties of temporality of which Bunge distinguishes three: time reversal (change of sign of "t" in equations); irreversibility (a property of a process, not that of time); and anisotropy (that durations are time-like intervals). Watanabe adds further warnings against confusing reversibility with retrodictability and irreversibility with non-retrodictability and develops his thesis accordingly while Costa de Beauregard prefers to speak of fact-like features (irreversibility) and law-like features (equations of physics).

---

[13] For references to Mircea Eliade in this paragraph see his *Time and Eternity in Indian Thought*. In: *Man and Time* (ed. by J. Campbell). New York: Pantheon Books 1957, p. 173.
[14] Grünbaum, op. cit. and O. Costa de Beauregard in *The Voices*, p. 417.
[14a] *Nat. Phil. Time*, 237 ff.

There is a dire need to clear up these concepts. It seems to me that irreversibility, anisotropy, time reversal or one-wayness of time are useful concepts only if applied to processes in time otherwise they amount to a *petitio principii*. Furthermore, anisotropy of time would be interesting only if time could also be regarded as sometimes isotropic, but this would be a contradiction in terms; that time, rather than a process in time is irreversible would be interesting if time could at least in principle be regarded as reversible, but there has been no such record in the history of man. The whole confusion is placed into perspective by Denbigh in the first half of his fine contribution.

When the career of thermodynamics began with the work of Sadi Carnot early in the 19th century, the view in mechanics was already entrenched in science that complete and ideal reversibility of processes is to be expected, hence, opinions of the unreality of time prevailed (Denbigh), illustrating the close connection between methodologies and philosophies. Working at physics includes idealizations and abstractions necessary for formulating laws of nature but what is legitimately negligible in solving physical problems is often quite essential to conceptual analysis. For instance, the validity of the Second Law of Thermodynamics demands isolated systems, yet no system totally isolated from the rest of the world can be considered seriously when one discusses time (Hund). This thought resembles the usual Mach Principle interpretation of inertia, which is understood as meaning that any reference to motion is but an abbreviated reference to the universe at large. The similarity of arguments brings to mind the difficulties of separating the concepts of time and motion.

Since both the microscopic and most of the macroscopic physical laws are formally timeless that is, physical development as represented by the equations of physics, is independent of the mathematical sign of the symbol "t", one may ask where and how is physics informed of the passing of time? Denbigh points to some conceptual subtleties in formulating our statements about physical experiments: observing anything excludes unobserving, putting together two bodies assumes that they are not taken apart. Watanabe seeks the sources of time in the way experiments are set up and interpreted. He believes that temporal development in physics derives from irretrodictability (not from irreversibility) which, in its turn, originates in the very notion of conditional probability used in the statistical descriptions, therefore, is introduced by the observer. As he sees it, human freedom of action is assumed in physics when one is permitted to establish prior conditions which lead to entropy increasing situations in terms of our prior knowledge of time.

The literature of time in physics stresses that our feeling of the passing of time coincides with the direction in which entropy of closed thermodynamic systems increases. But entropy can also be defined for open systems exemplified by living organisms or structures or behavior, systems which are believed to demand, on the average, a monotonic decrease in entropy if they are to function. It follows that association of time with increasing entropy can be justified only by the preponderance of non-self-organizing material. If our thinking were fundamentally organic rather than physical, we would associate time with the decreasing entropy of self-organizing systems. It seems, therefore, that time does not enter physics through the content of the Second Law of Thermodynamics.

A curious parallel may be drawn between the timelessness of the pre-Columbian Mayas (Whitrow) and that of the statistical interpretation of time in thermodynamics.

Although remarkably accurate, the periodicity embedded in the Maya imagery of calendrical time was so profound that in their views past and future events coalesced. History was searched for conditions preceding important events because the Maya priests held that if certain initial conditions prevailed, everything would start anew. Thus the memory of obscure pirates that hid out in northeastern Yukatan led to the identification of the Roman Catholic church with the Itza invaders, and of Protestantism with the old Maya religion. Bearded Spaniards were Toltec invaders, because their leaders also were bearded.[15]

This argument is isomorphic with the claim of Zermelo that since molecular processes should be cyclical and any initial configuration of particles in a closed box totally unconnected to the environment shall repeat itself, the statistical interpretation of entropy cannot be valid.[16]

When one cannot tell the individual Itza or individual molecules apart, if one admits the existence of strictly deterministic laws and if one refuses to regard the plurality of the single particles as a part of a larger universe, then the methods available to describe processes will obviously show a deterministic and timeless world. The S-matrix theory (Taylor) is of this nature. It describes relations between particles in the infinite future and past, leaving the intermediate, finite times inaccessible. But infinite past or future, like the religious and philosophical concepts of eternity, are essentially timeless. Whether the thermodynamics of finite systems, intermediate between the timelessness of eternity and that of single particle behavior can say anything about time remains to be seen.

It seems that temporality enters into microphysics by some more fundamental ways than a specific theory. I would like to mention two possible avenues: the employment of concepts embodying the one-and-the-many, and the construction of the "now" in microphysics.

The idea that the one, or identity, is essentially timeless, that it stands for permanence and is thus a being-like concept, has been amply debated by monists, dualists and pluralists. Plurality, on the other hand, whether in the sense of

---

15 J. E. S. Thompson: *The Rise and Fall of Maya Civilization*. Norman: U. Oklahoma Press 1966, p. 167.
16 *Nat. Phil. Time*, p. 279.

many things or many kinds of things, must postulate true change, because any configuration of things permanent through time would only constitute an identity, or unchanging one-ness. The many, therefore, is a becoming-like concept. The thought of an aggregate of elementary particles already combines in a rudimentary fashion being and becoming in time.

This antinomy may also be found embedded in the "now" of microphysical events. Because of their probabilistic nature, statistical theories cannot account for actual happenings but permit only potentialities; an event remains a surprise occurrence in a background of strict lawfulness. The physical "now" is seen, therefore, as the coexistence of "being-like" lawfulness and "becoming-like" unpredictability. Clockness itself may be defined on this basis: a process that exhibits identifiable, identical states of change according to a law of nature regarded as permanent.[17]

As I see it, the type of inquiry characteristic of the study of time in quantum theory, statistics and thermodynamics gives only some hints but does not reveal the sources of our sense of time, or even the means whereby our sense of time may derive or relate to some natural processes external to man. The contribution of quantum physical as well as relativistic approaches resides, rather, in the discovery of certain limitations to temporal processes heretofore unimagined and unexpected. Any universal theory of time, if it is to be taken seriously, must account for or at least accomodate these limitations. But ontological issues regarding time still cannot depend on epistemological questions about the verifiability of specific physical laws that do or do not involve time because this would amount to a metaphysical assertion that only those portions of temporal experience are valid, which can be dealt with within the limited methodologies of physical science.[18] On the contrary, in order to better understand such problems as the time vs. space-time syndrome of Relativity Theory, the relationship of the knower and the known in quantum physics, and other questions that relate to problems for which time-timeless relationships is essential, we must turn to an understanding of reality through ways other than the physical sciences.

Writing by the Yalu River in 1895, the French philosopher and poet, Paul Valéry, recorded an imaginary dialogue with a Buddhist sage. The monk reflected on Western ways of doing things: "You have neither the patience that weaves long lines, nor the feeling for the irregular... You are in love with intelligence until it frightens you... rage with desire for what is immediate and destroy your fathers and sons together."

These remarks apply uncomfortably well to most philosophical views of time which derive their essence from thermodynamics, quantum theory and relativity

---

17 *Time as a Hierarchy*... 607 ff., 688 ff.
18 This view is at variance with that taken by Henryk Mehlberg in: *Philosophical Aspects of Physical Time*. The Monist 53, 3 (1969) 340.

theory. The trend of what has been called the "elimination of time" in physics [19] that is the insistence that temporal flux is not an intrinsic feature of the world, seems always to be associated with a Platonic dislike of the irregular and the contingent. Although one may argue that the inveterate search for symmetry, stress on the noetic and the neglect of values all stem from the demands of the scientific method, it is important to keep in mind that the very same demands make physically based views of time unlikely sources for a universal theory of time and existence.

## 3. Time vs. Timelessness

This section deals with three interrelated problems: endings and beginnings; finitude and infinity, and the continuity vs. atomicity of time. What these problems have in common is that the failure of solving them comes from our inability to reach an acceptable interpretation of the time/timeless relationship. The epistemological assumption here is, of course, that a time/timeless relation does exist and that it is knowable. First I intend to show or at least imply that the three problem areas do depend on understanding time vs. timelessness, then I will speculate on possible ways that might lead to such an understanding.

The very essence of time, arguments of reversibility and virtuality notwithstanding, prohibits "time travel" (Park).[20] Instead, we must retrodict and predict past and future events through mental efforts in the present. This is generally done by applying to a law (a statement regarded as unchanging) certain findings or guesses independent of the truth of that law. The two sets of information together may then be used to predict or retrodict specific events. In a formal sense, using certain laws of physics, one may arrive at ideas of a timeless world if conditions can be found such that for them an operational time parameter cannot be designed. Examples are a primordial world of pure radiation or a world of "heat death" consisting only of ponderable mass.[21] Whether or not there is a general physical trend in the universe from a primordial to a final state of (operationally defineable) timelessness, can conceivably be scientifically investigated and eventually experimentally determined.

Yet, whatever forms such hypothetical proofs may take, perhaps photographs of the red shift of quasars, they will not assist us in answering the Augustinian question: What was before the beginning and what will be after the end of time? If we are to integrate the idea of a formally timeless universe into our experiential knowledge of time and existence, we must first make sense of the time/timeless interface.

---

19 *Nat. Phil. Time*, p. 1.

20 The impossibility of "time travel" suggests some fundamental connections between free will and time. *Time as Conflict*, p. 676 Note 116.

21 Regarding the former (a world of pure radiation) see J. T. Fraser: *The Interdisciplinary Study of Time*. In: *Interdisciplinary Perspectives of Time* (ed. by Roland Fischer). New York: N.Y. Acad. Sci., Annals, v. 138, Art. 2 (1967), p. 837.

It does not matter whether we speak about the beginning and end of the universe, with time "inside" as it were and timelessness outside, or about a chronon with time "outside" and timelessness "inside". The stumbling block is still the reconciliation of the timeless with the temporal. Čapek suspects that microphysical time has the same structure as the qualitative, temporal nature of experience. If this be the case, our problems of macroscopic time will be re-encountered in the microscopic world. In contrast to this view, Taylor believes that the microscopic properties of space and time will be found distinct but also different from the macroscopic properties of space and time. If this be the case, we must await the discovery of some conceptual transformation that will permit us to find continuity between our problems of time/timelessness and whatever corresponds to this problem in the microscopic world.

A thought experiment that illustrates the question of time/timelessness is Whitrow's ingenious paradox of the bouncing ball[22a], something of a Zeno's arrow with constant deceleration. As in the usual mathematical treatment of Achilles and the tortoise, the employment of finite limits of infinite series leads to a prediction that corresponds to experience; in our sense impression as well as in the mathematical imagery the bouncing ball does come to rest. However, as Whitrow demonstrates, taking the idealization as though it truly corresponds to physical nature leads to a paradox: the ball, according to mathematical description, both does and does not bounce an infinite number of times before it stops. The resolution of this paradox demands that we distinguish between limitations to natural processes and conceptual difficulties which derive from the simulacra (equations or pictures) that are used to describe such processes. The possible existence of minimal times in physics, or the quantization of perception (Efron, Pöppel) do not inform us of the way we ought to think about time and timelessness.

The question of finitude vs. infinity of time masks identical difficulties. Linguistic studies suggest that our thought-forming faculties associate the idea of infinite time with lawfulness and regularity, while finite time is associated with unique opportunity.

In English *always* translates time into space through the picture *all ways*, in complete analogy to the geometrical image of an infinite (spatial) extension.[22] Likewise, in Old English it was *allne waeg*, in Luther's German, *allewege*. Greek and Latin show a different approach; the formulators of those languages regarded continuity as embodied in the power of procreation. The Greek *aion* meant primarily "spinal marrow", the seat of generative power, a connotation which survives in everyday and in poetic usage.

---

22 Th. Thass-Tienemann: *The Subconscious Language*. New York: Washington Square Press 1967, p. 191.
22a G. J. Whitrow: *Time and Cosmical Physics*. Stud. Gen. 23 (1970) 231.

The Latin *aevum* unites the meanings of "eternity", "age", and "generation". The Old English *aew, awa*, and *ae* (preserved in "ever") meant "eternity" and "law" when used as nouns and "ever" as adverbs; the same root is preserved in the German "eternal", *ewig*. It is as though something in our articulating and thought-forming faculties associates infinity and eternity with lawfulness, that is, permanence.

Finitude also has its psycholinguistic counterpart in that meaning of time which identifies it with "opportunity", that is, something unique.[23] The Greek *kairos* means not only our "time" but also "temple", that is, the specific spot where the weapon can best penetrate the body to "cut off" the life time. We still "cut time" in English. The Latin *tempus* is not only "time" but "opportunity", as is the idea of time in Buddhism as the unexpected, the unique condition.

These observations suggest that beneath the debate of finite vs. infinite time there might be a deeper one: the interpretation of time vs. timelessness. It is my belief that if the time/timeless relation would be better understood, a resolution of this old and complex problem would also follow.

That continuity vs. atomicity also demand a prior understanding of time/timelessness has already been implied. Namely, the difficulty in incorporating the idea of minimal time intervals into the experiential and mathematically useful idea of continuous time resides in understanding how time may be constructed from elements of no-time. As mentioned above, a perceptual event of timelessness in a world of time is but an inside-out temporal universe in a timeless eternity.

Understanding time vs. timelessness is hampered by our heritage of spatial imagery of time in language. A line which ends can always be said to be adjacent to a non-line, hence the view arises that time and timelessness form a proper disjunction of mutually exclusive alternatives. I hold this to be incorrect and want to suggest that time/timelessness bear a hierarchical relationship on a scale of complexity of perception, somewhat as maturity bears to infancy. As infancy is a more primitive state than is maturity yet is nevertheless necessary before maturity is reached, so the atemporal or timeless description of the world shows access to existents more primitive than what we reach through our knowledge of time. It has been held that solutions to philosophical problems must be reached through the methods of philosophy, and that logical or mathematical interpretations of time should remain within the confines of logic and mathematics. But, the study of time is not within the confines of any of these ways of knowing, it calls instead for interdisciplinary understanding. An evaluation of the sources of the time/timeless relationship is most likely to come from behavioral sciences and from linguistics, and not necessarily from philosophy or logic where it has been of interest.

---

23 Th. Thass-Tienemann, *Symbolic Behavior*. (manuscript, private communication.)

In ordinary use "timelessness" refers to something outside the category of time and thereby it would seem to run up against the same difficulties as does time itself. Yet the experience of timelessness displays some distinct modes which are simpler to deal with than the many gradations of time perception and the sense of time.

Timelessness can mean the absence of ordering according to some principle, such as is often the case in the manifest content of dreams. There are some reasons to believe that such an absence of ordering reflects a pre-causal state of perception, resembling that of the child as distinct from that of the adult; events in the world of the child, or in the world of primitive man, are often connected by magic causation rather than by what we recognize in the mature waking state as causality. Timelessness is also often connected with states of ecstasy associated with religious meditation, or with dance, or with the ingestion of certain chemicals. These and similar types of experiences of timelessness are believed to represent a regression to a relatively undifferentiated ego, manifested by a sense of omnipotence and a loss of individual identity. In contrast, understanding the world in terms of future, past and present is possible only by a knower of a well differentiated ego that is, by a person of determined identity.[24]

The sketch just given suggests the following tentative conclusion. The logical difficulties that confuse such issues as the endings and beginnings of time, finitude and infinity of time, or continuity vs. atomicity of time stem from taking time and timelessness as mutually exclusive alternatives. Clarification of the position of time and timelessness in our perceptual world is likely to reveal that the two concepts are not mutually exclusive alternatives. The necessity of totally revising our methods of asking questions about the beginning and end of time and of the other two problem areas mentioned, would follow.

## 4. The Ontology of Bioclocks

Throughout the career of organic evolution the lives of man, animals and plants have been submerged in an unceasing variation of light and dark and a vast variety of other regularities; cyclicity in the living is probably coeval with life itself. Cyclical changes in life were regarded as manifestations of being in civilizations that understood themselves as participants in vast and uncontrollable order, but have only been lately subjected to scientific inquiry.[25]

---

24 For a discussion on the experience of timelessness see *Time as Conflict* . . . p. 647, and p. 684 ff. Whether what psychologists regard as regression in the service of the ego is a defense mechanism or a return to a healthy state is a veritable dilemma of civilizations, conjoining psychology, philosophy, sociology, anthropology and the study of time.

25 G. G. Luce: *Biological Rhythms in Psychiatry and Medicine*. Chevy Chase. Maryland: National Institute of Mental Health, (1970) is a valuable summary of its subject.

The overwhelming universality of biological rhythm, the realization that synchronism between the organism and its environment is imperative for survival and the possibility that spatial information may derive from temporal encoding suggests that life might more properly be regarded as a program than a structure. An elegant and impressive example of how temporal information may serve as the source of structuring of organizing systems is given by Goodwin when he shows how finite propagation velocities of disturbances can establish embryological fields. If this be so, then it should not be surprising that the nervous system which even in the adult retains the plasticity and developmental capacity of the embryological process, forms the basic temporal control of the mature adult through the numerous clocks of the central nervous system (Gooddy).

The capacity for rhythmic behavior is demonstrably independent of the environment, that is, physiological clocks can appear endogenous. Yet their rhythms cannot be divorced from ontogenetic imprinting and from the genetic memory of natural rhythm. The content of a past controversy, the Brown-Pittendrigh debate concerned precisely this question: are biological clocks exogenous or endogenous?[26] The generally accepted opinion is that the ultimate controlling system of rhythmic organization will be found in the biological clock, but with the need for occasional phasing supplied from the outside. This "occasional phasing", of course, has been part of the evolution of the physiological clock.

Similar ambivalence appears in the status of bioclocks as they are, or might be involved in time perception. The cortical representation of time perception (Pöppel) stressed by those who hold that the sense of time derives from endogenous processes cannot be separated from the evolution of the cortex. Those who stress the exogenous basis of time perception and employ data processing and computer language (Michon) cannot produce a completely exogenous scheme except at the expense of a totally non-material perceiver, a living point as it were, without extension.

It seems to me, therefore, that the sources of physiological rhythm are to be sought not in the biological clock or in the environment alone as though these two were separable, but in a totality that includes both. In conclusion, then, I feel that a critical examination of the epistemological status of the physiological clock is very much needed.

5. Time and Organic Evolution

In contrast to physiological rhythm stands the process of evolution recognized in the emergence of new forms of life. Statements about apparent directedness

---

26 For a statement and discussion of the controversy see J. L. Cloudsley-Thompson: *Rhythmic Activity in Animal Physiology and Behavior*. New York: Academic Press 1961, p. 5 and *passim*.

or teleology in evolution can usually be translated to mean that something has come about within a period which is presumed not sufficiently long to produce the same thing randomly. The coming about of an elephant in a certain period of time is not less likely, however, than any other combination of the particles making up the elephant. Randomly produced elephants are theoretically not impossible but, as Watanabe stresses, if one assumes an agent, whether a divine being or a principle of science, then the coming about of certain ordering is so much more probable that all other schemes appear unlikely. The question then remains, what are the constraints imposed on matter which make evolution appear directed?

Maynard Smith discusses one such constraint when he describes mutations which give rise to viable proteins. In his opinion and in the opinion of others, the important constraints on living matter are not the physical ones. It has been suggested that there may be regularities which include the physico-chemical laws of the physical world as well as certain principles unique to life.[27] As far as evolution and time is concerned, physical constraints on temporality are valid but not sufficient for an understanding of life. For principles that appear to us as directedness we will have to look to life itself.

If one desires, it is possible to find many biological arrows of time. The "one-way flow of time" in the living can be attributed to the averaging processes connecting the molecular and microscopic worlds (Landsberg). One may see the sources of time in the accumulation of random errors such as in certain processes of aging. Or, the famous central dogma of molecular biology might be thought of as the origin of the biological one-wayness of time: "DNA makes RNA makes Protein" or Crick's claim that "once genetic information has passed into protein it cannot get out again". With a philosophical stance of seeking the sources of the passing of time in life rather than in the physical world or in the functioning of the mind, irreversible biological process may be thought of as responsible for our idea that the world exhibits a temporal trend. But the very plurality of irreversible processes makes the validity of the approach suspect.

I believe that the problem lies deeper than biological irreversibility; it seems to reside somewhere in our views of lawfulness in life. Writing about organic evolution, George Gaylord Simpson speaks of determinism which is "historical and not mechanistic" and which permits "multiple solutions and not only a unique outcome".[28] We are too ignorant about biological invariance and cannot separate it clearly from biological contingencies. We are familiar only with life as it appears on earth, hence our sampling is not a fair one. All these work against our ability to write formal laws for life resembling the equations of

---

27 I am referring to the debate of "biotonic laws", a neologism coined by W. M. Elsasser: *The Physical Foundations of Biology*. London: Pergamon Press 1958. For details and references see *Time as Conflict* ... p. 630–1.

28 G. G. Simpson: *This View of life*. New York: Harcourt, Brace and World, 1964, p. 189.

physics. Goodwin already pointed to the great difficulties in applying the entropy concept to organisms[29], and suggested a more tractable entity, a measure of temporal organization of events.

Whether or not the nature of life and the nature of the responses of living things to stimuli can ever be fully understood in terms of physics alone (Landsberg) is not at all certain. Causality will have to be re-examined, perhaps as Watanabe has done regarding the nature of conditional probability, or perhaps as Whyte has done in connection with structural relaxation. Methodologies might have to change. We are likely to need a better understanding of the structure of temporal information (Meredith) and a flexible, modal logic (Prior). These are only a few examples of the effects of the Darwinian theory of evolution on our methods of dealing with the concept of time. As the arguments of Haber so clearly reveal, these effects are profound and far reaching and the end of the Darwinian revolution in the concept of time is not yet in sight.

## 6. Time Perception and the Sense of Time

In this section I am suggesting that there are two distinct modes in man's dealings with time as seen in the psychology of time: his time perception and his sense of time. Though these two modes sometimes blend one into the other and therefore a sharp line between them cannot be drawn, I believe that they reflect two distinct levels of integration in the temporal manifestations of the mind.

Much of the experimental work in the psychology of time deals with the simpler of these two, the perception of time. Specifically, I am thinking of the large number of tests of judgments of sequence and duration as functions of experimental variables, and the attendant concern of constructing mathematical models which can accommodate these findings. These and similar inquiries may be described with fair accuracy as being studies in the behavioral functions of man as a clock. Investigation of the sources and the experience of timekeeping I would regard as belonging in the psychology of time perception.

Under the concept of time sense I would subsume those behavioral functions in which the symbolic transformation of experience is preeminent: long term expectation and memory[30] including awareness of the individual's inevitable death, organization of personality in respect to the passage of time, temporal organization of communication in language, and the communal structuring of time.

Even though time perception is probably an older attribute of the mind than is the sense of time, one may speak of time perception only if the sense of

---

29 B. C. GOODWIN: *Temporal Organization in Cells.* New York: Academic Press 1963.
30 It is generally accepted that the physiology of short term memory processes is different from those which bring about long term memory traces. On this see E. Roy John: *Mechanism of Memory.* New York: Academic Press 1967.

time is first assumed as a primary datum of awareness. Only in terms of before and after, and future, past and present, can one critically examine the dependence of judgments of sequence and duration on various experimental conditions. A minimal period of time perception appears, then, as an experience of simultaneity with a before and after but without a temporal character. Such a happening may be conveniently designated an *event*.

Events conceived of as atoms of invariance, or unchanging identity may be identified in physics and in physiology, not only in psychology.[31] Efron found a minimal, timeless happening in perceptual duration forming, as it does, the lower limit of man's usefulness as a clock. In the reductionist view of the mind that is, if one holds that the functions of the mind are reducible to physicochemical processes, one would tend to agree with Efron's interpretation that mental processes are thus fundamentally measurable. If this be so, the upper limit of the mind is also measurable though it is not a universal quantity: it is a person's lifetime. In both cases the timeless portions of the happenings are outside our experience: we cannot experience anything within the small period determined by Efron, nor can we experience anything in the periods which preceed our birth or follow our death.

Michon supports the idea of quantized time perception on theoretical grounds. He assumes that temporal and non-temporal aspects of experience (such as time vs. size) are handled on equal footing by man as a processing system. Since the rate of information processing must be finite, and if the postulate holds, then some trade-offs ought to be observable between temporal and spatial perception and, in principle, such tradeoffs might be quantizable.

Opposing, as I see it, both of these views, Hamblin concludes from logical analysis that our elementary, primitive experience of time cannot be of the nature of atemporal instants but must be intervals. An instant, in terms of the definition of an event just given, is an event selected for usefulness. An interval, however, is a sample of time.

It is interesting to reflect on the unstated assumptions about time employed in the psychology of time perception as they relate to philosophical and physical ideas of time. For instance, the two models of time perception proposed by Michon, the pulse counter vs. the information theory model, correspond to the philosophical concepts of constitutive vs. relational time, respectively. The pulse counter model is one of constitutive time; the clocks tick away independently of what is happening in time, except for the ticking of the clock. The information theory model represents relational time: events are ontologically prior to time. Beneath both theories, as probably beneath most theories of time perception, one finds the Newtonian absolute time flowing "equably" and independently of the physiological goings on in time.

---

31 I have attempted to derive a universal and empirical definition of event from the spectrum of simultaneities one finds in physics, biology and psychology. In: *Time as Conflict...* p. 617 and passim.

The literature of time in psychology and in physics carries many assertions about the physical reversibility and the mind-independent irreversibility of time. Meredith believes that the opposite is true: psychological time is reversible in that we may travel back and forth in our imagination, but physical time is irreversible. He points out that both in kinematics and in dynamics the observer introduces the unpredictable element. The actual history of things is determined by the unpredictable acts of creation and annihilation together with predictable, lawful possibilities. But if we wish to think of such concepts as reversibility, lawfulness and unpredictability, being and becoming, and of the social and historical role, then in our dealings with time our attention must be transferred from concern with time perception to concern with the sense of time, for only within the broad purview of time sense can we include these features of temporal behavior.

I would like to define "mental activity" as pertaining to all those forms of behavior, whether available only through privileged access or observable by others, which make possible the accumulation and the passing on of knowledge to the descendents and the contemporaries of an individual. The great social continuities, especially the sciences, consist of such learning accumulated through time, thus partially overcoming the mortality of the individual. In this scheme of concepts the mind is not a spatial structure but a process of symbolic transformation of experience. The psychology of time perception, then, does not seem to contribute directly to our understanding of the sense of time for it does not deal with the origins of man's need for the creation of lasting and communicable forms, or records of his experience. For information on the roots of these needs we must turn elsewhere, perhaps to theoretical psychoanalysis.

In Freudian and neo-Freudian thought, the source of the sense of time is seen in the unresolvable separation of experience into life and death. While on the biological level life and death are regarded as forming an organic unity, in a mature ego they are analyzed into conflicting opposites.[32] If a man could exist without repression, without acknowledging death, sacrifice his individuality and become an undifferentiated part of the environment, the conflict would then disappear together with a sense of time. This time-ignorant unity of life and death is known to children, archaic man, saints and occasionally to each of us.

There is no need, or even possibility to explore here the detailed implications of the proposition that there are two integrative levels in the temporal manifestations of the mind. Enough has been said, however, to suggest that such an exploration, reaching out beyond the traditional confines of the psychology of time into philosophy and the social sciences would be both interesting and profitable for the study of time.

---

[32] For a detailed exposition of these views see the chapter on *Death, Time and Eternity* and passim in Norman O. Brown: *Life against Death*. New York: Random House, 1959. Also, *Time as Conflict* ... pp. 643–7.

## 7. The Self and the Other

The transformation of knowledge into modes which outlast the individual is a form of externalizing of inner experience; analysis of the self-other duality thereby created shows time at the root of this process (Sebba).

Modern Western man's search for identity reveals the paradoxical coexistence of the trends of destruction and creation. Knowledge of self-identity is an expression of the knowledge of time, because identity implies constancy through time and can be understood only through the concepts of being and becoming. Sebba's three schemes concerning the definition of the self in terms of self/non-self disjunction, suggest an epistemology of exact and non-exact knowledge. The first method would be that of rational speculation, the mind removes itself from change and looks at itself as an object in time: this, I believe, is the approach of the exact sciences. The self may analyze the memory of past experience and, thus, perceive in time a continuous now: this is a form of historical knowledge; or the self might resort to noetic expressions in the arts and letters depicting the existential anxiety and tyranny of time: this is the element of tragedy in the many forms of art.

Insofar as time is the essential ingredient of the identity process of the individual and of the race, the question of the modern self may be analyzed in terms of its relation to passing. Erickson and others look at it from the point of view of psychology, Sebba from literature, with somewhat similar results.[33]

In the mind of the protagonist of Samuel Beckett's *Murphy*, the world can be made "reversible" as in the timeless existence of the child, or in various types of ecstasies or in some philosophies of science. But since in such a world the clock goes both ways, it is a timeless world; the object-subject conflict is gone and with it vanishes the basis of our sense of time. Without the conflicts time changes to a timeless eternity of an "unbounded yet finite" prison, a gnostic metaphor describing the four-dimensional universe. That somewhere in these many ways of knowledge there is a common trend toward timelessness is fairly clear.

The symbolic inquiry into the self which Sebba finds in the works of Descartes, Rousseau and Beckett exposes some contemporary difficulties. These men sensed what was to become later the source of the crisis of man in re-evaluating his relationship to his environment. For much of natural science, but especially for its applied form, industrial technology, the division of self and non-self is essential for methodological reasons; philosophically, however, one must remember that the world of science is Watanabe's world to be acted upon, and not the world to be contemplated. When engineers and scientists set out to create utopias based on abstract understanding, their ideas are apt to

---

[33] See e.g. the work of Erik H. Erikson, especially *Identity and the Life Cycle*, New York: International University Press 1967 and *Identity, Youth and Crisis*. New York: Norton 1968.

lead to the difficulties experienced by Rousseau. "A self that steps away from the community of beings and feels itself to be wholly other gains access to operationally powerful modes of knowing and loses access to vital modes of participation ... Ignorance of its true state give this self that complete detachment from which enormous, cognitional and transformatory powers spring ..." (Sebba).

For Descartes, the essence of man is thinking. For Freud it is a drive to master and control the environment and other people as an externalization of sexual frustration. To strive and know time is the extroversion of the death instinct. For Hegel, the awareness of time which is the consciousness of death transforms into struggle to rule the lives of other men, a view that can be traced in current interpretations of dialectical materialism.

The self then, seems to be definable only by its unresolvable conflict with the non-self as manifested by its desire to reject temporality and to become timeless through the creation of knowledge. This awesome struggle finds expression in tragedy (Ungvari). While tragedy reflects a feeling of transcendental justice and is informed of hidden necessities, comedy speaks only of probabilities. Therefore, by observing that what had been tragic now tends to provoke laughter, Ungvari senses that the unresolvable conflicts that give rise to the sense of time are now regarded as solvable (by science and by communal actions) and are replaced by a harmonious but untrue view of existence. The divine design of Heilsgeschichte has been replaced by probabilistic laws of aggregates of particles. The interest in the immediate, prophesized by Paul Valéry and acted out in the now-ness of our subcultures, follows.

Hegel warned of the dangers of quantified thinking about time when he emphatically rejected the idea that pure mathematics, or reason, or logic may somehow be equated to time which deals with the sheer restlessness in life and its inherent differentiation. Since mathematics, in Hegel's view, deals with quantity, it can only represent arrested, quantitative aspects of our experience of time. Knowing time through geometry "degrades what is self-moving to the level of mere matter in order, thus, to get an indifferent, external, lifeless content."[34]

For the modern self, time has lost its aspects of becoming. The fear of the unknown and the unknowable has been repressed or relegated to the world of taboo and consequently, time became a commodity and not an arena of fear and hope. Exceptions to this are some of the existential thinkers such as Berdyaev, for whom history is a continuous attempt to resolve the conflict between free will and lawfulness, and also the contemporary romantic youth movements evincing anti-Weber principles. Consistently, the fate of the hero is a very thankless one in a society without feeling for tragedy, the sacrifice for the

---

34 G. W. F. Hegel: *The Phenomenology of Mind*. New York: Humanities Press 1966, p. 104.

community being that of the soldier in a forgotten war. As Voegelin clearly sees it, problems of alienation follow.[35]

Brandon finds the sources of our unease in the replacement of a teleological, optimistic and progressive view of history by a physical evaluation of time. Voegelin traces this same malaise to the ennui as a reaction to the loss of faith, while Sebba, as we have seen, diagnoses it as deriving from trying to define a "pure" self entirely divorced from the non-self. These three views concur in seeing the modern self as confused by a loss of identity in relation to a metaphysical scheme of things.

That personality and the view of time do correlate has been demonstrated (Knapp). Changes in behavior which accompany hypnotically suggested alterations in time stance were explored by Aaronson. A psychotypology based on views and attitudes to time has also been suggested[36] and personalities of societies based on temporal attitudes and time budgets have been studied.[37] Familiarity with attitudes to time, and with the organization of time as functions of ethnic and cultural divisions of man have been receiving increased attention in policy making of governments.

In the paper of Helen Green we have an example of usefulness of employing attitudes to time as ways of exploring cross-cultural relationship. In the Negro subcultures examined by her, the polarities which are so fundamental to industrial civilization, such as the separation of life and death, masculine and feminine, being and becoming, are not conceived of as warring dichotomies. Within the ideologies and cosmologies of these subcultures they form what Green calls a "dual unity". The idea of such a unity harks back in the history of ideas to the interpenetration of opposites of Friedrich Engels and even to Heraclitus. When the tension attendant to such dichotomies is absent, time itself loses in importance and one senses a unity of the self and non-self, representing the un-idealized version of Jean Jacque Rousseau's image of natural nobility. It is not by accident that Jean Jacques depreciated time budgeting: knowledge of time and the definition of the self are coemergent in ontogenesis and phylogenesis. The implications of the relationship between indentity and time are yet to be explored in detail through interdisciplinary studies.

II. *The Study of Time as an Intellectual Enterprise*

The sources of Western thought about time may be found in the reflections and practices of the people around the Mediterranean during the first millen-

---

35 E. Voegelin: *On Classical Studies*. (Paper prepared for the Jerome Lecture Committee Conference, Rome, 1971. Private Communication.)
36 J. E. Orme: *Time, Experience and Behaviour*. New York: American Elseviwer, 1969, Chapter 4.
37 See Part II of the article TIME in *International Enc. of the Social Sciences*. MacMillan and the Free Press, 1968. v. 16. Also, relevant publications of the Survey Research Center, Institute for Social Research, The University of Michigan.

nium B.C. Its broad lines, however, are set by processes of perception and modes of concept-formation that seem to be common to all mankind.

Thus, Hesiod in his *Works and Days,* which contains pronouncements such as are still to be found in Farmers' Alamanacs, advocated the conquest of want by proper attention to natural cycles. In his *Theogony* he gave an account of the origins of the world but remained quite unsuspecting of progression beyond the passing of individuals and generations. The restless Heraclitus argued for time as conflict and change, ceaseless becoming and perishing, and as a struggle between opposites, while the aristocratic Parmenides sought time's essence through the logic of permanence. The three metaphors: time as a carrousel, time as a struggle of opposites and time as a mirage of the essentially timeless substratum of the world may be traced through the history of Western ideas.

The unique position of time in the affairs of man and the vast sweep of thought which the idea of time commands are not evident, however, from the approaches in the exact sciences because these sciences are unhistorical. The sources of this unhistoricity and the attendant deemphasis on time are to be found in the idea of lawfulness in nature, a conviction of intricate history, reinforced by the success of science conducted on this basis. The power of the industrial revolution, based on the applied form of exact science, demonstrates the degree of influence on the environment which can be achieved through the derivation of timeless rules from a world of temporal experience. Unfortunately, through its emphasis on facts and invariability and through the neglect of values and the unpredictable, the sciences have created a chasm between the world views sanctioned by the intellect and our instinctual gropings for meaning and purpose. This chasm is not new.

Whitrow noted[38] that "The history of natural philosophy is characterized by the interplay of two opposing points of view", namely, the trend to eliminate time and the trend to regard time as fundamental and irreducible. The dichotomy did not disturb most of those who at various epochs have considered time because they usually became advocates of either one view or the other. It is only with the arrival of the age of specialization and the depreciation of value judgment that demands for a new and unified view of the world arose. As Brandon expressed it,

"... the hiatus that exists between the verdict of our science and that of our instincts, and of which we are becoming increasingly and disturbingly aware, is surely to be attributed in large measure to that spiritual malaise which afflicts our culture, and inevitably our personal lives."[39]

---

38 *Nat. Phil. Time,* p. 1.
39 *The Voices,* p. 157.

Whatever the historical causes of this hiatus might be, ours is an age of specialization with the scientific workers living in their commodious but bolted cells:

"In fact, there is a whole honeycomb of cells, each containing a section or two of human thought and experience, and the prison is so constructed that whatever opinion the inmates may have about each other, whenever they happen to open their mouths thereon they are miraculously struck dumb."[40]

The classic function of all philosophies is to remedy the centrifugal forces of thought by offering a comprehensive vision of the world, and thereby assisting and guiding man in his search for meaning and order in his life. Whitehead, Husserl, Merleau-Ponty, Sartre and some others whose views of time are examined by Mays followed this tradition and sought the outlines of universal philosophical systems. But to philosophize today in the tradition of wide concern, wisdom and balance is a prohibitively difficult task for two reasons. The first is the overwhelming amount of data that the sciences have revealed; critical evaluation and coordination of which appear to be beyond the power of any individual. Second, the spirit of the time rejects defensively, almost pathologically, any trend of thought which might claim universality reaching beyond the scientific horizons. Because of the success of analyticity in the exact sciences and in its applied form in industrial technology, academic philosophy tends to take refuge in positivistic arguments against the disturbing and awe inspiring dangers of transcendental speculation. The result, as I see it, is a great deal of formal but antiseptic argumentation.

Reflecting on the state of current academic philosophy at large, I sense an absence of ideals inspiring, intelligible and reasonable in terms of contemporary scientific and humanistic preparedness; I see our epoch for these reasons, as one essentially uninformed in spite of the spectacular achievements in man's control of his environment. It would follow that the search for universal principles among the complexity and multiplicity of the sciences and the variety of humanistic utterances should be the most important subject of reflective thought. For my own mind at least, the major motivating power behind the study of time has been a conviction that it is a suitable guide in the search for order and meaning in the contemporary chaos of scientific and humanistic knowledge.

When in the 1950's I took stock of the literature of time I was left with the same impression of disarray that is associated with the review of any unreduced data. It was clear that one had to operate on the assumption of discoverable truths common to the many ways of knowing time and remain motivated by intellectual passion. It became apparent that an added prerequisite for interdisciplinary studies is an intellectual climate where creativity regarding all

---

40 Joseph Needham: *Mechanistic Biology and the Religious Consciousness.* In: *Science, Religion and Reality* (ed. by J. Needham). New York: Braziller 1955, p. 225.

knowledge is permitted to flourish and aspects of reality previously separately understood are permitted to produce their synthesis. To bring this about, this Society was founded in 1966.

The intellectual history of time is characterized by the interplay of knowledge felt and knowledge understood. Reflection on the nature of time calls on experiential qualities such as those listed in the Introduction to this paper; consequently, issues concerning time have been many and varied, the attempts to understand them impressive and continuous. I regard the radically increased interest in the study of time as a reaction to the compartmentalization of knowledge and a call for truths more comprehensive that those revealed through the scientific method.

The study of time, unlike other fields of knowledge, is unlikely to be replaced by other themes even if our understanding of time were to increase substantially. Time occupies a privileged position among concerns of empirical and speculative nature because it pertains to man's search for his individual and social destiny. Therefore, the study of time as a process of clarification and re-definition of problems is likely to remain not only universal but also continuous.

The mainstream of contemporary philosophy has avoided embracing the vast speculative areas beyond science, mainly because its methodology is limited to the study of processes which can be mathematized in their representation. What seems to be called for is a free rationalism, subsuming but unencumbered by mathematical and logical thinking, one that can also accommodate the emotive and evaluational needs of homo sapiens. As demonstrated by the fine papers in this volume, there is a multitude of ways in which man encounters reality. The integration of discursive cognition and introspective experience around the idea of time might help us in grasping the roots common to knowledge felt and knowledge understood.

September 10, 1971

Brothers' Pool

Pocantico Hills, New York

*Acknowledgements*

The first draft of this essay was critically reviewed by Professors Haber, Müller, Park and Sebba. I am grateful to them for their detailed comments and for the value of the dialogues that ensued. For better or for worse, responsibility for the final content and form of the paper remains with the author.

# Special Session on Flight Dysrhythmia

# Introduction

J. T. Fraser

For want of a better name, I shall refer to the subject of these papers as that of "post-flight resynchronization". By these terms I intend to emphasize that although our scientific interest was in identifying the sources of a certain malaise which often accompanies long transmeridian flights, our practical concern was in finding efficient means for rapid resynchronization by the individual of his physiological clock and fast and complete adaptation to post-flight environment.

The state of knowledge of post-flight resynchronization shows the richness and confusion of all new understanding. I shall first give a sketch of the background problems, then speak of matters of operational interest and then point to some work which might lead to techniques for the enhancement of the rate of post-flight resynchronization and the reduction of its effects.

Rapid transmeridian travel produces in many travelers a sense of malaise. The complaints include gastric distress, fitful sleep, irritability, fatigue, apathy and a mild depression. The extent to which people seem to be affected varies from practically no reaction over and above tiredness, to some extremely uncomfortable depression.

Our present knowledge is summed up in the major contribution by Drs. Blatt and Quinlan. Following their article Dr. Reinberg gives useful information on some of the experimental work done on desynchronization under laboratory conditions and relates his results to conditions of transmeridian flights. Dr. Gooddy's paper is of a nontechnical nature, he speaks of those conditions of long distance transportation which in his judgment contribute to the sense of malaise. There are lists of references appended to the Blatt and Quinlan and to the Reinberg paper.

Contrary to early thought when it was believed that the flight malaise is caused solely by the organism being out of phase with its environment, it seems now that even on a physiological basis alone there are at least two groups of effects to deal with. One is *dysrhythmia*, which consists of the disparity between the various internal states of physiological clocks and of the external temporal referents; the second is *desynchronization*, that is the disturbance of the internal balance of the various physiological rhythms. In simple terms, the biological clocks in aggregate get out of phase with the environment during rapid transmeridian flights, while during the post-flight resynchronization period they also get out of phase with each other because their adjustment to the new environment proceed at different rates. Indeed, some functions acclimate within a day or so while some other functions tend to remain in phase with the original environment for several weeks.

The individual's daily rhythm is sometimes described by his activity curve. The higher and lower portions of the curves correspond roughly to the feeling of being "let down" versus the feeling of being "picked up". Depending on whether the traveler is a morning or evening person, on his personal habits and on numerous other conditions of a flight just terminated, he might find himself on a downward slope in a daily rhythm while the people around him are in the upward mode or vice versa. Some persons prefer long nights and short days, some others short nights and long days. Preferences such as these and the usual daily scheduling of the person determine his working and eating habits which are also tied in with the circadian rhythms of his original environment.

Conscious and unconscious apprehensions, the threatening aspects of strangely long or short days or nights compound the distress of flight malaise and the difficulties of resynchronization. The desynchronization among different physiological clocks adds to the lowering of the psychological stance while the conscious and unconscious anxieties hinted above influence the physiological rhythm of the traveler.

Adaptation times vary not only from function to function for a given subject, but also from subject to subject for a given physiological function, and for a given function with direction of flight. This latter problem is an interesting one for there is some evidence that west-east flights seem to be less disruptive than east-west travel. There is speculation that the differential effect between eastward and westward flights might stem from the fact that the free running periods of certain subjects are respectively less than or more than 24 hours. Accordingly, different individuals can resynchronize more easily to an artifically shortened or artificially lengthened day or night depending on the period of their circadian rhythms.

In addition to fatigue, dysrhythmia, desynchronization, the conscious and unconscious anxieties and the cultural dislocations often associated with transmeridian flights, there are problems of the digestive rhythm and added complications caused by sleep loss, again, with marked individual variability. If to these items one adds the elusive but psychologically important aspects of mood, fantasy and motivation, the conclusion one must draw is that a systematic assessment of post-flight resynchronization is a complex task.

Regardless of the outcome of any future study, however, the operational parameters under control of the carriers remain necessarily limited. They include flight rescheduling; scheduling of activity during flight; control of the psychological conditions on the ground and in flight; and the guidance and education of the passengers. There are some indications that programming of cabin temperatures and humidity; programming of feeding schedules; designing of entertainment programs and the use of psychological props might eventually be useful. At this time, however, there seems to be no understanding and not enough controlled data to decide whether such interference with the flow of traffic, cabin conditions and activity scheduling might be useful or indeed, desirable.

As will be seen from the following papers, the physiological and biological bases of the adaptation processes are obscure and ill understood even for strictly controlled laboratory tests, let alone for the extreme complexity of long distance travel. It seems to us, therefore, that we ought to concentrate on more immediate, though perhaps not total remedies, rather than imagine that a complete control of post-flight resynchronization is within reach. Setting aside the practicability of pharmaceutical interference, the direction to look is a study of altering passenger behavior patterns before, during and after flight and a review of personality patterns as they are involved in post-flight resynchronization.

If the personality factors now suggested by the papers prove to be valid correlates of the severity of effects, it is then envisaged that a personality inventory may be constructed in form of a self-assessing questionnaire. This would be designed to determine the flexibility of temporal organization of the individual and the degree to which he tends to respond to internal and external cues. Sample questions, offered only by way of speculations, might include those of personal habits, preferences of mornings, days or nights, management of time, and other related matters yet to be determined. An individual's score might then be used to suggest that in view of his personal preferences, and direction and length of flight, he should take an early morning, late morning, noon, evening or late night flight, and that he abide if he wishes by certain preflight, flight, and post-flight routines.

Returning now to the question of possible action, it is suggested that a study of personality with regard to post-flight resynchronization problems be conducted and the effects of changing behavioral patterns be investigated. Such research might lead to the discovery of some specific methods of increasing the pleasure of long distance air flights and safeguarding the efficiency and reliability of flying personnel.

In addition to the four following papers, I have received letter comments from Dr. Bernard S. Aaronson (Princeton, New Jersey), Robert Efron, M. D. (Boston, Massachusetts), Dr. B. C. Goodwin (Sussex, England), Professor B. Hess (Dortmund, Germany), Dr. med. K. E. Klein (Bad Godesberg, Germany), Dr. Robert Knapp (Middletown, Connecticut), Dr. John A. Michon (Soesterberg, The Netherlands) and Dr. Ernst Pöppel (Munich, Germany). The comments of Dr. Schaltenbrand, the letter comments as listed and my own introductory remarks have been delivered to our sponsor, Pan American World Airways Corporation.

# The Psychological Effects of Rapid Shifts in Temporal Referents*

SIDNEY J. BLATT** and DONALD M. QUINLAN***

Every day, thousands of travellers fly across many time zones arriving in new environments which are at a very different point in the daily routines of work and rest, sleep and wakefulness, and day and night. This rapid shift of temporal referents requires a change of psychological and physiological patterns so that the individual can accommodate to the new time schedule. This process of accomodation is frequently accompanied by malaises, including gastric distress, fitful sleep, irritability, fatigue, apathy, and mild depression. The occurrence of these symptoms highlights the close relationship that normally exists between an individual's psychological and physiological rhythms and environmental schedules.

The most fundamental coordination of psychological, physiological, and environmental rhythms is around the 24-hours cycle of night and day. It is in the attempts to adapt to an alteration of the 24-hour day-night cycle that we can observe how the various factors interweave to achieve a state of equilibrium. Since this equilibrium involves the total organism, the process of adaptation can be studied from a number of points of view. Thus far most of the research has investigated the physiological aspects of this process.

Many physiological functions have a rhythmic pattern of approximately 24 hours which has been referred to by Halberg (1960) as a "circadian rhythm." The typical pattern is for a function to rise shortly prior to, or shortly following, awakening, reaching a high point sometime in midafternoon, followed by a gradual decline through the evening with a "trough" between 3 and 6 a.m. Although many physiological functions, or more exactly, markers of physiological functions, follow a similar pattern, the various functions show different times of rise and fall. Therefore, it is more accurate to refer to circadian rhythms rather than a single rhythm. Arne Sollberger (1955) has suggested that the various rhythms may represent the body's distribution of available energy throughout the day. Generally, one can consider the organism at night and

---

\* We are indebted to Mr. Michael Millen, and Dr. Arne Sollberger for their assistance and comments.

\*\* Dr. Sidney J. Blatt, Associate Professor, Yale University, New Haven, Connecticut, U.S.A.

\*\*\* Dr. Donald M. Quinlan, Assistant Professor, Yale University, New Haven, Connecticut, U.S.A.

during sleep as a relatively closed system since input and output are essentially minimal. During the day and while awake, however, the organism can be considered as a relatively open system.

The various physiological rhythms are not fixed precisely at 24 hours, and in the studies using constant illumination of the Arctic day at Spitzbergen (Lewis and Lobban, 1957; Simpson and Lobban, 1967), most subjects were able to accommodate to artificial days of 21 and 28 hours. In those studies, various physiological functions adapted at different rates. Body temperature adapted almost immediately. Excretion of potassium and 17 OHCS adapted more slowly and in some individuals did not accomodate even after six weeks of environmental manipulation. This differential rate of physiological adaptation suggests that there are two phenomena caused by temporal alterations; a change in the phase relationships between the various physiological functions, and a disparity between the demands of the environment and the body's preparedness to work or rest (Aschoff, 1962, 1964). It may be well to reserve the term introduced by Strughold (1952) "desynchronization" for the disruption of the internal balance of the various physiological rhythms while reserving the term "dysrhythmia" for the disparity between the various internal states and the external temporal referents. There is not only wide variation in the rates at which various physiological functions accommodate to changes in the length of day, but findings (e.g., Lewis and Lobban) also suggest that there are significant differences between individuals in the rate and extent to which they adapt to changes in the length of the day.

In addition to the extensive literature on physiological circadian rhythms, there is also considerable evidence for a 24-hour periodicity in psychological functions. Freemann and Hovland (1934) found four different patterns of diurnal variation but all the tasks studied also showed evidence of a 24-hour periodicity. Klein, Wegmann, and Bruner (1968) found that performance on a number of complex psychomotor tasks had a similar efficiency curve in which maximal efficiency occurs in midafternoon and the period of minimal efficiency occurs between 2 and 4 a.m. Kleitman and his colleagues found similar circadian variations on a large number of tasks, including card sorting, mirror tracing, code transcription, multiplication, and complex reaction time. Highest body temperature, maximal performance, and lowest levels of fatigue all occurred at approximately the same time, in midafternoon (Kleitman, 1963). Kleitman reports a highly significant correlation between body temperature and psychomotor efficiency on complex tasks and he considers body temperature to represent the level of metabolic activity. The correlation between body temperature and psychomotor efficiency on simpler tasks is much lower (Kleitman, 1963).

Psychomotor performance usually reaches a trough in the early morning hours between 3 and 6 a.m. and the greatest number of errors and the highest accident rates have been found to occur during this time (Kleitman, 1963).

Psychomotor and cognitive efficiency on a variety of tasks, such as meter reading, automobile driving, problem solving, and radar tracking all rise and fall in the characteristic carcadian rhythm (Alluisi and Chiles, 1967; Kleitman, 1963). An extensive study of the circadian variability of psychological functions was recently conducted by Frazier, Rummel, and Lipscomb (1968). Vigilance performance of three subjects was observed during a 14-day period of confinement. Circadian periodicity was an important source of performance variation for all subjects and this was most apparent on complex information monitoring tasks and least apparent on simpler tasks. In addition to the evidence for circadian rhythms in complex vigilance performance, there were also important individual differences in the specifics of the circadian pattern.

Thus, there is evidence for the circadian periodicity of physiological and psychological functions. Recently there have been a number of studies which have examined the effects of disrupting these circadian patterns by rapid jet flight over multiple time zones. This disruption may be considered in terms of the typical west-east and east-west flights between the eastern time zone in North America and continental Europe. The most frequent departure from North America is midevening, on the downward slope of the activity curve. After a 7-hour flight, the traveler arrives in the early morning hours on his biological clock (around 2–3 a.m.), at the lowest point of his activity curve. But he deplanes midmorning European time when people around him are on the upward slope of their activity cycle, and the environment is most active and most likely to place the greatest demands on him. East-west flights, on the other hand, frequently begin around noon. After a flight through daylight hours, the traveler arrives in late evening according to European time, but in midafternoon on the North American continent, at the peak of activity. In both eastward and westward travel, the person arriving is at odds with the activity cycle of his environment. On the eastbound flight the traveler loses time, often missing a night's sleep, while the westbound traveler experiences an extension of his day.

In considering the effects of rapid transmeridian flight it should be remembered that the time zone changes occur in a stressful context. The traveler, exposed to a high level of continuous, monotonous, high-pitched sound, is relatively restricted in his movements, and is often plied with liberal quantities of rich food and alcohol. These unusual and stressful physical conditions, compounded by conscious and unconscious apprehensions about travel and particularly about air travel, create a stressful context and place the individual in a vulnerable position as he attempts to cope with the temporal alterations of his environment.

As early as 1931, Wiley Post experimented with his day-night cycle prior to his world flight (Post and Gatty, 1931). But it was primarily with rapid jet travel over multiple time zones, when travel became sufficiently brief, that travelers began to consider that the discomfort they experienced was more than just the fatigue and anxiety of a long flight. Recently there have been numerous accounts of the effects of dysrhythmia reported in the popular press and in scientific publication. The physical and psychological effects include reports of digestive disorders, fitful and unrefreshing sleep, impaired efficiency, and, most prominent, irritability, depression, fatigue and a vague dysphoric sense of something being wrong (e.g., Crane, 1963; Laverhne, Lafontaine, and Laplane, 1969; Strughold, 1965). Experienced pilots express recognition of the effects of dysrhythmia in their preference for north-south rather than east-west flights (Lodeesen and Crane, 1963). Airlines have been concerned about the effect of repeated shifting of time zones and the possibility that this leads to chronic anxiety and other disorders for the flight crews. There have been a number of recent recommendations to give crews additional credit for east-west flights (Buley, 1967) and several business firms recommend that executives travel one to several days in advance of critical meetings.

There have been a number of more systematic attempts to study the effects of dysrhythmia. Some of the earliest observations of the physiological effects of a time-zone change have been cited by Kleitman (1963) and include reports by Gibson (1905), who noted an inversion of temperature curves in two subjects who moved from New Haven to Manila, and by Osborne (1908) who confirmed these findings in a subject who moved from Melbourne to London. Much later, Burton (1956) reported a shift of body temperature following a change of time zones.

The first systematic observations of the disruptions caused by rapid trans-meridian air flight were reported by Strughold (1952). He observed that many people, particularly older people, were sensitive to the temporal shift that occurred in long distance flights and that they experienced physiological discomfort. They became hungry and sleepy, had to urinate during the night, and were awake at the wrong time. Some people who could sleep in almost any place under any circumstances were relatively free of the effects of temporal phase shift incurred in long distance flight. Pilots who had to cross and recross a number of time zones several times a month were prone to nervous stress and frequently required special attention (Strughold, 1965).

Flink and Doe (1959) and later Gerritzen (1962) measured 17 OHCS, sodium, and potassium levels in urinary samples following rapid time-zone displacement and their findings indicate that biochemical excretions are intially in phase with the original environment. The excretion patterns gradually become synchronized with the new environment. Flink and Doe (1959) and Sasaki (1964) suggest that this phase shift occurs at the rate of approximately one hour per day. Lafontaine, Laverhne, Courillon, Medvedeff, and Ghata (1967) have also recently reported similar findings. They studied urinary excretion of potassium and 17 OHCS after air travel from Paris to Anchorage, and return. Eight subjects flew to Anchorage and after 20 hours returned to Paris. Although the subjects were out of phase with Anchorage time in terms of urinary measures, upon return to Paris the urinary circadian rhythms were immediately concordant with Paris time. Two subjects remained in Anchorage for five days before returning to Paris and this 5-day stay in Anchorage led to definite modifications in the circadian variation of urinary secretion of potassium and 17 OHCS. The modification generally began on the third day and by the fifth day there was a complete reversal of the preflight phase cycle. Upon return to Paris after five days, these modifications in urinary excretion persisted and did not return to the prior cycle even after five days. Lafontaine et al. (1967), also noted that there were considerable differences between the two subjects who remained in Anchorage. Five days after their return to Paris, one subject was still completely reversed in his circadian patterns while the other subject tended to approach the pre-established base line curves.

In a recent attempt to assess the effects of transmeridian flight, the United States Federal Aeronautic Agency sponsored a series of studies in which a number of physiological functions were recorded several days prior to and subsequent to east-west, west-east, and north-south flights. Hauty and Adams (1965) measured rectal temperature, heart rate, critical flicker fusion, and palmar evaporation. In two studies of east-west travel, rectal temperature, taken the first day after flight, was still in phase with the preflight base line. It was not until the third or fourth day that the phase shift of rectal temperature appeared to have been complete and was in sequence with local time. The phasing of heart rate closely approximated the measure of rectal temperature, requiring 3 to 4 days for the shift to be complete. Palmar evaporation shifted more slowly and did not appear to be in phase until about the eight day.

In a subsequent study Hauty and Adams (1955 a, b) studied west-east travel in four subjects who were flown from Oklahoma City to Rome. Although the flight time to Rome was somewhat shorter than the flight to Manila and Tokyo (i.e., 15 1/2 hours as compared to 23 hours) the phase shifting of rectal temperature took longer and was not completed until about the sixth day. This longer delay in phase shifting was also apparent in other physiological measures. Also, heart rate and rectal temperature did not shift together as they had done previously in the east-west flights.

In their studies of east-west, west-east, and north-south flights, Hauty and Adams also obtained ratings of subjective fatigue and they measured simple and complex reaction time (decision making). On all flights subjects reported an increase in subjective fatigue on the first day of arrival. A significant increase in reaction time was noted only on the east-west flight and this decrement in the speed of reaction time lasted for only one day. Hauty and Adams concluded that rapid translocation through multiple time zones may affect a sense of well-being but "this is not accompanied by a commensurate change in the efficiency of basic psychological functions." Based on the lack of any significant psychological deficit on the west-east flight one might also conclude that west-east travel is less disruptive and causes less difficulty than east-west travel. Siegel,

Gerathewohl, and Mohler (1969), however, recently discussed the relative difficulty of west-east and east-west flight and they conclude that the data are contradictory and inconclusive.

There are limitations in the Hauty and Adams studies of psychological deficit after rapid time-zone dislocation. As Hauty and Adams suggest, reaction time tests can be vulnerable to practice effects. Since at least two of the four subjects in the west-east flight had already participated in the prior east-west flight they had considerable practice on the reaction time tasks. This may have obscured any evidence of psychomotor inefficiency in the subsequent west-east flight. The suggestion that west-east travel has little effect on psychological processes seems unwarranted in at least one other regard. The research on sleep deprivation demonstrates that individuals can compensate for the detrimental effects of stress for a brief time on relatively simple tasks and that the effects of stress are apparent only on more complex tasks assessed over a long time period. Circadian periodicity has been found in performance on complex information monitoring tasks but it is least apparent in reaction time tasks (e.g., Kleitman, 1963; Frazier, Rummel, and Lipscomb, 1968). Thus, psychomotor deficit may be more apparent in complex tasks rather than in the simpler psychological functions assessed in the Hauty and Adams (1965 a, b) studies.

A number of recent studies have indicated that considerable personal distress can occur in rapid transmeridian flights. Hartman and Cantrell (1967) observed the living and working schedule of pilots on extended flying missions. They comment that pilots try to adopt a number of strategies to cope with the problem but none of these strategies seem to solve the problem and pilots express dissatisfaction with extended flying missions and are eager to return home. Klein, Bruner, and Ruff (1966) found that the distress of a flight was greater when there was an increased interval between the departure time and the pilot's peak activity period and they infer that most performance failures and accidents are likely to occur during the hours of lowest resistance, during the trough of the circadian cycle.

As a result of the observations made on the effects of time-zone change on physiological and psychological functioning, airlines and civil aviation organizations have recently begun to take fuller account of the impact of time-zone stress on flight crews and passengers. The International Civil Aviation Organization (ICAO), for example, recently (Buley, 1967) developed a formula for calculating recommended rest stops in order to allow their personnel to accomodate to the stresses of air travel. The formula allows compensation for late night or early morning departures and arrivals and for the extent of the flight and the amount of time-zone change. In January 1968 the Civilian Aviation Department of the British Board of Trade attempted to adjust their policies to account for the adverse effects of time-zone changes. Because of the disturbances of bodily function, particularly during the first 24 hours of adaptation, and the difficulty in sleeping, they recommend that airlines brief their crews on the psychological effects of time-zone change so that crews can adjust their sleeping patterns accordingly. The British Airline Association also recommended that a pilot be allowed to have a maximum of 40 time-zone changes in any 28-day period.

Thus, increasing attention is being given to the disruptions caused by rapid shifts of time zone. There have been frequent reports by travellers and flight crews about the disruptions in living and the impairment of performance associated with dysrhythmia. But there have been relatively few studies conducted on the psychological effects of time-zone displacement. The few studies of the psychological effects of dysrhythmia have often studied personnel associated with aviation and the number of subjects in most studies has been very small. These studies have failed to take into account the greater variability of subjective reports, the vulnerability of performance measures to practice effects and the monotony inherent in repeated measures designs. Furthermore, most of the studies conducted thus far on psychological effects of time-zone displacement have assessed relatively simple psychological functions, even though prior research indicates that these simpler functions are least affected by circadian periodicity and by stress. On relatively simple psychological functions subjects can generally muster up sufficient energy to perform well for a relatively brief time. Our own research on the psychological effects of stress, for example, indicates that following a stressful experience subjects improve their performance by remembering more digits and by copying symbols more rapidly, but complex problem solving performance declines markedly (Blatt and Cohen, 1965; Sherman and Blatt, 1968). It is essential that more complex and demanding tasks be used in subsequent research on the psychological effects of time-zone dislocation.

One recent study in particular has provided a number of important suggestions about the psychological effects of time-zone shift. The Medical Department of Air France (Lafontaine et al., 1969) found that personnel on transatlantic flights frequently reported disorders of sleep and digestive functions following time-zone change. Sleep was reported to be shorter and less sound, and of 312 flight personnel, 72 % reported sleep disturbance. There were frequent reports that fatigue and nervous tension interfered with falling asleep. Younger pilots attempted to adopt local time schedules while older and more experienced pilots tried to maintain their original time schedules. All subjects, however, regardless of age, reported their sleep was heavy, sluggish, sometimes agitated, frequently interrupted by dreams and semi-awakenings, and followed by tiredness which made getting up difficult. Seventy-two percent of the subjects reported that they awaken at their normal French time, between 1 and 4 a.m. local time, and though many managed to go back to sleep again, this sleep was not sound and they kept awakening frequently for the remainder of the night. Some became fully awakened and could not return to sleep and they spent the final hours of the morning reading or doing personal work while awaiting breakfast. Thus, a large percentage of the subjects had sleep disturbances on the first night of arrival whether they went to bed early and stayed on French time or whether they went to bed late and tried to adopt local time. Even on the second night many subjects reported that they were awakened in the middle of the night about the time of their normal home-based waking time. In addition to these sleep disturbances, there were frequent reports of digestive difficulties (41 %).

The Air France study and a number of other studies clearly indicate that one of the major symptoms encountered in time-zone change is an impairment or loss of sleep. This sleep loss occurs because biological and psychological processes involved in sleep and wakefulness occur at inappropriate times. When sleep is attempted in the new environment the body may be prepared for activity and, therefore, the sleep may not be deep or refreshing. Thus, the research on sleep deprivation and sleep loss is a valuable source for suggesting some of the consequences of time-zone change. A study by Murray, Williams, and Lubin (1958), for example, indicates that there is a highly significant drop of body temperature after only night of sleep loss. Since Kleitman considers body temperature to represent the level of metabolic activity and since body temperature is highly correlated with psychomotor efficiency, the significant decrease in body temperature after one sleepless night may be very important.

Some of the other effects of sleep loss are anxiety, apprehension, nervousness, irritability, and apathetic depression, including a sense of lethargy and a dampening of emotional involvement and loss of energy (Murray, 1968). With sleep loss there is increasing difficulty in maintaining attention and concentration. Signal monitoring tasks are adversely affected by sleep loss. Subjects seemed to be able to pull themselves together briefly to perform adequately on simpler tasks but this performance cannot be sustained at an adequate level over a longer period of time. Thus there is a striking similarity between the effects of sleep loss and the effects reported after time-zone dislocation.

In time-zone dislocation the individual, already in a state of lower efficiency because of sleep loss, is then required to function in the new environment at a time when biologically and psychologically he is at a resting level. A number of studies have indicated that many psychomotor and cognitive capacities are at minimal efficiency after midnight. One report of night shift workers, for example, studied the errors made in recording meters in a large gas works over a period of 42 years (Bjerner, 1955). The plant operated on a rotating shift with workers on a different shift each week. The number of errors for each hour of the day closely paralleled the usual pattern of the various circadian rhythms; there were fewest errors in the afternoon and the greatest number of errors in the early morning hours. Error rates and error curves were similar for each day of the week, suggesting that adaptation to night shift occurs slowly since four nights were insufficient time for the workers to adapt to the new shift.

With some limitations, the study of night shift work can suggest some of the effects of shifts in time zone. In a study of the medical records of a large number of workers on fixed, nonrotating shift schedules, Aanonsen (1959) found a higher incidence of gastric and neurotic disorders in workers who began on the night shift but who requested to move to the day shift. Thus, in addition to the frequent inefficiency found on the night shift, there are some people who are unable to adapt to night shift schedules and experience considerable personal distress when they have to work at night. The reports of irritability and mood change in flight crews after time-zone dislocation are consistent with the higher incidence of emotional disorders in people beginning on night shifts (Aanonsen, 1959) and the apathetic depression observed after sleep loss (Murray, 1968).

It is interesting to note that not only are some of the symptoms of depression reported after time-zone displacement and after sleep loss but in some clinical states of depression there are suggestions of alterations of circadian cycles. Furthermore, depressed patients often have sleep disturbances, such as early morning awakening and hypersomnia and insomnia. Therefore it is not surprising that there is a similarity between the symptoms reported in depression, such as sleep disturbance, digestive difficulties, fatigue and irritability (Beck, 1968) and some of the malaises reported after time-zone dislocation. Murray's observations (1968) of apathy, lethargy, a lack of emotional involvement and a loss of energy may have important implications for the study of the psychological effects of time-zone displacement.

One additional factor in depression is the experience of some sort of loss or separation. This sense of loss is usually focused around the experience of a lack of environmental support such as in the loss of a person with whom there was strong emotional attachment, or the loss of a familiar setting, such as may occur when moving to a new home or taking a new position. Travel, if only temporarily, involves a similar loss; a separation from one's family and usual surroundings. Time-zone displacement can be experienced to some degree as a loss of environmental structure and the shift back to one's original environment may be facilitated physiologically by those biological rhythms which are unchanged and which may provide the basis for the reintegration of the more labile biological rhythms. In addition, the ease of the back shift after time-zone dislocation may be related to the feelings of safety and security in returning home.

Thus, the malaises often reported after rapid transmeridian flight seem to be the compounded effect of three different factors. First, the individual, after the physical and psychological stress of flying, is exposed to a phase shift in which there is a disparity between the activation patterns of the individual's physiological and psychological processes and the activity schedule of the environment. When physiological and psychological functions are at a resting level the individual is least prepared to process information and to make accomodations to the environment. During this time individual can experience an active environment as flooding him with stimuli. Conversely, when physiological and psychological functions are at the height of activation, a quiescent environment can be experienced as stimulus deprivation. This discrepancy between the activity cycle of the individual and environmental schedules can result in psychological distress and in less efficient performance.

The *dysrhythmia* between environmental schedules and the individual's activity curves also makes it difficult to rest. *Sleep loss*, the second factor in the malaises of rapid time-zone displacement, occurs either because the individual arrives at midmorning in the new setting and is unable to sleep because of the activity in the environment or, as occurs in east-west flight, the individual awakens at his biological awakening time, in the middle of the night.

The third factor in the reported malaises may be the effects of the *desynchronization* of the idividual's circadian rhythms which occurs as the physiological rhythms begin to shift to the new time zone. If the individual remains in the new environment his physiological functions begin to shift at a differential rate to the new time schedule. Because of these differential rates of adaptation there will be a desynchronization of the phase relationships among the various physiological rhythms. This desynchronization, the third factor in the malaises of time-zone displacement, probably occurs somewhat later, as the individual attempts to accomodate to the new time schedule. The psychological effects of this desynchronization of the various physiological circadian rhythms has not been as extensively studied as the effects of sleep loss or of the consequences of trying to function when one is normally at rest. But it seems likely that this desynchronization of circadian rhythms would lead to further psychomotor inefficiency and would exaggerate the feelings of dysphoria and irritability already present because of sleep loss.

Numerous references in the literature on circadian rhythms and the adaptation to changes in temporal referents indicate that there are considerable differences between individuals in the ease and rate at which they accommodate to shifts in time zone. These individual differences, we believe, are related to important aspects of personality and represent stable individual patterns of interacting with the environment. There are several aspects of inter-individual variation that are relevant to adaptation to time-zone shift. These dimensions, possibly interrelated, are: first, the time of maximal efficiency (the "morning" vs. "evening" type person); second, the rigidity or flexibility of the person's usual temporal structure and organization; and third, the degree to which the individual is responsive to internal physical states as opposed to the extent to which his behavior is determined primarily by environmental factors.

Kleitman (1963) stresses that there are individual differences in circadian patterns. He observed, for example, that though performance and body temperature are highly correlated there are at least two types of curves expressing the relationship between body temperature and performance efficiency. One type has its peak early in the working day and another has its peak later in the day. While the correlation between performance efficiency and temperature is high in both curves, the shapes of the curves are different. Kleitman related these two types of curves to the often noted difference between the "morning" and the "evening" type of people who reach their maximum efficiency at different times of the day.

In studying the effects of dysrhythmia, differentiation should be made between individuals in terms of their usual temporal organization and particularly the organization and flexibility of their sleeping patterns. Some individuals have a relatively fixed temporal organization and follow rigid time schedules. Others have little concern for time and they function in what appears to be a disorganized way. A third group could be characterized as more flexible with a general, but not rigid, sense of temporal organization. Research on punctuality and procrastination indicates that temporal organization is an important aspect of personality, for punctual and procrastinating individuals differ on a number of important personality dimensions (Blatt and P. Quinlan, 1966). Alpern, Feirstein and Blatt (1970) recently found that subjects with flexible sleep patterns were psychologically more stable and mature and better able to shift their cognitive functioning and adapt to a variety of different tasks than were subjects with a rigid or erratic sleep schedule. Kleitman also comments that there are considerable differences in the way people can accommodate to sleep loss. People vary in their ability to go to sleep early or to remain asleep beyond their customary waking time. This ability probably makes it easier for them to re-establish a sleep-wakefulness pattern. Flexible subjects, as compared to subjects with more rigid or erratic sleep patterns would probably more readily accommodate to changes in temporal referents following transmeridian flight. As suggested in the Air France study, many individuals awake at their usual biological time rather than immediately shifting their waking time to the new environmental time schedule. The more flexible individual who can adjust his sleeping patterns should make the accomodation to a new time zone more rapidly and with considerably less distress. Thus, sleeping patterns, as a specific aspect of the flexibility of temporal organization, and the particular patterns of performance efficiency are variables which probably affect the capacity to accommodate to rapid time-zone change.

Another important personality variable in the accommodation to time-zone change may be the degree to which the individual is responsive to internal physical cues. There is a growing body of literature about the differences between individuals in their tendency to respond to internal bodily cues as compared to the extent to which they are influenced by environmental cues. The relationship between appetite and eating habits is an instance of such a phenomena (Schachter and Singer, 1969; Schachter and Gross et al., 1969). There seem to be people who are "body oriented", whose eating is determined by feelings of hunger and stomach contractions and who under conditions of ordinary hunger eat about the same amount of food regardless of whether it is very attractive food or just merely palatable. Other individuals seem to be more environmentally determined and tend to eat more pleasant food than bland food. Their food consumption is determined more by "clock time" than by experiences of hunger and stomach contraction (Schachter, Goldman, and Gordon, 1969). The eating behavior of individuals primarily influenced by environmental factors is more responsive to a manipulation of clock time. The relevance of this dimension for adaptation to time shifts has been demonstrated by Schachter who found that in a study of Air France personnel, overweight individuals tend to report fewer incidents of gastrointestinal distress after time-zone change because they seemed better able to adapt their eating habits to environmental cues.

The degree to which internal bodily cues, as compared to cues in the external environment, influence behavior has been considered as a general dimension of cognitive perceptual style. This work has been articulated primarily by Herman Witkin and his colleagues (1962) who have examined many aspects of cognitive and perceptual functioning along this dimension which they call "field independence" and "field dependence". The field dependent person is influenced by environmental cues while the field independent person is more responsive to internal cues. We might expect that field independent and field dependent people would differ in their accommodations to changes in temporal referents. The field independent person, more attentive to his internal cues, is likely to be more affected by the disruption of internal states. The environmentally oriented field-dependent person, on the other hand, is more likely to be influenced by the temporal cues of the environment and to be relatively unaware of physical distress and therefore better able to make the accomodation to the change in temporal referents.

Research has begun on aspects of rapid transmeridian flight but the data on psychological effects are minimal and inconsistent. For adequate evaluation of the psychological effects of rapid time-zone change the following variables must be considered, either by systematic investigation or methodological control: first, the time of departure and arrival, and extent of time-zone displacement; second, the context of the flight, including pre- and postflight experiences and the conditions and length of the flight; and third, characteristics of the individual.

To illustrate the potential effects of times of arrival and departure, one can consider the following hypothetical flight. In a west-east flight of six hours with a six-hour time-zone displacement, earlier departure times result in less disruption of the normal sleeping schedule. The individual loses time on west-east travel, that is, his biological clock lags behind environmental time. Thus, under proper environmental conditions, he can go to bed early and sleep late because he will not be awakened by an increase in his biological activity until it is relatively late in the morning according to environmental time.

*Hypothetical Flights: 6 Hour Duration, 6 Time Zones Shifted.*[a] (Local times in parentheses)

| Time of Departure | Time of Arrival | Remain Awake | Time to Retire | Time of Awakening | Length of 1st Night of Sleep | ICAO Rest Allowance |
|---|---|---|---|---|---|---|
| A. West-East: (Day shortened-reference clock lags behind environmental time.) | | | | | | |
| 10 a.m. (ICAO=0) | 4 p.m. (10 p.m.) (ICAO=0.1) | 4 hrs. | 8 p.m. (2 a.m.) | 4 a.m. (noon) | 8 hrs. | 0.60 days |
| 4 p.m. (ICAO=0.1) | 10 p.m. (4 a.m.) (ICAO=0.4) | 0 | 10 p.m. (4 a.m.) | 4 a.m. (noon) | 6 hrs. | 1.0 days |
| 10 p.m. (ICAO=0.4) | 4 a.m. (10 a.m.) (ICAO=0.4) | No sleep entire day, full night's loss of sleep | | | | 1.3 days |
| B. East-West: (Day lengthened-reference clock precedes environmental time.) | | | | | | |
| 10 a.m. (ICAO=0) | 4 p.m. (10 a.m.) (ICAO=0.4) | 10 hrs. | 2 a.m. (8 p.m.) | 8 a.m. (2 a.m.) | 6 hrs. | 0.90 days |
| 4 p.m. (ICAO=0.1) | 10 p.m. (4 p.m.) (ICAO=0.2) | 4 hrs. | 2 a.m. (8 p.m.) | 8 a.m. (2 a.m.) | 6 hrs. | 0.80 days |
| 10 p.m. (ICAO=0.4) | 4 a.m. (10 p.m.) (ICAO=0.1) | 0 | 4 a.m. (10 p.m.) | 8 a.m. (2 a.m.) | 4 hrs. | 1.0 days |

[a] ICAO formula allowance 0.5 days.

A late evening west-east flight, however, encounters the most extensive problems. The individual, after a night flight, arrives when it is midmorning in the environment. Even though he may be able to sleep during the flight, for many people the sleep is usually not very deep, prolonged, or refreshing, and he arrives with virtually losing a night of sleep. If he is able to go to bed early the following night, however, and sleep longer than usual, he may be able to compensate for the sleep loss by the second day. The ICAO formula (Buley, 1967) clearly reflects the difficulty for the late evening departure and it recommends 1.30 days of rest for the west-east evening departure as compared to 0.60 days of rest for the morning departure. Morning or afternoon departures on eastward flights seem to be least demanding because they allow for more sleep. An examination of airline schedules, however, reveals that most flights from North America to Europe depart between 6 and 10 p.m.

On westbound flights, the individual gains time; his day is extended and his biological clock precedes environmental time. After a morning departure around 10 a.m., the individual arrives at his destination in midmorning, according to the environmental time, but at 4 p.m. on his biological reference time. If he adopts local time and goes to bed at 8 p.m. (which is the earliest practical time), his sleep is likely to be disturbed about six hours later, at his usual biological time for awakening. But this is at 2 a.m. local time. He can either remain awake waiting for morning and breakfast or else struggle to return to sleep. Though he may be able to get at least 6 hours of normal sleep, the awakening around 2 a.m. and the subsequent restless sleep until breakfast can cause distress. In late evening departures on westward flights, the individual arrives around 10 p.m. local time, but around 4 a.m. on his reference time. Even if the traveller retires immediately, his biological awakening time would occur at 2 a.m. local time, after only 4 hours of sleep.

In comparing the effects of eastward with westward flights, it seems that the degree of disruption is an interaction between the time of departure and the direction of flight. It is interesting to note that an estimation of potential sleep loss and the ICAO allowances for postflight recovery indicate the early morning eastbound departures are least demanding and the late eastbound departures are most demanding, with the westbound departures falling at intermediate levels.

In the preceding discussion, we have made the assumption of a consistent preflight and postflight routine. Obviously these can vary considerably. If the preflight routine or the trip itself is stressful, the individual would probably be more vulnerable to the adverse effects of dysrhythmia. Likewise, intemperate behavior during flight or after arrival could also have detrimental effects.

The context of the flight not only includes the comfort and conditions of the flight, but also includes the meaning of the flight. As discussed earlier, leaving home or returning home may have very different effects on individuals. The meaning of the flight also depends on previous experience in travelling. One limitation of most of the research to date has been that it has been conducted primarily on experienced travellers. In this regard, one might examine whether it is possible to learn how to cope with and accommodate to time-zone changes on one hand, or whether, on the other hand, the deleterious effects are cumulative.

Almost every report of circadian rhythms and alterations of them have commented on the extensive differences among individuals. However, no systematic evaluations of any aspect of these differences have been undertaken. Those aspects which seem most relevant are temporal organization, flexibility of sleep patterns, time of maximal efficiency, and sensitivity to internal cues. Our current research suggests that some of these characteristics of individuals may be interrelated and form a more general personality style or persistent mode of relating to the environment. Finally, the evaluation of the effects of time zone displacements should be conducted on a wide range of psychological functions, including simple and complex tasks, mood, affect, and fantasy.

In summary, the literature on physiological and psychological circadian rhythms suggests that the malaises reported subsequent to rapid jet flight across multiple time zones is a function of several factors. The individual after the physical and psychological stresses of flying has to adapt to an environment which is out of phase with his circadian rhythms. Being out of phase with the environment frequently results in a loss of sleep which has a detrimental effect on cognitive efficiency and contributes further to the feelings of irritability, dysphoria, lethargy, and apathy. As the individual's circadian rhythms begin to shift to the new time schedule, there is a desynchronization of the physiological circadian rhythms and this disruption of internal equilibrium may also contribute to the cognitive inefficiency, the physical distress and the psychological symptoms of depression and anxiety.

There is considerable evidence in the literature, furthermore, that there are marked differences among individuals in the relative rate and ease with which they can accommodate to changes in temporal referents. The capacity for accommodation to time-zone change in part may be a function of the time of the individual's maximal efficiency, the flexibility of the individual's temporal organization of his daily routine, and the degree to which his behavior is primarily determined by internal bodily cues or by the external cues of the environment. The further understanding of these factors which influence the capacity for accommodation to time-zone change is important not only for its obvious practical application but it is also important because it permits us to study the processes of psychological adaptation. Day-and-night is one of the most frequent, persistent and basic environmental changes that man encounters and many of man's biological and psychological functions are accommodated to it. Alterations of the day-night cycle require a shift in the physiological and psychological circadian patterns and it is in the study of these shifts that we can gain further understanding of how physiological and psychological processes interact to achieve a state of equilibrium and adaptation.

## *References*

Aanonsen, A.: *Medical problems of shift work.* Industrial Medicine and Surgery 28 (1959) 422–427.

Alluisi, E. A., Chiles, W. D.: *Sustained performance, work-rest scheduling, and diurnal rhythms in man.* Acta Psychologia 27 (1967) 436–442.

Alpern, M., Feirstein, A., Blatt, S. J.: unpublished report, 1970.

Aschoff, J.: *Timegivers of 24-hour physiological cycles.* In: K. E. Schaefer (Ed.), Man's dependence on the earthly atmosphere. New York: Macmillan 1962.

— Circadian clocks. Amsterdam: North-Holland Publishing Company 1964.

Beck, A. T.: Depression; clinical, experimental and theoretical aspects. New York: Harper & Row 1967.

Bjerner, B.: *Diurnal variation in mental performance.* British Journal of Industrial Medicine (1955) 103–110.

Blatt, S. J., Cohen, D. C.: *The effects of stress on the problem solving process.* Unpublished manuscript 1964.

— Quinlan, P.: *Punctual and procrastinating students: A study of temporal parameters.* Journal of Consulting Psychology 31 (2) (1967) 169–174.

Buley, L. E.: *The calculation of rest-stop entitlements on official travel by air, using an arithmetic formula.* Mimeographed manuscript of the International Civil Aviation Organization 1967.

Burton, A. C.: *The clinical importance of the physiology of temperature regulation.* Canadian Medical Association Journal 75 (1956) 715–720.

Crane, J. E.: *The time zone fatigue syndrome.* Flying Physician 7 (1963) 19–22.

Flink, E. B., Doe, R. P.: *Effects of sudden time displacement by air travel on synchronization of adrenal function.* Proceedings of the Society for Experimental Biology and Medicine 100 (1959) 498–501.

Frazier, T. W., Rummel, J. A., Lipscomb, H. S.: *Circadian variability in vigilance performance.* Aerospace Medicine 39 (1968) 383–395.

Freeman, G. L., Hovland, C. T.: *Diurnal variations in performance and related physiological processes.* Psychological Bulletin 31 (1934) 777–799.

Gerritzen, F.: *The diurnal rhythm in water, chloride, sodium, and potassium excretion during rapid displacement from east to west and vice versa.* Aerospace Medicine (1962) 697–701.

Gibson, R. B.: *The effects of transposition of the daily routine on the rhythm of temperature variation.* American Journal of Medical Science 129 (1905) 1048–1059.

Goldman, R., Jaffa, M., Schachter, S.: *Yom Kippur, Air France, dormitory food and the eating behavior of obese and normal persons.* Journal of Personality and Social Psychology 10 (1968) 117–123.

Halberg, F.: *Temporal coordination of physiologic function.* Cold Spring Harbor Symposium on Quantitative Biology 25 (1960) 289–310.

— *Physiologic rhythms.* In: J. D. Hardy (Ed.), Physiological problems in space exploration. Springfield, Illinois: Charles Thomas 1964.

Hartman, B. O., Cantrell, G. K.: *Sustained pilot performance requires more than skill.* Aerospace Medicine 38 (1967) 801–803.

Hauty, G. T., Adams, T.: *Phase shifting of the human circadian system.* In: J. Aschoff (Ed.), Circadian clocks 413–425. Amsterdam: North-Holland Publishing Company 1965 (a).

— — *Phase shifts of the human circadian system and performance deficits during the periods of transition: I. East-west flight; II. West-east flight; III. North-south flight.* Aerospace Medicine (b) (1965).

Klein, K. E., Bruner, H., Ruff, S.: *An investigation regarding stress on flying personnel in long distance jet flight.* Zeitschrift Flugwissenschaften 14 (1966) 109–121.

— Wegmann, H. M., Bruner, H.: *Circadian rhythm in indices of human performance, physical fitness, and stress resistance.* Aerospace Medicine 39 (1968) 512–518.

Kleitman, N.: *Sleep and wakefulness.* (Rev. ed.) University of Chicago Press 1963.

— Titelbaum, S., Feiveson, P.: *The effect of body temperature on reaction time.* American Journal of Physiology 121 (1938) 495–501.

Lafontaine, E., Laverhne, J., Courillon, J., Medvedeff, M., Ghata, J.: *Influence of air travel east-west and vice versa on circadian rhythms of urinary elimination of potassium and 17 Hydroxycorticosteriods.* Aerospace Medicine 38 (1967) 944–947.

Laverhne, J., Lafontaine, E., Laplane, R.: *An investigation on the subjective effects of time zone changes on flying staff in civil aviation.* Unpublished paper. Air France Medical Department 1969.

Lewis, P. R., Lobban, M. C.: *Dissociation of diurnal rhythms in human subjects living in abnormal time routines.* Quarterly Journal of Experimental Physiology 42 (1957) 371–386.

Lodeesen, M., Crane, J. E.: *Tired jet pilots.* Flying 72 (1963) 52.

Mohler, S. R., Dille, J. R., Gibbons, H. L.: *The time zone and circadian rhythms in relation to air craft occupants taking long-distance flights.* Paper presented at meeting of the American Public Health Association, Miami Beach. October 1967.

Murray, E. J.: *Sleep deprivation and personality adjustment.* Progress in clinical psychology, 44–62. New York: Grune & Stratton 1968.

—, Williams, H. G., Lubin, A.: *Body temperature and psychological ratings during sleep deprivation.* Journal of Experimental Psychology 56 (1958) 271–273.

Osborne, W. A.: *Body temperature and periodicity.* Journal of Physiology (London) 1908.

Post, W., Gatty, H.: *Around the world in eight days.* London: Hamilton 1931.

Quinlan, D.: *Body cues in perceptual adaptation.* Unpublished doctoral dissertation. Yale University 1968.

Sassaki, T.: *Effect of rapid transposition around the earth on diurnal variation of body temperature.* Proceedings of the Society for Experimental Biology and Medicine *115* (1964) 1129–1131.

Schachter, S., Goldman, R., Gordon, A.: *Effects of fear, food deprivation and obesity on eating.* Journal of Personality and Social Psychology *10* (1968) 91–97.

— Gross, L. P.: *Manipulated time and eating behavior.* Journal of Personality and Social Psychology *10* (1968) 98–106.

Sherman, A. R., Blatt, S. J.: *WAIS Digit Span, Digit Symbol, and Vocabulary performance as a function of prior experiences of success and failure.* Journal of Consulting and Clinical Psychology *4* (1968) 407–412.

Siegel, P. V., Gerathewohl, S. J., Mohler, S. R.: *Time zone effects.* Science *164* (1969) 1249–1255.

Simpson, H. W., Lobban, M. C.: *Effects of 21-hour day on the human circadian excretory rhythms of 17 OH corticosteriods and electrolytes.* Aerospace Medicine *38* (1967) 1205–1213.

Sollberger, A.: *Statistical aspects of diurnal biorhythm.* Acta Anatomia *23* (1955) 7–127.

Strughold, H.: *Physiological day-night cycle after global flight.* Journal of Aviation Medicine *63* (1952) 464–473.

— *The physiological clock in aeronautics and astronautics.* Annals of the New York Academy of Science *134* (1965) 413–422.

Williams, H. A., Lubin, A., Goodnow, J. J.: *Impaired performance with acute sleep loss.* Psychological Monographs *73* (1959) No. 14.

Witkin, H. A., Dyk, R. B., Faterson, H. F., Goodenough, D. R., Karp, S. A.: Psychological differentiation. New York: Wiley 1962.

# Evaluation of Circadian Dyschronism during Transmeridian Flights

ALAIN REINBERG[*]

*Summary.* In order to evaluate the effects of transmeridian flights upon human beings and other organisms, several facts should be kept in mind:

1. The most powerful synchronizer for animals seems to be the alternation of light and darkness; for man, under ordinary conditions, this role is probably played by the routine of the society in which he lives.

2. There will be a lag between a synchronizer phase-shift and corresponding shifts in the circadian rhythm acrophases of the physiologic functions studied. The time required to phase-shift the latter will vary according to function and subject and may range from a few days to several weeks.

3. Adaptation time after a transmeridian flight: a) varies from subject to subject for a given physiologic function; b) varies from function to function for a given subject; c) varies for a given function with direction of flight (being shorter after an East to West flight than after a West to East flight).

4. Experiments must be rigorously controlled and the data obtained thereby objectively analyzed (using modern electronic computer methods) in order to avoid differing interpretations of results.

The traveler going from one location to another may find it necessary to cross several time zones in a very short time and thus be submitted to a rapid shift of synchronizing environmental factors [33]. Rhythmometric methods [39] now in use in the study of chronobiology [28, 51] can detect and objectively quantify changes in parameters characterizing circadian (about 24-hour) rhythms of human biologic and physiologic functions detected after such "transmeridian" flights, i.e., flights crossing one or more time zones.

## I. Synchronizers

A number of environmental factors with cyclical variations of about 24 hours are instrumental in the synchronization of circadian rhythms [3–5, 14, 15, 17–38, 46–48, 52]. For animals and other organisms dependent upon natural lighting conditions, the alternation of light and dark probably plays the most

---

[*] Alain Reinberg, Maître de Recherches au CNRS (France), Laboratoire de Physiologie, Fondation A. de Rothschild, 29, rue Manin, Paris 19, France,
and Chronobiology Laboratories, University of Minnesota, Medical School, Department of Pathology, Minneapolis, Minn. 55455, USA.

important synchronizing role [3, 4, 6, 17–23, 28, 32, 33, 36, 38, 39, 48–53]. For human beings, who are able to manipulate the external environment, the societal routine in relation to the alternation of daily activity and nightly rest may be the dominant influence [2, 14, 15, 19–25, 28, 31–33, 36, 39, 49–51, 53].

After a "transmeridian" flight, when a subject must remain in a "new" geographic location – associated with a "new" societal routine – for any length of time, his circadian rhythms must undergo a phase-shift in order to accomodate his biologic functions to the new routine. The rate of phase-shift may be faster or slower, depending upon the direction of his flight [3, 34] (see below).

## II. Changes in Parameters Characterizing Circadian Rhythms are Detected if the Synchronizer is Phase-Shifted

From animal experiments the following facts have been established:

A) There is a delay between an abrupt phase-shift of the synchronizer and a "complete" phase-shift of the circadian acrophase (or peak of the function

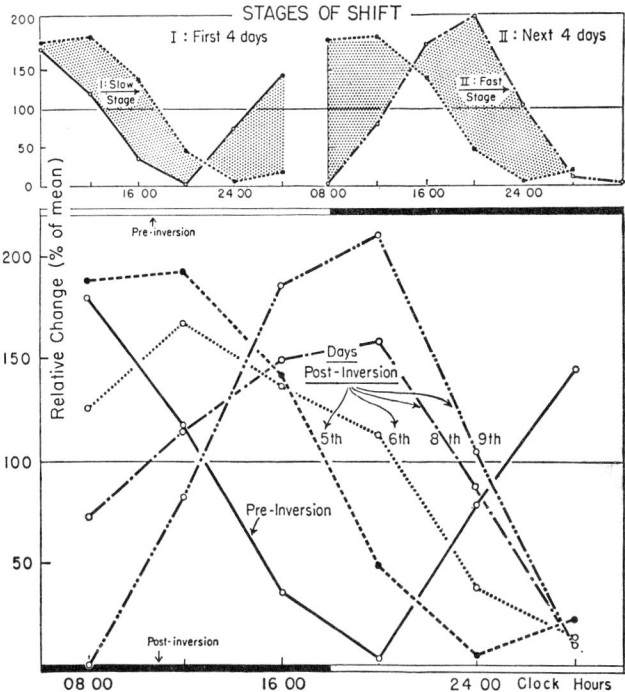

Fig. 1. Time course of phase shift in glycogen rhythm of intact mouse liver. Expressed as mg of glycogen/gm of liver; the values at times of peak and trough, respectively, were $54.1 \pm 3.10$ (at $\sim 12{:}00$) and $1.53 \pm 0.34$ (at $\sim 24{:}00$) on the 5th day after lighting inversion and $55.8 \pm 2.77$ (at $\sim 20{:}00$) and $0.04 \pm 0.11$ (at $\sim 08{:}00$) on the 9th day ($P$ of difference $< .01$) [29]

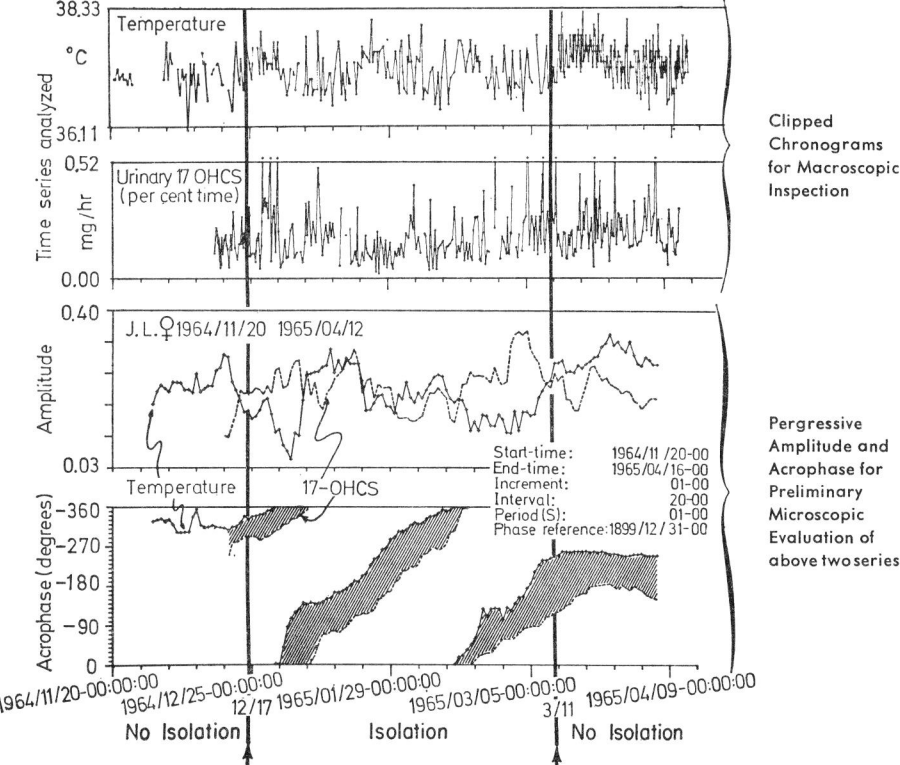

Fig. 2. Phase relations of circadian rhythms in the urinary excretion of 17-hydroxycorticosteroids and in rectal temperature of a woman during isolation in a cave and following resynchronization with a 24-hour-cyclic societal routine. Clipped chronogram of the time series shown on top. (For clipping, the mean and standard deviation were repeatedly computed and all values above and below mean ± 3 standard deviations were repeatedly equated to the nearer of these limits until the result of this iterative procedure was no longer associated with a change in mean and standard deviation to the nearest four decimal places. Extreme values thus "clipped" are indicated by a dot on top of the corresponding chronogram). – A macroscopic inspection of these time plots is barely contributory. Long-period changes of some regularity – corresponding presumably to the menstrual cycle – are apparent for rectal temperature in particular; changes with shorter period also are suggested by the record, yet it seems unjustified on the basis of inspection alone even to attempt to ascribe a precise period to a circadian rhythm and certainly it would be cumbersome, to say the least, to discuss the phase relations to each other of circadian components in the two time series. – By contrast, the display of acrophase in the bottom row – part of the microscopic approach – indicates first that the rhythms of both functions changed their period during isolation – only to be resynchronized with a 24-hour-cyclic routine thereafter; second, that the rectal temperature acrophase lagged behind that for 17-hydroxycorticosteroid excretion during isolation, as well as following resynchronization; and third, that resynchronization of body temperature occurred considerably faster than that of 17-hydroxycorticosteroid excretion. The latter finding may be related at least in part to the circumstance that the $\phi$ of rectal temperature was nearer its usual temporal placement in relation to the synchronizer than the $\phi$ of 17-OHCS, on the day of emergence from the cave. – From the pergressive amplitude diagram – third row from top – it can be seen that the amplitude, notably that of the rhythm in 17-OHCS excretion during isolation, showed no indication of damping as a conditioned reflex phenomenon might be anticipated to do. If there was a difference between the amplitude at the end of isolation and that upon resynchronization, this measure of the extent of circadian periodic change in 17-OHCS excretion indicated a more marked rhythm at the end of isolation than following resynchronization [15, 28]

used to approximate the rhythm) of each biologic function studied (Fig. 1) [29]. C. Pittendrich [47] expressed the same experimental evidence in the following terms: "Transients always precede attainment of a new steady-state ... This is true whether the former steadystate was disrupted by a single perturbation or by a phase-shift in the entraining cycle."

A phase-shift may be regarded as "complete" when the acrophase[1] has shifted approximately the same number of degrees as the synchronizer and has maintained the new value within reasonable limits for several days. In the case of transmeridian flights, a function my be said to be "resynchronized" when its acrophase* (after adjustment to the new time schedule) falls within the confidence limits of the acrophase obtained from the same function and subject prior to the phase-shift and remains steady. Reference standards for some human variables are already available for comparison [30, 36, 37, 45].

B) In a given animal species, the length of this delay varies from subject to subject for the same function and, what is more important, from function to function for the same subject.

Objective analyses of time series for several functions collected during and after isolation experiments indicate that this phenomenon can also be observed

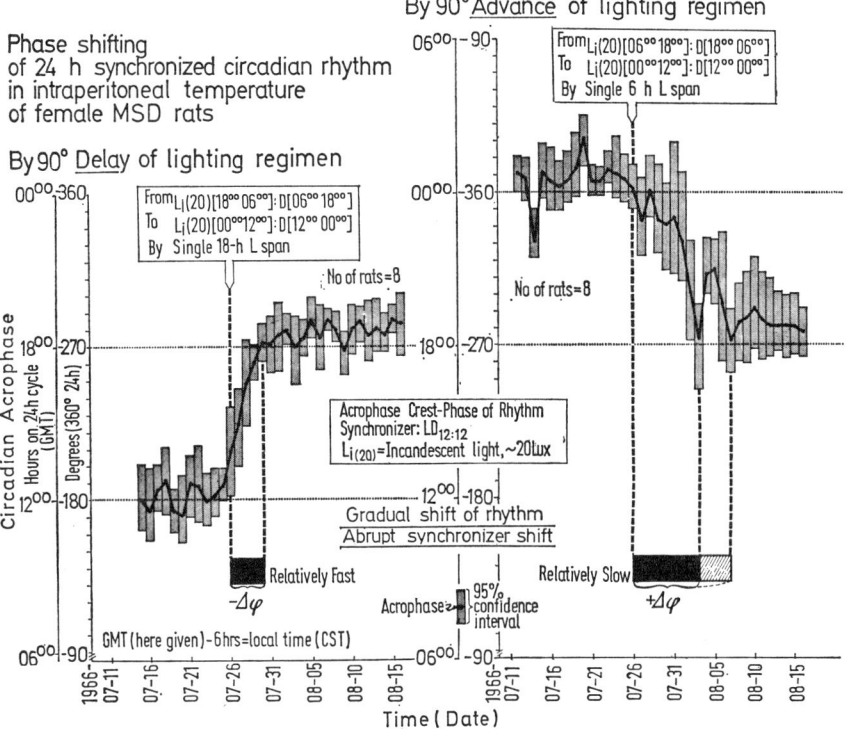

Fig. 3

---

[1] The peak of the function used to approximate the rhythm.

in human subjects (Fig. 2) [15, 28]. For instance, in comparing the circadian rhythm in urinary 17-hydroxycorticosteroid excretion – taken as an index of corticoadrenal activity – with that in rectal temperature from the same subject after a span of isolation, one notes a prolonged (greater than three weeks) desynchronization of the steroid rhythm, while temperature resynchronizes fairly rapidly. Thus, adrenal function is out of phase not only with the external environment but also internally in terms of its relationship to at least one other variable, if not more.

C) Circadian biorhythm parameters show different effects as a function of whether the synchronizer is advanced or delayed.

Examples of this phenomenon have been documented, among others, by J. Aschoff and R. Wever [6] for *Fringilla coelebs* (Chaffinches), and by F. Halberg et al. for the rat [34]. By macroscopic [27, 28] inspection of activity records, the former authors conclude that rhythm adaptation occurs faster when the synchronizer is advanced, that is, when either a light or a dark span is lengthened. On the other hand, F. Halberg and coworkers have reported [34], after microscopic analysis [27, 28] than more time is required to resynchronize the rat body temperature rhythm if the lighting regimen is advanced by 90° rather than delayed by the same amount (Fig. 3).

It may be noted here that transmeridian flights from East to West correspond to a synchronizer delay – from West to East they correspond to an advance.

*III. The Need for Rigorously Controlled Experiments and for Methods Permitting Objective Analyses of Results*

Many publications have been devoted to the effects of transmeridian flights upon biorhythms [1, 7–14, 16, 40–44, 54–57]. Conclusions from these studies are usually based on data displayed as chronograms and thus are subject to different interpretations by different viewers. The need for strictly controlled experimental designs and objective methods for analyzing results appears obvious.

The design should include a $\Delta t$ (the interval, whether equal or unequal, between measurements from a given subject on any given day) small enough and a T (the total number of days of study) at each "new" geographic location of a transmeridian flight large enough to provide sufficient data for analysis. It should be noted that for certain functions the length of time required for resynchronization can be longer than three weeks. One study has been carried out in which the subject collected data for about 30 days in France prior to an E to W flight, for about 70 days in the U.S.A. and finally for about 35 days in France after the (return) flight from W to E.

The study may be designed "transversely" or "longitudinally" [39] or a combination of the two (a "hybrid" design) depending upon the number of subjects and the length of time available [36].

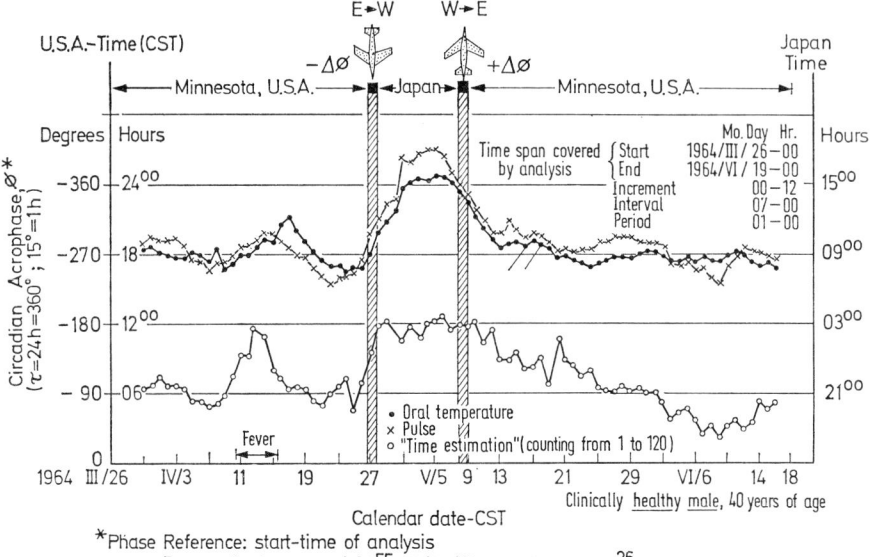

Fig. 4. Study of phase-shifts in three physiologic functions, in cooperative study between the University of Minnesota and the Federal Aviation Agency, Oklahoma City. Time estimation expressed as time elapsed during count from 1 to 120. High values indicate slow counts. Minimal counting rates are about 180° out of phase in relation to the crests of oral temperature and pulse. Maximum counting rates are roughly synchronized in phase with oral temperature and pulse, except for transient changes in system phase relations, e.g. during fever. Data collection started 06:55, III/26/1964 and ended 22:26, VI/19/1964 [28, 30, 35]

For purposes of analysis, special electronic computer programs have been developed and tested by the Chronobiology Laboratories at the University of Minnesota under the direction of Franz Halberg [24, 27, 28, 30, 32, 35, 36, 39]. Using such programs, objective estimates of rhythm parameters such as the rhythm-adjusted level, $C_0$, the amplitude, C (a measure of the extent of rhythmic change) and the acrophase, $\emptyset$ (the peak of the function used to approximate the rhythm), along with their respective (.95) confidence limits, are derived. Such "microscopic" [27] analytical methods permit the chronobiologic aspects of transmeridian flights to be considered as a part of quantitative biology.

## IV. Results

Results summarized herein are restricted to experiments carried out in accordance with the requirements discussed above, among others. All of them have been computer-analyzed in the Chronobiology Laboratories using the methods mentioned above [39, 28, 24, 25, 27, 30, 32, 35] (Fig. 4 and 5). From these experiments one can thus conclude that the time required for resynchronization after a transmeridian flight:

1. varies from subject to subject for a given function;
2. varies from function to function for a given subject (e.g., the sleep-wakefulness rhythm resynchronizes within a few days, while the rhytm in adrenal activity require several weeks to adjust);
3. varies with the direction of flight per subject per function: a slow adaptation usually corresponds to an advance of rhythm, i.e., a West-to-East transposition; while more rapid adaptation corresponds to East-to-West flights.

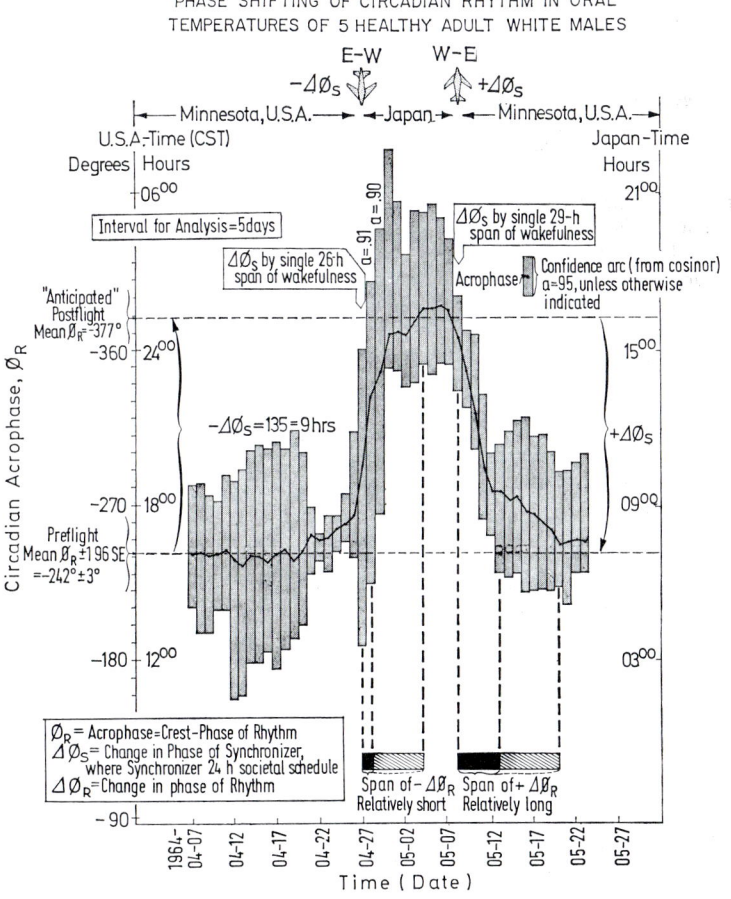

Fig. 5. Delay of circadian temperature acrophase occurs faster than advance following intercontinental flight from Minneapolis to Japan and back to Minnesota, respectively in the small sample of 5 subjects investigated. This finding must not be extrapolated to other groups or conditions; it does not apply to all individuals in the group here summarized. Genetic and other inter-individual differences deserve study in this connection as do factors complicating the difference between advance and delays of synchronizer – for instance, the rigor with which a given change in the social synchronizer schedule is enforced

Thus, a temporary dyschronism seems to occur – a transmeridian flight is followed by altered phase relations of circadian (and possibly other) rhythms – relations which normally characterize the circadian temporal structure of each organism and species.

The results obtained thus far cannot, however, be overly generalized – it is obvious that much work under standardized conditions with subsequent objective analysis remains to be done.

*References*

1. Alyakrinsky, B. S.: *Problems of latent desynchronosis.* XIIth Cospar meeting. Prague 1969.
2. Apfelbaum, M., Reinberg, A., Nillus, P., Halberg, F.: *Rhythmes circadiens de l'alternance veille-sommeil pendant l'isolement souterrrain de sept jeunes femmes.* Presse Médicale, 77 (1969) 879–882.
3. Aschoff, J.: *Exogenous and endogenous components in circadian rhythms.* In: Cold Spring Harbor Symp. Quant. Biol. Long Island Biol., Assoc., 25 (1960) 11–28.
4. — *Comparative physiology; diurnal rhythms.* Ann. Rev. Physiol., 25 (1963) 581–600.
5. — *Circadian rhythms in man: a self-sustained oscillator with an inherent frequency underlies human 24-hour periodicity.* Science, 148 (1965) 1427–1432.
6. — Wever, R.: *Resynchronisation der Tagesperiodik von Vögeln nach Phasensprung des Zeitgebers.* Z. f. verg. Physiol., 46 (1963) 321–335.
7. Burgard, P., Henry, M.: *Quelques aspects de la fatigue dans l'aviation de transport.* Presse Médicale, 89 (1961) 1903–1906.
8. Fischgold, H.: *La dette de sommeil.* Séance solennelle de l'Académie Internationale de Médecine Aéronautique et Spatiale. Athènes. Acad. Med. Aéron. Spat. Athènes: 1965
9. —, Lavernhe, J., Blanc, C.: *Sommeil, insomnie et dette de sommeil.* Presse Médicale, 75 (1967) 391–395.
10. Flink, E. B.: *Effect of sudden time displacement by air travel on synchronization of adrenal function.* Proc. Soc. Exp. Biol. Med., 100 (1959) 498–501.
11. Gerritzen, F.: *The diurnal rhythm in water, chloride, sodium and potassium excretion during a rapid displacement from east to west and vice versa.* J. Aerospace Med., 33 (1962) 697–701.
12. — *Methods for the study of the behaviour of circadian rhythms in kidney function, before, during and after global flights.* Réunion de Chronobiologie Appliquée à l'Hygiène de l'Environnement, Paris 30 juin–1 juillet, 1969.
13. —, Strengers, Th., Esser, S.: *Studies on the influence of fast transportation on the circadian excretion pattern of the kidney in humans.* Aerospace Med., 40 (1969) 264–271.
14. Ghata, J.: *Effets physiologiques des vols transméridiens.* Réunion de Chronobiologie Appliquée à l'Hygiène de l'Environnement. Paris 30 juin–1 juillet, 1969.
15. —, Halberg, F., Reinberg, A., Siffre, M.: *Rythmes circadiens désynchronises (17-hydroxy-corticostéroïdes température rectale, veille-sommeil) chez deux sujets adultes sains.* Ann. d'Endocrinol., 30 (1969) 245–260.
16. Gullett, C. C.: *Jet planes and the circadian cycle.* J. Amer. Med. Ass., 197 (1966) 935.
17. Halberg, F.: *Beobachtungen über 24-Stunden-Periodik in standardisierter Versuchsanordnung vor und nach Epinephrektomie und bilateraler optischer Enukleation.* Berichte über die gesamte Physiologie, 162 (1954) 354–355.
18. — *Temporal coordination of physiologic function.* In: Cold Spring Harbor Symp. Quant. Biol. Long Island Biol. Assoc., 25 (1960) 289–310.
19. — *The 24-hour scale: a time dimension of adaptive functional organization.* Persp. Biol. Med., 3 (1960) 491–527.
20. — *Symposium on "Some current research methods and results with special reference to the central nervous system". Physiopathologic approach.* Amer. J. Ment. Def., 65 (1960) 156–171.

21. — *Circadian temporal organization and experimental pathology.* VII Conf. Int. Soc. Studio Ritmi Biol. Siena (1960), Minerva Medica, Torino, 1962, pp. 20–26.
22. — *Physiologic 24-hour rhythms; a determinant of response to environmental agents.* In: Man's Dependence on the Earthly Atmosphere. K. E. Schaeffer (ed.) pp. 48—89. New York: Macmillan 1962.
23. — *Organisms as circadian systems; temporal analysis of their physiologic and pathologic responses, including injury and death.* In: Walter Reed Army Institute of Research Symposium, Medical Aspects of Stress in the Military Climate, 1964, pp. 1–36.
24. — *Circadian rhythms, a basis of human engineering for aero-space.* In: Psychophysiological Aspects of Space Flight. B. Flaherty (ed.) pp. 166–194. Columbia Univ. Press 1961.
25. — *Physiologic rhythms and bioastronautics.* In: Bioastronautics. K. E. Schaeffer (ed.), pp. 181–195. New York: Macmillan 1964.
26. — *Physiologic rhythms.* In: Physiological Problems in Space Travel. J. D. Hardy (ed.), pp. 298–322. Springfield, Ill.: Charles C. Thomas 1964.
27. — *Pouvoir de résolution des calculatrices électroniques en pathologie temporelle. Analogie avec la microscopie.* Scientia, 101 (1966) supplement, 172–179.
28. — *Chronobiology.* Ann. Rev. Physiol., 31 (1969) 675–725.
29. —, Albrecht, P. G., Barnum, C. P.: *Phase shifting of liver glycogen rhythm in intact mice.* Am. J. Physiol., 199 (1960) 400-402.
30. —, Engel, R., Swank, R., Seaman, G., Hissen, G.: *Cosinor Auswertung circadianer Rhythmen mit niedriger Amplitude im menschlichen Blut.* Physik. Med. Rehabil., 5 (1966) 101–107.
31. —, Engeli, M., Hamburger, C.: *The 17–ketosteroid excretion of a healthy man on weekdays and weekends.* Exp. Med. Surg., 23 (1965) 61–69.
32. —, —, —, Hillman, D.: *Spectral resolution of low-frequency, small-amplitude rhythms in excreted ketosteroid: probable androgen induced circaseptan desynchronization.* Acta Endocrinol. supplement 103 (1965) 1–54.
33. —, Halberg, E., Barnum, C. P., Bittner J. J.: *Physiologic 24-hour periodicity in human beings and mice, the lighting regimen and daily routine.* In: Photoperiodism and Related Phenomena in Plants and Animals. R. B. Withrow (ed.), pp. 803–878. AAAS Publ. 55, Washington, D. C., 1959.
34. —, Nelson, W., Runge, W., Schmitt, O. H.: *Delay of circadian rhythm in rat temperature by phase-shift of lighting regimen is faster than advance.* Abstract Fed. Proc., 26 (1967) 599.
35. —, Panofsky, H., Diffley, M., Stein, M., Adkins, G.: *Computer techniques in the study of biologic rhythms.* Ann. N. Y. Acad. Sci., 115 (1964) 695–720.
36. —, Reinberg, A.: *Rythmes circadiens et rythmes des basses fréquences en physiologie humaine.* J. Physiol. (Paris) 59 (1967) 117–200.
37. —, Reinhardt, J., Bartter, F. C., Delea, C., Gordon, R., Reinberg, A., Ghata, J., Halhuber, M., Hoffmann, H., Günther, R., Knapp, E., Peña, J. C., Garcia-Sainz, M.: *Agreement in endpoints from circadian rhythmometry on healthy human beings living on different continents.* Experientia, 25 (1969) 107–112.
38. —, Siffre, M., Engeli, M., Hillman, D., Reinberg, A.: *Etude en libre-cours des rythmes circadiens du pouls de l'alternance veille-sommeil et de l'estimation du temps pendant les deux mois de séjour souterrain d'un homme adulte jeune.* C. R. Acad. Sci., 260 (1965) 1259–1262.
39. —, Tong, Y. L., Johnson, E. A.: *Circadian system phase: an aspect of temporal morphology; procedures and illustrative examples.* In: The Cellular Aspects of Biorhythms. H. von Mayersbach (ed.) pp. 20–48. Berlin–Heidelberg–New York: Springer 1967.
40. Hauty, G. T., Adams, T.: *Phase shifts of the human circadian system and performance deficit during the periods of transition: 1—East-West flight.* J. Aerospace Med., 37 (1966) 664–668.
41. Klein, K. E., Brüner, H., Ruff, S.: *Untersuchungen zur Belastung des Bordpersonals auf Fernflügen mit Düsenmaschinen.* Z. f. Flugwissenschaften, 14 (1966) 109–121.
42. —, Wegmann, H. M., Brüner, H.: *Circadian rhythm in indices of human performance, physical fitness and stress resistance.* Aerospace Med., 39 (1968) 512–518.

43. Lafontaine, E., Ghata, J., Laverhne, J., Courillon, J., Bellenger, G., Laplane, R.: *Rythmes biologiques et décalages horaires. Etude expérimentale au cours de vols commerciaux long-courriers.* Concours Medical 189 (1967) 3731–3740 (n° 19) et 3963–3970 (n° 20).
44. —, Lavernhe, J., Courillon, J., Medvedeff, M., Ghata, J.: *Influence of air travel east west and vice versa on circadian rhythms of urinary elimination of potassium and 17-hydroxycorticosteroids.* Clin. Aviat. Aerospace Med., 38 (1967) 944–947.
45. Nelson, W., Halberg, F.: *Phase relations of circadian rhythms: animals.* In: Handbook of Environmental Biology. P. L. Altman and D. S. Dittmer (ed.) pp. 586–596. Bethesda: Amer. Soc. Exp. Biol. 1966.
46. Pittendrigh, C. S.: *Perspectives in study of biological clocks.* In: Symposium on Perspectives in Marine Biology, pp. 239—268. Berkeley: University of California Press 1958.
47. — *Circadian rhythms and the circadian organization of living systems.* In: Cold Spring Harbor Symp. Quant. Biol. Long Island Biol. Assoc., 25 (1960) 159–182.
48. —, Bruce, V. G., Kaus, P.: *On the significance of transients in daily rhythms.* Proc. Natl. Acad. Sci., 44 (1958) 965–973.
49. Reinberg, A.: *L'homme et les rythmes circadiens.* Cahiers Sandoz, 9 (1966) 1–50.
50. — *Rythmes biologiques (rappel de quelques données recentes).* Rev. Med. Aéronautique Spatiale, 7 (1968) 127–130.
51. — *Biorythmes et chronobiologie.* Presse Médicale, 77 (1969) 877–878.
52. —, Ghata, J.: *Les rythmes biologiques* (seconde edition), 128 pp. Paris: P.U.F. 1964.
53. —, Halberg, F., Ghata, J., Siffre, M.: *Spectre thermique (rythmes de la température rectale) d'une femme adulte saine, avant, pendant et après son isolement souterrain de trois mois.* C. R. Acad. Sci. Paris, 262 (1966) 782–785.
54. Sasaki, T.: *Effect of rapid transportation around the earth on diurnal variation in body temperature.* Proc. Soc. Exp. Biol. Med., 115 (1964) 1129–1131.
55. Siegel, P. V., Gerathewohl, S. J., Mohler, S. R.: *Time-zone effects. Disruption of circadian rhythms poses a stress on the long distance air traveler.* Science, 164 (1969) 1249–1255.
56. Strengers, T.: *The influence of intercontinental flights on the urinary excretion of steroid metabolites.* Réunion de Chronobiologie Appliquée à l'Hygiène de l'Environnement, Paris 30 juin-1 juillet, 1969.
57. Strughold, H.: *Day-night in atmospheric flight, space flight and other celestial bodies.* Ann. N. Y. Acad. Sci., 98 (1962) 1109–1115.

# Some Factors in the Production of Dysrhythmia and Disorientation Associated with Rapid Latitudinal Transfer

WILLIAM GOODDY*

*Summary.* There has been much neglect in the study of the factors causing malaise and illness resulting from rapid long-distance travel. Transmeridional travel causes a dysrhythmia because internal clock systems have to adjust to unaccustomed rhythms. The nature of neuronal action is described and the clock-like activity of the nervous system is outlined. Disturbances of brain rhythms (e.e.g.) are described. The important factors which influence health before and after flight are emphasised. A plea is made for wider publication of information about the causes and prevention of illness associated with rapid long-distance travel.

Only in the past ten years has there been an awakening interest in the medical and especially the neurological aspects of rapid travel as it affects the ordinary citizen. It is probable that there have been more or less elaborate studies for the armed services and for astronauts – how surprising it would be if there were not – but any such information, perhaps for security reasons, has not been released for the benefit of the multimillions of commercial route passengers who both keep the airlines in business and also pay the price for armed forces. Commercial airlines have been almost entirely silent about the various forms of flight malaise, and so have boards of aviation. However, there was an article on "Tummy Time" in a QANTAS passenger publication some years ago. The papers of Hauty and Adams (1966) are classical attempts to analyse the problem.

The average air passenger is almost completely uninstructed about the effects of rapid air travel and how to minimise the disadvantages. The times of his flights seem to be dictated by the convenience of the airlines; and alterations or delays have to be sustained as best they may. Some of the longest flights, such as those across the Pacific Ocean from Sydney, which start at 20.00 hours, are likely to cause the maximum discomfort to the passengers.

The neglect of *medical* aspects of everyday flying, an activity very closely related to the physiology and pathology of dysrhythmia, is the more surprising since non-medical disciplines, especially in biology and plant chemistry, have made such important and fundamental contributions in the field of observation of rhythmic activity in animals and plants. It is perhaps a

---

\* William Gooddy, MD, FRCP, National Hospital, Queen Square, London, W C 1, England.

sign of the urgency of our new problems that so many eminent scientists from many different fields of interest should have gathered at the German Institute of Mathematical Research in Oberwolfach in the summer of 1969.

If we are to make an accurate statement, *for general consumption*, for the benefit of all actual and potential passengers, about our psychological and physiological health before, during and at the end of rapid transmeridional travel, we must contribute our personal feelings as human beings; for such integrated human perceptions cannot be investigated adequately by analytical methods alone. In particular, human disorders and discomforts cannot be fully understood from the transfer of information derived from experiments on other animals.

The effects of high-speed travel are noted mainly after arrival at the new destination; and last for a period of from a few hours up to about ten days. We do not know what effects persist in astronauts. These effects are both psychological and physiological, commonly so closely integrated that it would be foolish to separate them; especially as it becomes obvious that the so-called psychological circumstances have their effects, very often, in purely physical, neurological, pharmacological ways.

One of the main points of this communication is that although we concern ourselves primarily with the health effects during and shortly after rapid transmeridional travel, we find that such effects have their origins days, weeks, months and even years before the flight begins, and, less important perhaps from the immediate health point of view, we *remember* the journey and its effects for a very long time after they are over – we are never the same again.

In seeking for a physiological unit by which the human being becomes aware of his own sense of time (personal time) in contrast with the non-human government or public time standards which we have devised from astronomical data (because we need a standard rather than a personal time for public business), the neurologist may well choose to study the basic unit of the nervous system as shown by anatomical, physiological and pathological studies, the *neuron*. We believe that behaviour is probably related to the integrative action of such units, of which there are supposed to be at least 10,000,000,000 in the human nervous system.

The neuron, with its rhythmic on-and-off, "all-or-none" form of signalling, has some of the properties of a clock-form. All that is required for a clock-form for the human observer is a regularly repeated natural phenomenon which is visible, audible, tactile, possibly olfactory (as in the incense clock). The end of each cycle must coincide with the beginning of the next identical cycle. By using any such mechanism to provide a counted number of cycles, the passage of time may be measured. Though we are familiar enough with the regular passage of stars by which public time is measured, and though we take for granted the perfection of our wrist-watches, it is perhaps a new idea for us

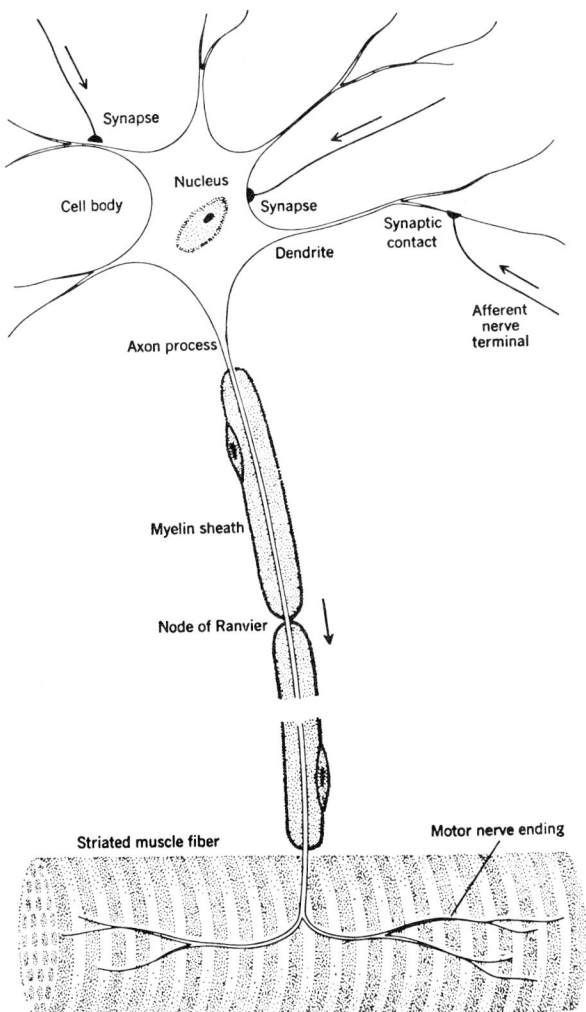

Fig. 1. Diagram of a neuron. (From Katz 1966)

that possibly the most important part of our internal structure, the nervous system, has innumerable examples of clock-forms, both in its single neurons and in the groups of neurons which form the nuclei, ganglia and tract pathways of the nervous system. The brain itself is merely a particularly large collection of such groups of cells and their fibre processes.

Since there may be some who are unfamiliar with this important cellular unit, the neuron, a brief description follows. The number of neurons is expressed by $10^9$ after various figures for different species. Man's remote ancestor, the *tarsier*, is estimated at $0.31 \times 10^9$, the *chimpanzee* at $5.5 \times 10^9$, and *man* at $6.9 \times 10^9$. The size of the cell body varies between less than 1 micron

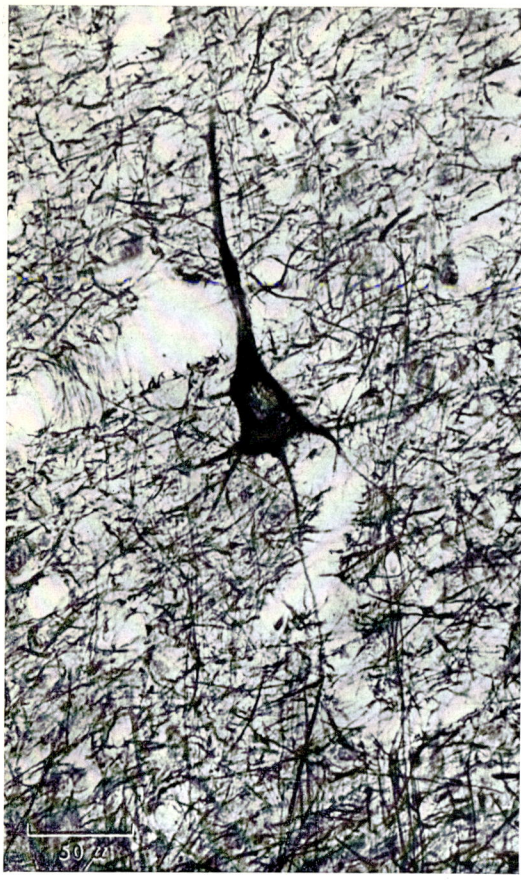

Fig. 2. Arrangement of cells in brain cortex, from Sholl 1956

and 100 microns. The length of fibre may be between 10 microns and a metre (1,000,000 microns). Each neuron may be in contact with about 1,000 other neurons directly (and with all other neurons indirectly).

The rate of conduction along a nerve fibre is between 1 metre per second and 100 metres per second, the rate of transmission depending on cross-sectional area, under standard conditions. Each neuron is capable of transmitting only at full power or not at all; and an average rate of transmission is at 50 times a second. Roughly speaking, a nerve unit may be said to transmit at 50 metres a second, and 50 times a second. This form of regularly intermittent release of energy, controlled by all those biochemical and biophysical processes which we term "normal physiology", is exactly comparable with the form of release of the escapement of a mechanical clock. Imagine the complexity and statistical uncertainty of such a system which is in an everchanging state. If we use as the smallest duration of scientific time, the *chronon*, as defined by Whitrow in his splendid *The Natural Philosophy of Time* (1961) at $10^{-24}$, then

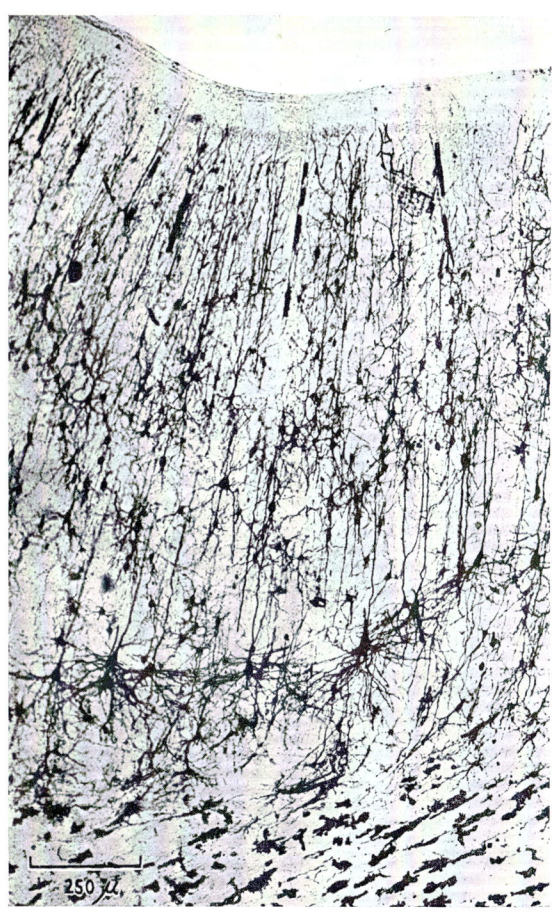

Fig. 3. Arrangement of cells in brain cortex, from Sholl 1956

we begin to realise the stupendous indeterminacy and fluctuations of such a system. Nevertheless, it is reasonable to use the neuronal basis of nervous activity as a method of understanding our sense of personal time (Gooddy, 1969).

The neuron transmits in an electrochemical fashion. The transmission of the nerve impulse has been studied by so many famous investigators that the mechanism is part of the everyday knowledge of any student of biology. Admirable expositions are to be found in *Nerve Muscle and Synapse* by Katz (1966) and *The Conduction of the Nervous Impulse* by Hodgkin (1964). The nerve cell is but a special example of a prototype cell; and it shares the properties of all cells in its general biochemical features. It is also a nervous *system* in miniature, as Katz points out. The transmission of the nerve impulse, the basis of nervous action, takes place as the result of a "wave of electrical negativity" passing without decrement along the nerve fibre. The wave of negativity results from the transport in or out of the neuronal membrane of sodium and potassium ions; that is to say, it is a highly specific chemical operation. When

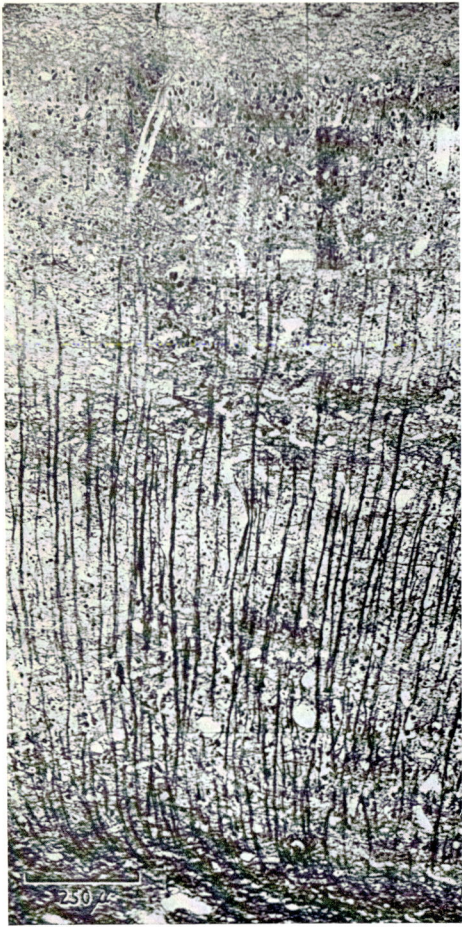

Fig. 4. Arrangement of cells in brain cortex, from Sholl 1956

the nerve impulse reaches the junction (synapse) with other nerve cells, transmission across the region takes place, again electrochemically, by means of transmitter substances. The first detected transmitters, with opposing action, were adrenaline and acetyl choline. Now at least nine transmitter substances are known. There are positive transmitter substances which facilitate the transmission across the junction: and there are also inhibitor substances.

The point under elaboration here is that at the very foundations of nervous activity there are highly developed clock-like forms of activity, taking place in elements which are dependent upon a variety of electrical and chemical influences. We shall not be surprised to learn that in this very labile biochemical system the administration of numerous drugs alters the pattern of nervous performance.

A general or local state of alteration in the mechanism or physiology of nervous activity may be accompanied by alteration of feeling or behaviour.

When we remind ourselves that many of the demonstrated shifts of circadian rhythms are found to be in the biochemical states which are produced by and directly affect cellular metabolism, we begin to realise how likely it is that the activity of the nervous system itself will be profoundly affected by travel-induced shifts. These shifts are gradually and constantly changing; and their effects are likely to be such as to ensure a long-drawn out state of discomfort, disorientation and dysrhythmia.

*Dysrhythmia.* This term, signifying any disturbance of any rhythm of which any observer is aware, has a special meaning in the world of neurology. It is important to show the neurological meaning here, since it is directly linked with the foregoing argument about the performance of single nervous units. As is well-known, it is possible, by the 1,000,000 times amplification of scalp potentials, to record rhythmic activity of the underlying brain. This is the electroencephalogram (e. e. g.): and it should never be forgotten that the detection of this brain activity was first shown by Berger in 1929. There are normal patterns of brain wave-forms, the rhythms extending over a spectrum of 1 to 30 cycles a second. The alpha-rhythm of 8–11 cycles a second is the best known one; and it is alleged to be highly specific (in the same way as fingerprints) for each person. The e. e. g. varies, as one would expect, with a large range of influences which are finally chemical in their mode of operation.

Such electroencephalographic changes can be seen in some records selected from the Sandoz Atlas of Electroencephalography. (Figs. 5–10)

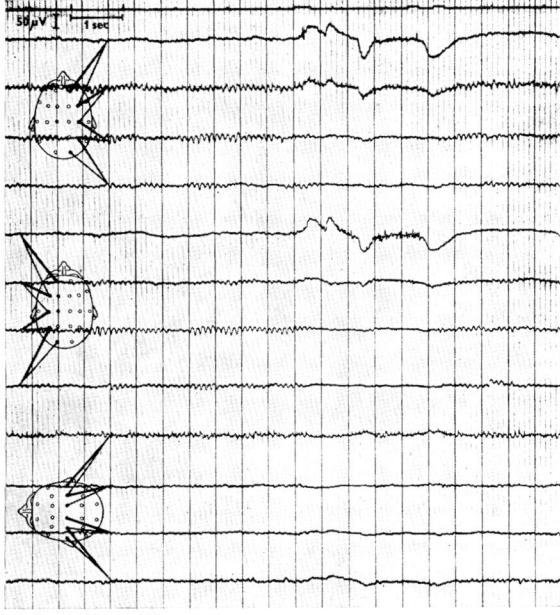

Fig. 5. Normal e.e.g.

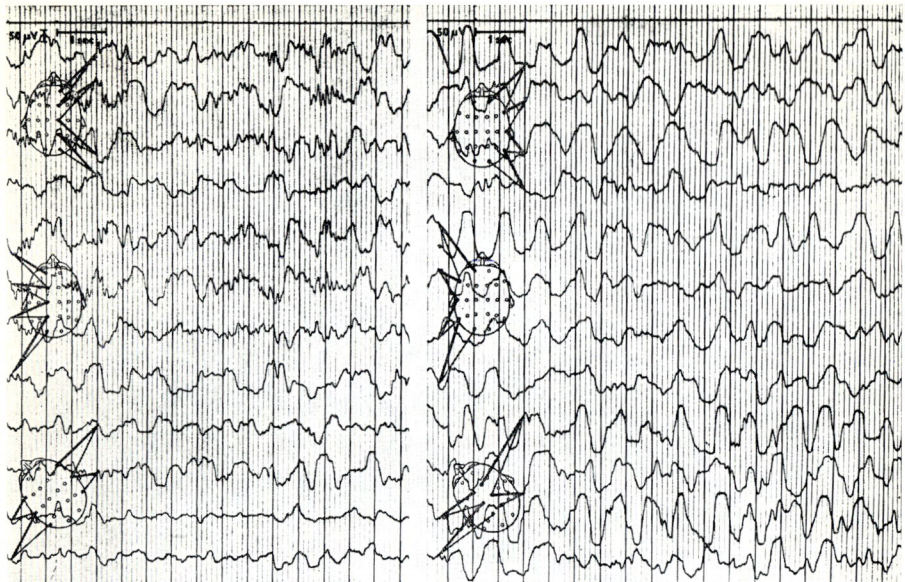

Fig. 6. E.e.g. pattern of increasing depth of sleep

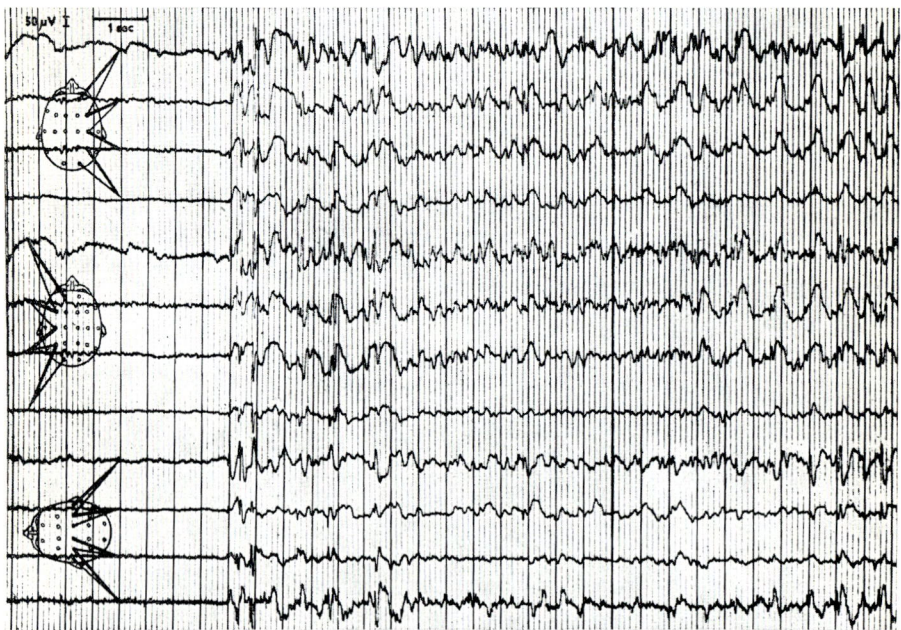

Fig. 7. E.e.g. showing effects after injection of cardiazol

These figures show how the performance of the brain varies under a variety of influences, of which the condition of sleep, drug administration and disease form an important part. Anxiety, the use of tobacco, alcohol and other drugs,

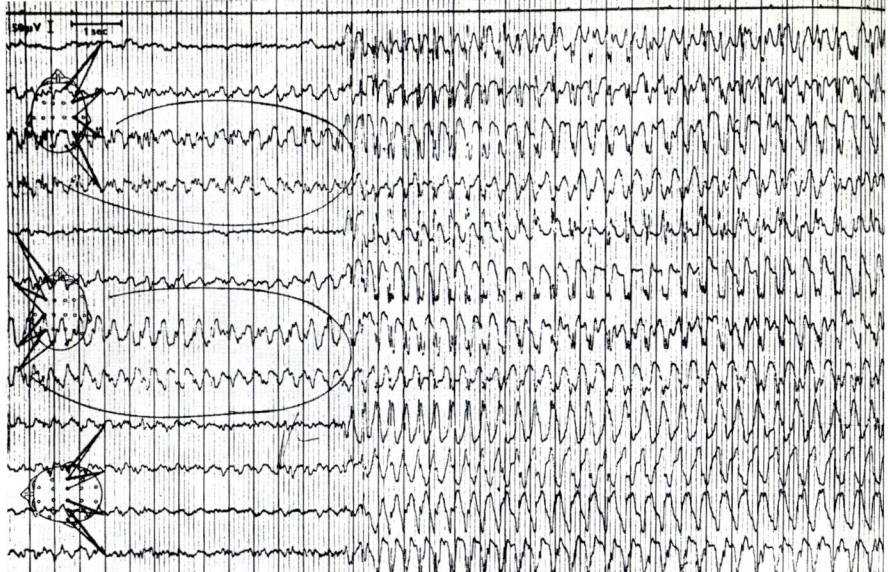

Fig. 8. E.e.g. showing burst of *petit mal* "spike-and-wave" epilepsy

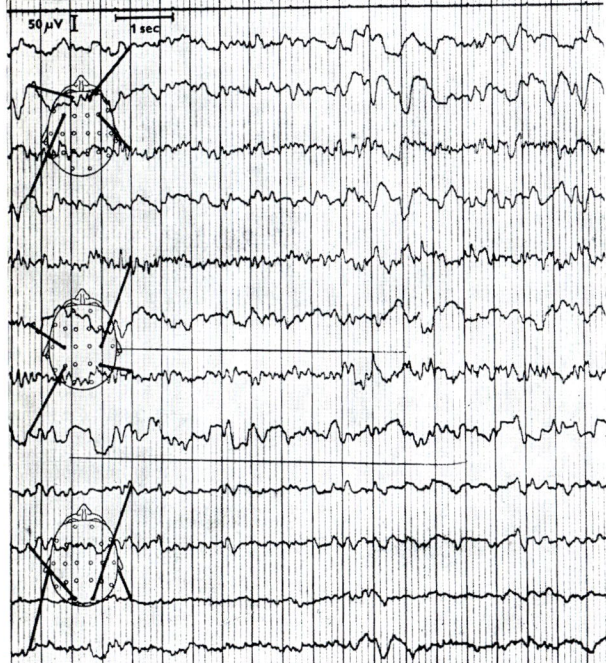

Fig. 9. E.e.g. indicating left-sided brain tumour

sleep or the prevention of sleep, the effects of flickering lights, are a few of the causes of alteration in a normal person's rhythm-producing mechanisms. Signs of dysrhythmia appear in the brain recordings.

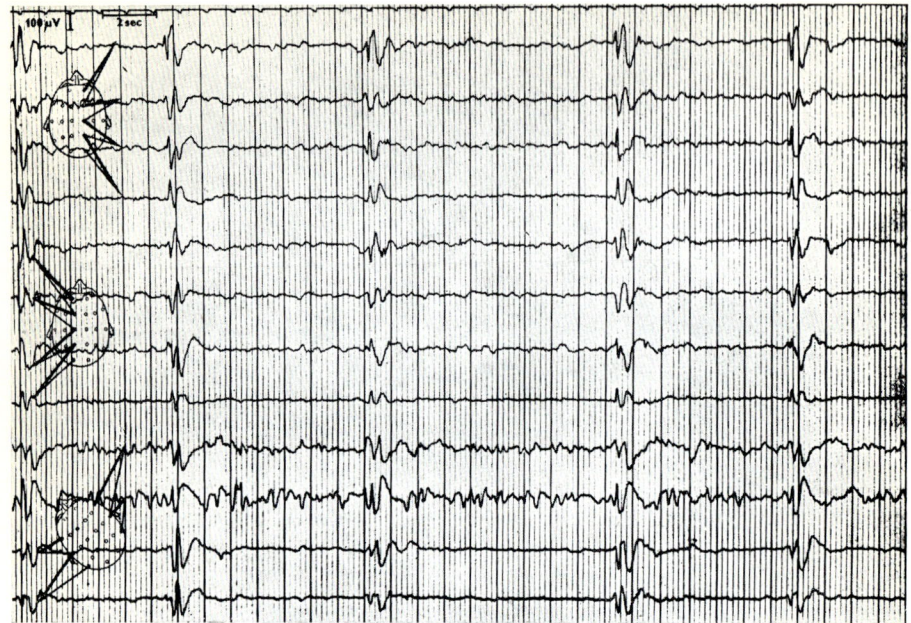
Fig. 10. E.e.g. showing rhythmic pattern in encephalitis

We must pass now briefly to consider the dysrhythmic influences of rapid travel, which may have such important effects upon neurophysiological and neuronal mechanisms. These influences, though apparently circumstantial or emotional in origin, are really pharmacological in effect. They do have, as we have already been told, a similarity to drug-induced states.

Mills (1969), referring to the work of Hauty and Adams (1966) has said:

"There is growing evidence that much of the discomfort and decrease in efficiency suffered by long-distance air travellers is due to the disturbance of physiological rhythms rather than to any fatigue induced by flying."

While the clinical observer agrees entirely with this important statement of Mills, it is here suggested that it may not be necessary to separate the notions of "disturbance of physiological rhythms" from "fatigue". The term "fatigue" applied to the feelings of the traveller is not just the tiredness of duration or hard work, but the mild, sometimes severe feeling of disturbed mental and physical health which many of us have noted: and is the single word we use to describe the sum of all biological derangements beyond the range of everyday adjustment.

The period of disordered function, the dysrhythmia itself, at the end of the journey is only partly related to the disturbance of the flight itself; for the journey is only the culmination, in a period of a few hours, of the plans of preceding weeks, months or even perhaps a year or so.

We may consider how these effects come about from a scheme simply shown by tables. First, the obvious three stages of the journey:

1. Pre-flight
   (*a*) and (*b*)
2. Flight
3. After-flight.

The first stage, before the flight begins, may be the most important of all in its disorientating effects. It is a plan for a prolonged dysrhythmia. It is obvious that pre-flight disturbances are the same whether the traveller goes across lines of longitude or along them.

## 1. *Pre-Flight (a)*

Special Occasion: "Once in a lifetime":
    Meeting, re-union:
    Important occasion not settled by post or telephone:
    New mode of life, transient or permanent:
    e.g. Meetings, lectures, speeches.

Hospitality:
Dress: (hot/cold)
Equipment:     related to 20 kg baggage

    Preparation of work (mental, written, spoken)

Documents:     Passport and visas:
    Inoculation and vaccination:
    Insurance.

Cabs and Porters.

Almost all of us will have the memories of pre-flight arrangements. I speak from the experiences of several journeys half way round the world: and a complete circuit last year. For that journey, during a Visiting Professorship in Australia, I had begun to make plans a year earlier, starting with the difficult decision whether I should go at all. Having decided to go, there was the complex situation of arranging for leave of absence to cover both professional and personal matters. There were the itinerary to plan, the lectures to prepare, inoculations, visas, clothes, photographic and other equipment; and other matters too numerous to mention. Some of us are harassed by the problem of having a weight allowance of 20 kg., whether we are going for the day or for half a year, whether we weigh 50 or 200 kg. The business of attending to these complicated matters on top of coping with current problems leads to the situation of a kind of double life. It appears to me to be disorientating and uncomfortable to such an extent that it is likely that many people require or actually take either tranquillisers or anti-depressant drugs.

As the time of departure approaches, the tempo of the double life increases to an almost unbearable degree; so that there is a physical feeling of severance as we leave our home for the start of the journey, with mouth dry, the pulse fast, the stomach unsettled, perhaps late at night or early in the morning.

*Pre-Flight (b)*

Journey to terminal: Wait
  Booking In
  Book seats
  Boarding Cards
  Baggage Tags

Journey to airport: Passports
  Customs
  Waiting time
  Boarding plane.

As soon as one leaves home the passage of personal time in relation to public time is out of our control. We have an anxious drive to the airport if traffic is dense or the weather bad. We are relieved to put ourselves into the hands of the flight machine when we hand in our tickets and baggage, to become mere passenger units of the airline. We are scrutinised by officials. We are passed to small, noisy public prisons, when we may not pass back to our friends or forward to our plane. We are in the limbo of transit. We may stay in this state for up to an hour; and sometimes we are last told of an unspecified hitch which means a delay of, usually, three hours in the first instance. We may have to return to our base and start again another day. We may have to stay in some strange accommodation until the delay is over. We may be in a quite disturbed state, anxious, sleepless and depressed, before we even fasten our seat-belts for take-off.

## 2. Flight

We must not neglect the anxiety of many passengers about the dangers of take-off and landing. Many is the time I have seen solid citizens rapidly mouthing passages from religious works whenever the plane was not flying straight and level. The highly abnormal condition of such people is often signified by their need for alcohol or other drugs. Much noradrenaline must be secreted as the plane accelerates, rotates and lifts off; and in the dangerous few moments after that, when a chance in the range of $10^6$ might cause multiple engine failure or control breakdown.

*Flight* Take-off
  Comfort: Confinement, lights, neighbours
  Diet: Alcohol, meals, excretion
  Entertainment: Reading, sleep, talk, music, films, drugs
  Arrival: Documents, passports, forms.

These are some of the factors affecting us during the hours of travel: and we have also to note the landing and take-off at intermediate stops, where we may have to disembark and re-embark, or else stay for an hour or more in the strange surroundings of the aircraft at rest.

## 3. *After-Flight*

After-flight: New internal state
              New external state
                  Meeting: needs for protection
                          currency and tips
                          phones, cabs
                          hotels,
                          professional duties
                          illness/over-entertainment.

After the aircraft has landed, we adjust ourselves as best we may to yet another set of unusual and often vaguely threatening circumstances. The emotions of reunion after a separation by space and time are powerful, the preparations for the work for which we travelled worry us, the adjustment to the new environment all take much out of us, affect us pharmacologically. We have immediately to adjust to different periods of light and darkness, while we know that our internal systems have not been able to adjust. Circadian rhythms persist, but they are circa-day which no longer exists for us. We are certainly not in a healthy state for sight-seeing, prolonged entertainment, work, or any serious decisions.

It is hoped that this brief exposition of the neurological unit of activity, the patterns of brain dysrhythmia, and the very numerous and longacting influences of a pharmacological and physiological nature derived from altered surroundings, will serve as an outline for discussion and further study. In a brief conclusion, a few suggestions for the improvement of well-being in future flights are made.

## *Conclusions*

Any high-speed long journey, especially over lines of longitude, produces mental and physical symptoms which are always important and sometimes severe. Such journeys are necessarily disturbing to the very important daily routine of our lives. *It is not only the journey itself, but also the preparations, over weeks or months as a rule; and the new circumstances on arrival which contribute to our feelings of abnormality.*

It is the duty of all health authorities, and especially of the airlines, to make the discomforts and hazards familiar to the would-be traveller. The airlines, only now becoming conscious of their responsibilities, should do everything possible to mitigate the unpleasant effects of pre-, during and post-flight procedures.

Passengers, including aircrew, need to be instructed and learn how they may become affected. They can then be more personally responsible for minimising the effects of those unavoidable stresses they have chosen to accept when they decide to travel by air. If they do not have to travel so fast, they may be wise to use other routes. Passengers should arrange their flights so that they have breaks of journey best suited to avoid the accumulation of flight stresses. At the end of their journeys, they should insist on a period of "convalescence", before undertaking any serious duties or engagements. They must insist on rest rather than entertainment, sight-seeing and important decisions, until they know from their own judgments that they have become "rhythmic" again.

Further investigation into dysrhythmic problems will be important, not only from the benefits to travellers, but also for the understanding of human chronometric activity.

*References*

Gooddy, W.: *Outside Time and Inside Time*. Perspectives in Med. and Biol. 12 (1969) 239–253.
Hauty, G. T., Adams, T.: *Phase shifts of the human circadian system and performance deficit during the periods of transition; east-west, west-east, north-south flight*. Aerospace Med. 37 (1966) 668, 1027, 1257.
Hess, R.: *E. E. G. Handbook*. Sandoz Monographs. Zürich 1966.
Hodgkin, A. L.: *The Conduction of the Nervous Impulse*. Liverpool: University Press 1964.
Katz, B.: *Nerve, Muscle and Synapse*. New York: McGraw-Hill 1966.
Mills, J. N.: *Man underground*. J. Roy. Coll. Phycns. Lond. 3 (1969) 329–332.
Sholl, D. A.: *The Organization of the Cerebral Cortex*. London: Methuen 1956.
Whitrow, G. J.: *The Natural Philosophy of Time*. London: Nelson 1961.

# Discussion Notes on the Lecture by Dr. Gooddy*

Georges Schaltenbrand**

These comments are intended as discussion notes on Dr. Gooddy's lecture; they complement his remarks from the point of view of anatomy, physiology and pathophysiology.

The problem of flight dysrhythmia is very closely linked with the problem of the internal clocks of the organism. Time signalling is one of the most important functions of living substance. Even single living cells act as time signallers for certain processes, for example, cell division, which proceeds according to an exact schedule, predetermined by a set of inherited chromosomes, yet modified by external stimuli. The time signals which are of greatest importance for the internal organs are always given by nerve cells. Complexes of such cells are found in a wide range of body regions, from the heart, whose rhythm is controlled by a small nerve-cell complex in the atrium, down to the peristaltic movement of the intestines, which is regulated by nerve cells in their walls. But these time signallers are always subject to modification by the constraints of the sympathetic and parasympathetic nervous systems which, by sending out stimulating or inhibiting impulses, can alter the rhythm of their timing. The more sophisticated time controls for the skeletal musculature are regulated by feedback circuits in the segments of the spinal cord. They govern, for example, the alternating movements of the extremeties in the fin motion of fishes, the alternating walking movements of mammals, or the tremors and cloni which occur in certain pathological conditions. Breathing is controlled from the brainstem by a special center.

The cerebellum is a reflex organ, producing in a highly differentiated manner temporal parameters for movements which are very precisely matched to the tasks which have to be performed under a great variety of conditions of gravity and inertia. The extreme regularity of the structure of the lamellae of this part of the brain, almost reminiscent of a vernier scale, illustrates the high degree of precision required. The highly differentiated performances of an ice-skating star or a concert pianist would be impossible without the innervation impulses to the periphery, synchronized down to minute fractions of a second. Yet the

---

\* Translated from the German by Barbara M. Crook, B. A. (Cantab.), c/o Springer-Verlag, BRD-69 Heidelberg, Postfach 1780, Germany.

\*\* Professor Dr. Georges Schaltenbrand, vorm. Direktor der Neurologischen Univ.-Klinik Würzburg, BRD-87 Würzburg, Lerchenweg 4, Germany.

function of this organ depends essentially upon the commands of its superior mechanism, the cerebrum, as well as upon the information received by it from numerous sensors located in the periphery. It is essentially a highly sophisticated reflex organ.

The system of internal clocks located deep within the cerebrum is quite a different matter. We are made aware of this system following certain illnesses, e. g. European sleeping sickness, which attack the grey matter of the third ventricle and interfere with the sleeping-waking rhythm. I first noticed this system when I was working with O. Girndt, experimenting on animals with the cerebrum removed. It was already known from Cannon's published work that cats after cerebrectomy show an intense motor restlessness with every sign of violent agitation. He called this state "sham rage". Girndt and I were able to establish that this strange state is not seen when the cerebrum is severed by a cut between the anterior end of the mesencephalon and the diencephalon, but that an even more excited and complex behavior is produced if the cut is made further forward, sparing most of the diencephalic cerebral portions of the brainstem ganglia. Animals in whom the thalamus is for the most part preserved, when protected from all external stimuli, show very regular periodic movement discharges with pauses of several minutes between them. Experiments of this kind can, of course, only be conducted in a completely noise-free room where the stimuli are minimal. We usually did them at night in the laboratory of R. Magnus (late Prof. of Pharmacology in Utrecht). Any external stimulus disrupts the spontaneous rhythm and hence introduces an irregularity.

Even at that time we suspected that this periodic restlessness was a pathological perversion of a physiological process; namely of the overriding organic regulation of the organism by the sleeping-waking system. And indeed, Hess has established in some very comprehensive experiments that electrical stimulation of certain regions of the thalamus can induce sleep, whereas other stimuli will wake the animal out of its sleep. We know from the investigations of Magoun (Los Angeles) that this system is the cephalic end of a very extensive reticular system which runs from the medulla oblongata through the midbrain to the diencephalon. This system is concerned with the excitation and damping of the nervous system as a whole. If it is destroyed within the pons Varoli, it provokes persistent sleep in mammals. However, while the lower parts of this system up as far as the midbrain are essentially stimulus-dependent in their reactions and hence induce a "poikilopoetic"[1] behavior in the animal, the diencephalic portion of the reticular system directs the "isopoetic"[1] time regulation according to the internal rhythms of the animal. Richter and Wang's studies of rodents have included a thorough treatment of this time-signalling function. They were able to show that a number of essential rhythms, for instance, stomach-intestine movements, the feeding drive, even the 4-day sexual cycle in females, are controlled from

---

[1] This expression is constructed on the lines of "poikilotherm" and "isotherm".

this region. These regions also match body rhythms to the day-night (circadian) rhythm. Careful studies and experiments carried out by Aschoff and coworkers have recently shown that, even when an animal is completely shielded from external influences, particularly noise, the circadian rhythm goes on "ticking" for a time, although the gradual accumulation of small errors causes it to get out of phase with astronomical time. Thus, the system of internal clocks regulates the organism in accordance with astronomical rhythms. In women, moon time also comes into the picture: this may have had some vital importance for the sexual cycle of our remote ancestors.

The effects of this system of internal clocks are complemented by a variety of reporting systems. Some of these consist of fibers which regulate the activity of the organs via the autonomic system, the sympathetic and parasympathetic nervous systems; others are hormonal controls, some of which still await investigation. Fairly well known now are the route from the hypothalamic centers via the nerve fibers to the posterior lobe of the hypophysis and others via the blood vessels to the anterior lobe of the hypophysis. These fibers do not just transmit nervous impulses, they also produce chemical substances which regulate hypophyseal function and this, in turn, releases into the blood special secretions which at the proper time set in train such processes as sexual maturation, birth, lactation, etc. Finally, the entire set-up has a feedback via the endocrine products of the organs it controls, i. e. the pituitary, the gonads and the adrenals.

While we are asleep, we are subject to a direct, vegetative influence exerted by the various parts of the brain via very fine nerve fibers which run from the "motor household centers" to all the pertinent parts of the brain; at the same time, the sleep-regulating areas apparently also produce a hormone, in a manner similar to that now assumed for the extrapyramidal system. However, the hormones from the "motor household centers" (waking or sleeping hormones) cannot influence one side of the brain at a time, like those transmitted via the extrapyramidal nerve pathways, instead they affect the entire body, as can be ascertained in the case of paraplegics, in whom even the lowest portions of the spinal cord "go to sleep", or in parabiotic animals. It follows that this hormone must be transported in the blood or cerebrospinal fluid, or both, to the distant parts of the nervous system.

The main difference between the pathway via the vegetative fibers of the autonomic system and the hormone transported in blood or cerebrospinal fluid lies in their time constants. This can be demontrated very effectively in attempts to control the heart rhythm: a very rapid effect can only be achieved by stimulating the vegetative fibers, i. e. the sympathetic or parasympathetic fibers; a rather longer-term effect, which begins to act within minutes, is set off by the release of adrenalin from the adrenal medulla; an even more long-term effect is due to the release of thyroid hormone. The two last are set in motion by the action of the vegetative fibers and the thyrotrophic hormone from the hypophysis.

Again, as with all connections of the nervous system, the connection between the center and the periphery is a two-way affair. We have already mentioned that the organs at the extremity of the system of internal clocks feed back to the diencephalon and the hypophysis. This mechanism is particularly impressive in the production of cortisone by the adrenal cortex: the cortisone level itself influences a person's well-being and periodicity. We neurologists know this only too well from our patients when we have to treat them with cortisone for certain conditions. Small doses induce optimism and good humor and mental and physical alertness but cut down sleeping time. Larger doses effect an extreme enhancement of excitability with complete sleeplessness, verging upon the psychotic. Conversely, when such a drug is withdrawn after it has been prescribed for some time, there is a neurasthenic lag, resembling a hangover.

What is important in the above for the problem we are discussing is the fact that physical and psychological troubles can upset this control mechanism to an extreme degree. These troubles we call "stress". It seems likely that it is precisely this mechanism which is profoundly affected by the dysrhythmias arising from air travel. The unaccustomed strains and the dislocation of the time rhythm induce nervous tension and irritability and these are accompanied by an excessive drain on the adrenal cortex, so that finally its production of cortisone is exhausted. This produces symptoms resembling those of the adaptation diseases described by Selje. However, it also points the way to treatment.

There are two possible ways: to avoid such a breakdown, or to correct it. It may be avoided by taking small amounts of tranquillizers which reduce the excitability of the reticular system; such drugs may be taken before or during the flight. Another simple possibility is to compensate for the dislocation of day-night rhythm by ensuring adequate sleep after the flight. Once exhaustion has occurred, a rapid recovery may be achieved by substitution with small quantities of cortisol. However, this is a two-edged weapon since it works like other stimulants and delays the time of recovery. Nevertheless, it can be necessary, particularly for aircrews, in order to maintain full performance during a critical period.